Lecture Notes in Computer Science

Lecture Notes in Artificial Intelligence **14196**

Founding Editor

Jörg Siekmann

The series Lecture Notes in Artificial Intelligence (LNAI) was established in 1988 as a topical subseries of LNCS devoted to artificial intelligence.

The series publishes state-of-the-art research results at a high level. As with the LNCS mother series, the mission of the series is to serve the international R & D community by providing an invaluable service, mainly focused on the publication of conference and workshop proceedings and postproceedings.

Murilo C. Naldi · Reinaldo A. C. Bianchi
Editors

Intelligent Systems

12th Brazilian Conference, BRACIS 2023
Belo Horizonte, Brazil, September 25–29, 2023
Proceedings, Part II

Springer

Editors
Murilo C. Naldi
Federal University of São Carlos
São Carlos, Brazil

Reinaldo A. C. Bianchi
Centro Universitario da FEI
São Bernardo do Campo, Brazil

ISSN 0302-9743 ISSN 1611-3349 (electronic)
Lecture Notes in Artificial Intelligence
ISBN 978-3-031-45388-5 ISBN 978-3-031-45389-2 (eBook)
https://doi.org/10.1007/978-3-031-45389-2

LNCS Sublibrary: SL7 – Artificial Intelligence

This Springer imprint is published by the registered company Springer Nature Switzerland AG
The registered company address is: Gewerbestrasse 11, 6330 Cham, Switzerland

Paper in this product is recyclable.

Preface

The 12th Brazilian Conference on Intelligent Systems (BRACIS 2023) was one of the most important events held in Brazil in 2023 for researchers interested in publishing significant and novel results related to Artificial and Computational Intelligence. The Brazilian Conference on Intelligent Systems (BRACIS) originated from the combination of the two most important scientific events in Brazil in Artificial Intelligence (AI) and Computational Intelligence (CI): the Brazilian Symposium on Artificial Intelligence (SBIA, 21 editions) and the Brazilian Symposium on Neural Networks (SBRN, 12 editions). The Brazilian Computer Society (SBC) supports the event, the Special Committee of Artificial Intelligence (CEIA), and the Special Committee of Computational Intelligence (CEIC). The conference aims to promote theoretical aspects and applications of Artificial and Computational Intelligence and exchange scientific ideas among researchers, practitioners, scientists, engineers, and industry.

In 2023, BRACIS took place in Belo Horizonte, Brazil, from September 25th to 29th, 2023, in the Campus of the Universidade Federal de Minas Gerais. The event was held in conjunction with two other events: the National Meeting on Artificial and Computational Intelligence (ENIAC) and the Symposium on Information and Human Language Technology (STIL).

BRACIS 2023 received 242 submissions. All papers were rigorously reviewed by an international Program Committee (with a minimum of three double-blind peer reviews per submission), followed by a discussion phase for conflicting reports. After the review process, 89 papers were selected for publication in three volumes of the Lecture Notes in Artificial Intelligence series (an acceptance rate of 37%).

The topics of interest included, but were not limited to, the following:

- Agent-based and Multi-Agent Systems
- Bioinformatics and Biomedical Engineering
- Cognitive Modeling and Human Interaction
- Combinatorial and Numerical Optimization
- Computer Vision
- Constraints and Search
- Deep Learning
- Distributed AI
- Education
- Ethics
- Evolutionary Computation and Metaheuristics
- Forecasting
- Foundation Models
- Foundations of AI
- Fuzzy Systems
- Game Playing and Intelligent Interactive Entertainment
- Human-centric AI

- Hybrid Systems
- Information Retrieval, Integration, and Extraction
- Intelligent Robotics
- Knowledge Representation and Reasoning
- Knowledge Representation and Reasoning in Ontologies and the Semantic Web
- Logic-based Knowledge Representation and Reasoning
- Machine Learning and Data Mining
- Meta-learning
- Molecular and Quantum Computing
- Multidisciplinary AI and CI
- Natural Language Processing
- Neural Networks
- Pattern Recognition and Cluster Analysis
- Planning and Scheduling
- Reinforcement Learning

We want to thank everyone involved in BRACIS 2023 for helping to make it a success: we are very grateful to the Program Committee members and reviewers for their volunteered contribution to the reviewing process; we would also like to thank all the authors who submitted their papers and laboriously worked to have the best final version possible; the General Chairs and the Local Organization Committee for supporting the conference; the Brazilian Computing Society (SBC); and all the conference's sponsors and supporters. We are confident that these proceedings reflect the excellent work in the artificial and computation intelligence communities.

September 2023

Murilo C. Naldi
Reinaldo A. C. Bianchi

Organization

General Chairs

Gisele Lobo Pappa	Universidade Federal de Minas Gerais, Brazil
Wagner Meira Jr.	Universidade Federal de Minas Gerais, Brazil

Program Committee Chairs

Murilo Coelho Naldi	Universidade Federal de São Carlos, Brazil
Reinaldo A. C. Bianchi	Centro Universitário FEI, Brazil

Steering Committee

Aline Paes	Universidade Federal Fluminense, Brazil
André Britto	Universidade Federal do Sergipe, Brazil
Anna H. R. Costa	Universidade de São Paulo, Brazil
Anne Canuto	Universidade Federal do Rio Grande do Norte, Brazil
Arnaldo Cândido Jr.	Universidade Estadual Paulista, Brazil
Felipe Meneguzzi	University of Aberdeen, UK
Filipe Saraiva	Universidade Federal do Pará, Brazil
Gina M. B. Oliveira	Universidade Federal Uberlandia, Brazil
Helida Santos	Universidade Federal do Rio Grande, Brazil
Leliane N. de Barros	Universidade de São Paulo, Brazil
Livy Real	B2W Digital, Brazil
Maria V. de Menezes	Universidade Federal do Ceará, Brazil
Marlo Souza	Universidade Federal da Bahia, Brazil
Renato Tinos	Universidade de São Paulo, Brazil
Ricardo Marcacini	Universidade de São Paulo, Brazil
Tatiane Nogueira	Universidade Federal da Bahia, Brazil
Thiago Pardo	Universidade de São Paulo - São Carlos, Brazil

Program Committee

Felix Antreich	Instituto Tecnológico de Aeronáutica, Brazil
Fernando Osório	Universidade de São Paulo, São Carlos, Brazil
Flavia Bernardini	Universidade Federal Fluminense, Brazil
Flavio Tonidandel	Centro Universitário FEI, Brazil
Flávio S. C. da Silva	Universidade de São Paulo, Brazil
Francisco Chicano	University of Málaga, Spain
Francisco De Carvalho	Universidade Federal de Pernambuco, Brazil
Gabriel Ramos	Universidade do Vale do Rio dos Sinos, Brazil
George Cavalcanti	Universidade Federal de Pernambuco, Brazil
Gerson Zaverucha	Universidade Federal do Rio de Janeiro, Brazil
Giancarlo Lucca	Universidade Federal do Rio Grande, Brazil
Gisele Pappa	Universidade Federal de Minas Gerais, Brazil
Gracaliz Dimuro	Universidade Federal do Rio Grande, Brazil
Guilherme Barreto	Universidade Federal do Ceará, Brazil
Guilherme Coelho	Universidade Estadual de Campinas, Brazil
Guilherme Derenievicz	Universidade Federal do Paraná, Brazil
Guilherme D. Pelegrina	Universidade Estadual de Campinas, Brazil
Guillermo Simari	Universidad Nacional del Sur in B. B., Argentina
Gustavo Giménez-Lugo	Universidade Tecnológica Federal do Paraná, Brazil
Heitor Gomes	Victoria University of Wellington, New Zealand
Helena Caseli	Universidade Federal de São Carlos, Brazil
Helida Santos	Universidade Federal do Rio Grande, Brazil
Heloisa Camargo	Universidade Federal de São Carlos, Brazil
Huei Lee	Universidade Estadual do Oeste do Paraná, Brazil
Isaac da Silva	Centro Universitário FEI, Brazil
Ivandré Paraboni	Universidade de São Paulo, Brazil
Ivette Luna	Universidade Estadual de Campinas, Brazil
Jaime S. Sichman	Universidade de São Paulo, Brazil
Jean Paul Barddal	Pontifícia Universidade Católica do Paraná, Brazil
João Papa	Universidade Estadual Paulista, Brazil
João C. Xavier-Júnior	Universidade Federal do RN (UFRN), Brazil
João Paulo Canário	Stone Co., Brazil
Jomi Hübner	Universidade Federal de Santa Catarina, Brazil
Jonathan Andrade Silva	Universidade Federal de Mato Grosso do Sul, Brazil
José Antonio Sanz	Universidad Publica de Navarra, Spain
José A. Baranauskas	Universidade de São Paulo, Brazil
Jose E. O. Luna	Universidad Católica San Pablo, Peru
Julio Nievola	Pontifícia Universidade Católica do Paraná, Brazil
Karla Roberta Lima	Universidade de São Paulo, Brazil

Karliane Vale	Universidade Federal do Rio Grande do Norte, Brazil
Kate Revoredo	Humboldt-Universität zu Berlin, Germany
Krysia Broda	Imperial College, UK
Laura De Miguel	Universidad Pública de Navarra, Spain
Leila Bergamasco	Centro Universitário FEI, Brazil
Leliane Nunes de Barros	Universidade de São Paulo, Brazil
Leonardo Emmendorfer	Universidade Federal do Rio Grande, Brazil
Leonardo Matos	Universidade Federal de Sergipe, Brazil
Leonardo T. Duarte	Universidade Estadual de Campinas, Brazil
Leonardo F. R. Ribeiro	Amazon, USA
Levy Boccato	Universidade Estadual de Campinas, Brazil
Li Weigang	Universidade de Brasilia, Brazil
Livy Real	B2W Digital, Brazil
Lucelene Lopes	Universidade de São Paulo - São Carlos, Brazil
Luciano Digiampietri	Universidade de São Paulo, Brazil
Luis Garcia	Universidade de Brasília, Brazil
Luiz H. Merschmann	Universidade Federal de Lavras, Brazil
Marcela Ribeiro	Universidade Federal de São Carlos, Brazil
Marcelo Finger	Universidade de São Paulo, Brazil
Marcilio de Souto	Université d'Orléans, France
Marcos Domingues	Universidade Estadual de Maringá, Brazil
Marcos Quiles	Universidade Federal de São Paulo, Brazil
Maria Claudia Castro	Centro Universitario FEI, Brazil
Maria do C. Nicoletti	Universidade Federal de São Carlos, Brazil
Marilton Aguiar	Universidade Federal de Pelotas, Brazil
Marley M. B. R. Vellasco	Pontifícia Universidade Católica do R. de J., Brazil
Marlo Souza	Universidade Federal da Bahia, Brazil
Marlon Mathias	Universidade de São Paulo, Brazil
Mauri Ferrandin	Universidade Federal de Santa Catarina, Brazil
Márcio Basgalupp	Universidade Federal de São Paulo, Brazil
Mário Benevides	Universidade Federal Fluminense, Brazil
Moacir Ponti	Universidade de São Paulo, Brazil
Murillo Carneiro	Universidade Federal de Uberlândia, Brazil
Murilo Loiola	Universidade Federal do ABC, Brazil
Murilo Naldi	Universidade Federal de São Carlos, Brazil
Myriam Delgado	Universidade Tecnológica Federal do Paraná, Brazil
Nuno David	Instituto Universitário de Lisboa, Portugal
Patrícia Tedesco	Universidade Federal de Pernambuco, Brazil
Paula Paro Costa	Universidade Estadual de Campinas, Brazil

Paulo Cavalin	IBM Research, Brazil
Paulo Pirozelli	Universidade de São Paulo, Brazil
Paulo Quaresma	Universidade de Évora, Portugal
Paulo Santos	Flinders University, Australia
Paulo Henrique Pisani	Universidade Federal do ABC, Brazil
Paulo T. Guerra	Universidade Federal do Ceará, Brazil
Petrucio Viana	Universidade Federal Fluminense, Brazil
Priscila Lima	Universidade Federal do Rio de Janeiro, Brazil
Priscila B. Rampazzo	Universidade Estadual de Campinas, Brazil
Rafael Giusti	Universidade Federal do Amazonas, Brazil
Rafael G. Mantovani	Universidade Tecnológica Federal do Paraná, Brazil
Rafael Parpinelli	Universidade do Estado de Santa Catarina, Brazil
Reinaldo A. C. Bianchi	Centro Universitario FEI, Brazil
Renato Krohling	UFES - Universidade Federal do Espírito Santo, Brazil
Renato Tinos	Universidade de São Paulo, Brazil
Ricardo Cerri	Universidade Federal de São Carlos, Brazil
Ricardo Marcacini	Universidade de São Paulo - São Carlos, Brazil
Ricardo Prudêncio	Universidade Federal de Pernambuco, Brazil
Ricardo Rios	Universidade Federal da Bahia, Brazil
Ricardo Suyama	Universidade Federal do ABC, Brazil
Ricardo A. S. Fernandes	Universidade Federal de São Carlos, Brazil
Roberta Sinoara	Instituto Federal de C., E. e T. de São Paulo, Brazil
Roberto Santana	University of the Basque Country, Spain
Rodrigo Wilkens	Université Catholique de Louvain, Belgium
Romis Attux	Universidade Estadual de Campinas, Brazil
Ronaldo Prati	Universidade Federal do ABC, Brazil
Rosangela Ballini	Universidade Estadual de Campinas, Brazil
Roseli A. F. Romero	Universidade de São Paulo - São Carlos, Brazil
Rosiane de Freitas R.	Universidade Federal do Amazonas, Brazil
Sandra Sandri	Instituto Nacional de Pesquisas Espaciais, Brazil
Sandro Rigo	Universidade do Vale do Rio dos Sinos, Brazil
Sílvia Maia	Universidade Federal do Rio Grande do Norte, Brazil
Sílvio Cazella	Universidade Federal de Ciências da S. de P. A., Brazil
Silvia Botelho	Universidade Federal do Rio Grande, Brazil
Solange Rezende	Universidade de São Paulo - São Carlos, Brazil
Tatiane Nogueira	Universidade Federal da Bahia, Brazil
Thiago Covoes	Wildlife Studios, Brazil
Thiago Homem	Instituto Federal de E., C. e T. de São Paulo, Brazil

Thiago Pardo	Universidade de São Paulo - São Carlos, Brazil
Tiago Almeida	Universidade Federal de São Carlos, Brazil
Tiago Tavares	Insper, Brazil
Valdinei Freire	Universidade de São Paulo, Brazil
Valerie Camps	Paul Sabatier University, France
Valmir Macario	Universidade Federal Rural de Pernambuco, Brazil
Vasco Furtado	Universidade de Fortaleza, Brazil
Vinicius Souza	Pontificia Universidade Católica do Paraná, Brazil
Viviane Torres da Silva	IBM Research, Brazil
Vladimir Rocha	Universidade Federal do ABC, Brazil
Washington Oliveira	Universidade Estadual de Campinas, Brazil
Yván Túpac	Universidad Católica San Pablo, Peru
Zhao Liang	Universidade de São Paulo, Brazil

Additional Reviewers

Alexandre Alcoforado
Alexandre Lucena
Aline Del Valle
Aline Ioste
Allan Santos
Ana Ligia Scott
Anderson Moraes
Antonio Dourado
Antonio Leme
Arthur dos Santos
Brenno Alencar
Bruna Zamith Santos
Bruno Labres
Carlos Caetano
Carlos Forster
Carlos José Andrioli
Caroline Pires Alavez Moraes
Cedric Marco-Detchart
Cinara Ghedini
Daiane Cardoso
Daniel da Silva Junior
Daniel Guerreiro e Silva
Daniela Vianna
Diego Cavalca
Douglas Meneghetti
Edilene Campos
Edson Borin
Eduardo Costa Lopes
Eduardo Max
Elias Silva
Eliton Perin
Emely Silva
Estela Ribeiro
Eulanda Santos
Fabian Cardoso
Fabio Lima
Fagner Cunha
Felipe Serras
Felipe Zeiser
Fernando Pujaico Rivera
Guilherme Mello
Guilherme Santos
Israel Fama
Javier Fumanal
Jefferson Oliva
João Fabro
João Lucas Luz Lima Sarcinelli
Joelson Sartori
Jorge Luís Amaral
José Angelo Gurzoni Jr.
José Gilberto Medeiros Junior
Juan Colonna

Leandro Lima
Leandro Miranda
Leandro Stival
Leonardo da Silva Costa
Lucas Alegre
Lucas Buzuti
Lucas Carlini
Lucas da Silva
Lucas Pavelski
Lucas Queiroz
Lucas Rodrigues
Lucas Francisco Pellicer
Luciano Cabral
Luiz Celso Gomes Jr.
Maëlic Neau
Maiko Lie
Marcelino Abel
Marcella Martins
Marcelo Polido
Marcos José
Marcos Vinícius dos Santos Ferreira
Marisol Gomez
Matheus Rocha
Mária Minárová
Miguel de Mello Carpi
Mikel Sesma
Murillo Bouzon
Murilo Falleiros Lemos Schmitt
Newton Spolaôr
Odelmo Nascimento
Pablo Silva
Paulo Rodrigues
Pedro Da Cruz
Rafael Berri
Rafael Gomes Mantovani
Rafael Krummenauer
Rafael Orsi
Ramon Abílio
Richard Gonçalves
Rubens Chaves
Sarah Negreiros de Carvalho Leite
Silvio Romero de Araújo Júnior
Tatiany Heiderich
Thiago Bulhões da Silva Costa
Thiago Carvalho
Thiago Dal Pont
Thiago Miranda
Thomas Palmeira Ferraz
Tiago Asmus
Tito Spadini
Victor Varela
Vitor Machado
Weber Takaki
Weverson Pereira
Xabier Gonzalez

Contents – Part II

Deep Learning Applications

Reinforcement Learning and GAN

Classification

Machine Learning Analysis

Transformer Applications

Transformer Model for Fault Detection from Brazilian Pre-salt Seismic Data

Letícia Bomfim[1,3], Oton Cunha[3], Michelle Kuroda[3], Alexandre Vidal[2,3], and Helio Pedrini[1](✉)

[1] Institute of Computing (IC), University of Campinas, Campinas, SP, Brazil
helio@ic.unicamp.br
[2] Institute of Geosciences (IG), University of Campinas, Campinas, SP, Brazil
[3] Center for Energy and Petroleum Studies (CEPETRO), Campinas, SP, Brazil
{mckuroda,vidal}@unicamp.br

Abstract. Carbonate reservoirs are known for their heterogeneity, which poses challenges in interpreting and defining geological models. The Brazilian reservoirs are formed mostly in highly faulted and fractured carbonate rocks, which can increase hydrocarbon transport and storage capacity. Strategies that permit the identification of these structures allow the optimization in the exploration of a reservoir. To fulfill this task, machine learning models have been able to provide an understanding of these environments through the use of data obtained by seismic methods. The use of convolutional neural networks (CNNs) has shown to be able to provide excellent abstractions in the field of semantic segmentation, including its use in seismic data. However, due to the highly heterogeneous formation of this type of data, the work of extracting information from these images remains challenging. From this, we investigate the potential of using Transformer models in this geological context focusing on the faults identification. As a technique to analyze this type of architecture, we use the TransUNet network, which combines the power of CNNs with the innovation brought by Transformers in a hybrid model of deep learning. To evaluate its performance, we made use of conventional CNNs to compare the results achieved. The results show that TransUNet outperforms conventional CNNs, achieving a Dice metric value of 88.34%, compared to 85.99% for U-Net, 83.41% for U-Net++, and 83.31% for SegNet, being also able to identify small structures beyond what is indicated in our target.

Keywords: Fault Detection · Transformer Models · Seismic Data

1 Introduction

Understanding and analyzing geological characteristics and their influence on reservoir properties is crucial for the success of hydrocarbon exploration and production. Moreover, automating these processes can result in the production of deliverables that are useful in well planning optimization, reservoir modeling,

M. C. Naldi and R. A. C. Bianchi (Eds.): BRACIS 2023, LNAI 14196, pp. 3–17, 2023.
https://doi.org/10.1007/978-3-031-45389-2_1

and risk analysis (Zheng et al. 2019). The Pre-Salt region in Brazil has enormous exploration potential, with reserve estimates suggesting that it contains approximately 70 to 100 billion barrels of oil. The majority of production comes from the Santos Basin, which is located in the southeastern part of the country and accounts for over 70% of Brazilian production. Therefore, geological and geophysical studies, in conjunction with technological tools, have been employed to investigate optimal exploration conditions and identify the structural and sedimentary characteristics of these complex basins that contribute to their potential for hydrocarbon accumulation.

Gaining a comprehensive understanding of the spatial distribution of structural, faciological, and petrophysical properties can aid in optimizing the characterization, prediction, and recovery phases of reservoirs. This is achieved by creating more reliable models. The petrophysical properties of interest include the porous medium, which comprises matrix systems, vugs, and fractures that define spaces for fluid accumulation in rock, as well as permeability, which determines the connectivity of pores and is critical for percolation and the recovery of subsurface fluids.

Faults and fractures are crucial features within the porous medium, as they play a vital role in engineering, geotechnical, and hydrogeological applications. They can exhibit dual behavior by either serving as pathways for fluid flow or barriers, depending on factors such as intensity, connectivity, dissolution, or cementing, which renders them impermeable. Numerous oil, gas, geothermal, and water supply reservoirs are formed in fractured rock. Therefore, an essential aspect of comprehending and predicting fracture behavior involves identifying and locating those that demonstrate hydraulically significant behavior (Council 1996). In order to analyze the properties of data from sources such as the pre-salt, which is located in ultra-deep waters below 5000m, and to identify fault regions, it is necessary to conduct seismic studies. This type of data offers insight into the structures present throughout the entire region of interest (Dondurur 2018) providing information about layer structures and their features. Upon obtaining this data, segmentation techniques are employed to extract the structures that will inform subsequent analyses.

Our aim is to incorporate the Transformer concept in the automated segmentation of seismic data. While convolutional networks have been used in the literature for this task, recent studies have highlighted the advantages of Transformers in computer vision tasks. According to Bi et al. [2021], Transformers exhibit a lower inductive bias, resulting in better performance due to fewer assumptions about the optimal approach. To this end, we have selected the TransUNet (Chen et al. 2021) model, which combines Transformers and U-Net, as a promising alternative for identifying faults through image segmentation.

2 Background

The analysis of seismic data is crucial for the progress of hydrocarbon exploration, and it is commonly done through the study of geological structures. Various techniques have been employed to analyze seismic data, including machine learning and image processing, which aim to automate and facilitate the interpretation process, using seismic data in the form of an image. Pepper and Bejarano [2005], for example, presented case studies on automatic fault interpretation using only seismic attributes that highlight faults. These attributes work similarly to filtering techniques used in image processing, and two of them, dip and azimuth, showed the best results in identifying fault regions that were extracted as connected components. Zhao and Mukhopadhyay [2018] explored the task of fault detection in synthetic and field data by using convolutional neural networks (CNNs) to develop prediction models. In particular, Zhao and Mukhopadhyay [2018] improved the final result by adding image processing algorithms, such as smoothing and sharpening, after the prediction step.

As the need for more robust models for seismic interpretation has become evident, researchers have turned to deep machine learning models for performing these tasks. Wu et al. [2019b] developed several models with a primary focus on fault prediction, including FaultNet3D and FaultSeg3D (Wu et al. 2019a). Using a single CNN, the FaultNet3D model aimed to estimate the probability of faults, cracks, and dips. Meanwhile, the FaultSeg3D model focused on fault delineation, with its output being a binary mask representing the seismic data, where 1 denotes the presence of faults and 0 represents the absence of faults.

Research on fault identification remains crucial in the geological context, as seismic data acquisition has significantly increased and deep convolutional neural networks have been successfully applied. Recent approaches, including (An et al. 2021), have created a large database labeled by experts to supplement synthetic data. A deep CNN based on edge detection has been proposed, producing a pixel-by-pixel binary classification of faults with superior results compared to commonly used CNNs.

3 Materials

The primary material for this work is the seismic data, which will be introduced along with its acquisition. Figure 1 shows the four main parts of this phase: data section, data split, data augmentation, and data normalization. Initially, two seismic cubes representing the input and target will be used to generate subsamples that will be augmented to provide the model with a diverse set of inputs.

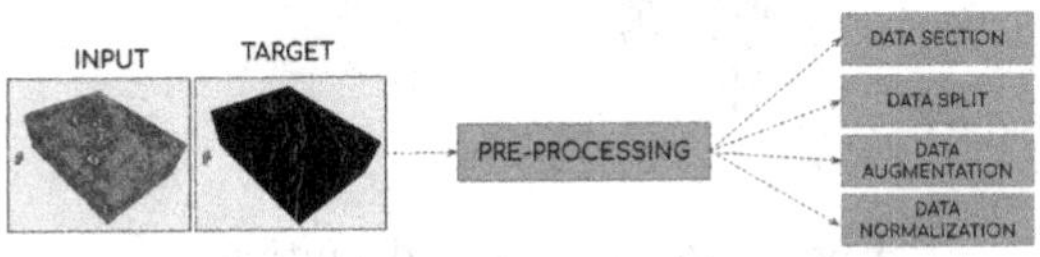

Fig. 1. Pre-processing workflow.

3.1 Seismic Data and Acquisition

The acquisition of seismic data using a marine approach is the foundation of this work, as it enables the extraction of information and characteristics from sedimentary basins, such as those in the Brazilian pre-salt. In order to conduct seismic surveys, a range of computational tools and systems are employed, with compatible real-time communication (Dondurur 2018). This process involves propagating elastic waves through the subsurface medium, which then reflect off interfaces and return to the surface, where they are detected by receivers.

Sound waves are generated by equipment called airguns, which penetrate the marine subsoil and, upon reflection, are detected by receivers equipped with vibrating coils that produce electrical signals. These receivers, such as hydrophones, are often positioned near the water surface, attached by cables, or towed by seismic vessels (known as streamers). The electrical signals are transmitted via cables to a seismograph recorder on the ship, resulting in representative images of subsurface structures. The data then undergo a careful processing step, in which they are grouped and the signal-to-noise ratio is enhanced to create images of subsurface structures.

The outcome of this acquisition process is a seismic volume, which is a three-dimensional function denoted by $a(x, y, z)$. It reflects the changes in seismic amplitude (a) along the three coordinates: x, y, and z. The seismic amplitude (a) can be expressed as a function of depth (z), a function of (x, z), or a function of (x, y, z) (Alsadi 2017).

The dataset utilized in this study comprises two seismic volumes: the input and the target. The dimensions of both volumes are $1401 \times 1481 \times 241$ pixels, and they cover an area of approximately 240 km^2, encompassing two pre-salt Santos basin fields. The vertical limit of the volumes is around 2000m within the area of interest. The input volume represents the amplitude seismic values of the area, while the target volume contains the faults interpreted from the amplitude seismic. The target volume was generated from 94 interpreted faults, as depicted in Fig. 2, and used to create a binary model of fault (1) and no-fault (0) scenarios, based on the proximity to faults.

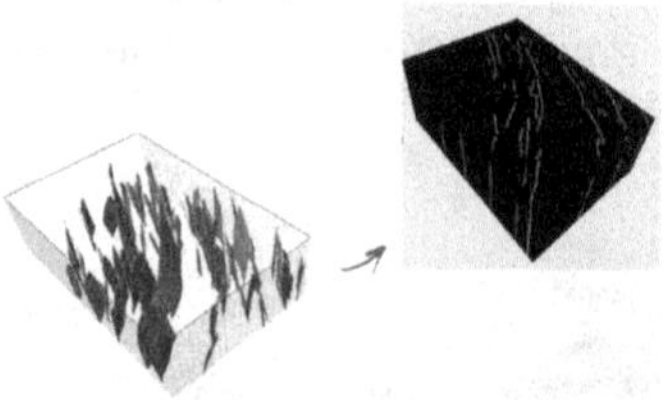

Fig. 2. Target seismic definition.

3.2 Pre-processing

We start by converting the seismic cube and mask into a NumPy array, which are structured in a three-dimensional form. Then, the data is manipulated in 2D sections to generate 2D sub-images that can be used to train the model. For this purpose, we extract smaller patches with dimensions $p \times p$ pixels from the seismic inlines, where $p = M$. This method provides a large image dataset with samples that have a conventional square dimensionality suitable for convolutional models. To avoid repetition of information, the inline region is sectioned into subimages with a $stride = M$, without overlapping, as shown in Fig. 3. This process generates $\lfloor (n/p) \rfloor$ subimages from each of the treated inline images.

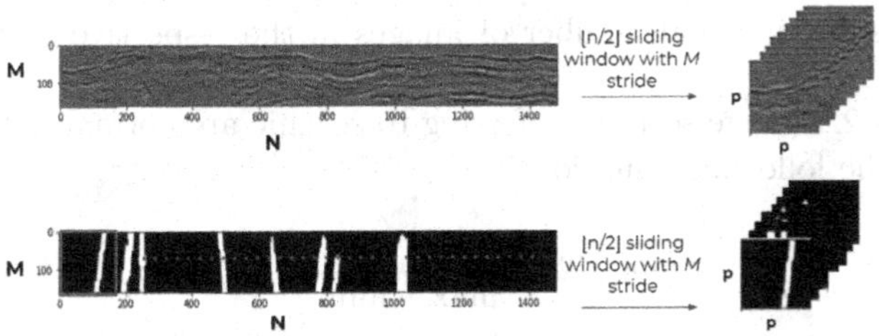

Fig. 3. Process of data sectioning.

Once the seismic data, input, and target have been prepared, the entire image dataset is randomized and divided into three separate sets: training, validation, and testing. These subsets are partitioned into a ratio of 70%, 15%, and 15% accordingly. The correspondence between the original seismic and the fault mask is maintained throughout the entire process. Therefore, once the data had been cropped and separated, we opted to apply data augmentation to the images that displayed fault presence. This was necessary since, in this type of task, the majority of the dataset contains images without faults, and classifying pixels one by one tends to be more unbalanced, with more pixels labeled as 0 than as 1 (non-fault/fault) (Wei et al. 2022).

The augmentation technique applied to the database was the flip transformation. Although there are various geometric transformations that can be used for augmentation, it's important to consider the data domain to ensure that the operations do not introduce errors in the learning process. In this case, flipping was chosen as it preserves the original orientation of faults, which is typically vertical. By rotating on the vertical axis, the flip operation doubles the number of images, as illustrated in Fig. 4.

Using the described procedure, we initially generated a database of 9369 sub-images from our seismic dataset, which had a dimension of $1401 \times 1481 \times 160$ pixels (inline/xlines/crossline). This was achieved by generating $\lfloor (n/p) \rfloor = 9$ images per inline, where $p = 160$. The resulting images were split into 6558 for training, 1405 for validation, and 1406 for testing. To increase the training and

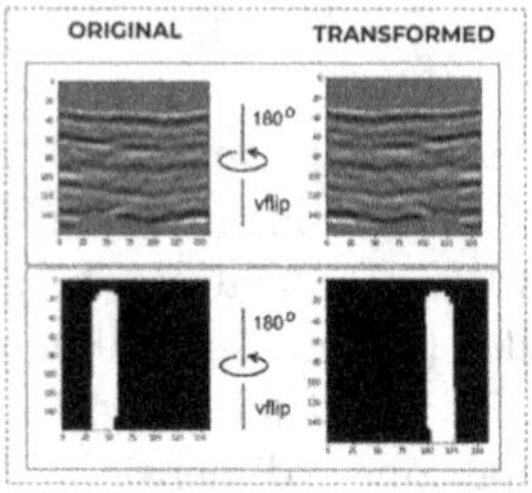

Fig. 4. Vertical flipping transformation.

validation sets, data augmentation was applied to all images containing faults, resulting in a 29% increase in the training data and a 36% increase in the test data. As a result, the total number of images in the respective groups became 9242 and 1914.

Finally, all 2D image sections referring to seismic are normalized between -1 and 1, using the following equation:

$$img = 2 \cdot \frac{img[:,:] - \min}{\max - \min} - 1 \tag{1}$$

where *img* is the 2D image resulting from the normalization, min is the minimum image value, and max is the maximum image value.

4 Methods

The TransUNet model, which uses Transformers, is utilized in this work. Given the success of this architecture in the realm of visual computing, our aim is to evaluate its efficacy for fault extraction in heterogeneous seismic fields and compare it with traditional models such as convolutional neural networks. The Fig. 5 shows the training input and the models used in this process.

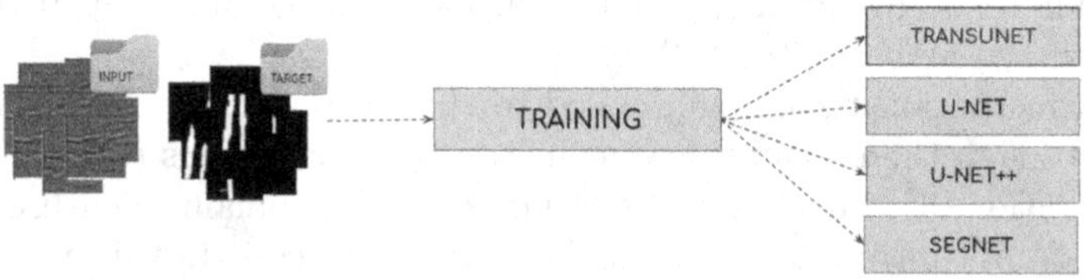

Fig. 5. Methodology workflow.

4.1 CNN Models

Image classification, a fundamental problem in computer vision, involves categorizing images into predefined classes and serves as the basis for other tasks such as region localization, detection, and segmentation. The CNNs are one of the most commonly used deep learning networks for this task, named after the linear mathematical operation called convolution between matrices (Albawi et al. 2017). CNNs are a type of feedforward neural networks, meaning that the information flows only in one direction, from the input to the output. Inspired by biological neural networks (Rawat and Wang 2017), CNNs differ from regular neural networks in that each unit in a CNN layer is a two-dimensional filter that is convolved with the input of that layer, enabling them to extract local features from images (Khan et al. 2018).

The convolutional layer of a CNN consists of a two-dimensional filter that convolves with the input feature map. The filter is an array of discrete numbers, where each element is a weight, learned during the training phase. In the beginning, these weights are randomly assigned until the learning process updates them. The CNN architecture includes layers for convolution, pooling to reduce the feature map size, and fully connected layers where each neuron is directly connected to the neurons in the previous and next layers. To gain an understanding of how convolutional networks can be constructed, we will explore three models from the U-Net family and subsequently utilize them to draw comparisons with the Transformer architecture.

U-Net. The original purpose of the U-Net network was to perform segmentation of medical images, and its architecture was an update and extension of the fully connected network. U-Net aimed to improve segmentation accuracy while minimizing the required amount of data (Long et al. 2015). The U-Net architecture, proposed by Ronneberger et al. [2015], consists of a contraction path (left side) and an expanding path (right side) for accurate segmentation of medical images. The contraction path has a typical convolutional network architecture, comprising two 3×3 convolutions, each followed by a rectified linear unit (ReLU), and a max-pooling operation with a 2×2 filter and stride 2 for downsampling. The number of feature channels is doubled at each downsampling step. The expansive path, on the other hand, involves increasing the resolution of the feature map, followed by a 2×2 convolution (half the number of channels), concatenation with the corresponding feature map of the contraction path, and two 3×3 convolutions (each followed by a ReLU). The last layer uses a 1×1 convolution to map the output to the desired number of classes.

U-Net++. Zhou et al. [2018] developed the U-Net++ architecture with the goal of improving the accuracy of medical image segmentation. This architecture is based on dense and nested skip connections, which provide a new approach to the segmentation task. The U-Net++ architecture was developed to enhance the performance of medical image segmentation by addressing limitations in previous models. Zhou et al. [2019] proposed an approach that employs multiple U-Nets with different depths, where the encoders and decoders are connected by dense and nested skip connections. The U-Nets share an encoder, while their decoders are interconnected, and deep supervision is used during training to simultaneously train all the constituent U-Nets while benefiting from a shared image representation. The redesigned skip connections in U-Net++ allow for variable-scale feature maps at a decoder node, enabling the aggregation layer to decide how attribute maps carried over the skip connections should be merged with the decoder feature maps.

SegNet. The SegNet (Badrinarayanan et al. 2017) is a convolutional neural network architecture designed for semantic pixel segmentation, comprising of an encoder network, a corresponding decoder network, and a pixel-wise classification layer. The encoder network architecture is similar to that of VGG-16 with 13 convolutional layers. The decoder network also has 13 layers, each corresponding to an encoder layer. The final output of the decoder is fed into a multiclass softmax classifier to generate class probabilities for each pixel. Unlike U-Net, SegNet does not reuse pooling indices but transfers the entire attribute map to the corresponding decoder and concatenates them into upsampled decoder feature maps through deconvolution.

4.2 Transformer Models

The Transformer (Khan et al. 2022; Vaswani et al. 2017) is a novel neural network that utilizes attention operations and was originally developed for natural language processing (NLP), where it has demonstrated remarkable success (Huang et al. 2020). In the field of computer vision, the Transformer has been increasingly employed to replace traditional techniques, resulting in various advantages (Bi et al. 2021).

The Transformer architecture includes an encoder and a decoder, both containing multiple attention blocks with the same architecture. The encoder produces encodings of the input, while the decoder takes these encodings and utilizes its contextual information to generate the output sequence (Han et al. 2022). Specifically, the Transformer encoder is composed of L layers of Multihead Self-Attention (MSA) and Multi-Layer Perceptron (MLP) blocks, alternating between the two. Before each block, Layer Normalization (LN) is applied, and residual connections are used after every block. Finally, the encoded feature representation is upsampled to full resolution to predict the dense output.

The success of the transformer in NLP has encouraged researchers to explore its potential in other areas. Consequently, similar models have been developed to learn useful image representations using the Transformer's concept. The Vision Transformer (ViT) (Dosovitskiy et al. 2020), for instance, has proven to be highly effective in several benchmarks, drawing inspiration from the self-attention mechanism in NLP, where word embeddings are substituted by patch embeddings (Fu 2022).

ViT has paved the way for the development of several other models based on attention mechanisms, which have brought about significant advances in various fields of computer vision. Surveys conducted by Guo et al. [2022] and Han et al. [2022] have shown that attention-based methods have been beneficial for tasks such as image classification, semantic segmentation, face recognition, few-shot learning, medical image processing, image resolution, 3D vision, among others.

TransUNet (Chen et al. 2021) is a model that harnesses the power of transformers for medical image segmentation. By combining CNN architectures, such as U-Net, which can extract low-level visual features to preserve fine spatial details, and Transformers, which excel in modeling global context, TransUNet creates a powerful hybrid architecture for accurate and efficient medical image segmentation.

TransUNet Architecture. TransUNet combines CNN and Transformer architectures to leverage the spatial details of CNN features and the global context captured by Transformers for medical image segmentation. The model follows a U-shape design, where Transformers establish self-attention mechanisms to encode the features in a sequence-by-sequence prediction perspective. The resulting self-attentive feature is upsampled and combined with high-resolution CNN features that were skipped during encoding, enabling precise localization.

Transformer is used as an encoder by transforming the input image into a sequence of flattened 2D patches through tokenization. To achieve this, the input image x is reshaped into N patches of size $P \times P$, where N is determined by the image's height and width (H and W) and the patch size (P), such that $N = \frac{HW}{P^2}$. A unique marker is assigned to each patch to preserve its positional information in the sequence, and the resulting sequence is fed as input to the encoder.

To recover the spatial order during upsampling, the encoded feature size is reshaped from $\frac{HW}{P^2}$ to $\frac{H}{P} \times \frac{W}{P}$, while the number of channels is reduced to the number of classes using 1×1 convolutions. Finally, the feature map is bilinearly upsampled to the full resolution of $H \times W$ to generate the final segmentation output. To address the issue of partial information loss resulting from using Transformer solely as an encoder, TransUNet utilizes a hybrid CNN-Transformer architecture that first leverages CNN to extract features from the input, followed by patch embedding of 1×1 patches extracted from the CNN feature map instead of the raw images. Therefore, the sequence of hidden features is reshaped to achieve full resolution from $\frac{H}{P} \times \frac{W}{P}$ to $H \times W$ by applying multiple cascades of upsampling blocks. Each block includes a $2\times$ upsampling operator, a 3×3 convolution layer, and a ReLU layer in sequence. This enables the aggregation of features at different resolution levels through skip connections.

4.3 Evaluation Metrics

As the task can be seen as a semantic segmentation task, it is crucial to rigorously evaluate the system's efficiency for it to be useful and produce effective contributions. To evaluate the effectiveness of a segmentation system used for extracting faults in images using machine learning methods, two essential metrics that assess the quality of segmentation were selected:

- Jaccard index: also known as the intersection over union (IoU), it quantifies the percentage overlap between the mask input (i.e., the target of the segmentation) and the predicted output. The Jaccard index (Shi et al. 2014) is calculated as the number of pixels that are common between the two images ($A \cap B$) divided by the number of pixels resulting from the union of both ($A \cup B$).
- Dice coefficient: is a measure frequently utilized in the computer vision field to evaluate the similarity between two images (Crum et al. 2006). It is similar to the IoU and computed as twice the overlap area divided by the total number of pixels in both images.

The performance assessment of the employed models can also be examined through a pixel-wise classification approach. This method employs metrics, including accuracy, recall, precision, and F1-score, to evaluate the prediction performance of positive samples. The matrix presents the predictions made for each class and provides a representation to measure the effectiveness of the classification model. In the case of fault detection applications, positive samples represent faults and their locations in a seismic volume (Huang et al. 2017).

5 Results and Discussions

The presentation and discussion of the results obtained from the trained models, seismic areas reconstruction, and performance metrics are presented using the previously described model.

5.1 Experiments

To assess the effectiveness of applying Transformer on seismic data, we utilized the TransUNet model on the database outlined in Sect. 3. Moreover, we employed three additional models that employ the convolutional neural network approach to compare the attained results.

The architectures used were parameterized with identical settings for the loss function, learning rate, and number of epochs to enable comparison of outcomes with the same initialization. An empirical value of $1e-4$ was chosen for the learning rate based on previous evaluations of different values. For the loss function, binary cross-entropy was chosen as it is commonly used for classification purposes and semantic segmentation is a pixel-level classification task. The number of epochs was set to 100 for all models, except for TransUNet where the batch size was reduced to 16 due to its higher memory complexity compared to the other networks.

The results obtained from the execution of all methods are presented in Table 1, where it can be observed that the TransUNet network surpasses the other architectures (U-Net, U-Net++, and SegNet) by 2.35%, 4.93%, and 5.03%, respectively, achieving an overall Dice score of 88.34%. Considering the IoU metric, this difference increases to 4.86%, 6.73%, and 7.23%, with TransUNet obtaining a value of 84.34%. It is worth noting that the difference between the values obtained by these two metrics is mainly due to the high penalty imposed by the IoU in cases where the classification results are poor. The evaluation of the other metrics confirms the superior performance of the TransUNet network.

Table 1. Quantitative comparison of the segmentation performance in fault detection task.

Models	Metrics					
	Dice	IoU	Accuracy	Precision	Recall	F1-Score
U-Net	0.8599	0.7948	0.9781	0.9258	0.9204	0.9230
U-Net++	0.8341	0.7761	0.9682	0.8852	0.8934	0.8892
SegNet	0.8331	0.7711	0.9570	0.9266	0.8312	0.8763
TransUNet	**0.8834**	**0.8434**	**0.9785**	**0.9303**	**0.9174**	**0.9238**

The comparisons between the predictions made by the previously presented models are illustrated in Fig. 6. The results indicate that U-Net and TransUNet produced images that are more similar to the target, while SegNet and U-Net++ exhibit a considerable amount of noise in their outputs. When comparing U-Net and TransUNet, a slightly more accurate border delimitation can be observed in U-Net. This may be due to the greater abstraction of global context extraction in the initial layers of TransUNet.

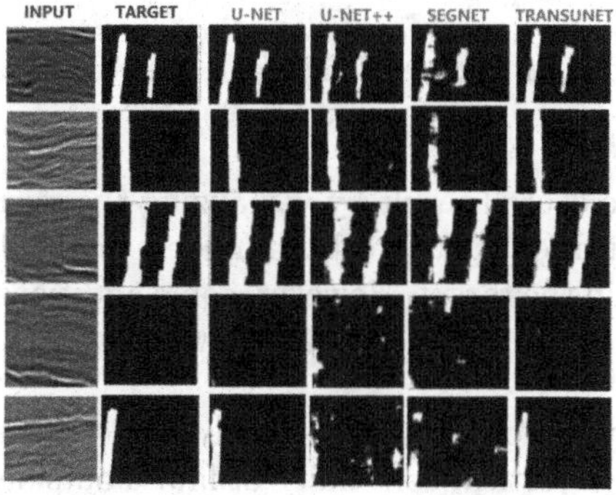

Fig. 6. Qualitative comparison of different models applied to seismic segmentation.

In order to conduct a more comprehensive analysis of the performance of the two networks, we reconstructed the slices that were formed by each predicted sub-image. This reconstruction was made possible by the fact that each input image to the network has a corresponding nomenclature that corresponds to the seismic inline and its cut order. Consequently, the training images were

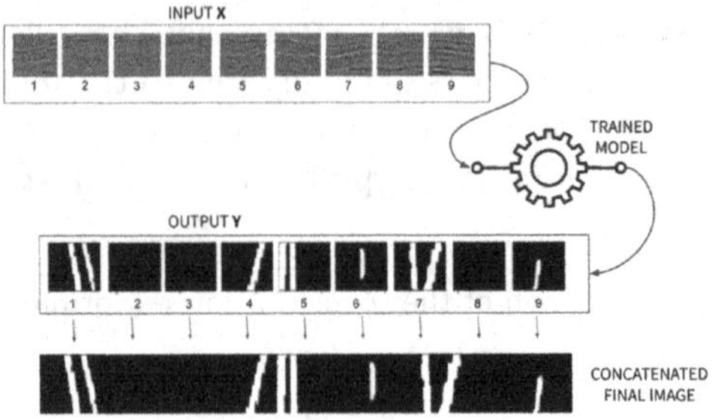

Fig. 7. Image prediction and concatenation process.

predicted in sequential order and their results were concatenated, as illustrated in an example shown in Fig. 7.

By applying this approach, we can reconstruct the entire seismic volume and examine the output of the two models. Figure 8 presents a comparison of the predictions made for two different slices, indicating that both models were able to detect the structures highlighted in the target, as well as some smaller regions, with TransUNet providing a more significant representation of them.

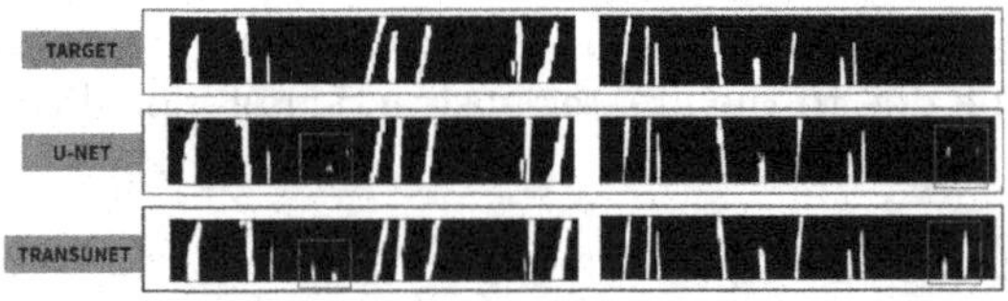

Fig. 8. Comparison between U-Net and TransUNet predictions.

Upon analyzing the overall result by visualizing the seismic cube prediction, as presented in Fig. 9, it is evident that both models were able to identify most of the structures present in the target. However, some additional small regions were also detected, which require detailed analysis when viewed in two dimensions, as depicted in Fig. 10.

The TransUNet prediction achieved slightly better results, with more indicated faults and greater vertical continuity of faults outside the regions present in the target binary cube. Nevertheless, the overall difference between the two methods was minimal.

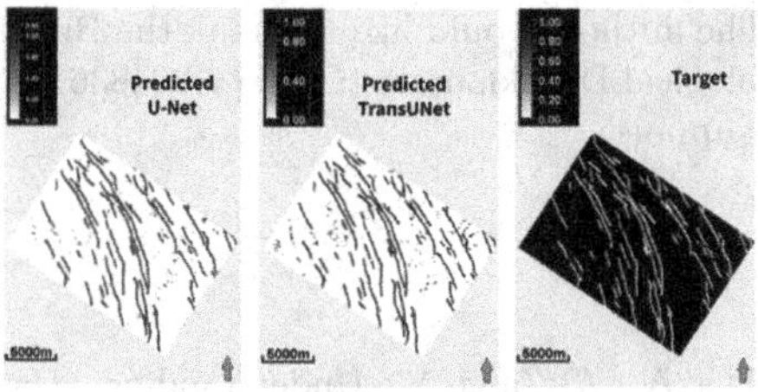

Fig. 9. Comparison between U-Net and TransUNet prediction in full 3D field.

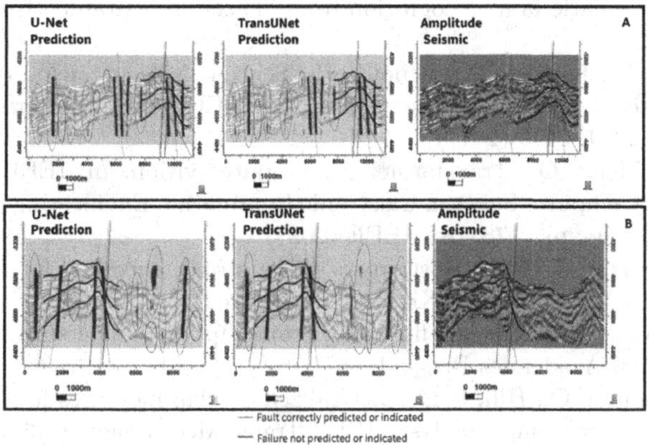

Fig. 10. Analysis of predicted failures in two different regions, *A* and *B*.

6 Conclusions

This work investigated the use of a hybrid model, TransUNet, which combines the strengths of convolutional networks and Transformer's content abstraction in the geological context. The results demonstrate the effectiveness of this approach in segmenting seismic images from a heterogeneous environment, such as the pre-salt layer, indicating potential applications of this architecture in various configurations for identifying and extracting geological structures in the field of seismic imaging.

The effectiveness of TransUNet in fault identification on seismic data was demonstrated by comparing it to conventional state-of-the-art methods. This study concludes that the incorporation of Transformer in this context has the potential to extract valuable information from seismic databases. The evaluation included both qualitative and quantitative approaches, suggesting that this new type of architecture could serve as a benchmark for other databases considering the use of TransUNet in the field of visual computing.

Acknowledgments. The authors would like to thank the Brazilian National Council for Scientific and Technological Development (CNPq #304836/2022-2) and Shell Brasil Petróleo Ltda for their support.

References

Albawi, S., Mohammed, T.A., Al-Zawi, S.: Understanding of a convolutional neural network. In: International Conference on Engineering and Technology, pp. 1–6. IEEE (2017)

Alsadi, H.N.: Seismic hydrocarbon exploration. In: 2D and 3D Techniques, Seismic Waves. Springer (2017)

An, Y., Guo, J., Ye, Q., Childs, C., Walsh, J., Dong, R.: Deep convolutional neural network for automatic fault recognition from 3D seismic datasets. Comput. Geosci. **153**, 104776 (2021)

Badrinarayanan, V., Kendall, A., Cipolla, R.: SegNet: a deep convolutional encoder-decoder architecture for image segmentation. IEEE Trans. Pattern Anal. Mach. Intell. **39**(12), 2481–2495 (2017)

Bi, J., Zhu, Z., Meng, Q.: Transformer in computer vision. In: IEEE International Conference on Computer Science, Electronic Information Engineering and Intelligent Control Technology, pp. 178–188. IEEE (2021)

Chen, J., et al.: Trans-UNet: transformers make strong encoders for medical image segmentation. arXiv preprint arXiv:2102.04306 (2021)

Council, N.R.: Rock fractures and fluid flow: contemporary understanding and applications. National Academies Press (1996)

Crum, W.R., Camara, O., Hill, D.L.: Generalized overlap measures for evaluation and validation in medical image analysis. IEEE Trans. Med. Imaging **25**(11), 1451–1461 (2006)

Dondurur, D.: Acquisition and Processing of Marine Seismic Data. Elsevier (2018)

Dosovitskiy, A., et al.: An image is worth 16×16 words: Transformers for image recognition at scale (2020). arXiv preprint arXiv:2010.11929

Fu, Z.: Vision Transformer: ViT and its Derivatives. arXiv e-prints, pages arXiv-2205 (2022)

Guo, M.-H., et al.: Attention mechanisms in computer vision: a survey. Comput. Visual Media, pp. 1–38 (2022)

Han, K., et al.: A survey on vision transformer. IEEE Trans. Pattern Anal. Mach. Intell. (2022). abs/2012.12556:1–20

Huang, L., Dong, X., Clee, T.E.: A scalable deep learning platform for identifying geologic features from seismic attributes. Lead. Edge **36**(3), 249–256 (2017)

Huang, M., Zhu, X., Gao, J.: Challenges in building intelligent open-domain dialog systems. ACM Trans. Inf. Syst. **38**(3), 1–32 (2020)

Khan, S., et al.: Transformers in vision: a survey. ACM Comput. Surv. **54**(10s), 1–41 (2022)

Khan, S., Rahmani, H., Shah, S.A.A., Bennamoun, M.: A guide to convolutional neural networks for computer vision. Synthesis Lectures Comput. Vision **8**(1), 1–207 (2018)

Long, J., Shelhamer, E., Darrell, T.: Fully convolutional networks for semantic segmentation. In: IEEE Conference on Computer Vision and Pattern Recognition, pp. 3431–3440 (2015)

Pepper, R.E.F., Bejarano, G.: PS advances in seismic fault interpretation automation. In: AAPG Annual Convention (2005)

Rawat, W., Wang, Z.: Deep convolutional neural networks for image classification: a comprehensive review. Neural Comput. **29**(9), 2352–2449 (2017)

Ronneberger, O., Fischer, P., Brox, T.: U-Net: convolutional networks for biomedical image segmentation. In: International Conference on Medical Image Computing and Computer-Assisted Intervention, pp. 234–241. Springer (2015)

Shi, R., Ngan, K.N., Li, S.: Jaccard index compensation for object segmentation evaluation. In: IEEE International Conference on Image Processing, pp. 4457–4461. IEEE (2014)

Vaswani, A., Shazeer, N., et al.: Attention is all you need. In: Advances in Neural Information Processing Systems, 30 (2017)

Wei, X.-L., et al.: Seismic fault detection using convolutional neural networks with focal loss. Comput. Geosci. **158**, 104968 (2022)

Wu, X., Liang, L., Shi, Y., Fomel, S.: FaultSeg3D: Using synthetic data sets to train an end-to-end convolutional neural network for 3D seismic fault segmentation. Geophysics **84**(3), IM35-IM45 (2019a)

Wu, X., Shi, Y., Fomel, S., Liang, L., Zhang, Q., Yusifov, A.Z.: FaultNet3D: predicting fault probabilities, strikes, and dips with a single convolutional neural network. IEEE Trans. Geosci. Remote Sens. **57**(11), 9138–9155 (2019)

Zhao, T., Mukhopadhyay, P.: A fault detection workflow using deep learning and image processing. In SEG International Exposition and Annual Meeting, OnePetro (2018)

Zheng, Y., Zhang, Q., Yusifov, A., Shi, Y.: Applications of supervised deep learning for seismic interpretation and inversion. Lead. Edge **38**(7), 526–533 (2019)

Zhou, Z., Siddiquee, M.M.R., Tajbakhsh, N., Liang, J.: U-Net++: a nested U-Net architecture for medical image segmentation. In: Deep Learning in Medical Image Analysis and Multimodal Learning for Clinical Decision Support, pp. 3–11. Springer (2018)

Zhou, Z., Siddiquee, M.M.R., Tajbakhsh, N., Liang, J.: U-Net++: redesigning skip connections to exploit multiscale features in image segmentation. IEEE Trans. Med. Imaging **39**(6), 1856–1867 (2019)

Evaluating Recent Legal Rhetorical Role Labeling Approaches Supported by Transformer Encoders

Alexandre Gomes de Lima[1,2(✉)], José G. Moreno[3], Taoufiq Dkaki[3], Eduardo Henrique da S. Aranha[2], and Mohand Boughanem[3]

[1] Instituto Federal do Rio Grande do Norte, Natal, Brazil
alexandre.lima@ifrn.edu.br
[2] Universidade Federal do Rio Grande do Norte, Natal, Brazil
eduardoaranha@dimap.ufrn.br
[3] Institut de Recherche en Informatique de Toulouse, UMR 5505, CNRS, 31000 Toulouse, France
{Jose.Moreno,Taoufiq.Dkaki,Mohand.Boughanem}@irit.fr

Abstract. Pre-trained Transformer models have been used to improve the results of several NLP tasks, which includes the Legal Rhetorical Role Labeling (Legal RRL) one. This task assigns semantic functions, such as *fact* and *argument*, to sentences from judgment documents. Several Legal RRL works exploit pre-trained Transformers to encode sentences but only a few employ approaches other than fine-tuning to improve the performance of models. In this work, we implement three of such approaches and evaluate them over the same datasets to achieve a better perception of their impacts. In our experiments, approaches based on data augmentation and positional encoders do not provide performance gains to our models. Conversely, the models based on the DFCSC approach overcome the appropriate baselines, and they do remarkably well as the lowest and highest improvements respectively are 5.9% and 10.4%.

Keywords: Indian legal system · Sentence classification · Sentence encoding · DFCSC

1 Introduction

One of the main goals of Legal AI, the application of Artificial Intelligence (AI) in the legal domain, is task automation, which increases the productivity of legal professionals and makes the law more accessible. Several works related to Legal AI are about tasks concerning judgments such as judgment prediction [14,24],

This work was supported by NPAD/UFRN, by the Coordenação de Aperfeiçoamento de Pessoal de Nível Superior - Brasil (CAPES, Finance Code 001) and by the LawBot project (ANR-20-CE38-0013), granted by ANR the French Agence Nationale de la Recherche.

M. C. Naldi and R. A. C. Bianchi (Eds.): BRACIS 2023, LNAI 14196, pp. 18–32, 2023.
https://doi.org/10.1007/978-3-031-45389-2_2

and legal summarization [7,9], which rely on judgment documents as the main source of data. A judgment document is a record of a lawsuit authored by the Court that sets forth its decision about the case. The document includes such as statements of the facts, analysis of relevant law, rights and liabilities of the parties, and rulings, among others.

Specific information is required or can be exploited to boost Legal AI tasks: facts for fact-based search; facts and violated statutes for judgment prediction; facts, arguments, and judicial reasoning for legal summarization. The *Legal Rhetoric Role Labeling* (Legal RRL) task assigns semantic functions to sentences in judgment documents (e.g., fact, argument, ruling). Thus, Legal RRL is a way of finding information that is useful in itself as well as for downstream tasks. The Legal RRL task is not trivial, even for humans. One of the limiting factors is the heterogeneity in the structure of documents. It is common for a lawsuit to be judged by different courts (e.g., lower courts, supreme court), which may adopt different standards or guidelines to write judgment documents. Another issue is the subjectivity related to the interpretation of different rhetorical roles [3,15]. Inter-annotator agreements reported from works about the development of Legal RRL datasets are good evidence of such subjectivity. In [15], the authors report a low inter-agreement score (Kappa index is 0.65) among three annotators on judgments from the income tax domain over 1,600 sentences. Similarly, in [8] a low inter-agreement score (Kappa index is 0.59) is reported for 40,305 sentences and 30 annotators.

Legal RRL automation can be implemented as a sentence classification task, where sentences are fed into a machine learning model that assigns a label to each sentence. The model consists of two core components, a sentence encoder and a classifier. The sentence encoder is responsible to produce numeric representations (i.e., vectors), commonly known as sentence embeddings, of text sentences while the classifier assigns labels to each representation as illustrated in Fig. 1.

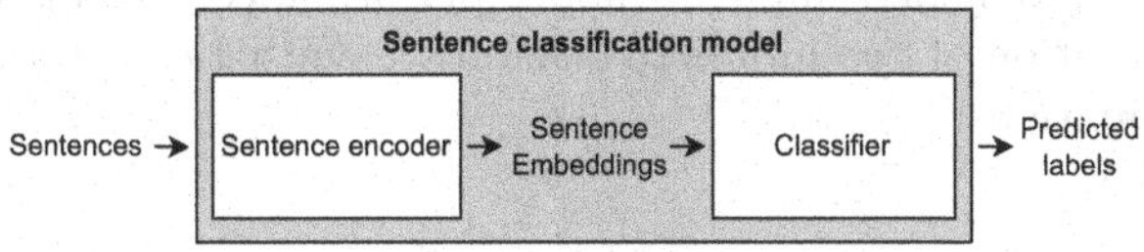

Fig. 1. Sentence classification task.

The representation vectors directly impact the performance of the model since richer representations lead to more precise classifications [21]. The exploitation of pre-trained Transformer models [20] is the current state-of-the-art in many NLP tasks and this includes the sentence encoding one. Several pre-trained Transformer models are publicly available, such as BERT [5] and RoBERTa [13]. When generating the vector representation of a word w in a sentence, a Transformer model takes w and its neighbor words into account. Therefore, it yields different vector representations for w when it occurs in different sentences or different positions of the same sentence. This context exploitation ability allows the model to yield rich text representations.

Fine-tuning is a common procedure for exploiting pre-trained Transformers that adjusts the weights of a pre-trained model for a specific task and dataset. This is a general procedure and it can be employed to tackle the Legal RRL task [1,3,8,18]. Some recent works [10–12,15] propose additional procedures as a way of improving the performance of models handling the Legal RRL task but, despite their potential, such works employ different datasets what hinders a fair comparison. In this context, the goal of this work is to perform an equitable comparison among approaches proposed in recent works [10–12,15]. To achieve this, we implement models based on three of such approaches and evaluate them over two datasets. Only the DFCSC-based models overcome the appropriate baselines, and they do so remarkably well: the lowest and highest improvements respectively are 5.9% and 10.4%.

2 Recent Approaches to Handle Legal RRL

Pre-trained Transformer models have achieved remarkable results in many NLP tasks. Thus, it is not a surprise the publication of Legal RRL works that rely on such models [1,3,8,18]. Fine-tuning is a common training procedure for one working with pre-trained models, whose goal is the adjustment of the model weights to the task at hand. A pre-trained Transformer model can be employed out-of-the-box (i.e., without fine-tuning) but since the task at hand is mostly different from the task used in its pre-training, the fine-tuning improves the performance of the model. Other training procedures can be employed together with the fine-tuning one as a way of improving the performance of a model. Regarding the Legal RRL task, there are a few recent works that follow this strand.

Mixup [23], a data augmentation technique, is exploited in [10] to augment the training set and hence improve model performance. Mixup relies on the weighted interpolation of two inputs to generate a synthetic vector as defined by the following equations:

$$\begin{aligned} \tilde{\mathbf{x}} &= \lambda \mathbf{x}_i + (1-\lambda)\mathbf{x}_j, & \mathbf{x}_i, \mathbf{x}_j &\in \mathcal{R}^e \\ \tilde{\mathbf{y}} &= \lambda \mathbf{y}_i + (1-\lambda)\mathbf{y}_j, & \mathbf{y}_i, \mathbf{y}_j &\in \mathcal{R}^{|L|}, \end{aligned}$$

where $\mathbf{x}_i$ and $\mathbf{x}_j$ are sentence embedding vectors, e is the embedding dimension, $\mathbf{y}_i$ and $\mathbf{y}_j$ are one-hot label encodings, and $|L|$ is the number of labels or classes. $(\mathbf{x}_i, \mathbf{y}_i)$ and $(\mathbf{x}_j, \mathbf{y}_j)$ are two examples drawn at random from the training data and λ is a value in the interval $[0, 1]$ drawn at random from a Beta(α, α) distribution, where $\alpha > 0$. In their setup [10], the authors exploit Sentence BERT [17] as an encoder without fine-tuning. The exploited dataset comprises 60 judgments, 10,024 sentences, and seven rhetorical roles. As a result, the Mixup-based models overcome the respective baselines in terms of F1 score. For example, a linear classifier achieves a gain of 4.3% in the worst case and 6.8% in the best case.

Another work [12] leverages positional information in addition to encoded sentences. Each sentence is represented by a sentence embedding and a positional embedding, the latter indicating the position a sentence occupies in its source document. Positional embeddings are computed by the same sinusoidal positional encoding method exploited in the original Transformer architecture [20]:

$$PE_{(pos,2k)} = \sin(pos/m^{2k/e})$$
$$PE_{(pos,2k+1)} = \cos(pos/m^{2k/e})$$

where pos is an integer indicating the position of a sentence, k is a dimension of a **PE** vector where $0 \leq k < (e-1)/2$, e is the embedding dimension, and m is the max number of sentences in a document. The sentence and positional embeddings must be combined before being fed to the models' classifier and to do this the authors employ vector sum and vector concatenation. In their setup [12], they experiment with four pre-trained Transformers as sentence encoders. The exploited dataset comprises 369 judgments, 54,244 sentences, and two rhetorical roles. The authors report gains from 3.3% to 12.6% achieved by fine-tuned models.

The approaches presented in [10, 12] rely on the single sentence classification paradigm: a sentence is encoded in isolation and so its context (neighbor sentences) is ignored. Although, Transformers are notable for exploiting context to encode sentences and thus the work of [11] leverages this ability to produce richer sentence embeddings as a way to achieve better results. The encoder is fed with chunks of sentences and a special character is employed to signal sentences' limits. The authors propose a chunk layout and an assembly procedure, named DFCSC (Dynamic-filled Contextualized Sentence Chunk), that avoids the use of padding tokens, the truncation of core sentences, and that defines at run time the number of core sentences and edge tokens in each chunk. Core sentences are the sentences taken into account by the classifier and edge tokens are tokens inserted at the edges of the chunks as a way to provide context for the first and last core sentences (see Fig. 2). The procedure requires two hyperparameters: c_{len} (the number of tokens in a chunk) and m_{edges} (the minimum desired number of tokens in the edges of a chunk). In their setup [11], the authors exploit a dataset with 277 judgments, 31,865 sentences, and 13 rhetorical roles. They experiment with four pre-trained Transformers and two additional approaches based on sentence chunks besides the DFCSC one. DFCSC-based models perform the best and the reported gains are respectively of 3.76% and 8.71% in the best and worst cases.

3 Experimental Setup

3.1 Datasets

We employ two datasets to evaluate the models which we refer to as 7-roles and 4-roles. Both are derivations of the original dataset developed in [15][1] that

[1] Available at https://github.com/Exploration-Lab/Rhetorical-Roles.

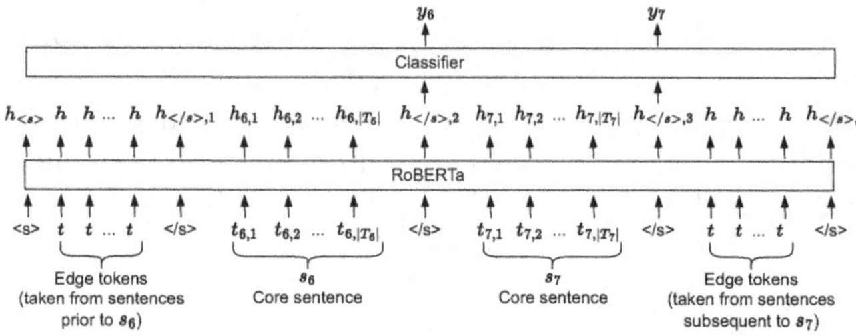

Fig. 2. Illustration of a DFCSC. Remark that the classifier takes into account only the tokens chosen as core sentence representations. s_i means the i-th sentence, y_i is the label assigned to s_i, h represents a token embedding, t represents a token, $|T_i|$ is the number of tokens in s_i, <s> and </s> are special tokens.

comprises 100 judgment documents written in English from the Indian legal system split into training set (80 documents and 16,845 sentences), validation set (10 documents and 2,142 sentences), and test set (10 documents and 2,197 sentences). A team of legal experts annotated the documents and each sentence is labeled with one of the following 13 rhetorical roles: *Fact* (FAC), *Issues* (ISS), *Argument Petitioner* (ARG-P), *Argument Respondent* (ARG-R), *Statute* (STA), *Dissent* (DIS), *Precedent Relied Upon* (PRE-R), *Precedent Not Relied Upon* (PRE-NR), *Precedent Overruled* (PRE-O), *Ruling By Lower Court* (RLC), *Ruling By Present Court* (RPC), *Ratio Of The Decision* (ROD), and *None* (NON).

We follow the same procedure from [15] to create the 7-roles dataset. The *Issue* labels are renamed as *Fact*, the *Argument Petitioner* and *Argument Respondent* labels are renamed as *Argument* (ARG), and the *Precedent Relied Upon*, *Precedent Not Relied Upon* and *Precedent Overruled* (PRE-O) labels are renamed as *Precedent* (PRE). Also, the *Dissent* and *None* sentences are ignored for classification purposes. Although, some models rely on document-level data to produce feature vectors and discarding such sentences would result in a data loss degree. Regarding chunk-based models, *Dissent* and *None* sentences are part of the model input but their predicted labels are discarded when we compute the loss function value. Regarding PE-based models, such sentences are not part of the model input but they are exploited when computing the positional embeddings to preserve the actual sentence positions. By discarding the *Dissent* and *None* sentences, the 7-roles dataset is composed of 16,264, 2,109, and 2,094 sentences in the training, development, and test sets, respectively.

To be effective, PE-based models require some degree of correlation between sentence positions and rhetorical roles [12]. Thus, we designed the 4-roles dataset to perform a fairer evaluation of PE-based models. To achieve this goal, we modified the original dataset from [15] by maintaining the roles occurring often at the same positions (i.e., the *Fact*, *Ruling By Present Court* and *RatioOfTheDecision*) and by renaming the remaining roles as *Other*. We do not discard sentences in this process and hence the 4-roles dataset has the same number of sentences as the original one.

The original dataset is highly unbalanced and this is a characteristic of the 7-roles and 4-roles datasets as well. The distribution of sentences by rhetorical role in the three datasets is shown in Fig. 3.

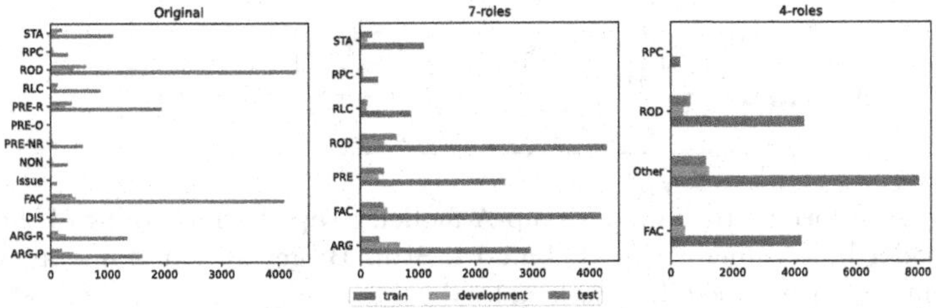

Fig. 3. Number of sentences per rhetorical role in the original, 7-roles, and 4-roles datasets.

3.2 Models

We implement models relying on the approaches from [10,12], and [11]. For each approach, we experiment with two or three Transformer encoders. Details of our results and source code are available at https://github.com/alexlimatds/bracis_2023.

Mixup-A: These are models whose training includes mixup vectors as proposed by [10]. At first, a Transformer encoder is fine-tuned on the training dataset in the same fashion as the SingleSC approach described ahead (step A in Fig. 4). In the following, the fine-tuned Transformer is employed to yield sentence embeddings which serve as source data of the mixup procedure (step B in Fig. 4). Finally, the mixup vectors and the sentence embeddings are employed to train a Mixup-A model (step C in Fig. 4). The training of a Mixup-A model updates only the weights of its classification layer. Sentences are represented by the last embedding vector of the token marking the beginning of the input (`[CLS]` or `<s>`). We employ InCaseLaw [16] and RoBERTa-base [13] as sentence encoders. InCaseLaw is a BERT model pre-trained with corpora from the Indian legal system.

The generation of mixup vectors is oriented to selected labels and an augmentation rate γ. For each selected label l_i, we generate $n_{l_i} \cdot \gamma$ mixup vectors, where n_{l_i} means the number of training sentences labeled with l_i. A mixup vector is generated from a sentence embedding labeled with l_i and a sentence embedding labeled with l_j where $i \neq j$ (the embeddings are selected at random). When experimenting with the 7-labels dataset, we set $\gamma = 0.5$, $\alpha = 0.1$, the selected labels are *Argument*, *Statute*, *Precedent*, *RulingByLowerCourt*, and *RulingByPresentCourt* which results in 3,881 synthetic vectors. For the 4-labels dataset, we set $\gamma = 0.5$, $\alpha = 0.1$, the selected labels are *Fact*, *Ratio Of The Decision*, and *Ruling By Present Court* which results in 4,408 synthetic vectors.

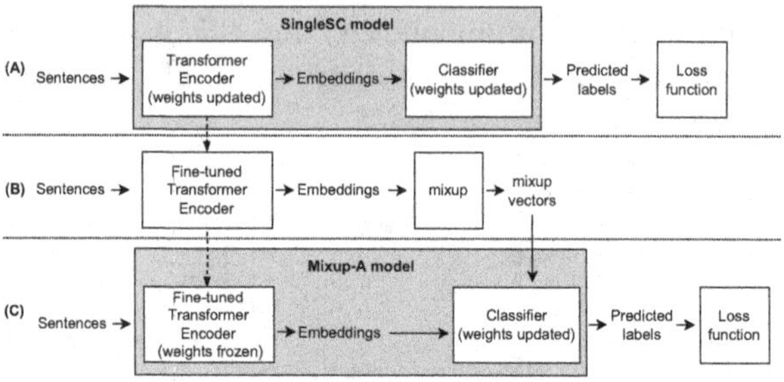

Fig. 4. Steps performed to train a Mixup-A model. **Step A:** fine-tuning of the Transformer encoder by training a SingleSC model. **Step B:** generation of mixup synthetic vectors from sentence embeddings yielded by the fine-tuned Transformer encoder. **Step C:** employing sentence embeddings and mixup vectors to train the classifier of the Mixup-A model.

Mixup-B: These are models similar to the Mixup-A ones but without a prior encoder fine-tuning step. The training of a Mixup-B model updates the weights of the encoder and of the classification layer (i.e., the Transformer is fine-tuned during the training of the model). Since the sentence embeddings change during the fine-tuning, the mixup procedure yields a new set of synthetic vectors at each training epoch as shown in Fig. 5. As before, we exploit InCaseLaw and RoBERTa as sentence encoders. When experimenting with the 7-labels dataset, we set $\gamma = 1.0$, $\alpha = 0.1$, and select all seven labels, which results in about 14,800 synthetic vectors per training epoch. For the 4-labels dataset, we set $\gamma = 1.0$, $\alpha = 1.0$, the selected labels are *Fact*, *Ratio Of The Decision*, and *Ruling By Present Court* which results in about 8,470 per training epoch.

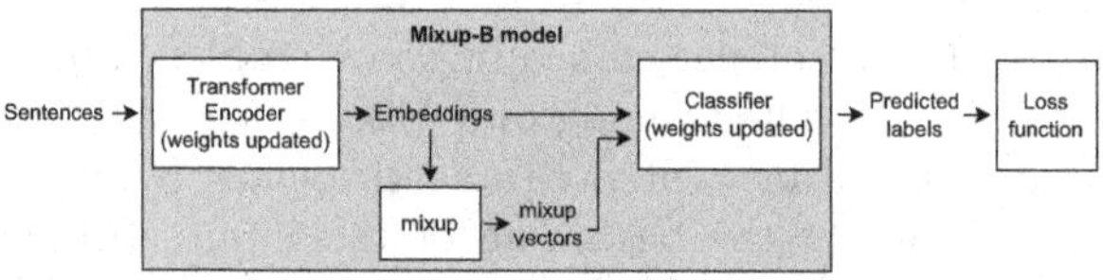

Fig. 5. Training procedure of a Mixup-B model.

PE-S: These are models that rely on the sum of sentence embeddings and positional embeddings as proposed in [12]. We exploit InCaseLaw and RoBERTa as sentence encoders and sentences are encoded in isolation.
PE-C: Like PE-S models, but we employ concatenation instead of sum to combine sentence embeddings and positional embeddings.
DFCSC-SEP: These are models based on the DFCSC approach [11]. We employ InCaseLaw, RoBERTa-base, and Longformer-base [2] as sentence encoders. Longformer is a model that relies on a sparse self-attention mechanism, which allows it to handle longer sentences than the BERT-based models.

In all models, core sentences are represented by the last hidden states of the respective separator tokens (the `[SEP]` or `</s>`token to the right of a core sentence). We set $m_{edges} = 250$ for the three models, $c_{len} = 512$ for the models based on InCaseLaw and RoBERTa, and $c_{len} = 1,024$ for the model based on Longformer.

DFCSC-CLS: Like DFCSC-SEP models, but each core sentence is represented by the concatenation of the last hidden states regarding the special token marking the beginning of the chunk (`[CLS]` or `<s>`) and the corresponding separator token (`[SEP]` or `</s>`).

We also implement the following baselines:

SingleSC: These are single sentence classification models, i.e. models that do not rely on chunks. Each sentence is fed to the model one by one and is represented by the hidden state of the token that marks the beginning of the input (`<s>` or `[CLS]`). SingleSC models are the reference baselines of Mixup and PE models, since such approaches do not consider context to encode sentences. The exploited pre-trained encoders are InCaseLaw and RoBERTa-base.

Cohan: These are models that follows the chunk design from [4], that is, chunks that do not share sentences or tokens. Cohan models are the reference baselines of DFCSC models since such approaches rely on chunks to encode sentences. The exploited pre-trained encoders are InCaseLaw, RoBERTa, and Longformer. Models based on InCaseLaw and RoBERTa have the following hyperparameter values: 85 as the maximum sentence length, 512 as the chunk length, and 7 as the maximum number of sentences in a chunk. The model based on Longformer has the following hyperparameter values: 85 as the maximum sentence length, 1,024 as the chunk length, and 14 as the maximum number of sentences in a chunk. Each sentence is represented by the hidden state of the following separator token.

All models use a single fully-connected layer as their classifier. Except for the Mixup-A models, the encoder is fine-tuned together with the classifier during the training of the respective model. For each Mixup-A model, the Transformer encoder is tuned with the same hyperparameters as its SingleSC counterpart (e.g., SingleSC-InCaseLaw and Mixup-A-InCaseLaw). For their classifiers, we adopt 200 training epochs, 10^{-2} as initial learning rate, and 0.2 as dropout rate. The remaining models are fine-tuned for four epochs, and we adopt 10^{-5} as initial learning rate and 0.2 as dropout rate of the classification layer. The pre-trained Transformer models are used with default parameters, except for the Longformer, whose global attention is set to consider `<s>` and `</s>` tokens. For all models we employ the cross entropy loss function and the Adam optimizer ($\beta_1 = 0.9$, $\beta_2 = 0.999$, $\epsilon = 10^{-8}$, $weight\_decay = 10^{-3}$). The learning rate is scheduled linearly from an initial value to zero, and there are no warm-up steps. We adopt a batch size of 4 for Longformer-based models and a batch size of 16 for the other models. Python 3.8.13, PyTorch 1.13.0, and Hugging Face Transformer 4.28.1 are the programming language and main libraries we use to develop all models. Hugging Face is mainly used to load pre-trained Transformer models. A single 16GB Tesla V100 SXM2 GPU is used in all experiments.

3.3 Evaluation

All models are trained on the training set and evaluated on the test set. For each model, the training and evaluation procedure is repeated five times with a different random seed for each repetition (the same five seeds for all models). The evaluation metric is the macro F1 score.

To compare the results obtained by the implemented models, we use the Almost Stochastic Order (ASO) test [6] with Bonferroni correction as implemented by [19]. ASO is a statistical significance test designed to compare two score distributions produced by deep learning models. The test analyzes two distributions and quantifies the amount of stochastic order violation by computing a value ϵ_{min}. Given the scores over multiple random seeds from two models A and B, A is declared superior if $\epsilon_{min} < \tau$, where τ is 0.5 or less (or more precisely, we confirm that A is stochastically dominant over B in more cases than vice versa). We can also interpret ϵ_{min} as a confidence score: the lower it is, the stronger the evidence that A is superior to B. In this context, and following the recommendation of $\tau = 0.2$ for multiple comparisons, we formulate our hypotheses as

$$H_0 : \epsilon_{min} \geq 0.2 \qquad (A \text{ is not superior to } B)$$
$$H_1 : \epsilon_{min} < 0.2 \qquad (A \text{ is superior to } B)$$

In our experiments, we compare all pairs of models based on five random seeds each using ASO with a confidence level of $\alpha = 0.05$ (before adjusting for all pairwise comparisons using the Bonferroni correction). We only compare scores obtained on the same dataset.

4 Results and Discussion

Table 1 presents the results achieved by the models described in Sect. 3 and by the MTL-LSP-BiLSTM-BERT model reported in [15]. We confirm the statistical power of the samples (scores) by employing Bootstrap power analysis [22].

Baselines: All Cohan models achieve higher scores than SingleSC ones and the differences are statistically significant in both datasets ($\epsilon_{min} \leq 0.05$ in all comparisons). This is expected since Cohan models exploit more contextual data to yield sentence embeddings. The differences between SingleSC-InCaseLaw and SingleSC-RoBERTa models are significant in both datasets as well ($\epsilon_{min} \leq 0.08$). We also have significant differences for Cohan-Longformer vs. Cohan-InCaseLaw and Cohan-Longformer vs. Cohan-RoBERTA ($\epsilon_{min} = 0.01$ in both comparisons) for the 7-roles dataset, and for Cohan-InCaseLaw vs Cohan-RoBERTa ($\epsilon_{min} = 0.04$) and Cohan-InCaseLaw vs Cohan-Longformer ($\epsilon_{min} = 0.00$) for the 4-roles dataset. The MTL-LSP-BiLSTM-BERT model remarkably outperforms all baselines. Although we cannot perform ASO tests among this model

Table 1. Results achieved by the baseline models and the models of interest. The F1 scores are averages of five evaluations and refer to the best fine-tuning epoch (i.e., the epoch that gives the highest F1 score). The **Epoch** column reports the best fine-tuning epoch (e.g., 1/4 indicates the 1st epoch out of a total of 4 epochs). For Mixup-A models, we report the encoder fine-tuning epoch followed by the Mixup-A classifier training epochs. σ denotes the standard deviation. The scores in bold are the best for each dataset.

Approach	Encoder	7-roles			4-roles		
		Epoch	F1	σ	Epoch	F1	σ
SingleSC	InCaseLaw	1/4	0.604	0.010	3/4	0.628	0.006
	RoBERTa	4/4	0.577	0.007	4/4	0.617	0.007
Cohan	InCaseLaw	4/4	0.638	0.011	4/4	0.672	0.011
	RoBERTa	4/4	0.636	0.008	4/4	0.648	0.015
	Longformer	4/4	0.666	0.008	4/4	0.653	0.008
MTL-LSP[a]	BiLSTM-BERT	–	0.700	0.010	–	–	–
Mixup-A	InCaseLaw	1/4 200	0.601	0.006	3/4 200	0.630	0.004
	RoBERTa	4/4 200	0.572	0.008	4/4 200	0.613	0.004
Mixup-B	InCaseLaw	2/4	0.599	0.009	4/4	0.632	0.006
	RoBERTa	4/4	0.577	0.006	4/4	0.616	0.008
PE-S	InCaseLaw	4/4	0.597	0.004	4/4	0.631	0.009
	RoBERTa	4/4	0.563	0.010	4/4	0.624	0.010
PE-C	InCaseLaw	4/4	0.614	0.008	4/4	0.637	0.017
	RoBERTa	4/4	0.572	0.017	4/4	0.623	0.018
DFCSC-SEP	InCaseLaw	4/4	0.659	0.006	3/4	0.657	0.016
	RoBERTa	4/4	0.702	0.008	4/4	0.696	0.011
	Longformer	4/4	**0.726**	0.004	4/4	**0.712**	0.007
DFCSC-CLS	InCaseLaw	4/4	0.652	0.011	4/4	0.648	0.010
	RoBERTa	4/4	0.698	0.008	4/4	0.686	0.012
	Longformer	4/4	0.721	0.010	4/4	0.707	0.018

[a] Results from [15].

and our baselines because the scores of MTL-LSP-BiLSTM-BERT are not available, the differences are likely significant because the differences are large and the reported standard deviation is small.

Mixup Models: Regarding the two datasets, the score differences between each Mixup-A or Mixup-B model and its SingleSC counterpart (e.g., Mixup-A-InCaseLaw vs. SingleSC-InCaseLaw) are really small, and according to the ASO tests, these differences do not have statistical significance. Hence, the Mixup approaches do not harm or improve the base SingleSC models. Also, there are no significant differences among Mixup-A and Mixup-B models, considering both datasets and the same encoder. Previous work [10] achieves performance gains by exploiting mixup vectors: the F1 score achieved by a linear classifier improves

by 4.3% in the worst case and by 6.8% in the best case. So why do our mixup-based models not outperform their baselines? This may be due to the size of the datasets. In [10], there is little data, so the additional data provided by mixup is beneficial. Conversely, the dataset of the current work is larger, leaving no room for improvements based on data augmentation.
PE Models: Concerning the 7-roles dataset, the PE-S models perform worse than their respective SingleSC counterparts, but the difference is only significant for SingleSC-RoBERTa vs. PE-S-RoBERTa ($\epsilon_{min} = 0.03$). This means that the PE-S approach harms the base SingleSC-RoBERTa model instead of improving it. For the 4-roles dataset, the PE-S models outperform their respective SingleSC counterparts, but the differences are not significant. Also, there are no significant differences among the PE-C models and their respective SingleSC counterparts for the two datasets. We expected poor results (i.e., no gain or decrease in performance) from the PE models with respect to the 7-roles dataset, since most of its labels do not have a clear correlation with sentence positions. In the case of performance degradation, PE vectors should be working as noise data. Conversely, we expected PE vectors to provide gains in the experiments with the 4-roles dataset, which does not happen. We believe that the positional embeddings employed do not represent positions in a way that can enrich the input data of the models.
DFCSC models: Regarding the 7-roles dataset, all DFCSC models outperform their respective Cohan counterparts. The differences are significant with $\epsilon_{min} = 0.14$ for DFCSC-CLS-InCaseLaw vs. Cohan-InCaseLaw, and $\epsilon_{min} = 0.01$ for the remaining comparisons. For the 4-roles dataset, only DFCSC-SEP-InCaseLaw and DFCSC-CLS-InCaseLaw do not outperform their respective Cohan counterparts. The ASO test results in $\epsilon_{min} = 0.23$ for Cohan-InCaseLaw vs. DFCSC-SEP-InCaseLaw, and in $\epsilon_{min} = 0.00$ for Cohan-InCaseLaw vs. DFCSC-CLS-InCaseLaw, meaning that only the latter comparison is significant. The DFCSC models using RoBERTa and Longformer encoders achieve remarkable improvements: the lowest and highest gains are 5.9% and 10.4%, respectively. This behavior is also reported in [11], and thus we consider this additional evidence that the RoBERTa and Longformer encoders exploit contextual data in a better way than the InCaseLaw one (RoBERTa and Longformer rely on the same training procedure, which is different from the one used for InCaseLaw). Comparing the performance of the DFCSC models using Table 2, we come to some conclusions. First, regarding the 7-roles dataset, we see that all Longformer-based models are superior to those based on RoBERTa and InCaseLaw; that all RoBERTa-based models are superior to those based on InCaseLaw; and that there is no significant difference between the DFCSC-SEP and DFCSC-CLS approaches when we compare models that employ the same encoder. Regarding the 4-roles dataset, we find that DFCSC-SEP-Longformer is superior to all models using RoBERTa and InCaseLaw; that DFCSC-CLS-Longformer is superior to DFCSC-CLS-RoBERTa, DFCSC-SEP-InCaseLaw, and DFCSC-CLS-InCaseLaw; that the two RoBERTa-based models are superior to those based on InCaseLaw; and that there is no significant difference between

Table 2. Confidence scores (ϵ_{min}) from ASO tests comparing DFCSC models. Comparisons follow the row-to-column direction, and significant differences (i.e., $\epsilon_{min} < 0.2$) are indicated by underlined values.

	7-roles dataset						**4-roles dataset**					
		CLS-Longformer	SEP-RoBERTa	CLS-RoBERTa	SEP-InCaseLaw	CLS-InCaseLaw		CLS-Longformer	SEP-RoBERTa	CLS-RoBERTa	SEP-InCaseLaw	CLS-InCaseLaw
	F1	0.721	0.702	0.698	0.659	0.652	**F1**	0.707	0.696	0.686	0.657	0.648
SEP-Longformer	0.726	0.45	0.01	0.01	0.01	0.01	0.712	0.61	0.07	0.01	0.01	0.01
CLS-Longformer	0.721	–	0.00	0.01	0.00	0.01	0.707	–	0.45	0.16	0.00	0.00
SEP-RoBERTa	0.702	–	–	0.57	0.01	0.01	0.696	–	–	0.40	0.01	0.01
CLS-RoBERTa	0.698	–	–	–	0.00	0.01	0.686	–	–	–	0.02	0.01
SEP-InCaseLaw	0.659	–	–	–	–	0.42	0.657	–	–	–	–	0.42

the DFCSC-SEP and DFCSC-CLS approaches when we compare models that employ the same encoder. When comparing DFCSC and MTL-LSP-BiLSTM-BERT models, we find that RoBERTa-based models perform similarly and that Longformer-based models perform better.

5 Final Remarks

In this work, we compare approaches from recent proposals to handle the Legal-RRL task. All approaches rely on pre-trained Transformers to encode sentences and employ different strategies to improve reference baselines. Models based on the Mixup and PE approaches are unable to overcome the SingleSC baselines. For the PE models, we conclude that the 7-role dataset does not have a sufficient correlation between labels and sentence positions, which nullifies the usefulness of the models. For the 4-role dataset, we believe that the positional embeddings employed are not powerful enough to enrich the input data of the models.

Models based on the DFCSC approach, on the other hand, show remarkable performance. Most of them overcome strong baselines based on sentence chunks. DFCSC models that exploit RoBERTa and Longformer encoders perform similarly or better than a model based on the MTL-LSP approach proposed in [15].

From the obtained results, we propose a future work that combines MTL-LSP and DFCSC approaches, since the MTL-LSP-BiLSTM-BERT model does not rely on sentence chunks to encode sentences with BERT. Such work would provide interesting evidence about the ability of LSP signals to boost sentence chunk-based models.

We also say that the sentence classification approaches evaluated in this work do not rely on knowledge specific to the legal domain. Therefore, more studies should be conducted to evaluate the efficiency of such approaches in other domains. Despite this, we believe that incorporating domain-specific knowledge into models based on pre-trained Transformers is a valuable research direction.

References

1. Aragy, R., Fernandes, E.R., Caceres, E.N.: Rhetorical role identification for Portuguese legal documents. In: Britto, A., Valdivia Delgado, K. (eds.) BRACIS 2021. LNCS (LNAI), vol. 13074, pp. 557–571. Springer, Cham (2021). https://doi.org/10.1007/978-3-030-91699-2_38
2. Beltagy, I., Peters, M.E., Cohan, A.: Longformer: the long-document transformer. CoRR abs/2004.05150 (2020). https://arxiv.org/abs/2004.05150
3. Bhattacharya, P., Paul, S., Ghosh, K., Ghosh, S., Wyner, A.: Deeprhole: deep learning for rhetorical role labeling of sentences in legal case documents. Artificial Intelligence and Law, pp. 1–38 (2021)
4. Cohan, A., Beltagy, I., King, D., Dalvi, B., Weld, D.: Pretrained language models for sequential sentence classification. In: Proceedings of the 2019 Conference on Empirical Methods in Natural Language Processing and the 9th International Joint Conference on Natural Language Processing (EMNLP-IJCNLP), pp. 3693–3699. Association for Computational Linguistics, Hong Kong, November 2019. https://doi.org/10.18653/v1/D19-1383. https://aclanthology.org/D19-1383
5. Devlin, J., Chang, M.W., Lee, K., Toutanova, K.: BERT: pre-training of deep bidirectional transformers for language understanding. In: Proceedings of the 2019 Conference of the North American Chapter of the Association for Computational Linguistics: Human Language Technologies, Volume 1 (Long and Short Papers), pp. 4171–4186. Association for Computational Linguistics, Minneapolis, Minnesota, June 2019. https://doi.org/10.18653/v1/N19-1423. https://aclanthology.org/N19-1423
6. Dror, R., Shlomov, S., Reichart, R.: Deep dominance - how to properly compare deep neural models. In: Korhonen, A., Traum, D.R., Màrquez, L. (eds.) Proceedings of the 57th Conference of the Association for Computational Linguistics, ACL 2019, Florence, Italy, July 28–August 2, 2019, Volume 1: Long Papers, pp. 2773–2785. Association for Computational Linguistics (2019). https://doi.org/10.18653/v1/p19-1266. https://doi.org/10.18653/v1/p19-1266
7. Feijo, D., Moreira, V.: Summarizing legal rulings: comparative experiments. In: Proceedings of the International Conference on Recent Advances in Natural Language Processing (RANLP 2019), pp. 313–322. INCOMA Ltd., Varna, Bulgaria, September 2019. https://doi.org/10.26615/978-954-452-056-4_036. http://aclanthology.org/R19-1036
8. Kalamkar, P., et al.: Corpus for automatic structuring of legal documents. In: Proceedings of the Thirteenth Language Resources and Evaluation Conference, pp. 4420–4429. European Language Resources Association, Marseille, France, June 2022. https://aclanthology.org/2022.lrec-1.470
9. Li, D., Yang, K., Zhang, L., Yin, D., Peng, D.: Class: a novel method for chinese legal judgments summarization. In: Proceedings of the 5th International Conference on Computer Science and Application Engineering. CSAE 2021. Association for Computing Machinery, New York (2021). https://doi.org/10.1145/3487075.3487161

10. de Lima, A.G., Boughanem, M., da S. Aranha, E.H., Dkaki, T., Moreno, J.G.: Exploring SBERT and mixup data augmentation in rhetorical role labeling of Indian legal sentences. In: Tamine, L., Amigó, E., Mothe, J. (eds.) Proceedings of the 2nd Joint Conference of the Information Retrieval Communities in Europe (CIRCLE 2022), Samatan, Gers, France, July 4–7, 2022. CEUR Workshop Proceedings, vol. 3178. CEUR-WS.org (2022). https://ceur-ws.org/Vol-3178/CIRCLE_2022_paper_29.pdf
11. de Lima, A.G., Moreno, J.G., da S. Aranha, E.H.: IRIT_IRIS_A at SemEval-2023 Task 6: legal rhetorical role labeling supported by dynamic-filled contextualized sentence chunks. In: Proceedings of the 17th International Workshop on Semantic Evaluation (SemEval-2023). Association for Computational Linguistics, Toronto, Canada, July 2023
12. de Lima, A.G., Moreno, J.G., Boughanem, M., Dkaki, T., da S. Aranha, E.H.: Leveraging positional encoding to improve fact identification in legal documents. In: First International Workshop on Legal Information Retrieval (LegalIR) at ECIR 2023, pp. 11–13 (2023). https://tmr.liacs.nl/legalIR/LegalIR2023_proceedings.pdf
13. Liu, Y., et al.: Roberta: a robustly optimized BERT pretraining approach. CoRR abs/1907.11692 (2019). https://arxiv.org/abs/1907.11692
14. Ma, L., Zhang, Y., Wang, T., Liu, X., Ye, W., Sun, C., Zhang, S.: Legal judgment prediction with multi-stage case representation learning in the real court setting. In: Proceedings of the 44th International ACM SIGIR Conference on Research and Development in Information Retrieval, pp. 993–1002. SIGIR '21. Association for Computing Machinery, New York (2021). https://doi.org/10.1145/3404835.3462945
15. Malik, V., Sanjay, R., Guha, S.K., Hazarika, A., Nigam, S., Bhattacharya, A., Modi, A.: Semantic segmentation of legal documents via rhetorical roles. In: Proceedings of the Natural Legal Language Processing Workshop 2022, pp. 153–171. Association for Computational Linguistics, Abu Dhabi, United Arab Emirates (Hybrid), December 2022. https://aclanthology.org/2022.nllp-1.13
16. Paul, S., Mandal, A., Goyal, P., Ghosh, S.: Pre-training transformers on indian legal text. CoRR abs/2209.06049 (2022). https://doi.org/10.48550/arXiv.2209.06049
17. Reimers, N., Gurevych, I.: Sentence-bert: sentence embeddings using siamese bert-networks. In: Inui, K., Jiang, J., Ng, V., Wan, X. (eds.) Proceedings of the 2019 Conference on Empirical Methods in Natural Language Processing and the 9th International Joint Conference on Natural Language Processing, EMNLP-IJCNLP 2019, Hong Kong, China, November 3–7, 2019, pp. 3980–3990. Association for Computational Linguistics (2019). https://doi.org/10.18653/v1/D19-1410
18. Savelka, J., Westermann, H., Benyekhlef, K.: Cross-domain generalization and knowledge transfer in transformers trained on legal data. In: Ashley, K.D., Atkinson, K., Branting, L.K., Francesconi, E., Grabmair, M., Walker, V.R., Waltl, B., Wyner, A.Z. (eds.) Proceedings of the Fourth Workshop on Automated Semantic Analysis of Information in Legal Text held online in conjunction with the 33rd International Conference on Legal Knowledge and Information Systems, ASAIL@JURIX 2020, December 9, 2020. CEUR Workshop Proceedings, vol. 2764. CEUR-WS.org (2020). https://ceur-ws.org/Vol-2764/paper5.pdf
19. Ulmer, D., Hardmeier, C., Frellsen, J.: Deep-significance-easy and meaningful statistical significance testing in the age of neural networks. arXiv preprint arXiv:2204.06815 (2022)

20. Vaswani, A., et al.: Attention is all you need. In: Guyon, I., von Luxburg, U., Bengio, S., Wallach, H.M., Fergus, R., Vishwanathan, S.V.N., Garnett, R. (eds.) Advances in Neural Information Processing Systems 30: Annual Conference on Neural Information Processing Systems 2017, December 4–9, 2017, Long Beach, CA, USA, pp. 5998–6008 (2017). https://proceedings.neurips.cc/paper/2017/hash/3f5ee243547dee91fbd053c1c4a845aa-Abstract.html
21. Wang, C., Nulty, P., Lillis, D.: A comparative study on word embeddings in deep learning for text classification. In: Proceedings of the 4th International Conference on Natural Language Processing and Information Retrieval, NLPIR 2020, pp. 37–46. Association for Computing Machinery, New York(2021). https://doi.org/10.1145/3443279.3443304
22. Yuan, K.H., Hayashi, K.: Bootstrap approach to inference and power analysis based on three test statistics for covariance structure models. Br. J. Math. Stat. Psychol. **56**(1), 93–110 (2003). https://doi.org/10.1348/000711003321645368. https://bpspsychub.onlinelibrary.wiley.com/doi/abs/10.1348/000711003321645368
23. Zhang, H., Cissé, M., Dauphin, Y.N., Lopez-Paz, D.: mixup: beyond empirical risk minimization. In: 6th International Conference on Learning Representations, ICLR 2018, Vancouver, BC, Canada, April 30 - May 3, 2018, Conference Track Proceedings. OpenReview.net (2018). https://openreview.net/forum?id=r1Ddp1-Rb
24. Zhong, H., Xiao, C., Tu, C., Zhang, T., Liu, Z., Sun, M.: How does NLP benefit legal system: a summary of legal artificial intelligence. In: Proceedings of the 58th Annual Meeting of the Association for Computational Linguistics. pp. 5218–5230. Association for Computational Linguistics, Online, July 2020. https://doi.org/10.18653/v1/2020.acl-main.466. https://aclanthology.org/2020.acl-main.466

Dog Face Recognition Using Vision Transformer

Victor Hugo Braguim Canto[1(✉)], João Renato Ribeiro Manesco[1], Gustavo Botelho de Souza[2], and Aparecido Nilceu Marana[1]

[1] Sao Paulo State University (UNESP), Bauru, Sao Paulo, Brazil
{victor.canto,joao.r.manesco,nilceu.marana}@unesp.br
[2] Bank of Brazil Brasilia, Federal District, Brazil
gustavo.botelho@bb.br
http://www.unesp.br, https://www.bb.com.br

Abstract. The demand for effective, efficient and safe methods for animal identification has been increasing significantly, due to the need for traceability, management, and control of this population, which grows at higher rates than the human population, particularly pets. Motivated by the efficacy of modern human identification methods based on face biometrics features, in this paper, we propose a dog face recognition method based on vision transformers, a deep learning approach that decomposes the input image into a sequence of patches and applies self-attention to these patches to capture spatial relationships between them. Results obtained on DogFaceNet, a public database of dog face images, show that the proposed method, which uses the EfficientFormer-L1 architecture, outperforms the state-of-the-art method proposed previously in literature based on ResNet, a deep convolutional neural network.

Keywords: Animal Biometrics · Dog Identification · Facial Recognition · Convolutional Neural Network · Visual Transformer · ArcFace

1 Introduction

According to Pet Brazil Institute [9], in 2018, there were 54.2 million dogs, 39.8 million birds, 23.9 million cats, 19.1 million fishes, and 2.3 million reptiles and small mammals, considered pets, in the homes of Brazilian families. Compared to the 2013 census data, the cat and dog populations have grown in households at 8.1% and 5%, respectively, with an overall average growth of 5.2%. In addition, data from the World Health Organization (WHO), published by São Paulo University (USP) [14], indicate that in Brazil there are approximately 30 million lost or abandoned animals, with 10 million cats and 20 million dogs.

The uncontrolled proliferation and the inappropriate management of animals in society can be factors with a very negative impact on public health, as animals can transmit many diseases and parasites. Therefore, it is of paramount importance to carry out monitoring, control and also traceability of animals through

M. C. Naldi and R. A. C. Bianchi (Eds.): BRACIS 2023, LNAI 14196, pp. 33–47, 2023.
https://doi.org/10.1007/978-3-031-45389-2_3

complete medical records and effective identification mechanisms to enable monitoring and evaluation of the animal population in the country.

Some countries have already approved mandatory pet identification registration, such as South Korea, where all dogs must be monitored through RFID (Radio Frequency Identification), tags on collars or other devices [10].

As occurs in human identification, animal identification is also susceptible to fraud. In the UK, for example, pet insurance fraud cases have been increasing by 400% per year, reaching 2 million, annually, in claims, according to Youtalk-Insurance [18]. Therefore, in addition to issues of effectiveness and efficiency, animal identification methods also need to be robust enough to discourage, stop or even prevent fraud.

According to Kumar et al. [11], animal identification approaches can be classified into invasive or non-invasive. Examples of invasive approaches are ear tags, ear tattoos and microchip implantation in the animal's body. Examples of non-invasive approaches are RFID (Radio Frequency Identification) devices, collars with GPS (Global Positioning System), Internet of Things (IoT) devices, Bluetooth devices, and biometric features (e.g. face, iris, retina and muzzles). Among these approaches, animal identification based on biometric features has proved to be the best choice, since it is effective, efficient and does not require special devices, which would introduce more costs to the identification process, being also important to discourage or prevent fraud.

Among several human biometric traits, the face is one of the most popular due to its high coverage, high acceptability, easiness of capture at a distance, convenience of use and lower costs since digital cameras are cheap and popular sensors. Motivated by the success of modern human face recognition methods, in this paper, we propose a dog face recognition method based on vision transformers, a deep learning approach that decomposes the input image into a sequence of patches and applies self-attention to these patches to capture spatial relationships between them.

2 Biometric Dog Identification

Biometric dog identification approaches have been reported in the literature by using snout and facial features, among others [10,12,17,21]. This section presents some approaches proposed so far, their identification rates, as well as their advantages and disadvantages.

2.1 Identification Using Snout Features

Snout is the region of an animal's face consisting of its nose, mouth, and jaw. In many animals, this structure is called muzzle. The muzzle information can be used as a biometric feature for dog identification, as reported by Jang et al. [10].

Figure 1 shows, on the left, a device used to capture images of dog snouts and, on the right, a captured image, with the region of interest for biometric identification highlighted.

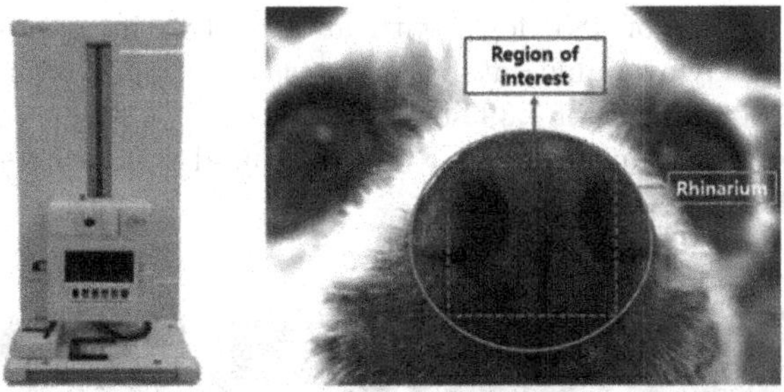

Fig. 1. Device used to collect images of dog snouts (on the left) and the region of interest of a collected image, used for feature extraction (on the right) [10].

Jang et al. [10] have proposed the use of some feature extraction methods, such as SIFT (Scale-Invariant Feature Transform), SURF (Speeded Up Robust Features), BRISK (Binary Robust Invariant Scalable Keypoints) and ORB (Oriented FAST and Rotated BRIEF), and the method that presented the best performance was the ORB with EER (Equal Error Rate) of 0.35%, when evaluated with 55 dog muzzle pattern images acquired from 11 dogs and 990 images augmented by the image deformation (i.e., angle, illumination, noise, affine transform). However, despite this low error rate, the technique proposed by these authors has some disadvantages:

- Even though it is not an invasive technique, it is necessary to immobilize the animal to capture the image and extract features from the snout;
- It is important to avoid light reflection and blur in muzzle images;
- It is necessary to have good hygiene of the muzzle.

Figure 2 shows examples of dog muzzle images that were discarded in the work by Jang et al. [10] due to light reflection and blurring caused by dog movements.

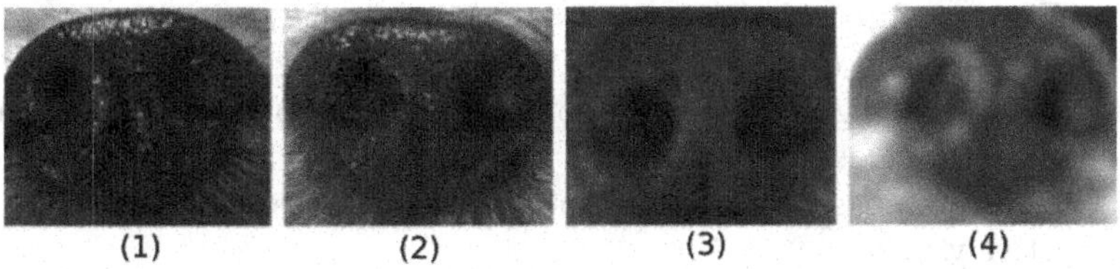

Fig. 2. Dog muzzle images with light reflection (1–2) and blurring (3–4) [10].

2.2 Identification Using Facial Features

Recent works on dog biometric identification based on face features have used the DogFaceNet database, proposed by Mougeot, Li and Jia [17]. This database contains 3,148 images of 485 dogs, with at least two images per individual. However, the publicly available database that was developed after the published work contains 8,363 images of 1,393 dogs.

Figure 3 shows four images of four dogs captured from different viewpoints that compose the DogFaceNet dataset.

Fig. 3. Samples of four dog images from the DogFaceNet dataset [17].

The images of the dog's faces were all aligned, using the animal's eyes as a reference. Figure 4 shows the process of detecting the fiducial points (eyes and snout) on the face of a dog (on the left), used for the alignment of the images, and the result of the alignment and cropping processes applied on the six images of one dog from the DogFaceNet dataset (on the right). The dog face images of the DogFaceNet database were divided into training and test sets, with 90% and 10%, respectively.

Dog identification through facial features is a challenging task, given the high interclass similarity and the high intraclass variability that can occur. Figure 5 shows, on the left, six pairs of images captured from distinct pairs of individuals that present faces with high similarity (false positive pairs of canine image faces according to the method proposed by Mougeot, Li and Jia [17]), and, on the right, six pairs of images, each pair from the same individual, that present high variability (false negative pairs of canine image faces according to the method proposed by Mougeot, Li and Jia [17]).

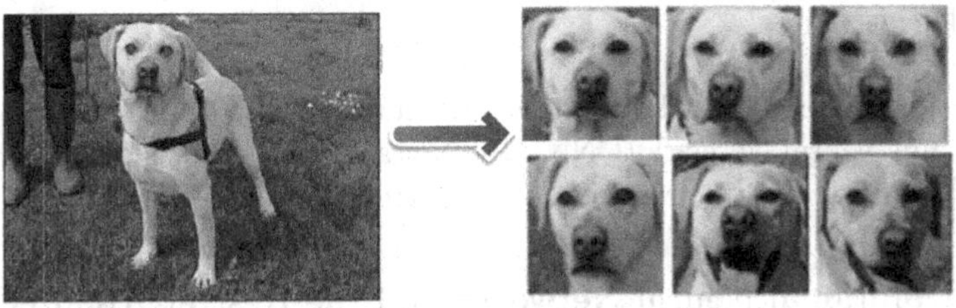

Fig. 4. Detection of the animal's eyes and snout (left) and the results obtained after the aligning and cropping processes, applied to six images of one dog from the DogFaceNet dataset [17].

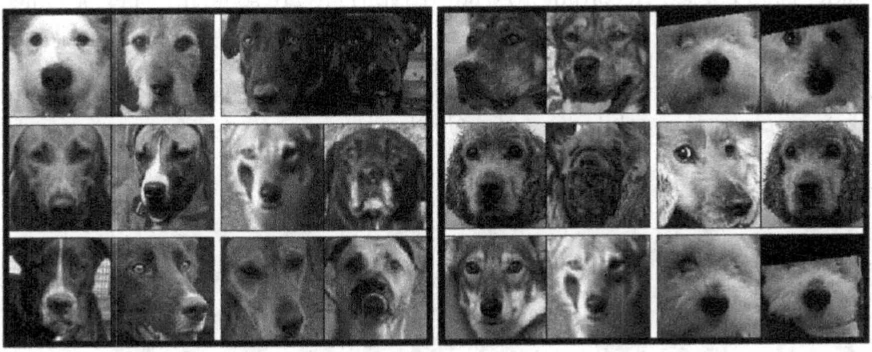

Fig. 5. Examples of six false positive pairs of canine faces (left) and six false negative pairs of canine faces (right), according to the method proposed by Mougeot, Li and Jia [17].

Mougeot, Li and Jia [17] proposed a Deep VGG-like and a ResNet-like Convolutional Neural Network (CNN) for the dog face verification task containing 48 dogs from an open-set of test data of DogFaceNet, and concluded that the ResNet-like CNN performed better with 92% of accuracy. For the identification task, the best performance was also obtained by the ResNet-like CNN, with 60.44% of accuracy for rank-1 and 92% for rank-5.

Yoon, So and Rhee [21], based on the studies of Mougeot, Li and Jia [17], proposed a methodology for using vector spaces to improve the performance in the dog identification process. Thus, in the scenario of dog facial identification, the authors proposed, using the triplet loss, the removal of the L2 norm, for comparing feature vectors from the images of the dogs in a new space within a sphere of radius 1. With this, they obtained a mean accuracy of 88.8% with ArcFace+VL in the verification task.

3 Background

In this section, we describe the fundamental concepts used in this work, such as Convolutional Neural Network (CNN), ResNet architecture, Vision Transformers (ViT), EfficientFormer-L1 architecture and ArcFace function.

3.1 Convolutional Neural Networks and the ResNet Architecture

Convolutional Neural Networks (CNN) have been widely used in solutions for Computer Vision applications, in problems such as object detection, image recognition, and Biometrics [1,2]. Their architectures are similar to the ones of the human visual cortex. According to Chollet [1], there are some high-performance CNN architectures, such as AlexNet, VGG, GoogleNet and Resnet.

Convolutional Neural Networks are organized into layers, such as the convolutional layers, the pooling layers and the fully connected layer. In Fig. 6 there is an example of the architecture of a CNN.

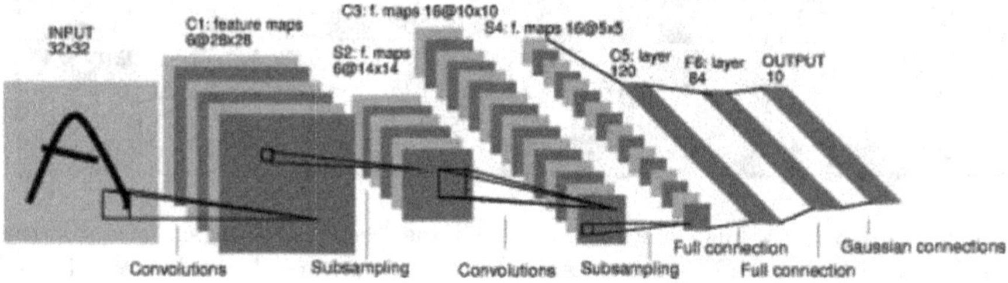

Fig. 6. Simplified architecture of a Convolutional Neural Network [13].

In CNNs, to limit the range of a result of a convolution layer, activation functions are used, such as the linear function ReLU, the sigmoid function, the hyperbolic tangent function, etc.

Thus, according to LeCun et al. [6], the convolutional layers have the objective of performing the extraction of features through the convolution operation. Furthermore, the adjustments of the backpropagation algorithm weights are performed in this layer.

Also according to LeCun et al. [6], the pooling layer aims to reduce dimensionality by combining neighbouring pixels into a single value. Consequently, there is a reduction in time and a minimization of the use of computational resources for training.

The fully connected layer aims to perform image classification. For example, in a neural network for classifying dogs, the result of the fully connected layer expresses which dog is associated with the input image.

In addition to conventional CNN architectures, ResNet is a CNN architecture proposed by He et al. [8] and its main characteristic, as its own name indicates - Residual Networks - is the use of shortcut connections, i.e., residual information is passed from previous layers to posterior ones, which improves and, consequently, change the training process (avoiding data vanishing or attenuation as well as saving computational resources, for instance). In Fig. 7, there is an example of using shortcut connections and routing to the output layer of the neural network in order to optimize its training process.

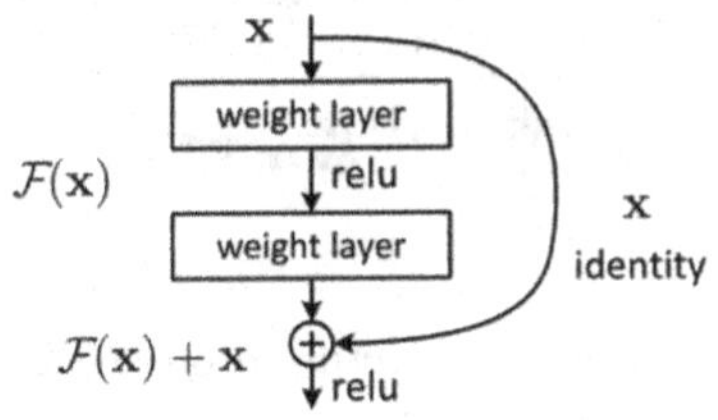

Fig. 7. ResNet features [7].

According to Li et al. [15], several improvements were made to the ResNet architecture to increase its performance. Among them, Zhang et al. [22] proposed multi-level shortcut connections and Targ, Almeida and Lyman [19] proposed using more convolution layers and data flow between layers, among others. Also, different benchmark architectures of ResNet neural networks were proposed, however, one of the most used and effective for many Computer Vision tasks is the ResNet-50 (with 50 layers).

3.2 Vision Transformers and the EfficientFormer Architecture

Transformers were proposed by Vaswani et al. [20] to solve a problem involving natural language processing (NLP) and have become state-of-the-art for many types of applications.

For image processing applications, Dosovitskiy et al. [5] proposed the Vision Transformers (ViT) models, in which the input image is divided into patches of fixed size, transforming it into an input sequence of image patches. Continuing through the initial patches, feature vectors are projected across an initial linear layer and position references are added directly to a standard transformer. In Fig. 8 there is an example of the architecture of a Vision Transformer and its operations.

According to [5], in the benchmarks performed, the Vision Transformers architecture outperformed state-of-the-art image recognition architectures, specifically with the use of Convolutional Neural Networks. For example, in the ImageNet database, the Vision Transformer architecture had an accuracy of 88.5%, and the others, using CNNs, were below 88%.

Even with the great progress caused by ViTs in the Computer Vision area, models based on transformers are still very heavy and present a problem of high latency, which hinders both the learning and the use of this type of model in less capable environments. EfficientFormer-type networks arise to mitigate

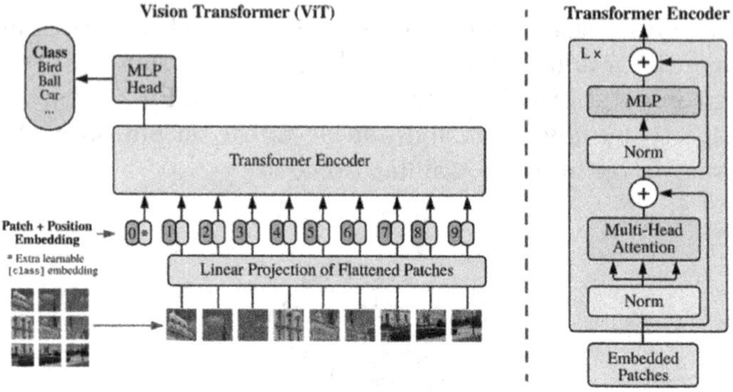

Fig. 8. Vision Transformer architecture proposed by Dosovitskiy [5].

this problem through a pure transformer, with consistent data dimensionality, capable of running in more diverse environments. Figure 9 shows an overview of the EfficentFormer proposed by Lie et al. [16]. The network starts with a convolution stem as patch embedding, followed by MetaBlock (MB). The MB^{4D} and MB^{3D} contain different token mixer configurations, that is, local pooling or global multi-head self-attention, arranged in a dimension-consistent manner.

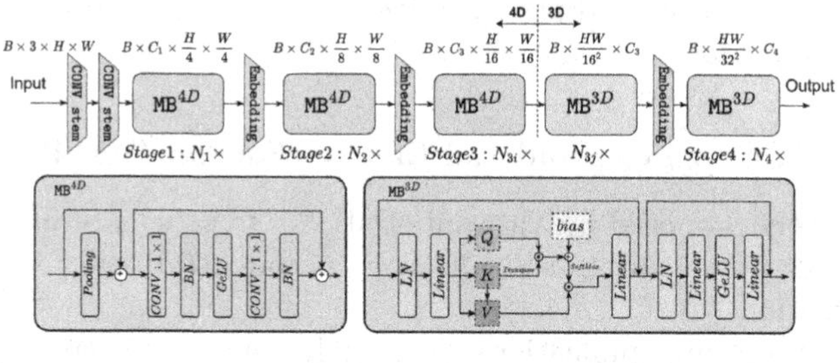

Fig. 9. Overview of the EfficientFormer [16].

3.3 ArcFace

In the ArcFace [4] error function, the objective of learning becomes jointly minimizing the angular distance between elements of the same class while maximizing the intra-class similarity through Eq. 1.

$$L = -\frac{1}{N}\sum_{i=1}^{N} log \frac{e^{s(\cos(\theta_{y_i}+m))}}{e^{s(\cos(\theta_{y_i}+m))} + \sum_{j=1, j\neq y_i}^{n} e^{s\cos(\theta_j)}} \tag{1}$$

In this case, the class samples are positioned in a hypersphere of radius s, and the objective of the error function becomes to minimize the angular distance between elements of this hypersphere, in such a way that the features extracted in the last layer before the softmax, represent a sample in this space, whose elements similar to this sample always end up being positioned at an angular distance in the interval m.

The error function can be extended to perform an analysis on subspaces by adding an additional parameter of subcenter when extracting features, this parameter consists of a matrix of weights of dimensions $C \times k$, where C is the number of classes and k the number of subspaces, such that the features are mapped to the k subspaces, so that the ArcFace error function can be applied to each of them [3].

4 Proposed Method

In order to improve the state-of-the-art results in the task of dog facial recognition, in this work we propose a method that uses EfficientFormer-L1, a Vision Transformer architecture proposed by Lie et al. [16].

In the proposed method, the EfficientFormer-L1 is coupled with the ArcFace error function [4], a popular loss function in biometrics which aims to minimize the interclass angular distance while the intraclass angular distance is maximized.

In addition, an SVM classifier will be used in the output of both architectures, since in biometrics calculation is used using distances, such as the cosine distance.

Figure 10 demonstrates the architecture of the proposed method with the architecture of EfficientFormer-L1.

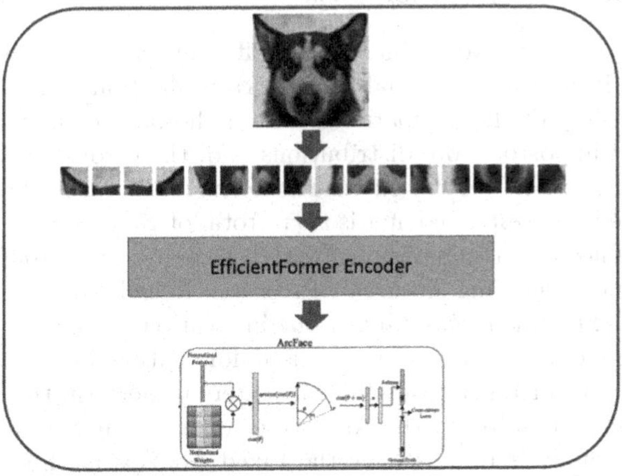

Fig. 10. Diagram of the proposed method, describing both the transformer step and the ArcFace technique used to position the face descriptors from the same dog near each other and the face descriptors from distinct dogs far away, in the latent space.

5 Results and Discussion

In order to assess the proposed method for dog face recognition, using visual transformers, more specifically by using the EfficientFormer-L1 architecture, and ArcFace, two experiments were carried out on the DogFaceNet dataset. As the subset of data used by Mougeot, Li and Jia [17] is different than the one publicly shared, a protocol, inspired by them and using standard practices for open-set biometric authentication was used. The results obtained were compared with a baseline method based on the ResNet-50 architecture [8] proposed by Mougeot, Li and Jia [17].

It is important to point out that the experiments are not identical to those carried out by Mougeot, Li and Jia [17], as the publicly disclosed base is different from the original base.

In addiction, in Table 1 there are details of the servers that were used to run the experiments. Being the implementation of the ResNet architecture on server 1 and the EfficientFormer-L1 architecture on server 2.

Table 1. Server Settings

#	Linux Version	vCPU	RAM Memory	GPU	HD
1	Ubuntu 20.04	Intel Xeon @ 8x 2.4 GHz	24 GB	2x NVIDIA GEFORCE RTX 3080 12 GB	90 GB (SSD)
2	Ubuntu 20.04	Intel Xeon @ 8x 2.4 GHz	24 GB	1x NVIDIA GEFORCE RTX 3090 24 GB	90 GB (SSD)

5.1 Experiment - Verification Task

In the verification task, two architectures, EfficientFormer-L1 and ResNet-50, were employed in a standard facial biometrics evaluation protocol. This protocol involves using all the comparisons within the dataset in order to obtain the genuine and impostor score distributions and, then, construct the Receiver Operating Characteristic (ROC) curve. Furthermore, to ensure balanced evaluation, we selected all positive comparisons (a total of 2561 pairs of dogs) and an equal number of negative comparisons. From these pairs, we computed additional evaluation metrics, including Accuracy, Precision, Recall, and F-1 Score.

Initially, an experiment was carried out in relation to the verification task, an EfficientFormer-L1 architecture was used for feature extraction combined with the ArcFace error function with 3 subcenters to perform the learning. The hyperparameters in relation to the ArcFace error function were obtained using as reference the analysis proposed by the Fixed AdaCos method, proposed by [23], in which the margin parameter is fixed at $m = 0.5$ and the scale parameter is obtained dynamically by $s = \sqrt{2} \cdot \log(C-1)$, where C is the number of classes.

Figure 11 shows the genuine comparison score distributions and the imposter comparison score distributions obtained using ResNet-50 and EfficientFormer-L1 features. One can observe that the EfficientFormer-L1 features generated more

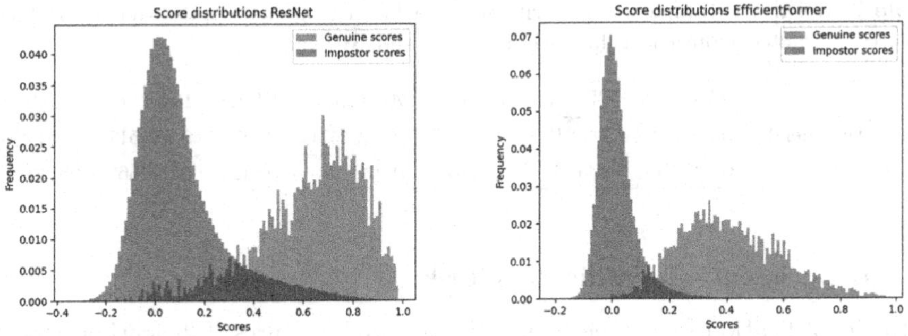

(a) ResNet-50 (b) EfficientFormer-L1

Fig. 11. Distributions of similarity scores obtained through features extracted using ResNet-50 and EfficientFormer-L1.

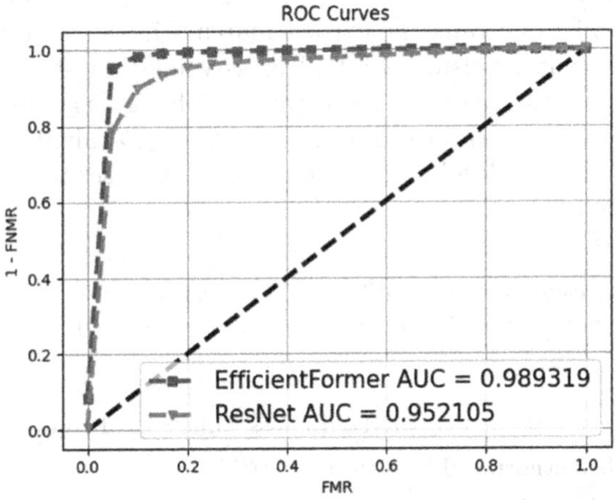

Fig. 12. ROC curves obtained on DogFaceNet dataset by using features extracted from the ResNet-50 and EfficientFormer-L1.

separated distributions than the Resnet-50 features, leading to a lower intersection between the distributions and, consequently, lower error rates. EfficientFormer-L1 obtained a better AUC value, equal to 0.989319, while the Resnet-50 obtained an AUC value equal to 0.952105, as shown in Fig. 12.

Table 2 presents the values of Area Under the ROC Curve (AUC), Equal Error Rate (EER), Accuracy, F1-Score, Precision and Recall, obtained in the verification task, on the DogFaceNet dataset, using face features extracted from the EfficientFormer-L1 and ResNet-50. One can observe that EfficientFormer-L1 features performed better than the ResNet-50 features in all metrics.

Table 2. Results obtained in the verification task, on DogFaceNet dataset, using face features extracted from the EfficientFormer-L1 and ResNet-50.

	AUC	EER THR	EER	Accuracy	F1-Score	Precision	Recall
EfficientFormer-L1	0.989319	0.145812	0.048873	0.961147	0.960757	0.970517	0.951190
ResNet-50	0.952105	0.333500	0.100762	0.922100	0.920247	0.942669	0.898867

5.2 Experiment - Identification Task

In the identification task experiment, the two architectures, EfficientFormer-L1 and ResNet-50, were used again. To evaluate their performance, we employed two distinct settings with two different types of data split.

In the first setting, we computed the Cumulative Matching Characteristic (CMC) curve by comparing the cosine distance between every element in the probe set and every element in the gallery set. This analysis provides insights into the model's ability to rank the correct matches in the gallery.

The data was split by using a k-fold cross-validation strategy, with k set to 10, where 90% of the open-set data was used for gallery creation, and the remaining 10% was reserved for evaluation. To further evaluate the models, we trained an SVM using the k-fold cross-validations with k set to 10. Table 3 shows the results obtained.

Table 3. Accuracy results in the identification task using face features extracted from EfficientFormer-L1 and ResNet-50 and an SVM classifier in the K-fold cross-validation protocol.

	K-Fold Cross Validation Protocol
EfficientFormer-L1	0.8834 ± 0.0356
ResNet-50	0.4653 ± 0.0485

For the second split, a random sample approach was employed, as proposed by Mougeot, Li and Jia [17], where m samples for each class are randomly selected from the dataset and included in the gallery set, while the rest of the samples are used for evaluation. For this data split approach, a k-NN classifier was used, with the k value being chosen also as proposed by Mougeot, Li and Jia [17]. Figure 3 presents the rank-10 cumulative match characteristic (CMC) curve in each of the data separation protocols for the identification task. One can observe that the EfficientFormer-L1 features obtained very superior performances when compared with the ResNet-50 features, in the identification task. The best rank-1 accuracy value, 91.4%, was obtained by EfficientFormer, with $m = 1$, while the best rank-1 accuracy value obtained by ResNet-50 was only 63.3%, also with $m = 1$.

Regarding the related works presented in Sect. 2.2, using the smaller initial version of the DogFaceNet, Mougeot, Li and Jia [17], obtained 92% of accuracy in the verification task, and 60.44% of rank-1 accuracy (best case) in the identification task, while Yoon, So and Rhee [21], using the same dataset that we have used, obtained a mean accuracy of 88.8% with ArcFace+VL in the verification task (Fig.13).

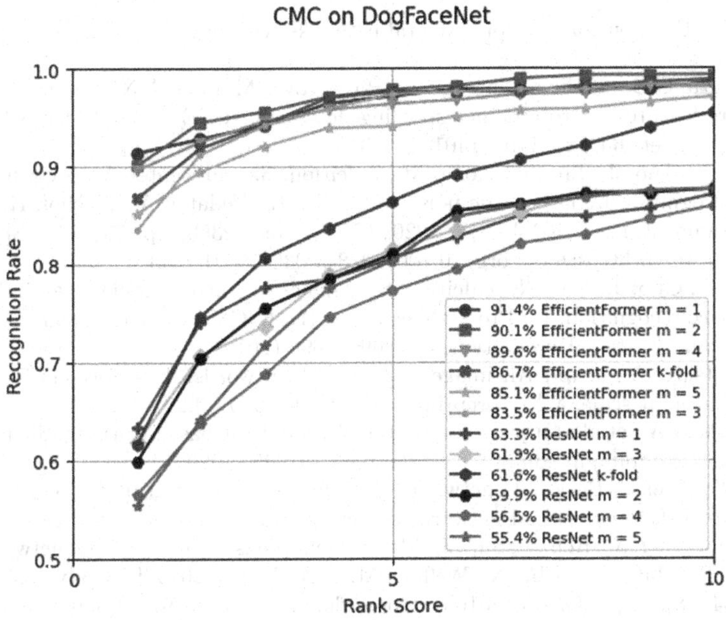

Fig. 13. CMC curves obtained on DogFaceNet dataset by using face features extracted from EfficientFormer-L1 and ResNet-50.

6 Conclusion

With the advancement of technology, especially in the scenario of smart cities and Internet of Things devices, there is an improvement and, consequently, sophistication in biometric applications. With this, it is possible to explore the field of animal biometrics, as there are still few applications and studies in the area applied to dogs, according to the study by Mougeot, Li and Jia [17]. In the case of dogs specifically, biometric identification through faces has shown to be very promising, as it does not require specific hardware resources and can be used through smartphone cameras. As for the recognition techniques, it is noted that the application of architectures based on transformers, especially with EfficientFormer-L1 used in this work, brings excellent results compared to the state of the art in the area of pattern recognition in images that are the Convolutional Neural Networks, in this case from ResNet-50. Still, the application of biometric techniques in dogs is a vast area to be explored and its results add positively to the population, because, through this, it provides ways of monitoring these animals in cities, searching for lost animals, reducing abandonment, carrying out disease control more effectively and help prevent or detect fraud.

References

1. Chollet, F.: How convolutional neural networks see the world. The Keras Blog 30 (2016)
2. De Souza, G.B., da Silva Santos, D.F., Pires, R.G., Marana, A.N., Papa, J.P.: Deep texture features for robust face spoofing detection. IEEE Trans. Circuits Syst. II Express Briefs **64**(12), 1397–1401 (2017)
3. Deng, J., Guo, J., Liu, T., Gong, M., Zafeiriou, S.: Sub-center ArcFace: boosting face recognition by large-scale noisy web faces. In: Vedaldi, A., Bischof, H., Brox, T., Frahm, J.-M. (eds.) ECCV 2020. LNCS, vol. 12356, pp. 741–757. Springer, Cham (2020). https://doi.org/10.1007/978-3-030-58621-8_43
4. Deng, J., Guo, J., Xue, N., Zafeiriou, S.: Arcface: Additive angular margin loss for deep face recognition. In: Proceedings of the IEEE/CVF Conference on Computer Vision and Pattern Recognition, pp. 4690–4699 (2019)
5. Dosovitskiy, A., et al.: An image is worth 16x16 words: Transformers for image recognition at scale. arXiv preprint arXiv:2010.11929 (2020)
6. Forsyth, D.A., et al.: Object recognition with gradient-based learning. Shape, contour and grouping in computer vision, pp. 319–345 (1999)
7. GeeksforGeeks: Residual networks (resnet) - deep learning. https://www.geeksforgeeks.org/residual-networks-resnet-deep-learning/. Accessed 18 June 2022
8. He, K., Zhang, X., Ren, S., Sun, J.: Identity mappings in deep residual networks. In: Leibe, B., Matas, J., Sebe, N., Welling, M. (eds.) ECCV 2016. LNCS, vol. 9908, pp. 630–645. Springer, Cham (2016). https://doi.org/10.1007/978-3-319-46493-0_38
9. Institute, P.B.: Pet census (2019). https://institutopetbrasil.com/imprensa/censo-pet-1393-milhoes-de-animais-de-estimacao-no-brasil. Accessed 18 June 2022
10. Jang, D.H., Kwon, K.S., Kim, J.K., Yang, K.Y., Kim, J.B.: Dog identification method based on muzzle pattern image. Appl. Sci. **10**(24), 8994 (2020)
11. Kumar, S., Singh, S.K.: Visual animal biometrics: survey. IET. Biometrics **6**(3), 139–156 (2017)
12. Lai, K., Tu, X., Yanushkevich, S.: Dog identification using soft biometrics and neural networks. In: 2019 International Joint Conference on Neural Networks (IJCNN), pp. 1–8. IEEE (2019)
13. LeCun, Y., Bottou, L., Bengio, Y., Haffner, P.: Gradient-based learning applied to document recognition. Proc. IEEE **86**(11), 2278–2324 (1998)
14. Lemos, S.: The number of adoptions and abandonment of animals in the pandemic (2021). https://jornal.usp.br/atualidades/cresce-o-numero-de-adocoes-e-de-abandono-de-animais-na-pandemia. Accessed 18 June 2022
15. Li, S., Jiao, J., Han, Y., Weissman, T.: Demystifying resnet. arXiv preprint arXiv:1611.01186 (2016)
16. Li, Y., Yuan, G., Wen, Y., Hu, J., Evangelidis, G., Tulyakov, S., Wang, Y., Ren, J.: Efficientformer: vision transformers at mobilenet speed. Adv. Neural. Inf. Process. Syst. **35**, 12934–12949 (2022)
17. Mougeot, G., Li, D., Jia, S.: A deep learning approach for dog face verification and recognition. In: Nayak, A.C., Sharma, A. (eds.) PRICAI 2019. LNCS (LNAI), vol. 11672, pp. 418–430. Springer, Cham (2019). https://doi.org/10.1007/978-3-030-29894-4_34

18. Software, A.: Pet insurance fraud increases (2018). https://youtalk-insurance.com/broker-news/400-rise-in-pet-insurance-fraud-highlights-need-for-new-approach. Accessed 18 June 2022
19. Targ, S., Almeida, D., Lyman, K.: Resnet in resnet: Generalizing residual architectures (2016). arXiv preprint arXiv:1603.08029
20. Vaswani, A., et al.: Attention is all you need. Advances in neural information processing systems 30 (2017)
21. Yoon, B., So, H., Rhee, J.: A methodology for utilizing vector space to improve the performance of a dog face identification model. Appl. Sci. **11**(5), 2074 (2021)
22. Zhang, K., Sun, M., Han, T.X., Yuan, X., Guo, L., Liu, T.: Residual networks of residual networks: multilevel residual networks. IEEE Trans. Circuits Syst. Video Technol. **28**(6), 1303–1314 (2017)
23. Zhang, X., Zhao, R., Qiao, Y., Wang, X., Li, H.: Adacos: adaptively scaling cosine logits for effectively learning deep face representations. In: Proceedings of the IEEE/CVF Conference on Computer Vision and Pattern Recognition, pp. 10823–10832 (2019)

Convolutional Neural Networks

Convolutional Neural Networks for the Molecular Detection of COVID-19

Anisio P. Santos Jr.[1], Anage C. Mundim Filho[1], Robinson Sabino-Silva[2], and Murillo G. Carneiro[1](✉)

[1] Faculty of Computing, Federal University of Uberlandia, Uberlandia, MG, Brazil
{anisio4322,anage.mundim,mgcarneiro}@ufu.br

[2] Department of Physiology, Institute of Biomedical Sciences, Federal University of Uberlandia, Uberlandia, MG, Brazil

Abstract. The ongoing COVID-19 pandemic caused an unprecedented overburning of healthcare systems and still represents a global health issue with the emergence of COVID-19 variants. The relevance of mass testing for COVID-19 in the find-test-trace-isolate-support strategy suggested by the World Health Organization (WHO) is imperative to reduce COVID-19 transmission. Although real-time polymerase chain reaction (RT-PCR) is considered a reference standard for COVID-19 detection, it is an expensive, lengthened, and laborious process, and problems in RNA extraction can reduce the sensitivity. In this context, the Raman spectroscopy analysis in biofluids is a label-free method performing a suitable cost-benefit application for COVID-19 detection. We propose a Convolutional Neural Network (CNN) architecture that processes spectra images generated by the Raman spectrum and returns the COVID-19 diagnosis of the spectrum sample. The predictive performance of the CNN was compared against several other algorithms widely adopted in the literature. The CNN architecture discriminates COVID-19 with Raman spectroscopy of blood samples with 96.8% accuracy, 95.5% sensitivity, and 98.2% of specificity, representing the best results as well as a promising alternative to distinguish samples. Moreover, we also present a model explanation analysis that contributes to clarifying the salient features taken into account by our CNN.

Keywords: COVID-19 · Raman Spectroscopy · Convolutional Neural Networks · Deep Learning

1 Introduction

The COVID-19 pandemic caused an unprecedented global effect on private and public healthcare systems and social and economic impacts. Among the COVID-19 containment measures is vaccination. However, the distribution of COVID-19 vaccines continues to be applied with low vaccination coverage in underdeveloped countries, especially adapted vaccines to prevalent COVID-19 variants. Therefore, mass testing for COVID-19 is still a strategic preventive measure to reduce its transmission.

M. C. Naldi and R. A. C. Bianchi (Eds.): BRACIS 2023, LNAI 14196, pp. 51–62, 2023.
https://doi.org/10.1007/978-3-031-45389-2_4

There is a need for new techniques developed to prevent it from mass propagating again and also to prevent possible pandemic scenarios. Locations such as airports, schools, workplaces, and remote communities would benefit from faster and reagent-free screening techniques.

Vibrational spectroscopy is considered a promising method for biological analyses. It enables label-free extraction of biochemical information and images toward diagnosis and the assessment of the biochemical and molecular composition of biofluids [1]. Vibrational spectroscopy techniques are bioanalytical approaches with great potential to discriminate between healthy and pathological conditions, like Diabetes mellitus [4] and Zika virus [15]. In addition, it allows the analysis of different bio-fluids, such as blood, urine, saliva, and tissues [2,14,18].

Among several vibrational spectroscopy technologies, Raman is known by its ability to determine molecular components of various types of materials quickly and without the need for the use of reagents. It uses inelastic light scattering based on a monochromatic laser source. The changes in the spectra represent parts of specific molecules. Each Raman shift is unique in each wavenumber frequency, representing a specific functional group [8].

The application of biofluids in Raman spectroscopy for diagnostic or screening fields permits simple sample preparation without the insertion of reagents and thus can be considered a green technology due to a significant reduction in environmental waste [8]. Furthermore, its usage has been investigated in the diagnosis of several diseases like breast, colon, lung, and prostate cancers [9].

In this paper, we are interested in the analysis of Raman spectra obtained from blood serum samples for the detection of COVID-19. The ongoing COVID-19 pandemic caused an unprecedented overburning of healthcare systems and still represents a global health issue with the emergence of COVID-19 variants.

Due to the relevance of the topic, there are already some related works in the literature. In [18] the authors presented a support vector machine (SVM) model to detect COVID-19 using Raman spectrum of serum samples, and in [19] the authors extended the work of [18] by combining the SVM with an extreme gradient boosting (XGB) technique. Despite the Raman spectroscopy data used in [18,19] being the same considered in our research, our work goes a step further by investigating deep learning techniques, like convolutional neural networks (CNNs).

Different from most Raman spectra analysis, which adopts simple statistical analysis [6], and also of some machine learning initiatives with Raman, which are focused on traditional classifiers [18,19], we hypothesize that convolutional neural networks [2] can contribute in the detection of COVID-19 from Raman spectra of serum samples by efficiently learning robust representations of such a problem and also improve model explainability. Despite some few works have already applied 1D CNN to classify breast cancer, minerals, and pancreatic cancer by Raman spectroscopy, e.g., [11,13,16], the investigation of such architectures for COVID-19 detection from Raman spectra is yet a barely explored topic in the literature.

Therefore, the work most related to ours in the literature is the study presented in [5], which investigated traditional and deep learning techniques for the diagnosis of COVID-19 using Raman spectroscopy. In that work, the authors evaluated SVM, XGB, Random Forest (RF), and CNN, which achieved good predictive performance over several groups of experiments. However, our work differs from this for some reasons, such as: we consider Raman spectra of serum samples instead of saliva ones; we consider a publicly available Raman data set instead of a private one, which benefits reproducibility aspects; and besides the predictive performance, we also are interested in the explainability analysis.

The main obstacle to differentiate each class based on the Raman spectrum from control and COVID-19 samples is the similarity of both spectra. The complexity of the dataset is a critical feature to solve issues for discriminating samples of each class. Truncation in regions with organic compounds can be performed at specific wavelength regions. The Raman dataset analyses and interpretation require expertise due to the potential overlap of peaks corresponding to distinct chemical components. The Raman spectra differentiation can be especially problematic in complex biofluids with thousands of molecular components, featuring variability which is an issue for discrimination. This spectral variability can impact the accuracy and discrimination associated with high-dimensional data, with a large number of wavelengths contributing to the spectrum. The management of high-dimensional data is a challenge for computational complexity, feature selection, and generalization of the classification models for spectral datasets.

This paper aims to contribute to the development and evaluation of CNNs for the detection of COVID-19 using Raman spectra of serum samples, in order to obtain more accurate models. To be specific, the study presents a one-dimensional (1D) CNN architecture to process Raman spectra as well as an explainability analysis of its salient features. In summary, the main contributions of our work are listed as follows:

- Development and evaluation of the proposed 1D CNN method for detection of COVID-19 from Raman spectra of serum samples;
- Comparative evaluation of several traditional and deep learning techniques for the problem;
- Explainability analysis regarding salient Raman shifts to detect and discriminate COVID-19 and healthy groups.

The remainder of the paper is organized as follows. Section 2 presents the Raman spectroscopy data as well as the steps of data preparation and preprocessing. Section 3 describes the 1D CNN proposed in this paper. Section 4 shows the experimental results and explainability analysis regarding our CNN model. Section 5 concludes the paper.

2 Data Set Description and Preparation

The data adopted in this work was publicly available by [18]. Blood samples were collected between February 10th and May 10th, 2020. Blood samples were taken from COVID-19 patients and suspected cases upon admission between February 10 and May 10, 2020. The serum was isolated by centrifuging at 3000 rpm for 10 min after a one-hour rest. The serum samples were stored at 4°C and measured within 36 h. For measurement, approximately 0.5 ml of serum sample was prepared in cryopreservation tubes and strictly sealed for Raman scan. Additional spectral data were collected from cryopreservation tubes with saline solution [18].

The data were obtained from blood serum samples in a total of 309 spectra, of which 159 received a positive diagnosis for COVID-19 (COVID-19 group) and 150 received a negative test result for COVID-19 (Healthy group). The serum samples were processed by a Raman spectrometer. For each sample, the equipment is responsible for generating a corresponding spectrum.

Figure 1 shows the average spectrum of the Healthy group (a), the average spectrum of the COVID-19 group (b), the average spectrum of both groups (c) and all spectra from both groups (d). One can see some interesting regions in Figs. 1(c) and 1(d), in which samples from different groups seem better separated.

Regarding data preparation and preprocessing, two steps were conducted in this study: Savitzky-Golay filtering and vector normalization:

- The Savitzky-Golay (SG) is a filter for smoothing and differentiation that optimally fits a set of data points to a polynomial in the least-squares sense [17]. This process is important to reduce noise and smooth the signal while maintaining higher-order moments of the original spectrum. SG has been used to normalize and preprocess spectral data for FTIR, Raman, and other spectroscopy equipments due to the inherent presence of noises during sample collection, manipulation, and so on.
- Vector normalization is the process in which the intensity values of all spectra are normalized by the Euclidean norm.

3 Model Description

This section describes the one-dimensional CNN proposed in this study for the detection of COVID-19 using Raman spectra of serum samples. The proposed architecture is illustrated by Fig. 2. There are three main types of layers: convolutional, pooling, and fully connected. In the following, each one of them is briefly described.

Convolutional Layer. The convolutional layer aims to learn representative features of the input data. The convolution layer is made up of several convolution kernels, also known as filters, which are used to compute different feature maps.

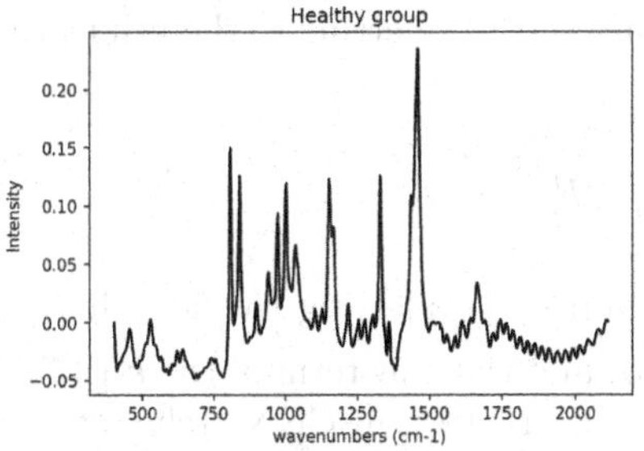

(a) Average spectrum of healthy group

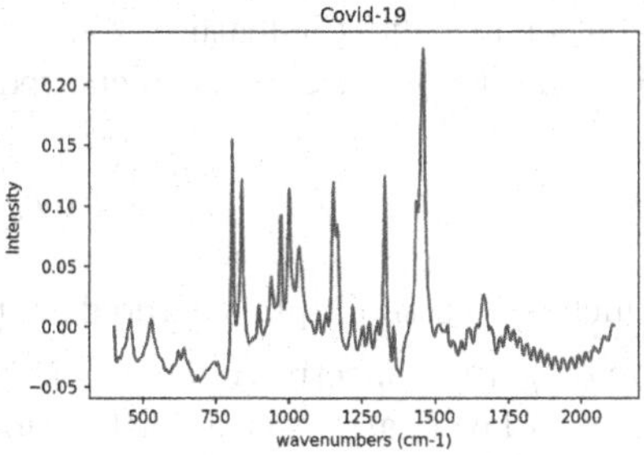

(b) Average spectrum of COVID-19 group

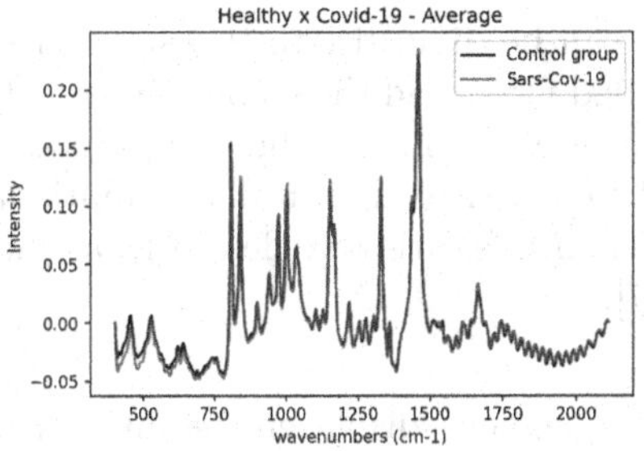

(c) Average spectrum of COVID-19 and healthy groups

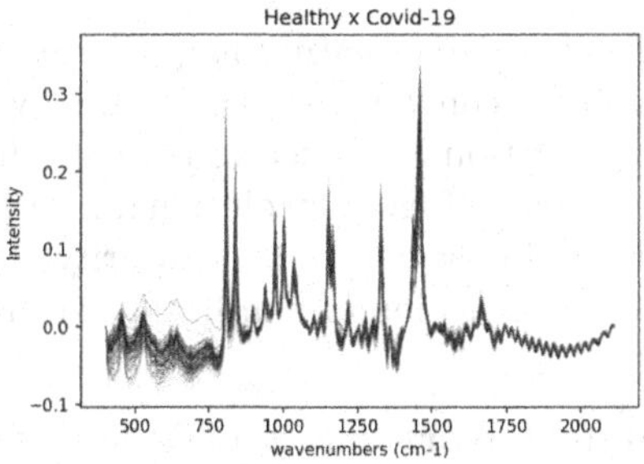

(d) Samples of COVID-19 and healthy groups

Fig. 1. (a) Average of healthy group spectra, (b) Average of COVID-19 group spectra, (c) Average of healthy and COVID-19 spectra, (d) Total COVID-19 and healthy spectral plot

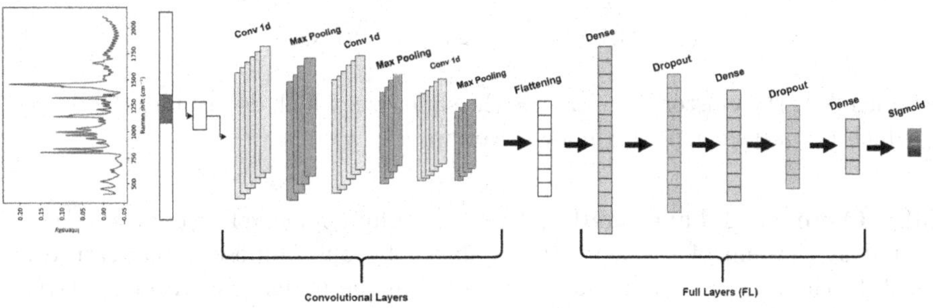

Fig. 2. The architecture of the one-dimensional CNN proposed.

Specifically, each neuron in a feature map is connected to a region of neighboring neurons in the previous layer. Such a neighborhood is referred to as the receptive field of the neuron.

Mathematically, the feature value at location (i, j) referring to the kth feature map of the lth layer, $z^l_{i,j,k}$ is calculated as:

$$z^l_i = \sigma\left(w^{l^T}_k x^l_{i,j} + b^l_k\right) , \quad (1)$$

in which $x^l_{i,j}$ is the input centered at position (i, j) referring to the l-th layer, $w^{l^T}_k$ and b^l_k are respectively the weight vector and the bias term of the k-th filter of the l-th layer, and σ is the activation function which in CNN architecture is usually ReLU (Rectified Linear Units):

$$ReLU(x) = max\{0, x\} . \quad (2)$$

To compute a feature map, the process can be divided into two steps: first, the input is convoluted with an already learned filter, and then a non-linear activation function is applied to the convolution results. This non-linearity is not an integral part of the convolutional layer, but in practice, almost all convolutional layers are followed by a non-linearity. These non-linearities are desirable, as they allow the detection of non-linear features [7].

Pooling Layer. The pooling layer aims to achieve displacement invariance, reducing the spatial resolution of feature maps. This spatial reduction is achieved through a non-unitary stride. In this way, only one activation at each n (stride size) position of the input is calculated in each spatial dimension.

Each feature map of a pooling layer is connected to the corresponding feature map of the previous convolutional layer. The output generated by the pooling function for the feature map $z^l_{i,j,k}$ is represented by the expression:

$$p_{m,k} = \max_{n=1}^{N}\left(z^l_{i-1} \times q + n\right) , \quad (3)$$

in which N is the cluster size; and q is the slide size, which determines the degree to which the adjacent pool windows overlap.

Fully Connected Layer and Dropout. Fully connected layers are usually located at the end of the network, as shown by our proposed architecture in Fig. 2. In these layers, the features extracted in the previous convolution layers are used to obtain the network classification output.

A fully connected layer takes all the neurons in the previous layer and connects them to every neuron in the current layer. It also establishes which high-level attributes most closely relate to the object's class.

Dropout is a technique usually adopted in this layer, in which some neurons are randomly turned off, along with their connections, during training only. During test, all neurons are kept active. The reason for doing this is to avoid overfitting in training. In our architecture, we adopted a dropout rate of 20% between the dense layers.

Finally, after the fully connected layers and the dropouts come to the output layer. For the classification problem, it is common to use as many neurons as there are classes to predict and the output has a sigmoid activation function that can show the result of the classification. The sigmoid is represented in the following equation:

$$\sigma(x) = \frac{1}{1+e^{-x}} \ . \tag{4}$$

This function only returns values between 0 and 1. It is responsible to classify the output and is located in the final step of our model shown in Fig. 2.

The CNN architecture shown by Fig. 2 receives a Raman spectrum as input. It then extracts features of the spectrum through three one-dimensional convolutional layers. At the end of each convolutional layer, ReLU is adopted as the activation function. Between each convolutional layer, there is a MaxPooling layer. By stacking multiple convolutional and pooling layers, one can gradually extract features with a higher (complex) level of abstraction. After extracting the features of the input, the Flattening step transforms the arrays into an one-dimensional vector to start the classification method. The Dense layers, which are also composed by Dropout functions, learn the weights to accurately adjust the classification of each group, and the Sigmoid function is used to return the prediction.

4 Experimental Results

In this section, we present the experiments and results obtained by our CNN in comparison with other techniques in the problem of COVID-19 detection from Raman spectra. In addition, we also present an explainability analysis regarding the salient features considered by CNN in its classification using Shapley additive explanations (SHAP) [3].

The experiments were conducted using a k-fold cross-validation process. This method partition the dataset samples in k disjoint subsets, with k-1 adopted as a training set and the remaining one as a test set. A total of k executions are performed, which means each subset is adopted as the test set once. The predictive performance of the techniques is then obtained over k executions in which performance metrics like accuracy, sensitivity, and specificity are calculated. In this study, we considered 5-fold cross-validation.

We compared our CNN technique against widely known supervised learning techniques like Naive Bayes (NB), Random Forest (RF), Extreme Gradient Boosting (XGB), Multi-Layer Perceptron (MLP), and Support Vector Machine (SVM) as well as against other deep learning techniques like Fully Convolutional Network (FCN), Multi-Channel Deep Convolutional Neural Network (MCDCNN) and Residual Networks (ResNet). All models were trained using the parameters recommended in [10,19].

Regarding the computational simulations, the experiments were done using Python on two machines: a laptop with Core i7 9th Gen processor, 32 GB of Ram, and Geforce GTX 1660Ti GPU, and a desktop with Ryzen 9 processor, 32 GB of Ram, and Titan V GPU.

Table 1. Predictive performance of the techniques under analysis for the COVID-19 detection from Raman spectra.

Model	Accuracy	Sensitivity	Specificity	Mean(Se, Sp)	F1(Se, Sp)
CNN	**0.968**	0.955	0.982	<u>**0.973**</u>	<u>**0.968**</u>
NB	0.890	0.883	0.903	0.893	0.892
RF	0.952	0.966	0.943	0.955	0.954
XGB	0.916	0.898	0.930	0.914	0.913
MLP	0.952	0.912	**1.000**	0.956	0.953
SVM	0.952	**1.000**	0.912	0.956	0.953
FCN	0.813	0.875	0.753	0.814	0.809
MCDCNN	0.942	0.912	0.978	0.945	0.943
RESNET	0.932	0.983	0.860	0.922	0.894

4.1 Predictive Results for COVID-19 Detection

Table 1 shows the predictive results of the techniques under analysis in terms of accuracy, sensitivity, and specificity. As sensitivity and specificity covers different groups, in order to support our analysis we also evaluate the arithmetic and harmonic mean between both metrics, respectively named Mean(Se, Sp) and F1(Se, Sp). The best result obtained by each metric is boldfaced in the table. Among traditional techniques, RF, SVM and MLP achieved the best results. Indeed, SVM and MLP have respectively the highest sensitivity and specificity in the table. However, the best overall results were achieved by the CNN which classified samples correctly more than any other technique. On the other hand, worse results were obtained by NB and FCN. While NB is the unique linear technique under analysis, FCN is a deep learning technique essentially based on the convolutional layer, which means it does not have fully connected layers to help in the learning of complex mapping functions.

Now we analyze each one of the executions of the 5-fold cross-validation in order to better understand the performance of the techniques under analysis. Figure 3 shows the predictive performance of each model in terms of accuracy, specificity, sensitivity, and Mean(Se, Sp). One can see FCN really had troubles in some executions. One can also see that ResNet, a state-of-the-art deep learning technique, had its performance oscillating among the executions. For example, in Fig. 3 ResNet achieved the worst (execution 1) and the best predictive result (execution 3), indicating that the model may be overfitted in some scenarios, which will be subject of our further investigations. Regarding the CNN performance, it was very stable in terms of accuracy, specificity, and Mean(Se, Sp), indicating that such a technique can contribute in the detection of COVID-19 from serum Raman spectroscopy samples. In order to better understand the salient features that contributed to the almost 97% of accuracy, we next investigate the explainability of the CNN model.

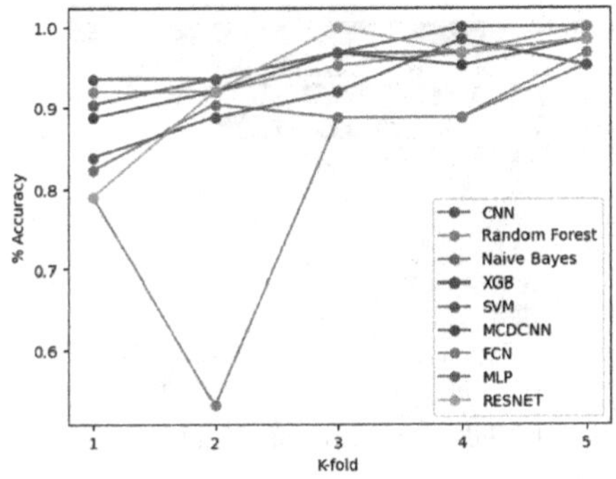

(a) Accuracy analysis

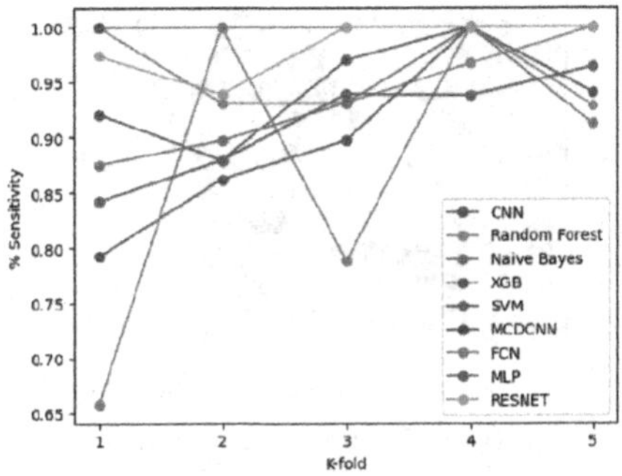

(b) Sensitivity analysis

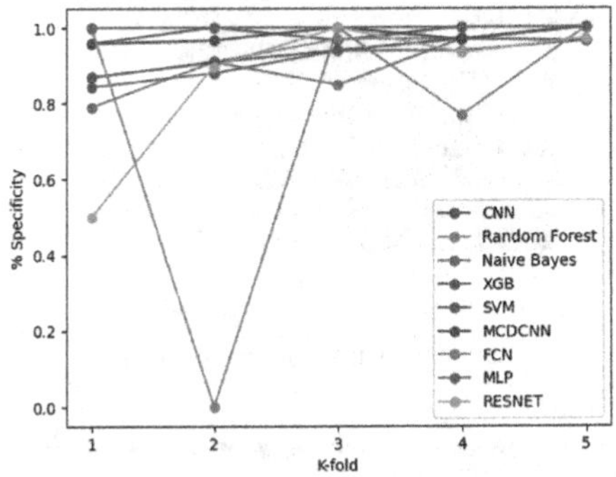

(c) Specificity analysis

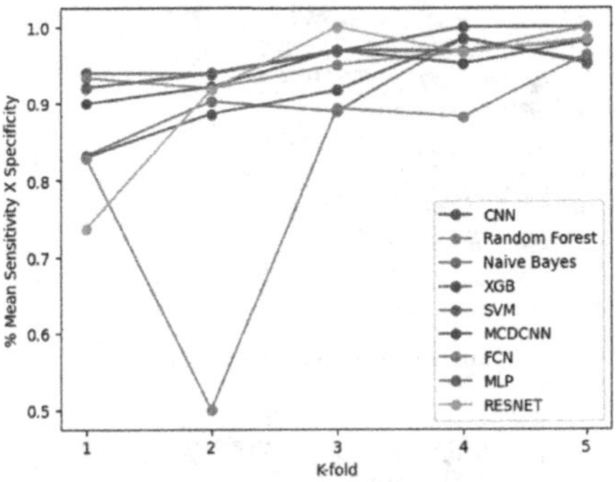

(d) Mean(Se,Sp) analysis

Fig. 3. Analysis of the predictive performance of the techniques under comparison over the 5-fold cross-validation.

4.2 Explainability Analysis

The SHAP method is an approach to explain the output of machine learning models. It connects optimal credit allocation with local explanations using the classic Shapley values from game theory and their related extensions [12]. With that in mind, to understand a bit more about how the CNN model created its inductive mechanism for COVID-19 detection, some SHAP analysis were conducted in order to provide insights regarding the CNN most relevant features.

Figure 4 shows the SHAP features of the main Raman shift capable to discriminate both groups of samples. The main Raman shifts between 1605–1607 cm^{-1} can be attributed to nucleoproteins. The Raman shifts between 1448–1450 cm^{-1} and 1271–1273 cm^{-1} can be attributed to (CH2/CH3) from lipids and the proline from collagen, respectively. Besides, the Raman shift at 831 cm^{-1} can be attributed to the amino acid Tyrosine. Furthermore, the Raman shifts between 427–432 cm^{-1} and 402-410 cm^{-1} were attributed to Phosphate and the heme deformation of proteins. Lastly, the Raman shift at 1657 cm^{-1} can be attributed to Amide I in proteins [19] . In general, the main Raman shifts used to

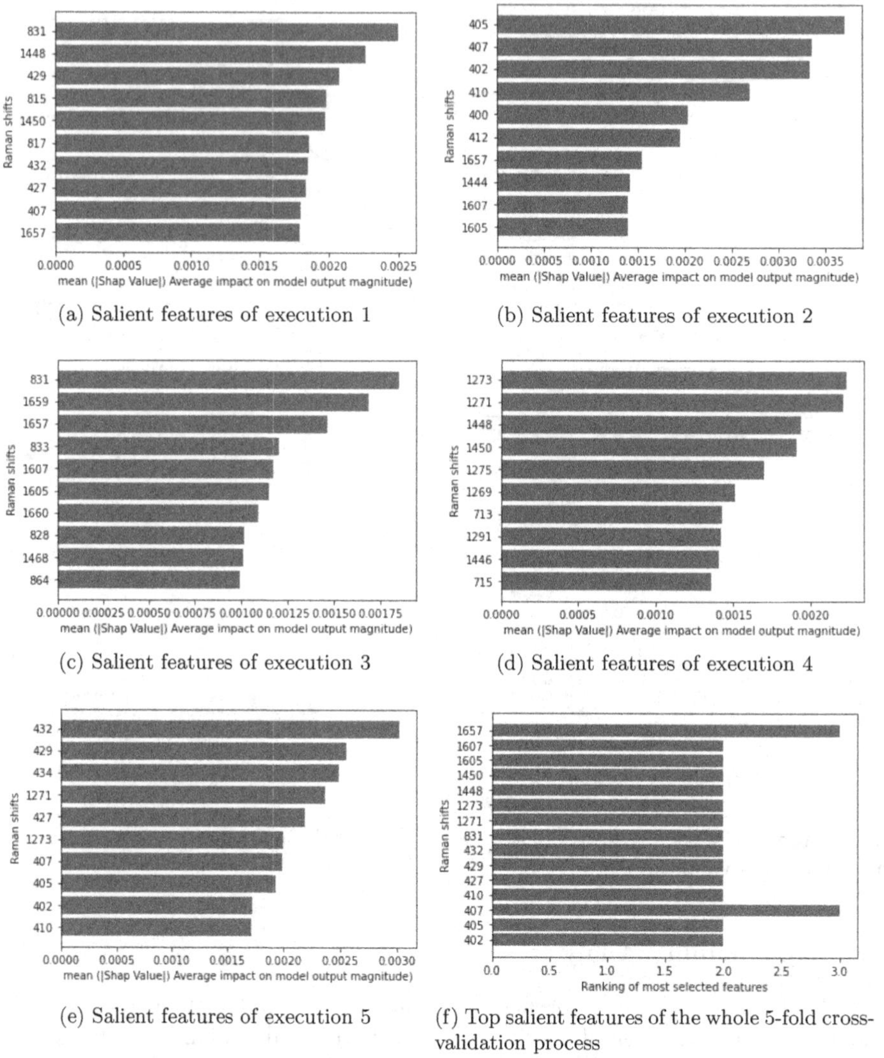

(a) Salient features of execution 1

(b) Salient features of execution 2

(c) Salient features of execution 3

(d) Salient features of execution 4

(e) Salient features of execution 5

(f) Top salient features of the whole 5-fold cross-validation process

Fig. 4. Explainability analysis of the CNN model considering the 5-fold cross-validation for COVID-19 detection.

discriminate control from COVID-19 samples were related to proteins, indicating changes in the proteomic profile in the serum of COVID-19-infected subjects.

5 Conclusions

This paper investigated deep learning solutions based on CNNs for the detection of COVID-19 using Raman spectra of serum samples in order to obtain more accurate models. To be specific, an 1D CNN was proposed and comparatively evaluated against several traditional and deep learning techniques.

Experimental results showed that the CNN model achieved the highest predictive performance in comparison with other techniques, with 96.8% accuracy, 95.5% sensitivity, and 98.2% of specificity, outperforming even other robust deep learning solutions like ResNet.

In addition, explainability analysis regarding salient Raman shifts to detect COVID-19 and healthy groups were also conducted, revealing some proteins which probably are associated with the spectral differences between the groups.

The investigation conducted here contributes to the Raman spectroscopy field and may also impact the society in the future by providing regular, and financially accessible diagnosis for the population.

The fast and accurate results achieved by the CNN method proposed can make the system suitable for use in screening and routine examinations, for example, contributing to the promotion and regular testing of the population.

Forthcoming works will focuses on the development of even better deep learning architectures as well as in addressing other applications based on spectroscopy technologies.

Acknowledgment. Authors thank the financial support given by Google (through the 2020 and 2021 Google Latin America Research Awards), Minas Gerais Research Foundation - FAPEMIG (grants number APQ-00410-21), Brazilian National Council for Scientific and Technological Development - CNPq (grants number 402196/2021-0 and 408216/2022-0), and National Institute of Science and Technology in Theranostics and Nanobiotechnology - INCT-Teranano (grant number CNPq-465669/2014-0). RS-S also thanks the CNPq for the productivity fellowship. We also thank NVIDIA Corporation by the donation of a Titan V GPU used in this research.

References

1. Baker, M.J., et al.: Using fourier transform ir spectroscopy to analyze biological materials. Nat. Protoc. **9**(8), 1771–1791 (2014)
2. Barauna, V.G., et al.: Ultrarapid on-site detection of sars-cov-2 infection using simple atr-ftir spectroscopy and an analysis algorithm: high sensitivity and specificity. Anal. Chem. **93**(5), 2950–2958 (2021)
3. Van den Broeck, G., Lykov, A., Schleich, M., Suciu, D.: On the tractability of shap explanations. J. Artif. Intell. Res. **74**, 851–886 (2022)
4. Caixeta, D.C., et al.: Salivary atr-ftir spectroscopy coupled with support vector machine classification for screening of type 2 diabetes mellitus. Diagnostics **13**(8), 1396 (2023)
5. Carlomagno, C., et al.: Covid-19 salivary raman fingerprint: innovative approach for the detection of current and past sars-cov-2 infections. Sci. Rep. **11**(1), 4943 (2021)
6. Desai, S., et al.: Raman spectroscopy-based detection of RNA viruses in saliva: a preliminary report. J. Biophotonics **13**(10), e202000189 (2020)
7. Géron, A.: Hands-on machine learning with Scikit-Learn, Keras, and TensorFlow. "O'Reilly Media, Inc" (2022)

8. Giamougiannis, P., et al.: A comparative analysis of different biofluids towards ovarian cancer diagnosis using Raman microspectroscopy. Anal. Bioanal. Chem. **413**, 911–922 (2021)
9. Hanna, K., Krzoska, E., Shaaban, A.M., Muirhead, D., Abu-Eid, R., Speirs, V.: Raman spectroscopy: current applications in breast cancer diagnosis, challenges and future prospects. Br. J. Cancer **126**(8), 1125–1139 (2022)
10. Ismail Fawaz, H., Forestier, G., Weber, J., Idoumghar, L., Muller, P.A.: Deep learning for time series classification: a review. Data Min. Knowl. Disc. **33**(4), 917–963 (2019)
11. Li, Z., et al.: Detection of pancreatic cancer by convolutional-neural-network-assisted spontaneous Raman spectroscopy with critical feature visualization. Neural Netw. **144**, 455–464 (2021)
12. Lundberg, S.M., Lee, S.I.: A unified approach to interpreting model predictions. In: Guyon, I., Luxburg, U.V., Bengio, S., Wallach, H., Fergus, R., Vishwanathan, S., Garnett, R. (eds.) Advances in Neural Information Processing Systems 30, pp. 4765–4774. Curran Associates, Inc. (2017)
13. Ma, D., Shang, L., Tang, J., Bao, Y., Fu, J., Yin, J.: Classifying breast cancer tissue by Raman spectroscopy with one-dimensional convolutional neural network. Spectrochim. Acta Part A Mol. Biomol. Spectrosc. **256**, 119732 (2021)
14. Naseer, K., Ali, S., Qazi, J.: Atr-ftir spectroscopy as the future of diagnostics: a systematic review of the approach using bio-fluids. Appl. Spectrosc. Rev. **56**(2), 85–97 (2021)
15. Oliveira, S.W., et al.: Salivary detection of zika virus infection using atr-ftir spectroscopy coupled with machine learning algorithms and univariate analysis: A proof-of-concept animal study. Diagnostics **13**(8), 1443 (2023)
16. Sang, X., Zhou, R.g., Li, Y., Xiong, S.: One-dimensional deep convolutional neural network for mineral classification from Raman spectroscopy. Neural Processing Letters, pp. 1–14 (2022)
17. Savitzky, A., Golay, M.J.: Smoothing and differentiation of data by simplified least squares procedures. Anal. Chem. **36**(8), 1627–1639 (1964)
18. Yin, G., et al.: An efficient primary screening of covid-19 by serum Raman spectroscopy. J. Raman Spectrosc. **52**(5), 949–958 (2021)
19. Zeng, W., Wang, Q., Xia, Z., Li, Z., Qu, H.: Application of xgboost algorithm in the detection of sars-cov-2 using Raman spectroscopy. J. Phys. Conf. Seri. **1775**, 012007. IOP Publishing (2021)

Hierarchical Graph Convolutional Networks for Image Classification

João Pedro Oliveira Batisteli(✉), Silvio Jamil Ferzoli Guimarães, and Zenilton Kleber Gonçalves do Patrocínio Júnior

Image and Multimedia Data Science Laboratory (IMSCIENCE), Pontifícia Universidade Católica de Minas Gerais (PUC Minas) Belo Horizonte, Minas Gerais, Brazil
joao.batisteli@sga.pucminas.br,
{sjamil,zenilton}@pucminas.br
http://www.imscience.icei.pucminas.br/

Abstract. Graph-based image representation is a promising research direction that can capture the structure and semantics of images. However, existing methods for converting images to graphs often fail to preserve the hierarchical information of the image elements and produce sub-optimal or poor regions. To address these limitations, we propose a novel approach that uses a hierarchical image segmentation technique to generate graphs at multiple segmentation scales, capturing the hierarchical relationships between image elements. We also propose and train a **H**ierarchical **G**raph **C**onvolutional Network for **I**mage **C**lassification (HGCIC) model that leverages the hierarchical information with three different adjacency setups on the CIFAR-10 database. Experimental results show that the proposed approach can achieve competitive or superior performance compared to other state-of-the-art methods while using smaller graphs.

Keywords: Deep Learning · Graph Neural Networks · Image Classification

1 Introduction

Graph-based image representation is an emerging research area that leverages the spatial relationships between image elements to model image content more effectively [16]. Graph-based approaches can enhance the understanding of image semantics and context by incorporating domain-specific knowledge into the learning process. Moreover, these approaches can provide multi-scale representations of the same image [17,22,23], capturing local and global information about its structure. These advantages make graph-based image representation appealing for various image analysis tasks.

Code is available at: https://github.com/IMScience-PPGINF-PucMinas/HGCIC.

M. C. Naldi and R. A. C. Bianchi (Eds.): BRACIS 2023, LNAI 14196, pp. 63–76, 2023.
https://doi.org/10.1007/978-3-031-45389-2_5

Applying machine learning algorithms to graph data presents unique challenges due to its irregularity, variable sizes, and diverse neighbor relationships [24]. Graph Neural Networks (GNNs) have emerged to address these challenges, adapting neural network architectures to process graph-structured data effectively. GNNs capture the intricate relationships and dependencies between vertices in a graph, leveraging information from neighboring vertices and edges to encode local and global structural information. This approach enables accurate predictions and improved generalization, making GNNs ideal for tackling various graph-based machine-learning problems.

One of the main applications of GNNs is graph-based image analysis, which requires representing images as graphs. However, this is a challenging task. One common possibility in literature is to apply image segmentation methods for partitioning the image into regions and representing each region by a vertex in a graph [2,9,17,19,22,23]. Despite the simplicity of this approach, it has two important drawbacks: (i) the dependency on the quality and quantity of the regions produced by the segmentation algorithm; and (ii) the hierarchical relationship between image elements are not captured. Here we argue that this property is essential for effective image reasoning, which a graph representation must capture. It is worth mentioning that hierarchical information could be seen as a multi-scale representation. Figure 1 illustrates failures in image classification when the hierarchical structure is disregarded (all cases were classified as airplanes, using [2] representation).

(a) Deer (b) Automobile (c) Ship

Fig. 1. Examples of images classified as airplanes, according to [2].

To overcome the limitations of existing methods, we propose a **novel approach that leverages hierarchical segmentation techniques to generate graph vertices for image classification from graph representation**. Hierarchical segmentation is a well-established technique in computer vision and image processing [6,7,13], enabling the identification of objects and regions within an image based on their visual characteristics. Additionally, our method allows incorporating hierarchical relationships between image elements into the resulting graph, facilitating image analysis tasks. More specifically, the idea of hierarchical segmentation is similar to generating a set of image segmentations at different levels of detail with respect to the principles of multi-scale image analysis [12]. By adopting this approach, we argue that we can effectively capture the rich structural information in the image and encode it in the resulting graph. Thus, our proposal for image classification uses a hierarchical segmentation method to capture hierarchical information for representing image data features. However, in some cases, while working on raw images, noise may produce poor results in image segmentation, for that, in this work, instead of using raw images, we generate superpixels from them to produce homogeneous and concise regions in conjunction with good object delineation. From these superpixels, region adjacency graphs are computed for representing superpixel images.

To evaluate the effectiveness of our proposed representation, we conducted experiments using images from the CIFAR-10 dataset, leveraging our hierarchical segmentation technique from the superpixel images to generate graphs. We also introduce a novel model, the Hierarchical Graph Convolutional Network for Image Classification (HGCIC), that can effectively learn from the hierarchical graphs and classify images.

Experiments demonstrated promising results, outperforming some state-of-the-art methods, highlighting the model and the graph representation effectiveness. Furthermore, the graphs used to train our model have a much smaller number of edges and vertices than those utilized in other works, reducing the computational complexity of GNNs and speeding up the learning process.

We can briefly describe the two major contributions of this work to graph-based image analysis: (i) the proposition of a novel graph representation method that leverages hierarchical image segmentation to capture hierarchical representations of the underlying image structure; and (ii) the introduction of a novel graph convolutional network (GCN) architecture that can extract and use the essential information from our new graph representation.

This work is organized as follows. Section 2 presents some related works. Section 3 describes the most important concepts needed for understanding the proposed method. Section 4 presents a hierarchical graph convolutional network for classifying images based on superpixel graphs. Section 5 describes the experiments and some comparative analysis of the proposed approach to the state-of-the-art methods. And, finally, in Sect. 6, we have drawn some conclusions and present some further work.

2 Related Works

Superpixel image segmentation could be seen as a fundamental task in image processing that divides an image into homogeneous regions. This process can reduce the complexity of an image and provide a more efficient representation for further analysis. By treating these regions or segments as vertices, we can transform the image into a graph structure, which can facilitate the use of graph-based algorithms.

A common strategy in literature for representing images as graphs is based on considering vertices as superpixels [2,9,11,17,20,22]. Superpixels are groups of pixels into perceptually meaningful regions based on similarity criteria like color or location. This approach leverages the flexibility of graph neural networks, which can handle irregular graphs with different shapes and sizes. Other methods include splitting the images into fixed patches viewed as vertices [14] or interpreting each pixel as a vertex [8,20]. Some works also explore multi-scale graph (from image) representations to achieve better graph classification performance.

Multiscale graph-image representations can capture different levels of details and features from the image. For instance, in [22], the authors proposed a novel superpixel algorithm that produces segments with a wide size distribution, allowing a more flexible representation of an image, as it can capture fine and coarse details. The authors in [23] model hyperspectral images as multiple graphs with different neighborhood scales and propose a dynamic graph convolution operation that updates the similarity measures among vertices by fusing feature embeddings. To the best of our knowledge, the work in [17] is the only one that has used a multiscale graph segmentation, in which superpixels are obtained at several scales and different types of relations between vertices are explored to improve the model expressiveness.

The edges of the graph are also crucial for constructing the graph representation, as they enable the message-passing mechanism of GNNs. The message-passing mechanism updates the feature representations of vertices by exchanging information with neighboring vertices through edges. One natural way to build adjacency is using a region adjacency graph (RAG), in which edges are created between spatial neighbors [2,22]. Another strategy is constructing k-nearest adjacency based on the vertex's spatial and/or feature distances, as used in [9,14]. Both methods have shown promising results, and their choice mainly depends on the problem requirements and the properties of the data.

3 Theoretical Background

3.1 Graph Neural Networks

Graph neural networks (GNNs) can be divided into two types: spatial or spectral. Spatial GNNs work directly on the graph structure and compute vertex (and maybe edge) representations using information from their neighbors. Spectral GNNs, on the other hand, function on the graph spectral domain. They utilize the eigenvectors of the graph Laplacian matrix as the foundation for representing the graph data. This work focuses solely on spatial GNNs.

The input of each GNN layer are the vertex feature vectors $\{h_u \in \mathbb{R}^d \mid u \in \mathcal{V}\}$, the set of edges $\mathcal{E}$, and optionally edge feature vectors (maybe seen as weights) $\{w_{uv} \in \mathbb{R}^d \mid (u,v) \in \mathcal{E}\}$. The result of each layer is a new vertex representation $\{h_u^{'} \in \mathbb{R}^{d^{'}} \mid u \in \mathcal{V}\}$, in which the same parametric function is applied to each vertex given its neighbors $\mathcal{N}_u = \{v \in \mathcal{V} \mid (u,v) \in \mathcal{E})\}$ and on the edges incident to it, generically given by:

$$h_u' = f_\theta\left(h_u, aggregate(h_v, w_{uv}' \mid v \in \mathcal{N}_u)\right) \tag{1}$$

in which *aggregate* is a permutation invariant function (like max, min, sum), f_θ is the parametric function, and w_{uv}' is the updated edge feature defined by:

$$w_{uv}' = g_\theta(h_u, h_v, w_{uv}) \tag{2}$$

in which g_θ is a distinct parametric function. Each vertex update step is also called the message passing step since vertices send information to their neighbors.

3.2 Residual Gated Graph Convolutional Network

According to [4], GatedGCN is a fusion of the vanilla GCN and edge gating mechanism. In [9], the authors suggested modifications to the GatedGCN architecture by introducing residual connections and batch normalization [15]. The vertex update is given by:

$$h'_u = h_u + \text{ReLU}(\text{BN}(U^\ell h_u + \sum_{v \in N_v} \alpha_{uv} \odot V^\ell h_v)) \tag{3}$$

in which U^ℓ, V^ℓ are linear transformations, $\odot$ denotes Hadamard product, ReLU stands for Rectified Linear Unit, BN represents batch normalization, and α_{uv} are the edge gates defined by:

$$\alpha_{uv} = \frac{\sigma(\hat{w}_{uv})}{\sum_{v' \in N_u} \sigma(\hat{w}_{uv'}) + \varepsilon} \tag{4}$$

$$\hat{w}_{uv} = A^\ell h_u^\ell + B^\ell h_v^\ell + C^\ell w_{uv} \tag{5}$$

in which A^ℓ, B^ℓ, C^ℓ are linear transformations, σ is the sigmoid function, and ε is a small-fixed constant for numerical stability. The edge gate in Eq. 4 works as a soft attention mechanism [9], allowing the model to learn the importance of different vertices in a neighborhood. Finally, the edge features are updated as follows:

$$w'_{uv} = w_{uv} + \text{ReLU}(\text{BN}(\hat{w}_{uv})) \tag{6}$$

3.3 Hierarchical Segmentation

Hierarchical image segmentation is a set of image segmentations at different detail levels [13]. The segmentations with lower levels of detail can be created by merging regions from segmentations at higher levels of detail.

Hierarchical approaches must obey the principles of multi-scale image analysis. These principles ensure that the segmentation is consistent across different levels of detail. The causality principle defines that a contour presented at a scale k_1 should be present at any scale $k_2 < k_1$. The location principle defines that contours should be stable because they neither move nor deform from one scale to another [12].

Hierarchical image segmentation organizes image segments into a tree structure where each vertex represents a different level of detail or abstraction. The highest level of the tree represents the entire image, while lower levels correspond to smaller and more specific sub-regions or sub-segments. This structure provides a way to represent the image at different levels of resolution, allowing for a better understanding of the image's contents.

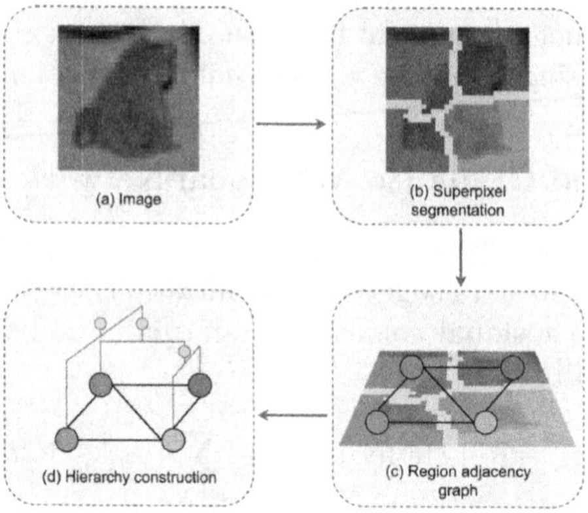

Fig. 2. Pipeline for computing a hierarchy from the original image.

4 Hierarchical Graph Convolutional Networks by Using Hierarchy of Superpixels

Given a finite set V, a *partition* of V is a set $\mathbf{P}$ of nonempty disjoint subsets of V whose union is V. Any element of $\mathbf{P}$, denoted by $\mathbf{R}$, is called a *region* of $\mathbf{P}$. Given two partitions $\mathbf{P}$ and $\mathbf{P}'$ of V, $\mathbf{P}'$ is said to be a (total) refinement of $\mathbf{P}$, denoted by $\mathbf{P}' \preceq \mathbf{P}$, if any region of $\mathbf{P}'$ is included in a region of $\mathbf{P}$. Let $\mathcal{H} = (\mathbf{P}_1, \ldots, \mathbf{P}_\ell)$ be a set of ℓ partitions on V. $\mathcal{H}$ is a hierarchy if $\mathbf{P}_{i-1} \preceq \mathbf{P}_i$, for any $i \in \{2, \ldots, \ell\}$.

4.1 Graph Construction

Let $G = (V, E)$ be a RAG computed from the superpixels in which the set V represents the superpixels and the set E the adjacency relation between the superpixels. Let $\mathcal{H} = (\mathbf{P}_1, \ldots, \mathbf{P}_\ell)$ be a hierarchy computed from the graph G. Let $\mathcal{R}_j$ be the set of regions in the partition $\mathbf{P}_j$ of the hierarchy $\mathcal{H}$. Let $\mathcal{R}$ be set containing all regions belonging to all partition $\mathbf{P}_j \in \mathcal{H}$.

Figure 2 illustrates our proposal for computing the hierarchy from the original image. In the following, we describe how to compute three different graphs from a given hierarchy. These graphs will be used as input in the learning step of our method.

Hierarchy-Based Graph. This graph, denoted by $G_h = (V_h, E_h)$, is a graph computed from the hierarchical structure $\mathcal{H}$ in which the set of vertices V_h is equal to the $\mathcal{R}$. The set of edges E_h is defined by $E_h = \{(r_i, r_j), (r_j, r_i) \mid r_i, r_j \in \mathcal{R},\ r_i \neq r_j,\ \text{inwhich}\ r_j\ \text{isthesmallestregionthatcontains}\ r_i\}$.

kNN-Based Graph. This graph, denoted by $G_k = (V_k, E_k)$, is a graph computed from the k-nearest neighborhood of regions in the hierarchical structure $\mathcal{H}$

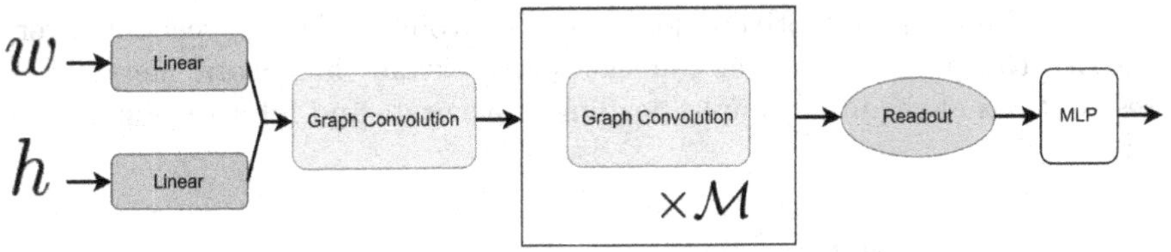

Fig. 3. Model Architecture. The input edges and vertices features are h and w, respectively.

in the feature space, in which the set of vertices V_k is equal to the $\mathcal{R}$. Let $W(V) = \{w(v),\ \forall v \in V\}$ be the set of feature vectors related to the vertex set V. The set of edges E_k is defined by $E_k = \{(r_i, r_j) \mid w(r_j) \text{ isoneofthek} - \text{nnof } w(r_i)\}$.

Complete-Based Graph. This graph, denoted by $G_c = (V_c, E_c)$, is a graph computed from hierarchical structure $\mathcal{H}$, in which the set of vertices V_c is equal to the $\mathcal{R}$. Let $W(V) = \{w(v),\ \forall v \in V\}$ be the set of feature vectors related to the vertex set V. The set of edges E_c is defined by $E_c = \{(r_i, r_j) \mid r_i, r_j \in \mathcal{R},\ r_i \neq r_j\}$.

It is important to mention that each region is represented by the following set of features: color (color channels mean and 2-bin color histogram), texture (contrast, dissimilarity, homogeneity, energy, correlation, and angular second moment), region (orientation, bounding box area, solidity, area, eccentricity, convex area, perimeter, mean intensity, Euler number, and Hu moments), position (X and Y mean position) and the vertex altitude in the segmentation tree. The extracted features were then used to create a vertice feature vector $h_u \in \mathbb{R}^{104\times 1}$ for each $u \in V$.

4.2 Architecture

Figure 3 shows the proposed GCN architecture for the image classification task. To embed the input edge and vertices features, two linear layers were applied to produce D-dimensional embeddings. The dimension of the edge and vertices embeddings remained the same across all layers. Inspired by [5], this work adopted $\mathcal{M}+1$ convolutions, in which $\mathcal{M}$ is chosen at inference time. All $\mathcal{M}$ convolutional layers share the same weights, which improved the results and made the proposed architecture parameter-efficient [5].

An adaptive architecture that adjusts its depth can capture important features and patterns that lead to better accuracy and performance. However, balancing model capacity and complexity is crucial to avoid sub-optimal results. If the model is too shallow, it may not be able to capture complex patterns, while a model that is too deep may suffer from overfitting or be computationally expensive. After the graph convolution layers, we employ a *readout layer* that generates a fixed-size vector representation from the graph features. The output of the readout layer is then fed into a multi-layer perceptron (MLP) that learns to make class predictions based on the graph features.

By combining the adaptive depth graph convolution layers, readout layer, and MLP, the proposed model can effectively extract and learn hierarchical representations of the input graphs, leading to accurate and robust classification results.

5 Experimental Results

To analyze our proposal for image segmentation, we have applied our strategy to a well-known database. We have considered the three different graphs computed from the hierarchy. Also, we have trained all the models for 1,000 epochs with an initial learning rate of 10^{-3}, which is reduced by half if the validation accuracy does not improve after ten epochs until reaching a stopping learning rate of 10^{-5}, Cross Entropy loss, batch size of 64 and Adam optimizer with $\beta_1 = 0.9$, $\beta_2 = 0.98$, and $\epsilon = 10^{-9}$. We saved the weights on the epoch with the best validation accuracy.

We evaluated the proposed model in the CIFAR-10 database [18], comprising 60,000 32×32 color images across ten classes, with 6,000 images each. The database is split into 45,000 training images, 5,000 validation images, and 10,000 test images. It is important to observe that we have followed the procedure described in [9] and randomly sampled 5,000 images from the training set for validation. The same splits were used for all experiments.

5.1 Implementation Details

Graph Construction. We adopt the SLIC [1] as the superpixel segmentation method since it is simple, fast, and memory efficient and to make a fair comparison since almost all work in our comparative analysis uses it. The target number of superpixels is typically 20 but may vary for each image. We have used the *watershed by area* [6] as the hierarchical segmentation method, which outperformed other methods based on different attributes in our experiments. Thus, the hierarchy-based, kNN-based, and complete-based are constructed from the hierarchy computed using watershed by area, which is applied to the RAG of the superpixels obtained by the SLIC. The number of nearest neighbors k was set to 8 on the kNN-based graph setup.

Architecture. To implement our model, we use PyTorch Geometric [10], which batches multiple graphs into a single graph with multiple subgraphs for mini-batch training. Following [5], we could set the variable depth $\mathcal{M}$ as half of the number of vertices in each graph. However, this is not feasible for graphs with different sizes in the same batch. Instead, we use $\mathcal{M} = \lfloor max(|\mathcal{V}|_{minibatch})/2 \rfloor$, in which $max(|\mathcal{V}|_{minibatch})$ is the maximum number of vertices among the graphs in the batch.

We use GatedGCN as the graph convolution since it preserves and updates the edge features w_{uv} between vertices u and v at each layer [9], and the soft attention mechanism in Eq. 5 enables the model to learn how important each

Table 1. HGSIC Architecture Details

	Input Size	Output Size	# of Parameters	Activation Function
Edge Embedding	($\lvert\mathcal{E}\rvert$ x 1)	($\lvert\mathcal{E}\rvert$ x 70)	140	–
Vertex Embedding	($\lvert\mathcal{V}\rvert$ x 104)	($\lvert\mathcal{V}\rvert$ x 70)	7,350	–
GatedGCN	($\lvert\mathcal{E}\rvert$ x 70)	($\lvert\mathcal{E}\rvert$ x 70)	25,270	ReLU
	($\lvert\mathcal{V}\rvert$ x 70)	($\lvert\mathcal{V}\rvert$ x 70)		
GatedGCN	($\lvert\mathcal{E}\rvert$ x 70)	($\lvert\mathcal{E}\rvert$ x 70)	25,270	ReLU
	($\lvert\mathcal{V}\rvert$ x 70)	($\lvert\mathcal{V}\rvert$ x 70)		
Linear	140	256	36,096	ReLU
Layer Norm.	256	256	512	–
Linear	256	10	2,570	–

neighbor v is for vertex u. Furthermore, we changed Eqs. 3–6 by replacing batch normalization with layer normalization and adding nonlinearity and normalization before calculating the attention coefficients, resulting in Eqs. 7–10.

$$h'_u = h_u + \text{ReLU}(\text{LN}(U^\ell h_u + \sum_{v \in N_v} \alpha_{uv} \odot V^\ell h_v)) \tag{7}$$

$$\alpha_{uv} = \frac{\sigma(\hat{w}_{uv})}{\sum_{v' \in N_u} \sigma(\hat{w}_{uv'}) + \varepsilon} \tag{8}$$

$$\hat{w}_{uv} = \text{ReLU}(\text{LN}(A^\ell h_u^\ell + B^\ell h_v^\ell + C^\ell w_{uv})) \tag{9}$$

$$w'_{uv} = w_{uv} + \text{ReLU}(\text{LN}(\hat{w}_{uv})) \tag{10}$$

Distinct from other GCN models, this work used a readout layer that concatenates the global mean and the max pooling, capturing the average and maximum values of the vertex features from the entire graph. Combining two permutation invariant functions was motivated by the better results obtained in the initial experiments compared to using only one. It is worth mentioning that the MLP includes two linear layers with layer normalization that help improves the model's training stability and generalization performance. Table 1 shows the details of each layer in our proposed architecture.

5.2 Quantitative Analysis

Table 2 shows the results for the HGCIC model and other state-of-the-art methods. The HGCIC model was trained using three different graphs: kNN-based (HGCIC$_k$), hierarchy-based (HGCIC$_h$), and complete-based (HGCIC$_c$). Interestingly, the HGCIC$_h$ model, based on the hierarchy structure, has a much smaller number of edges and outperforms the other two graphs.

This result is noteworthy since the number of edges in a graph directly impacts the message-passing step of graph neural networks, which are designed

Table 2. Accuracy of the proposed model and state-of-art methods in the CIFAR-10 dataset. * means the target number of nodes since the authors do not report the average value, and the cells with – refer to data that have not been reported.

Model	# params	# nodes	# vertices	Accuracy
Spatial GNN				
HGCIC$_k$ (ours)	97,208	47.13	377.04	0.6043
HGCIC$_h$ (ours)	97,208	47.13	92.26	**0.6313**
HGCIC$_c$ (ours)	97,208	47.13	2,175.11	0.6186
GAT [2]	55,364	75*	–	0.4593
GatedGCN [9]	104,357	117.63	941.04	**0.6731**
SplineCNN [22]	139,178	197 ± 82	–	0.5869
Spectral GNN				
H-L Cheby-net [17]	200,000	253*	–	**0.7318**

to learn from the graph's structural information. However, our findings suggest that the relationships captured by edges in the graph are more critical than the graph's number of edges. This is consistent with prior work showing that incorporating hierarchical relationships between vertices and edges can help improve GNN performance [17]. Overall, test results demonstrate the effectiveness of the proposed hierarchical adjacency method in enhancing our model's performance.

Despite the superior performance of the HGCIC$_h$ model, our work did not achieve the best accuracy in image classification. However, we achieved a competitive result without resorting to complex strategies that other methods used, such as recurrent neural networks [5], multiple relations [17], positional embeddings [21], or vector fields to define the directions of information propagation [3]. These techniques could also be incorporated into our model in the future to boost its performance further. Another factor that could affect our results is the size of our graphs, which were the smallest among the compared methods. The only model with fewer parameters than ours was the one proposed by [2] but with a lower result compared to the proposed method. Therefore, our work demonstrates a promising approach for graph-based image analysis and shows the potential of hierarchical segmentation for creating effective graph representations of images.

A drawback of the methods proposed in [17] is that they require computing the eigenvalues and eigenvectors of the graph Laplacian matrix. This step is essential for learning filters that depend on the Laplacian eigenbasis and capture the graph structural information, but it can be costly. Furthermore, since the Laplacian eigenbasis is specific to each graph, models trained on one graph may not generalize well to others. This limitation can restrict the scalability and applicability of these methods to a broader range of graph data. Using spatial GNNs, our model can capture the geometric features of the graph without relying on the Laplacian eigenbasis and avoid the problem of generalization to new graphs.

Table 3. Examples of predictions of the proposed models. Images have been enlarged for easy viewing.

	Groundtruth	Hierarchy-based	kNN-based	Complete-based
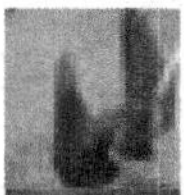	Bird	Bird	Dog	Horse
	Deer	Deer	Bird	Ship
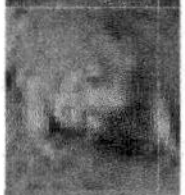	Frog	Frog	Frog	Frog
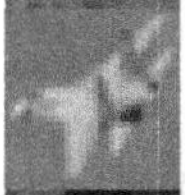	Airplane	Airplane	Airplane	Ship
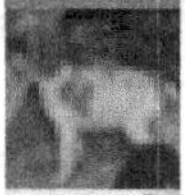	Dog	Dog	Bird	Dog
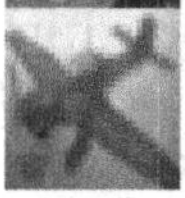	Airplane	Bird	Frog	Frog

5.3 Qualitative Analysis

In Table 3, we present the qualitative results of our models. We observe that the model trained with the hierarchy-based adjacency can classify not only simple images but also images that are challenging for humans to classify due to their low resolution (only 32×32).

5.4 Ablation Study

We conducted an ablation study to assess the performance impact of various components in our proposed method. We compared our proposed architecture with a baseline model based on [9], which has four GatedGCN layers, a global average pooling for the readout layer, and an MLP similar to ours, but without normalization as in the original paper. Additionally, we tested the proposed architecture and graph creation method with a simple feature vector as used in [2,9,17], which consists of the concatenation of the average value for each color channel and the geometric centroid, to show that a more expressive feature set can enhance the performance of the task.

Table 4. Performance of the ablation study to assess how changes in the proposed method affect performance. $HGCIC^{5F}_{\bullet}$ are the models with the proposed architecture trained with the simpler 5D feature vector, and $G4GCN_{\bullet}$ are trained with the full set of features and four GatedGCN layers.

Model	# params	# vetices	# edges	Accuracy
$HGCIC_k$	97,208	47.13	377.04	**0.6043**
$HGCIC^{5F}_k$	90,278	47.13	377.04	0.546
$G4GCN_k$	129,316	47.13	377.04	0.4716
$HGCIC_h$	97,208	47.13	92.26	**0.631**
$HGCIC^{5F}_h$	90,278	47.13	92.26	0.548
$G4GCN_h$	129,316	47.13	92.26	0.4658
$HGCIC_c$	97,208	47.13	2,175.11	**0.6186**
$HGCIC^{5F}_c$	90,278	47.13	2,175.11	0.5698
$G4GCN_c$	129,316	47.13	2,175.11	0.4803

Table 4 shows the result of our ablation study. All modified models showed degradation in accuracy compared to their originals, proving the benefits of both the proposed architecture and used features.

Out of all the models tested, the ones using architecture proposed in [9] showed the most significant decrease in accuracy. This was due to the limited capacity of the architecture to learn complex features from the data. Specifically, the four layers in this architecture were deemed insufficient for the graph-learned embeddings to converge, resulting in poor representations in the final GNN layer.

Among the models trained with the simpler 5D feature vector, the one with the highest accuracy was the model trained with complete graph adjacency. The superior performance of this model was because the distance between vertices is not a reliable indicator of segment differences. As a result, the model gave similar weightage to all neighbors in the message-passing, consequently prioritizing more neighbors to perform the feature aggregation. Similar behavior also occurred in experiments with the [9] architecture, where the hierarchy adjacency caused relevant vertices to be farther away from others. Again, the four layers are insufficient for the information to propagate throughout the graph.

6 Conclusion

This work introduces a new approach for constructing graphs from images using hierarchical segmentation methods. Additionally, it presents a new model called HGCIC, which has been trained using graphs obtained from hierarchical segmentation in three different adjacency setups. The proposed model has demonstrated remarkable results by incorporating variable depth, hierarchical relationships through edges, and well-defined features. The implications of this research are exciting and could have a far-reaching impact on various applications that rely on graph-image analysis.

Although the results were slightly inferior, they still showed promise. The proposed approach utilizes significantly smaller graphs than those in previous works, and the proposed GCN architecture contains fewer parameters yet still delivers promising results. Moreover, the ablative study confirmed the hypothesis that the choice of architecture and features positively impacted the model's overall performance. These findings highlight the potential of the proposed approach as a more efficient and effective means of graph construction and analysis.

In future works, we plan to investigate the impact of attention mechanisms on our approach. Additionally, we aim to conduct a more in-depth analysis of the relationship between the number of vertices and model performance while exploring multiple graph relations.

Acknowledgements. The authors would like to thank Conselho Nacional de Desenvolvimento Científico e Tecnológico - CNPq - (Universal 407242/2021-0 and PQ 306573/2022-9), and Fundação de Amparo Pesquisa do Estado de Minas Gerais - FAPEMIG - (Grants PPM- 00006-18). This study was also financed in part by PUC Minas and by the Coordenação de Aperfeiçoamento de Pessoal de Nível Superior - Brasil (CAPES) - Finance Code 001.

References

1. Achanta, R., Shaji, A., Smith, K., Lucchi, A., Fua, P., Süsstrunk, S.: Slic superpixels compared to state-of-the-art superpixel methods. IEEE Trans. Pattern Anal. Mach. Intell. **34**(11), 2274–2282 (2012)
2. Avelar, P.H., Tavares, A.R., da Silveira, T.L., Jung, C.R., Lamb, L.C.: Superpixel image classification with graph attention networks. In: 2020 33rd SIBGRAPI Conference on Graphics, Patterns and Images (SIBGRAPI), pp. 203–209. IEEE (2020)
3. Beaini, D., Passaro, S., Létourneau, V., Hamilton, W., Corso, G., Liò, P.: Directional graph networks. In: International Conference on Machine Learning, pp. 748–758. PMLR (2021)
4. Bresson, X., Laurent, T.: Residual gated graph convnets. arXiv preprint arXiv:1711.07553 (2017)
5. Corso, G., Cavalleri, L., Beaini, D., Liò, P., Veličković, P.: Principal neighbourhood aggregation for graph nets. In: Advances in Neural Information Processing Systems, vol. 33, pp. 13260–13271 (2020)
6. Cousty, J., Najman, L.: Incremental algorithm for hierarchical minimum spanning forests and saliency of watershed cuts. In: Soille, P., Pesaresi, M., Ouzounis, G.K. (eds.) ISMM 2011. LNCS, vol. 6671, pp. 272–283. Springer, Heidelberg (2011). https://doi.org/10.1007/978-3-642-21569-8_24
7. Cousty, J., Najman, L., Kenmochi, Y., Guimarães, S.: Hierarchical segmentations with graphs: quasi-flat zones, minimum spanning trees, and saliency maps. J. Math. Imaging Vis. **60**(4), 479–502 (2018)
8. Defferrard, M., Bresson, X., Vandergheynst, P.: Convolutional neural networks on graphs with fast localized spectral filtering. In: Advances in Neural Information Processing Systems, vol. 29 (2016)
9. Dwivedi, V.P., Joshi, C.K., Luu, A.T., Laurent, T., Bengio, Y., Bresson, X.: Benchmarking graph neural networks. J. Mach. Learn. Res. **24**(43), 1–48 (2023)

10. Fey, M., Lenssen, J.E.: Fast graph representation learning with PyTorch Geometric. In: ICLR Workshop on Representation Learning on Graphs and Manifolds (2019)
11. Fey, M., Lenssen, J.E., Weichert, F., Müller, H.: SplineCNN: fast geometric deep learning with continuous b-spline kernels. In: Proceedings of the IEEE Conference on Computer Vision and Pattern Recognition, pp. 869–877 (2018)
12. Guigues, L., Cocquerez, J.P., Le Men, H.: Scale-sets image analysis. Int. J. Comput. Vision **68**(3), 289–317 (2006)
13. Guimarães, S., Kenmochi, Y., Cousty, J., Patrocinio, Z., Najman, L.: Hierarchizing graph-based image segmentation algorithms relying on region dissimilarity. Math. Morphol.-Theory Appl. **2**(1), 55–75 (2017)
14. Han, K., Wang, Y., Guo, J., Tang, Y., Wu, E.: Vision GNN: an image is worth graph of nodes. In: NeurIPS (2022)
15. Ioffe, S., Szegedy, C.: Batch normalization: accelerating deep network training by reducing internal covariate shift. In: International Conference on Machine Learning, pp. 448–456. PMLR (2015)
16. Johnson, J., et al.: Image retrieval using scene graphs. In: Proceedings of the IEEE Conference on Computer Vision and Pattern Recognition, pp. 3668–3678 (2015)
17. Knyazev, B., Lin, X., Amer, M., Taylor, G.: Image classification with hierarchical multigraph networks. In: Sidorov, K., Hicks, Y. (eds.) Proceedings of the British Machine Vision Conference (BMVC), pp. 223.1-223.13. BMVA Press (2019). https://doi.org/10.5244/C.33.223
18. Krizhevsky, A., Hinton, G., et al.: Learning multiple layers of features from tiny images (2009)
19. Liu, Q., Xiao, L., Yang, J., Wei, Z.: CNN-enhanced graph convolutional network with pixel-and superpixel-level feature fusion for hyperspectral image classification. IEEE Trans. Geosci. Remote Sens. **59**(10), 8657–8671 (2020)
20. Monti, F., Boscaini, D., Masci, J., Rodola, E., Svoboda, J., Bronstein, M.M.: Geometric deep learning on graphs and manifolds using mixture model CNNs. In: Proceedings of the IEEE Conference on Computer Vision and Pattern Recognition, pp. 5115–5124 (2017)
21. Rampášek, L., Galkin, M., Dwivedi, V.P., Luu, A.T., Wolf, G., Beaini, D.: Recipe for a general, powerful, scalable graph transformer. In: Advances in Neural Information Processing Systems, vol. 35, pp. 14501–14515 (2022)
22. Vasudevan, V., Bassenne, M., Islam, M.T., Xing, L.: Image classification using graph neural network and multiscale wavelet superpixels. Pattern Recogn. Lett. (2023)
23. Wan, S., Gong, C., Zhong, P., Du, B., Zhang, L., Yang, J.: Multiscale dynamic graph convolutional network for hyperspectral image classification. IEEE Trans. Geosci. Remote Sens. **58**(5), 3162–3177 (2019)
24. Wu, Z., Pan, S., Chen, F., Long, G., Zhang, C., Yu, P.: A comprehensive survey on graph neural networks. IEEE Trans. Neural Netw. Learn. Syst. (2020)

Interpreting Convolutional Neural Networks for Brain Tumor Classification: An Explainable Artificial Intelligence Approach

Dieine Estela Bernieri Schiavon(✉), Carla Diniz Lopes Becker, Viviane Rodrigues Botelho, and Thatiane Alves Pianoski

Federal University of Health Sciences of Porto Alegre,
Porto Alegre, RS 90050-170, Brazil
{dieineb,carladiniz,vivianerb,tathiane}@ufcspa.edu.br
https://www.ufcspa.edu.br/english/presentation

Abstract. Brain tumors pose a complex medical challenge, requiring a specific approach for accurate diagnosis and effective treatment. Early detection can significantly improve outcomes and quality of life for patients with brain tumors. Magnetic resonance (MRI) is a powerful diagnostic tool, and convolutional neural networks (CNNs) are efficient deep learning algorithms for image analysis. In this study, we explored using two CNN models for brain tumor classification and applied hyperparameter optimization and data augmentation techniques to achieve an accuracy of up to 96%. In addition, we use Explainable Artificial Intelligence (XAI) techniques to visualize and interpret the behavior of CNN models. Our results show that CNN models accurately classified MRI images with brain tumors. XAI techniques helped us to identify the patterns and features used by the models to make predictions. This study supports the development of more reliable medical diagnoses for brain tumors using CNN models and XAI techniques. The source code is available at (https://github.com/dieineb/Bracis23). The repository also has images generated during the experiments.

Keywords: Brain Tumor · Convolutional Neural networks · XAI Techniques

1 Introduction

The Central Nervous System (CNS) tumors are due to the abnormal growth of cells in the tissues of these locations. According to the National Cancer Institute, CNS cancer represents 1.4% to 1.8% of all malignant tumors worldwide. About 88% of CNS tumors are in the brain [1]. These tumors cause profound deficits and are fatal in patients whose initial symptoms are often not perceived.

Malignant-based tumors are classified by the World Health Organization (WHO) [2] into four stages (I-IV) according to aggressiveness. Tumors classified

M. C. Naldi and R. A. C. Bianchi (Eds.): BRACIS 2023, LNAI 14196, pp. 77–91, 2023.
https://doi.org/10.1007/978-3-031-45389-2_6

in stages III and IV are more aggressive and need immediate treatment and can lead the individual to death in less than two years. The classification of Magnetic Resonance Imaging (MRI) is an important part of the processing of medical imaging and assists physicians in making diagnoses and more accurate treatments. The MRI is one of the leading imaging tools for studying brain issues. It is often the modality of choice for diagnosis and treatment planning for tumors brain due to the high resolution and significant contrast of soft tissues [3].

The first methods of characterization of cancer brains were predominantly based on modeling neuroimaging data statistics. Driven by advances in computing, deep learning has become standard in image classification medical. Integrated statistics and deep learning methods have recently emerged as a new direction in the automation of medical practice, unifying knowledge multidisciplinary in medicine, statistics, and artificial intelligence [4].

Convolutional neural networks (CNNs) are deep learning algorithms, very efficient for image analysis [5]. They are frequently used in health, mainly for classifying, locating, detecting, and segmenting tumors on magnetic resonance imaging. The architecture typical of CNN for image processing consists of a series of layers of convolution filters interspersed with a series of reduction layers or data grouping.

Convolutional layers in CNNs serve as feature extractors, learning feature representations from input images. Neurons in these layers are organized into feature maps, and filters connect them to neurons in the previous layer. The convolution filters detect relevant image features, progressively capturing more complex patterns. Additionally, pooling layers are commonly employed after convolutional layers to reduce image dimensions. Multiple convolutional layers are often stacked together to create a deep model, enabling the extraction of abstract feature representations [6]. The fully connected layers interpret these extracted features and perform high-level reasoning tasks.

Deep learning algorithms are having an increasing impact on our everyday lives. As machine learning algorithms are increasingly applied to high-impact, high-risk tasks like medical diagnoses, it is critical that researchers can explain how these algorithms arrived at their predictions [7]. However, neural network architectures have a natural propensity for opacity in understanding data processing and prediction.

Explainable artificial intelligence (XAI), used in machine learning, is an artificial intelligence approach that aims to create transparent and interpretable models providing human-understandable explanations for their decisions or predictions. XAI aims to improve the reliability of AI systems and facilitate collaboration between humans and machines. These techniques include methods for visualizing model internals, feature importance analysis, and rule extraction, among other methods. This paper focuses on the model visualization technique: feature visualization by layers and class activation maps (CAMs).

Visualization and CAMs techniques are used to interpret and understand the behavior of machine learning models. Feature visualization generates images that activate specific neurons in a neural network, allowing researchers to understand the patterns and features that the network uses to make predictions. CAMs identify the important regions of an input image that contribute the most to the output of a model, which helps understand which features the model uses to make predictions and diagnose performance issues. Both techniques are valuable tools for identifying areas for improvement in a model's performance.

By explaining the predictions made by the model, XAI techniques increase transparency and reliability, allowing medical professionals to understand the reasoning behind the diagnosis better.

This study aims to evaluate two CNN models that, after using hyperparameter optimization techniques and data augmentation, presented accuracy results greater than 96% in all stages (training, validation, and testing) to classify resonance images with brain tumors and demonstrate the characteristics of the networks considered for image classification using the XAI technique. For this purpose, the databases Br35H - Brain Tumor Detection [8] and The Brain Tumor Classification (MRI) by Sartaj [9], both available in Kaggle, will be used.

This paper is structured as follows. Section 2 presents related works in deep learning on medical image analysis and explainable artificial intelligence. Then, Sect. 3 presents the dataset, pre-processing, data augmentation, and the networks' structure used for training. Section 4 presents the results, the discussion, and the extracted XAI features for enhancing interpretability. Section 5 concludes the paper.

2 Related Works

Explainable Artificial Intelligence is a technique used to understand an AI model and explain how it arrived at its results or decisions. XAI is particularly important in deep learning, as deep neural networks can be very complex, making it difficult to understand how they arrived at a particular decision or prediction.

In deep learning, XAI provides insight into a neural network's internal workings. This can be achieved through various techniques, such as visualizing the intermediate layers of a network or generating heatmaps to show which parts of an input image were most important in making a particular prediction.

XAI techniques are becoming increasingly important in medical diagnosis using deep learning models, as they can help improve the interpretability and trustworthiness of these models and improve patient outcomes. Medical diagnostic support has shown great promise with them in identifying and classifying various medical conditions based on medical imaging data, such as X-rays, MRI, and Computed Tomography scans. However, they can be highly complex.

This section reports some published research on the question of this article: Are the results obtained through CNNs reliable for supporting the medical diagnosis?

The paper [10] discusses the need for a more scientific approach to interpretability in machine learning. They propose a framework for characterizing different types of interpretability and identify research challenges in developing a rigorous science of interpretable machine learning. The authors discuss potential future directions for research in this area, such as developing new evaluation metrics and exploring the ethical implications of interpretability. The paper emphasizes the importance of interpretability as a critical component of responsible and trustworthy machine learning systems.

In [11], the authors propose a method that uses a combination of deep neural networks (DNNs) and multiclass support vector machines (SVMs) to improve the accuracy and interpretability of brain tumor diagnosis. To improve the interpretability of their model, the authors use explainable artificial intelligence (XAI) techniques, including feature visualization and salience maps, to identify the key features and regions of the input images most relevant for tumor detection. They also compare the performance of their model to other state-of-the-art brain tumor detection methods. The authors demonstrate that the proposed method achieves high accuracy in detecting brain tumors greater than 97% and that using XAI techniques improves the interpretability of their model and provides information about the underlying characteristics of brain tumors. The authors tested the model using two databases (figshare and BraTS 2018). The paper highlights the potential of combining deep learning, machine learning, and XAI techniques to improve the accuracy and interpretability of brain tumor diagnosis.

In the study [12], seven CNN models are evaluated for the classification of brain tumors. The models considered include a generic CNN model and six pre-trained CNN models. The dataset used for this research is called Msoud, which comprises three existing datasets: Fighshare, Sartaj, and Br35H. This combined dataset consists of 7023 MRI images. The study results indicate that the InceptionV3 model achieves the highest accuracy among all evaluated CNN models. It achieves an average accuracy of 97.12% on the Msoud dataset. The authors used the three banks together (80% for training and 20% for testing). The paper focused on showing the performance of the models through metrics based on values, even though it is a recent study published in 2023. However, there has been a growing tendency to incorporate explainability techniques for greater transparency of the models, in addition to the traditional metrics in the last few years.

In [13], the authors propose an explanation-driven DL model that utilizes a convolutional neural network, local interpretable model-agnostic explanation (LIME), and SHAP for predicting discrete subtypes of brain tumors (meningioma, glioma, and pituitary) using an MRI image dataset. Unlike previous models, the proposed model uses a dual-input CNN approach to overcome classification challenges posed by low-quality images with noise and metal artifacts by adding Gaussian noise. The CNN training results demonstrate 94.64% accuracy, outperforming other state-of-the-art methods. To ensure consistency and local accuracy for interpretation, the authors employ Shapley Additive exPlanations (SHAP), which examines all future predictions applying all possible combinations of inputs, and LIME, which constructs sparse linear models around each prediction to illustrate how the model operates in the immediate area.

The study [14] proposes a Convolutional Neural Network (CNN) model with a new form of Grad-CAM called numerical Grad-CAM (numGrad-CAM) to provide a user-friendly explainability interface for brain tumor diagnosis.

The numGrad-CAM-CNN model was evaluated using technical and physician-oriented (human-side) evaluations and achieved an average accuracy of 97.11% a sensitivity of 95.58% and a specificity of 96.81% for the targeted brain tumor diagnosis setup. Moreover, the integration of numGrad-CAM provided an accuracy of 90.11% compared to other CAM variations in the same CNN model. Physicians who used the numGrad-CAM-CNN model provided positive feedback on its usefulness in providing an explainable and safe diagnosis decision-making perspective for brain tumors. This study demonstrates that the proposed method can effectively enhance the interpretability of deep learning models for brain tumor diagnosis, making them more trustworthy and accessible to healthcare professionals.

The study [15] demonstrated the performance of the commonly used convolutional neural network (CNN) models: VGG16, ResNet50, and MobileNet. These models were applied to brain magnetic resonance imaging to identify tumor cells. The reported accuracy of these models was 97%, 94.5%, and 99%, respectively. The authors aimed to develop a modified CNN model comparing the proposed model with the three known networks. The model proposed in the study achieved an overall classification accuracy of 98.5%. All models were trained and tested using only one database, the Br35H. The authors did not explicitly mention the use of explainable techniques in their study. However, they expressed their intention to increase the model's reliability by implementing similar medical imaging datasets.

Using a single dataset in a study, as mentioned in some papers presented in this section, may limit the generalizability of the results. Robust conclusions typically require validation across multiple independent datasets. While the studies presented provide insights into the performance of specific CNN models for medical diagnosis, further research, and validation on diverse datasets are essential to establish the reliability and generalizability of these models.

In addition to evaluating the performance of the models with statistical metrics and testing in more than one database, an excellent current practice is to evaluate through interpretability techniques so that the models are not seen as black boxes. These XAI techniques offer different approaches to explain the decision-making processes of deep learning models. Visualizing important regions, assigning feature importance values, or calculating gradients provide insight into the model's inner workings and help users understand why specific decisions or predictions were made. These techniques contribute to the reliability, acceptance, and responsible application of deep learning models in various domains, including medical diagnosis.

3 Experiments

For the development and execution of codes necessary for manipulating data and implementing the model, was used Google Collaboratory Pro. The machine learning library used was Keras via the TensorFlow platform.

Two CNN architectures were used for the experiments, a standard multi-layered deep CNN (Fig. 1), Conv2D developed with three convolutional layers, and a deeper one, the Xception model (Fig. 2) with 36 and separable convolutional layers.

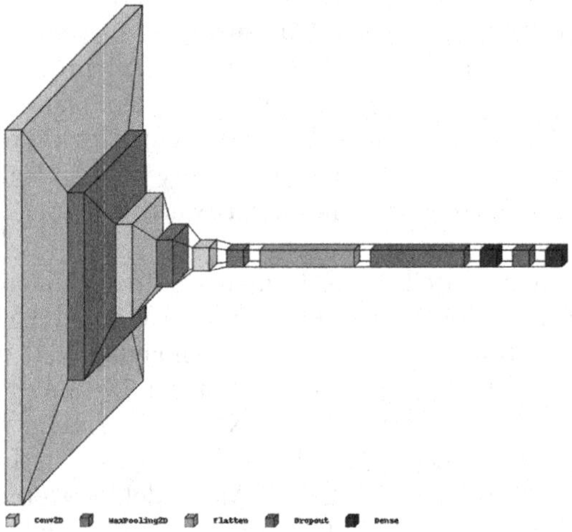

Fig. 1. The Basic Model layout [own author].

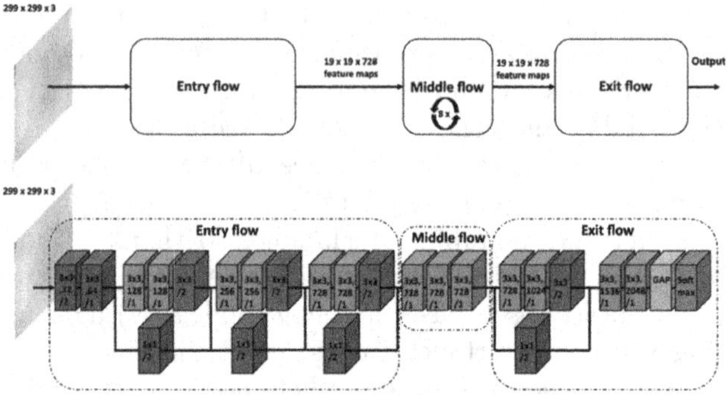

Fig. 2. Xception Model layout [16]

Datasets. This study utilized two databases for brain tumor detection and classification. The first database used was the Br35H [8] Brain Tumor Detection database, which consists of 3000 magnetic resonance images (MRI) representing both images with brain tumors and normal images. The database is well-balanced, containing 1500 images depicting brain tumors and an equal number of 1500 normal images. The available interpretation categories for this database are "no" (without a brain tumor) and "yes" (with the presence of a brain tumor).

The Brain Tumor Classification (MRI) (Sartaj) [9] database from the Kaggle platform was employed to assess the generalization of the models. This database consists of 2870 images for training and 394 for testing, divided into four classes: without tumor, meningioma, pituitary tumor, and glioma. Among the test folder images, there are 105 images without tumors, 115 with meningioma, 100 with glioma, and 74 with pituitary tumors.

Since the research paper focuses on binary classification (with and without tumor), without classifying the tumor types, the authors decided to conduct an experiment using the meningioma class and the no-tumor class due to their balanced distribution. As a result, only 220 images were selected for this experiment: 105 files representing "No Tumor" and 115 files representing "Meningioma Tumor".

Data Preprocessing and Data Augmentation. The images were resized to (224, 224) for the first model and (299,299) for the Xception pre-trained model.

Data augmentation generates more training samples by augmenting the samples through some random transformations that produce believable-looking images. The goal is for the model to see the same image only once during training.

The following transformations were applied in the training of both models:

RandomFlip("horizontal"): this layer performs random horizontal flips on the input images. It randomly mirrors images horizontally, which helps the model learn invariant features and improve its generalizability.

RandomRotation(0.1): this layer applies a random rotation to the input images. Parameter 0.1 indicates the maximum angle of rotation in radians. In this case, the images will be rotated by a random angle within the -0.1 to 0.1 radians range.

RandomZoom(0.2): this layer randomly zooms in on the input images. Parameter 0.2 specifies the maximum zoom range. This means that images can be enlarged or reduced by a random factor within 0.8 to 1.2.

Architectures. Below we present the details of the two architectures used in the experiments.

Basic Model. The basic model was developed with a lean architecture, with three convolutional layers, three Max Pooling layers, a Dense layer with 128 neurons, and the final layer with one neuron for the binary classification problem. A dropout of 40% was added after the Flatten and Dense(128) layers to avoid overfitting.

Xception. Inspired by the assumptions of Inception architecture, the Xception model, which means "Extreme Inception," was named because, according to the author [17], this hypothesis is a stronger version of the hypothesis underlying Inception architecture. The author proposed a convolutional neural network architecture based entirely on depthwise separable convolution layers. In effect, they make the following hypothesis: mapping cross-channel and spatial correlations in the feature maps of convolutional neural networks can be entirely decoupled. The Xception architecture has 36 convolutional layers forming the feature extraction base of the network. The convolutional layers are structured into 14 modules with linear residual connections around them, except for the first and last modules [17].

Metrics. The statistical metrics used to evaluate the models were: Accuracy, Precision, Recall, Specificity, and Receiving Operational Characteristics (ROC Curve). These metrics are often employed in similar works [11, 14].

XAI Techniques. The feature visualization technique and Gradient-weighted Class Activation Mapping (Grad-CAM) were used in the basic CNN model and the CAM with LIME (Local Interpretable Model-Agnostic Explanation) techniques in the Xception model.

The general idea behind visualizing CNN layers is to identify which parts of the input image activate the different filters at each layer and how these activations are combined to form the final prediction [18]. This information can be used to understand how the network is making its predictions and identify potential improvement areas.

The heatmap technique [18], denominated "class activation maps" (CAMs), involves taking the output of the last convolutional layer and obtaining a vector feature by applying a global average pooling operation. The vector feature is then used to compute the final prediction and the class activation map, which is obtained by weighting the output of the last convolutional layer with the corresponding weights of the final prediction (Eq. 1).

$$M^{(c)} = \sum_{k} w_k^{(c)} \phi_k(x) \tag{1}$$

where $w_k^{(c)}$ is the weight of the k-th feature map in the last convolutional layer for class c, and $\phi_k(x)$ is the activation map of the k-th feature map in the last convolutional layer for the input image x.

Grad-CAM [19] is an XAI technique used in CNNs to identify the most important regions of an image that contributed to a particular class prediction. It uses the gradient information flowing into the last convolutional layer of the network to generate a heatmap that highlights these regions. Grad-CAM is an extension of CAM that can be used with any CNN architecture and produces more accurate and finer-grained heatmaps than CAM. These techniques help interpret CNN predictions and identify potential biases or limitations in the model.

3.1 Training, Validation, and Test

The Br35H [8] database was used in both models for training and validation using the holdout technique (80% for training and 20% for validation), and in the first step, 10% of the entire dataset, 300 samples, was saved for testing. Afterward, the models were trained for 60 epochs and compiled using the Adam learning rate optimizer by default 0.001. The metrics for monitoring the convergence of the models were accuracy and binary cross-entropy for loss. In the final phase, to test the models and generate the confusion matrix, 220 images from the other database [9] were used.

4 Results and Discussion

This section will present and discuss the performance of the models in the Br35H and Sartaj datasets. Afterward, information regarding the XAI techniques applied to the models will be presented.

The hyperparameters used were selected based on previous research and empirical evaluation and were considered effective for the classification task of brain tumor images.

The Adam optimizer was used for both models, with a learning rate of 0.001. The Adam optimizer is a widely used optimization algorithm that adapts the learning rate for each parameter during training, making it suitable for various tasks. The learning rate determines the step size in each iteration during optimization. The Batch size was 32 for the base model and 64 for the Xception model. Batch size affects the data used to update model parameters on each iteration. Larger batch size can lead to more stable convergence and increase memory requirement. Finally, both models were trained for 60 epochs.

The learning curves shown in Fig. 3 and Fig. 4 demonstrate the training and validation accuracy and loss over the 60 epochs for the Basic Model CNN and Xception Model CNN, respectively.

In the case of the Basic Model CNN, the training accuracy gradually increases and plateaus at around 95% after about 20 epochs. The validation accuracy also increases but plateaus at around 90% after about 15 epochs. The loss decreases sharply in the first few epochs, then decreases at a slower rate until it plateaus after about 30 epochs. These results suggest that the Basic Model CNN can achieve high accuracy on the training data but may be overfitting after 20 epochs, as the validation accuracy is a bit lower than the training accuracy.

For the Xception Model CNN, the training accuracy increases more rapidly than the Basic Model CNN and plateaus at around 98% after ten epochs. The validation accuracy also increases and plateaus at around 96% after about 15 epochs. The loss decreases more smoothly and plateaus after about 40 epochs. These results suggest that the Xception Model CNN can achieve even higher accuracy than the Basic Model CNN and is less prone to overfitting.

The hyperparameters used appear to have been effectively training both models. However, the Xception Model CNN performs better than the Basic Model CNN, likely due to its more complex architecture.

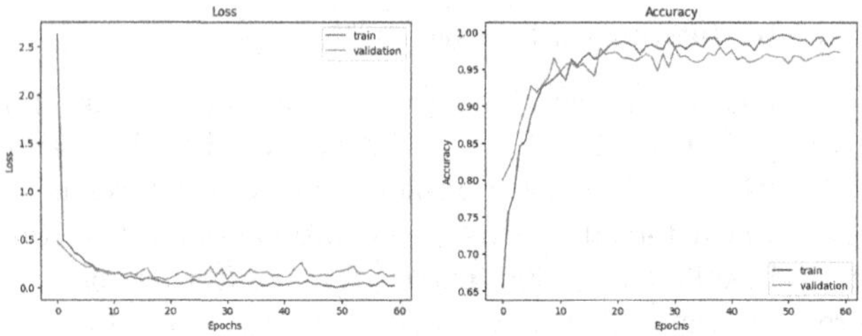

Fig. 3. Learning curves of Basic Model CNN

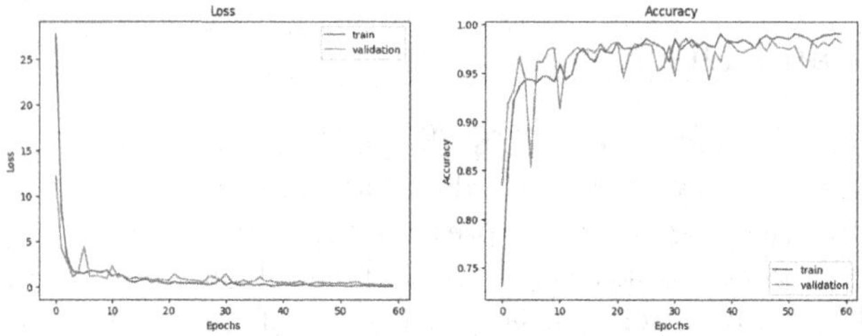

Fig. 4. Learning curves of Xception Model CNN

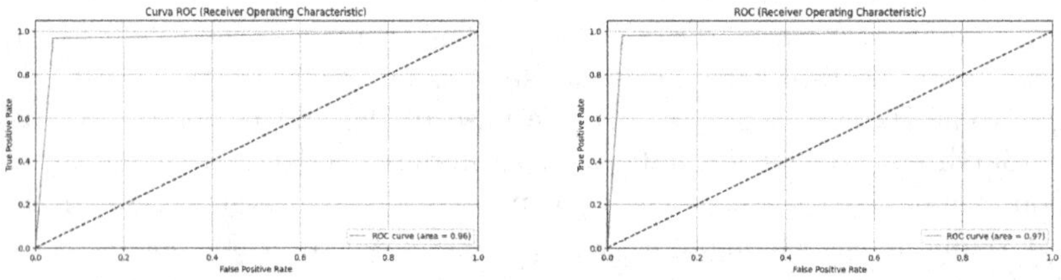

Fig. 5. ROC Curve Basic and Xception models on Br35H dataset.

The experimental results in Table 1 indicate that the basic model and Xception perform well on the Br35H dataset, achieving high accuracy and AUC (Area Under the Curve) values (Fig. 5). However, Xception outperforms the basic precision, recall, and specificity model. Specifically, Xception achieves a precision of 96.71%, a recall of 98%, and a specificity of 96.67%, compared to the basic model's precision of 96.03%, recall of 96.67%, and specificity of 96%. These results suggest that Xception can better distinguish between positive and negative instances in the Br35H dataset.

On the other hand, the test data did not perform well in the Sartaj dataset. The basic model and Xception achieve moderate accuracy and AUC values (Fig. 6), but Xception performs better in precision, recall, and specificity. Xception achieves a precision of 80.15%, a recall of 94.78%, and a specificity of 74.28%, compared to the basic model's precision of 77.60%, recall of 84.35%, and specificity of 73.3%. Despite these improvements, the overall performance of the Sar-

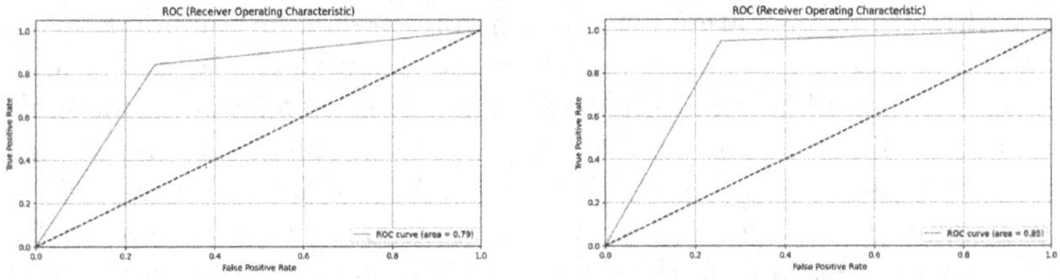

Fig. 6. ROC Curve Basic and Xception models on Sartaj dataset.

taj dataset is still relatively low. One hypothesis for this low performance could be due to the variability of the Sartaj dataset, which may contain variations in image features. It could be due to differences in imaging capture techniques, image quality, or specific characteristics of the tumors in the dataset. The models were trained on a different dataset than the Sartaj dataset and may struggle to generalize effectively.

The results demonstrate the importance of evaluating a model on multiple datasets to ensure its generalization performance. Although the models are promising on the Br35H dataset, it has yet to show good generalizability on new images. More investigations are needed to improve its performance on the Sartaj and in other datasets.

Table 1. Test Results on Br35H and Sartaj Datasets

Dataset	CNN	Loss	Acc. (%)	Precision (%)	Recall (%)	Specificity (%)	AUC
Br35H	Basic	0.1802	96.33	96.03	96.67	96.00	0.96
	Xception	0.6124	97.33	96.71	98.00	96.67	0.97
Sartaj	Basic	0.7762	79.09	77.60	84.35	73.33	0.79
	Xception	3.2430	85.00	80.15	94.78	74.28	0.85

The following topic aims to apply the XAI techniques, following the interpretable evaluation of the CNNs' performance. Our investigation employed a triad of XAI methodologies to extract insights regarding the CNN models utilized for brain tumor classification.

The first technique was visualizing images in convolutional layers of the Basic Model, which showed how the input image (Fig. 7) was processed in each of the three convolutional layers (Fig. 8). This technique aimed to provide an understanding of how the model made its predictions and identify which parts of the input image were used by the model for classification. These images allow us to observe the features learned by the model at each layer and the increasing complexity of the representations as we move deeper into the network.

The Grad-CAM technique is applied to the basic model's first, second, and third convolutional layers (Fig. 9). For each layer, the input image is passed through the model, and the gradient information flowing into the layer generates a heat map highlighting the most important regions of the image that contributed to the classification. The heatmap is overlaid on the input image to show the image regions that contributed to the ranking. The results show that Grad-CAM can provide valuable information about the model's decision-making process. For the first convolutional layer, the heatmap highlights the edges and contours of the object in the image. The second convolutional layer highlights more complex features, such as textures and patterns. The heatmap highlights even more complex features in the third convolutional layer, such as shapes and structures.

The third technique was the Class Activation Map, applied to the Xception model. Figure 10 illustrates a false positive output generated by the Xception model, showing its prediction with 100% confidence. The true class of the image is identified as "no tumor", while the model predicts it as "yes tumor". In addition to correctly interpreting classified classes, it is also essential to understand the factors contributing to false positives and negatives to improve model performance, reduce classification errors, and increase confidence in model predictions. Interpretability techniques offer a unique advantage, allowing humans to study models and actively look for flaws. When flaws are exposed and understood, we can better understand the potential risks associated with these models and design safeguards to mitigate them. It could ensure that deployed systems meet transparency requirements and are more reliable.

Visualizing the CNN layers, the CAM and Grad-CAM techniques provide an understanding of the inner workings of the CNN models used to classify medical images, improving interpretability and consequently making the network more transparent to human eyes. These visualization techniques help explain the model's decision-making process and can provide insights into improving model performance.

5 Conclusion

This paper evaluated the performance of two CNN models, the Basic Model and Xception, for brain tumor classification on two datasets, Br35H and Sartaj. The accuracy values of the models reached 96.33% and 97.33% in the test set of the database in which they were trained. The results obtained here are similar to those found in recent research [11,12,14,15] and relatively higher than in the study [13]. The results also indicate that Xception outperforms the basic model on both datasets. However, the overall performance of the Sartaj dataset remains low, suggesting the need for further investigation to improve model generalization.

Using different databases in research is crucial to support robust conclusions about the generalization of a model. It helps assess a model's ability to generalize diverse data, identify biases, increase reliability and reproducibility, and validate real-world applicability. It is a common approach that researchers in deep learning and medical imaging must follow to ensure the validity and usefulness of their discoveries.

Furthermore, the paper introduces explainable artificial intelligence (XAI) techniques, the visualization of CNN layers, CAM, and Grad-CAM techniques to enhance the CNN models' interpretability. The visualization of CNN layers demonstrates the learned features by the model at each layer and the increasing complexity of the representations. These visualization techniques improve the understanding of the CNN models' decision-making process and could provide valuable insights into enhancing model performance.

XAI techniques have several benefits, including increased trust and reliability of the model's output, better communication between medical professionals and patients, and potential improvements in model performance. Additionally, the visualizations generated by XAI techniques can assist radiologists in identifying the most important features in medical images, leading to more accurate diagnoses.

However, there are also some limitations to consider. XAI techniques can be computationally expensive, requiring significant processing power and time. Furthermore, the visualizations generated by these techniques can be subjective. Therefore, it is crucial to perform thorough validation and testing of the models using XAI techniques to ensure their generalization to new data.

XAI techniques have demonstrated significant potential for improving the interpretability of CNN models in medical image classification tasks. Their use is likely to become increasingly important in the future. Nevertheless, further research and development are necessary to overcome the limitations and challenges associated with these techniques to ensure their reliability and usefulness in real-world medical applications.

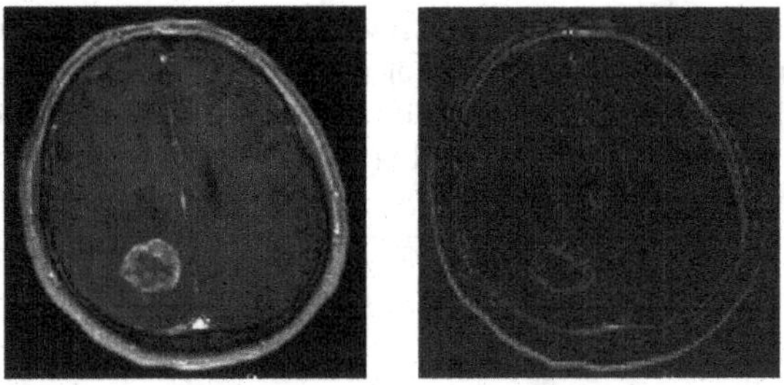

Fig. 7. Basic Model - The original image input is on the left, and the right is in the first convolutional layer.

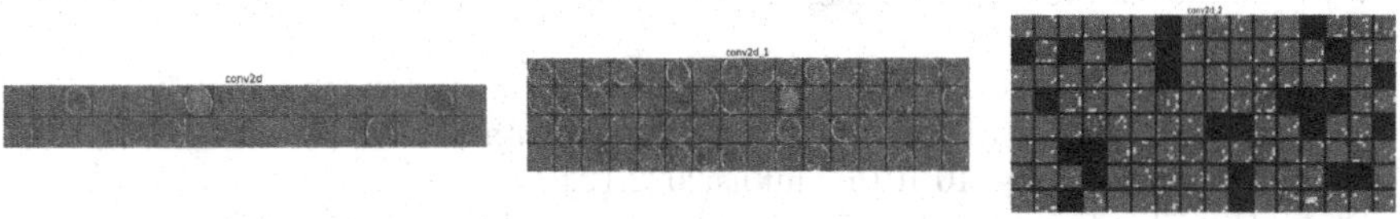

Fig. 8. Basic Model visual layers technique - 1st, 2nd and 3rd convolutional layers showcasing the learned features and their growing complexity within the network.

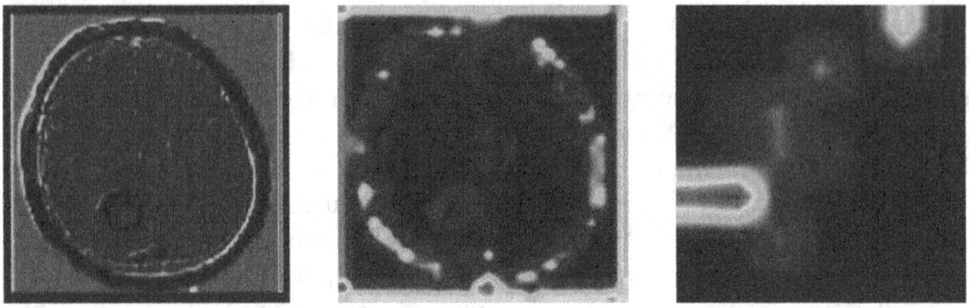

Fig. 9. Grad-CAM analysis for Basic Model (1st, 2nd, and 3rd convolutional layers) reveals distinct insights: edges and contours (1st), complex textures and patterns (2nd) and shapes and structures(3rd).

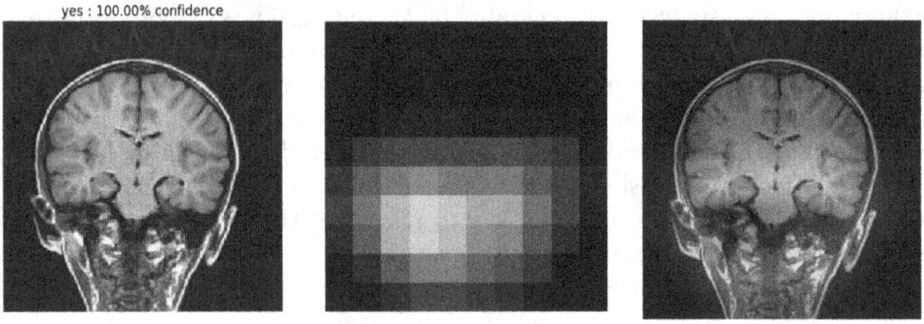

Fig. 10. False Positive output and after CAM in Xception Model. The image on the left shows the classification of the Xception model with a 100% of confidence. True class: "no tumor"; Predicted class: "yes". The heatmap is shown in the central image. In the image on the right, the area colored is those that increase the probability that the image belongs to the yes class (tumor).

References

1. Câncer do sistema nervoso central. Instituto Nacional de Câncer-INCA (n.d.). https://www.gov.br/inca/pt-br/assuntos/cancer/tipos/sistema-nervoso-central
2. Brain and central nervous system cancer-IARC. (n.d.).https://www.iarc.who.int/cancer-type/brain-and-central-nervous-system-cancer/. Accessed 16 May 2023
3. Magadza, T., Viriri, S.: Deep learning for brain tumor segmentation: a survey of state-of-the-art. J. Imaging **7**(2), 19 (2021). https://doi.org/10.3390/jimaging7020019
4. Fernando, K.R.M., Tsokos, C.P.: Deep and statistical learning in biomedical imaging: state of the art in 3D MRI brain tumor segmentation. Inf. Fusion **92**, 450–465 (2023). https://doi.org/10.1016/j.inffus.2022.12.013
5. DSA, E.: Chapter 43 - Pooling layers in convolutional neural networks. In: Deep Learning Book, 10 December 2022. https://www.deeplearningbook.com.br/camadas-de-pooling-em-redes-neurais-convolucionais
6. Alzheimer's disease diagnosis using deep learning techniques. Int. J. Eng. Adv. Technol. **9**(3), 874–880 (2020). https://doi.org/10.35940/ijeat.c5345.029320

7. Lipton, Z.C.: The mythos of model interpretability. Commun. ACM **61**(10), 3643 (2018). https://doi.org/10.1145/3233231
8. Br 35H: Brain Tumor Detection 2020 (n.d.). www.kaggle.com, https://www.kaggle.com/datasets/ahmedhamada0/brain-tumor-detection
9. Brain Tumor Classification (MRI) (n.d.). www.kaggle.com, https://www.kaggle.com/datasets/sartajbhuvaji/brain-tumor-classification-mri
10. Doshi-Velez, F., Kim, B.:Towards a rigorous science of interpretable machine learning. arXiv:1702.08608 [cs, stat]. https://arxiv.org/abs/1702.08608 (2017)
11. Maqsood, S., Damaševičius, R., Maskeliūnas, R.: Multi-modal brain tumor detection using deep neural network and multiclass SVM. Medicina **58**(8), 1090 (2022). https://doi.org/10.3390/medicina58081090
12. Gómez-Guzmán, M.A., et al.: Classifying brain tumors on magnetic resonance imaging by using convolutional neural networks. Electronics **12**(4), 955 (2023). https://doi.org/10.3390/electronics12040955
13. Gaur, L., Bhandari, M., Razdan, T., Mallik, S., Zhao, Z.: Explanation-driven deep learning model for prediction of brain tumour status using MRI image data. Front. Genet. **13**. https://doi.org/10.3389/fgene.2022.822666
14. Marmolejo-Saucedo, J.A., Kose, U.: Numerical grad-cam based explainable convolutional neural network for brain tumor diagnosis. Mob. Netw. Appl. (2022). https://doi.org/10.1007/s11036-022-02021-6
15. Islam, Md.A., et al.:A low parametric CNN based solution to efficiently detect brain tumor cells from ultrasound scans (2023). https://doi.org/10.1109/ccwc57344.2023.10099302
16. Westphal, E., Seitz, H.: A machine learning method for defect detection and visualization in selective laser sintering based on convolutional neural networks. Additive Manuf. **41**, 101965 (2021). https://doi.org/10.1016/j.addma.2021.101965
17. Chollet, F.: Xception: deep learning with depthwise separable convolutions (2017). arXiv:1610.02357 [Cs]. https://arxiv.org/abs/1610.02357v3
18. Chollet, F.: Deep learning with Python. Manning Publications, New York (2017)
19. Selvaraju, R.R., Cogswell, M., Das, A., Vedantam, R., Parikh, D., Batra, D.: Grad-CAM: visual explanations from deep networks via gradient-based localization. Int. J. Comput. Vision **128**(2), 336–359 (2020). https://doi.org/10.1007/s11263-019-01228-7

Enhancing Stock Market Predictions Through the Integration of Convolutional and Recursive LSTM Blocks: A Cross-market Analysis

Filipe Ramos(✉), Guilherme Silva, Eduardo Luz, and Pedro Silva

Computing Department, Federal University of Ouro Preto, Ouro Preto, Brazil
filipe.rss@aluno.ufop.edu.br,
{guilherme.lopes,eduluz,silvap}@ufop.edu.br

Abstract. This study explores convolutional and recursive LSTM blocks within a singular architecture for forecasting stock prices. We propose a method that integrates convolutional networks, which learn to process signals through filters, with recursive LSTM blocks to account for critical temporal information often overlooked in convolutional approaches. Our investigation primarily revolves around two research questions: (1) Can integrating convolutional and recursive LSTM blocks within a singular architecture enhance prediction accuracy? and (2) What is the impact of training and testing with disparate data distributions? The latter question arises from our experiment of training the model using data drawn from the Indian Stock Market and testing the predictions with New York Stock Market data, thus deviating from the traditional focus on uniform stock market distributions. Our results reveal a notable improvement in prediction accuracy (MAPE reduction of 2.22%), strongly suggesting that pre-processing data via Convolutional Neural Networks (CNN) benefits LSTM blocks and can enhance the performance of stock market prediction methodologies.

Keywords: Stock Prince · Machine Learning · CNN · LSTM

1 Introduction

A stock market is where companies can raise funds for their business by selling fractional shares. Many individuals buy and sell these shares regularly to generate profits, a strategy commonly referred to as day trading.

Day trading involves purchasing and selling shares on the same day to profit from price differences. Traders who use this strategy often rely on statistical models and candlestick patterns to predict future prices, even though the market may be unpredictable [6].

In contrast to traders who use statistical models or candlesticks, some investors perform fundamental analysis. These individuals base their actions on

Supported by Universidade Federal de Ouro Preto, FAPEMIG (APQ-01518-21), CAPES and CNPq (308400/2022-4).

M. C. Naldi and R. A. C. Bianchi (Eds.): BRACIS 2023, LNAI 14196, pp. 92–106, 2023.
https://doi.org/10.1007/978-3-031-45389-2_7

various factors within the company, such as its profit, market value, return on equity, and other indicators.

According to Vijh et al. [24], the most challenging aspect of predicting stock prices is the influence of external factors, such as the global economy and political conditions. Despite this challenge, previous works in this field often employ classical algorithms, such as Linear Regression, Random Walk Theory [11], Moving Average Convergence/Divergence [2], Autoregressive Moving Average [22], and Autoregressive Integrated Moving Average [3] to predict stock prices.

Traditional machine learning techniques, such as Support Vector Machines (SVMs) [14] and Random Forests [19], can be used to enhance stock market prediction. Additionally, many researchers are exploring artificial neural networks (ANNs) to predict stock prices and automate trading for passive income. For example, Lu et al. [17] and Singh and Srivastava [21] utilize ANNs for this purpose. Currently, certain types of ANNs, such as Recurrent Neural Networks (RNNs) [8], are showing promising results. These networks analyze current data and use previous data to make a single prediction.

In [9], the authors demonstrated that deep learning models, specifically multilayer perceptron, recursive (RNNs), and convolutional networks, can be more effective than traditional statistical methods, like ARIMA, for stock prediction. The authors trained the models on data from a single company listed on India's National Stock Exchange (NSE) and evaluated the approach on five stocks listed on the New York Stock Exchange (NYSE).

Many works in the literature explore machine learning using sequential models for forecasting stock prices [8,9], often incorporating filters such as moving averages to smooth time series data. Some studies employ convolutional networks that learn to process signals via filters, yet these networks tend to neglect temporal information. This observation steered us towards a research question: Can the integration of convolutional and recursive LSTM blocks within a singular architecture bolster the results?

Simultaneously, this work aims to probe a further research question: What is the impact of training and testing with different data distributions? A majority of the existing studies prioritize predicting stock trends from a more uniform data distribution, often concentrating on the same stock market. To heighten the level of challenge, we ran experiments to predict from different distributions: specifically, training with data drawn from the Indian Stock Market and testing the predictions using data from the New York Stock Market.

Our experiments indicate a substantial improvement and strongly suggest that pre-processing data through Convolutional Neural Networks (CNN) benefits LSTM blocks for the task in question. This approach underscores the potential efficacy of incorporating convolutional layers in sequence prediction tasks, thereby contributing to the broader conversation around stock market prediction methodologies

This manuscript is divided as follows. Section 2 shows similar papers written by different researchers and the reported results. Section 3 presents the data and models used and how they were built. Section 4 addresses the experiments made and compares the results obtained and the literature results Hiransha et al. [9]. Finally, conclusions are presented in Sect. 5.

2 Related Works

In this section, we present several works that use Machine Learning and Deep Learning for stock prediction. Besides, it is also presented the kind of data used in these works.

Stock prediction is a challenging and popular research area that, if successful, can yield profitable results. Jiang [12] presents several papers on this topic, where most utilize standard data (open, close, high, low, and volume) to train models and compare their performance with real-market results. For example, Vijh et al. [24] collected ten years of data on five USA-based companies using Yahoo Finance and trained a low-complexity ANN model, which achieved an RMSE of 0.42 and a MAPE of 0.77% in the best result.

In addition to predicting stock prices, models can be built to determine when to buy or sell stocks, as demonstrated by Santuci et al. [20], who compared SVM and Random Forest models to predict such events. In the best case, their work yielded a profit of 17.74%.

Another internal topic for discussion is the accuracy of different models in predicting stock prices, which is demonstrated by Huang et al. [10] using two models: Feed-forward Neural Network and Adaptive Neural Fuzzy Inference System.

Pang et al. [18] used two different types of LSTM to make predictions based on Shanghai A-share data, achieving an MSE of 0.017 in the best test. Many external factors can influence the stock price, and Jin et al. [13] utilized an external indicator called sentiment to indicate how investors feel about the economy or a specific stock. Das et al. [4] also used sentiment data extracted from tweets on Twitter to predict stock prices. These papers show that the market responds to people's actions, and using this information can better represent the market than numbers.

Agrawal et al. [1] utilized Stock Technical Indicators to predict not the stock price but the trend of the price, i.e., whether the price will increase or decrease. Their work achieved an accuracy of 63.59% in the best case.

Despite the works presented in this section using different datasets, the objective is stock price prediction. These works were selected due to the made of traditional machine learning and different neural network architectures applied to the stock price prediction context.

3 Proposed Methodology

s This section shows how the research was driven and the specific models definition. Subsection 3.1 describes the data used and how it was collected. Subsection 3.2 explains the baseline technique used for comparison purposes and also the proposed architecture, combining deep learning architectures. The metrics used to evaluate the proposed approaches are described in Subsect. 3.3.

3.1 Dataset

The first step is data collection by downloading the stock information from the website infomoney.com. The collected raw data contains:

- Date: The days to which data refer. This data is a string with the month with three letters, followed by the day and year.
- High: The highest share price on a specific day. This data is a float number.
- Close: The share price when the stock market closes on a specific day. This data is a float number too.
- Open: The share price when the stock market opens on a specific day. This also is a float number.
- Low: The lowest share price on a specific day. And this is a float number.
- Volume: The number of transactions that happen on that day. This data is a integer number, but in the case of a thousand volumes, it is represented with the number divided by 1000 followed by a letter "K". And if is a million or more the number is divided by one million and followed by the letter "M".
- Change: The difference in price, in percentage, that happens on a specific day. This data wasn't used in the model but is presented because it was collected along with others.

This data is a time series, so the previous samples directly influence the following sample. The close price rarely moves from a low value to a high value (or vice versa) without the other data that has changed drastically.

Based on Hiransha et al. [9], we chose Tata Motors Ltd. (which uses the token TAMO) stock as training data and used Axis Bank (AXBK), Maruti (MRTI), and HCL (HCLT) as testing and the close price as to be predicted. Table 1 shows the datasets used in this work and the period collected. The data was collected respecting the date range used in [9] in order to directly compare the results.

Table 1. Data collected of the stocks.

Stock	Start Date	End Date
Tata Motors (TAMO)	Jan 01, 1996	Jun 30, 2015
Axis Bank (AXBK)	Oct 05, 2007	Jun 30, 2017
Maruti (MRTI)	Jan 02, 2008	Jun 30, 2017
HCL (HCLT)	Jun 05, 2007	Jun 30, 2017

The second step consists of preprocessing and cleaning the data due to the nature of the raw data being different from the one requested by a machine learning model. The objective is to delete null data and convert the fields to float numbers. We removed the "date" and "change" columns as they are unnecessary. The "date" is due to its nature which is an index, and the "change" column because it is a linear combination of the other fields. And last, before the training, it is mounted frames with a window of several days for training and testing according to the machine learning model needs. Besides, the letters "M" and "K" are also converted to "000000" and "000" respectively. For example, the string "14k" would be converted to 14,000 and "2M" would be converted to 2,000,000.

Finally, the data needs to be normalized due to the high differences in the features scale. By normalizing the data, we ensure that all data has the same impact on training. The normalization strategy adopted was the min-max, which can be defined as:

$$x_{norm} = \frac{(x - x_{min})}{x_{max} - x_{min}}, \tag{1}$$

where x is a unit of data, x_{max} is the largest value in the dataset, x_{min} is the smallest value in the dataset, and x_{norm} is the return that will replace x.

3.2 Model Definition and Training

Two strategies are evaluated, one base on a traditional machine learning strategy and another based on deep learning architectures.

The first model is based on the Support Vector Regression (SVR) [16]. The SVR works like SVM (Support Vector Machine), but instead of making a classification based on the hyperplane division, it calculates and returns the distance between the point and the hyperplane as output. In this work, the SVR model uses information from a single day to predict the close price of the next day.

As in trade operations, recent data is just as important as current ones since the current stock value depends on the previous ones. RNN works well because it uses the current data and previous ones. So the input is a frame with the several days including the one right before the day that will be predicted. One example of RNN is the long short-term memory (LSTM) architectures [7]. We made a sequence of empirical tests with a different number of layers to compare different LSTM architectures and try to find the best fit for the problem treated. We fix the number of 75 cells in each layer, a dropout of 20% (avoid overfitting [23]), and the last layer as a dense layer with one node for the regression.

The LSTM model learns the time series. Towards to extract more features from the time series, a Convolutional Neural Network (CNN) [15] is applied. So, instead of the LSTM learning directly from the raw data, it could learn from more complex features learned by convolution layers of a CNN. This strategy was observed in other works that use a 1D signals [5]. The CNN-LSTM model was constructed by using the best-performing LSTM model, adding a one-dimensional convolutional layer with 32 filters, ReLU activation, followed by a max-pooling layer of size 4. The values for the CNN and max-pooling layer were determined empirically.

3.3 Evaluation Metrics

To evaluate the effectiveness of the proposed approaches, we used three different evaluation metrics, all available on the *scikit-learn* library. The first one is the root mean square error (RMSE), which is the mean error squared to turn positive the metric, and the square root to move back to the original proportion. Besides, it also focuses on emphasizing large outlier errors. The metric is defined as

$$RMSE = \sqrt{\frac{\sum (y_i - y_p)^2}{n}}, \quad (2)$$

where y_i represents the right value, y_p is the value predicted by the model, and n is the number of predictions made in tests.

The second is mean absolute error (MAE), which is the module of the mean error and it is represented by the formula

$$MAE = \frac{\sum |(y_i - y_p)|}{n}, \quad (3)$$

in which y_i, y_p and n are the same presented for RMSE metric.

The last metric used was the mean average percentage error (MAPE), which is represented by the formula

$$MAPE = \frac{\sum |y_i - y_p|/|y_p|}{n} \quad (4)$$

in which y_i, y_p and n are the same presented for RMSE and MAE metrics.

4 Experiments and Results

This section is about showing, explaining, and discussing the results. All the training and testing were made in a computer with 128 GB RAM, Titan X 12 GB VRAM, and Intel(R) Core(TM) i7-5820K CPU @ 3.30 GHz. The framework used was Tensorflow 2 and Scikit-learn 1.1 for Python3.

We aim to answer the two researches questions of this work: (1) Can integrating convolutional and recursive LSTM blocks within a singular architecture enhance prediction accuracy? and (2) What is the impact of training and testing with disparate data distributions? To correctly answer the second question, first we propose baseline results using the SVR method. To answer the first question, first we evaluate a proper LSTM architecture and then, add CNN layer.

All the models were trained for 10,000 epochs. The SVR model has an RBF kernel and a regularization parameter of 1. The LSTM and CNN-LSTM models were trained with a batch size of 2,048, Adam optimizer, and mean squared error as a loss. The models were trained with the TAMO dataset and tested using the HCLT, MRTI, and AXBK datasets, following the pattern of the work carried out by Hiransha et al. [9].

4.1 SVR Model

The first model evaluated was the SVR. In all three datasets tested, there is a gap between the predicted close price and the real one. This can be seen in Figs. 1, 2 and 3 and the error measured in Table 2.

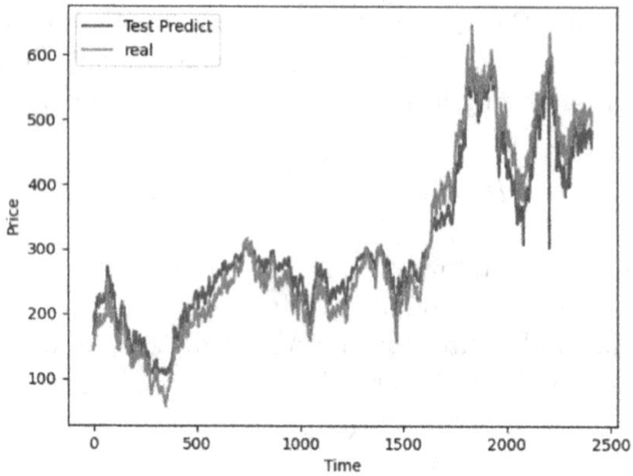

Fig. 1. Predictions made by SVR model for the Axis Bank dataset. Time in days.

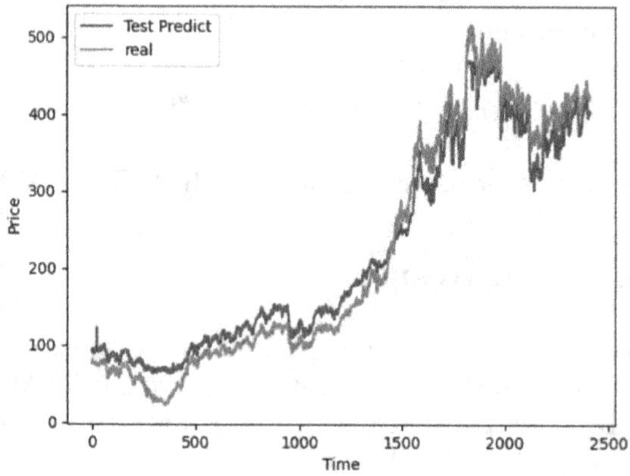

Fig. 2. Predictions made by SVR model for the HCL Techonologies dataset. Time in days.

Table 2. Metrics of result in SVR model

Stock	RMSE	MAE	MAPE
Axis Bank (AXBK)	0.0498	0.0446	10.8372
HCL (HCLT)	0.0525	0.0489	21.8596
Maruti (MRTI)	0.0469	0.0436	22.1453

It is possible to observe that the predicted close prices from SVR have a smoother line compared to the real one. As the SVR only uses the data from today to predict tomorrow, the information from the previous days are not used, and the model cannot follow the real close price. The greatest gap is observed in the HCL Technologies dataset, which has a bigger variance as the range of values is greater.

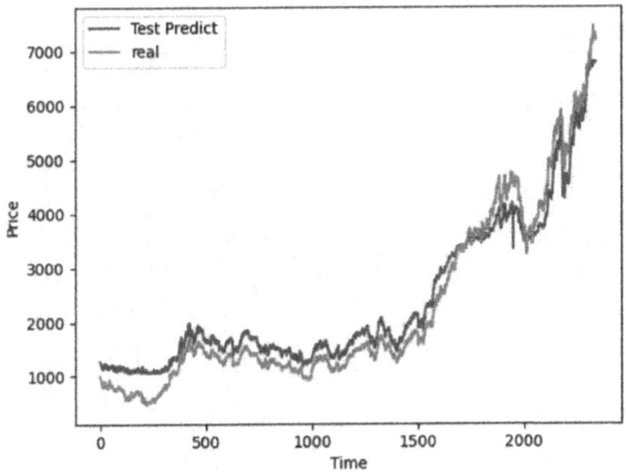

Fig. 3. Predictions made by SVR model for the Maruti dataset. Time in days.

4.2 LSTM Model

We conduct an empirical study to define the best LSTM architecture to reach the highest MAPE in the TAMO dataset. Different models were evaluated by varying the size of the layer. After that, the model with the best results in the TAMO data was chosen.

As shown in Table 3, the model with the slowest result has six layers. All these layers were intercalated with Dropout layers of 20% to minimize the chance of overfitting. The best LSTM architecture can be seen in Fig. 4.

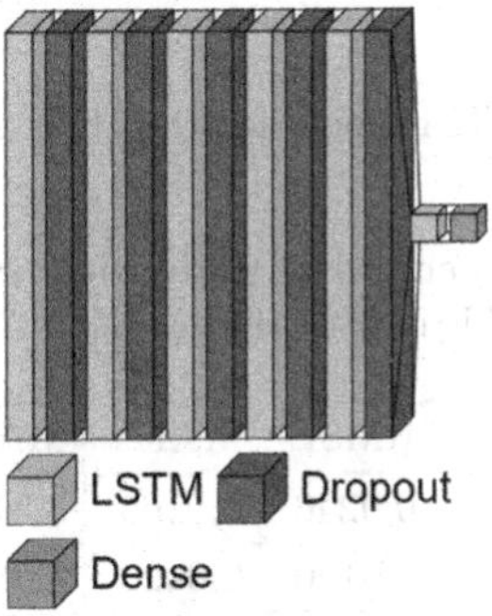

Fig. 4. LSTM architecture with six layers with 75 unities in each one.

As seen in Figs. 5, 6 and 7, the LSTM model could predict the closest stocks value compared to the SVR one. This scenario can be seen in Table 4, where the reported error are below the ones reported in Table 2. MAPE below three shows that the distance between the predicted close price and the real one is below 3.

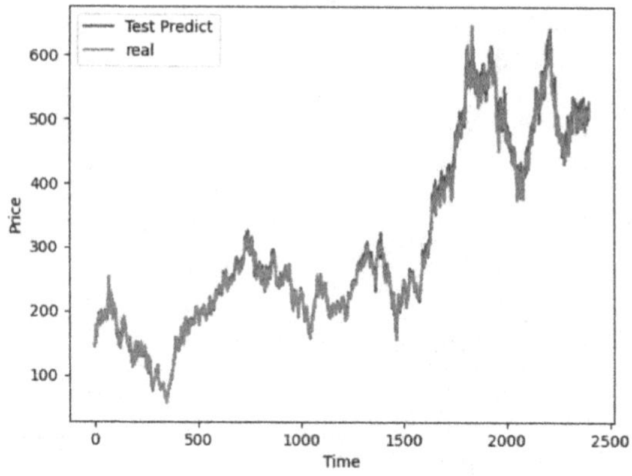

Fig. 5. Predictions made by LSTM model for the Axis Bank dataset. Time in days.

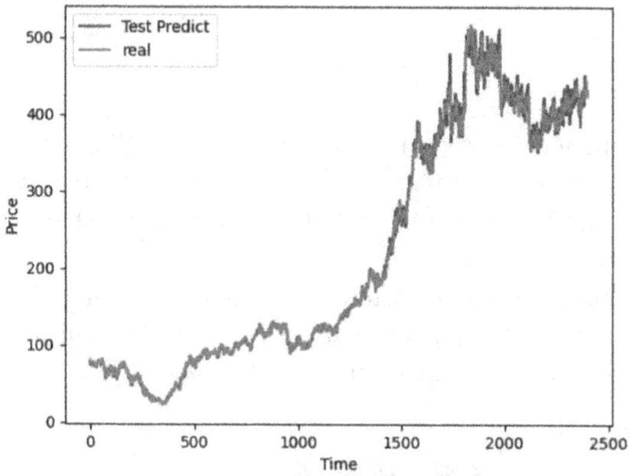

Fig. 6. Predictions made by LSTM model for the HCL Technologies dataset. Time in days.

Table 3. Reported metrics of an empirical study with LSTM models. The model with smaller MAPE, RMSE, and MAE is with six layers.

Model	RMSE	MAE	MAPE
1 Layer	0.0246	0.0165	3.3173
2 Layer	0.0190	0.0120	2.3882
3 Layer	0.0173	0.0118	2.4162
4 Layer	0.0191	0.0130	2.6081
5 Layer	0.0165	0.0114	2.3327
6 Layer	**0.0165**	**0.0114**	**2.3182**
7 Layer	0.0169	0.0119	2.4022

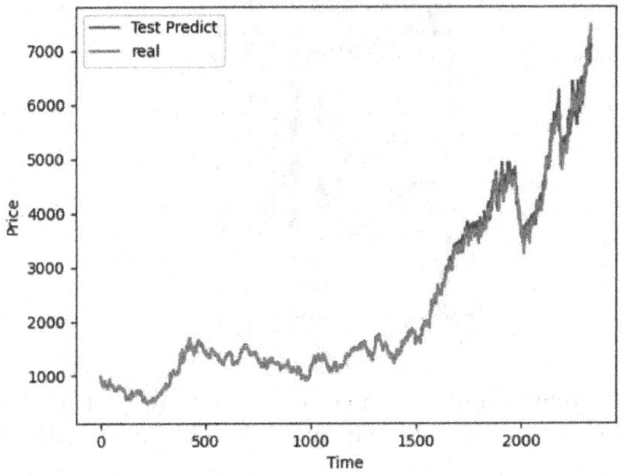

Fig. 7. Predictions made by LSTM model for the Maruti dataset. Time in days.

Table 4. Metrics of result in LSTM model

Stock	RMSE	MAE	MAPE
Axis Bank (AXBK)	0.0165	0.0114	2.3182
HCL (HCLT)	0.0146	0.0089	2.1197
Maruti (MRTI)	0.0302	0.0236	2.5932

Compared to the SVR models, the LSTM models predict the closed price near the real one. One of the reasons is the number of days used for each model: while the SVR uses only one day, the LSTM uses 12 days. Similar to the SVR model, HCL Technologies also presented the highest gap. The LSTM model could follow the real close price curve when the variance is small, although, when it is high, the model could not follow the real curve and presented a smoother behavior.

4.3 CNN-LSTM Model

The last model build was with a combination of CNN and LSTM. This model has two 1D convolution layers and was defined with empirical tests, stacking the convolutional layers before the LSTM with six layers and dropout defined in the experiments presented in Table 3. The CNN-LSTM architecture is seen in Fig. 8.

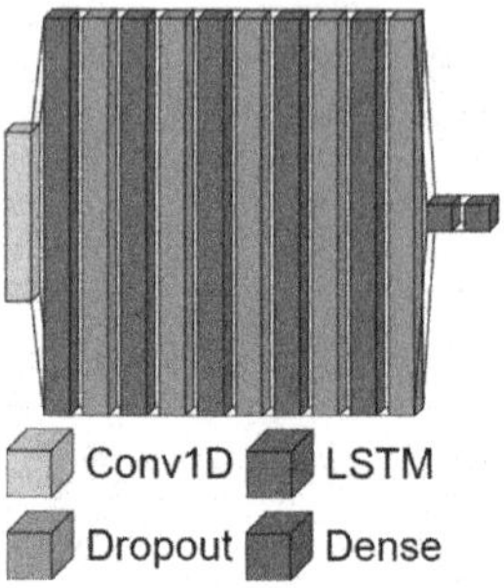

Fig. 8. CNN-LSTM architecture with a convolutional layer with 32 filters, kernel size equal to 8 and ReLU activation followed by six LSTM layers with 75 unities in each one.

As can be seen in Figs. 9, 10 and 11, the line that represents the prediction almost can not be seen, which means that the predicted stock price is very close to the actual value. Some parts of the graphic have a peak, but this does not affect the result in general. The errors can be seen in Table 5.

Table 5. Metrics of result in CNN-LSTM model

Stock	RMSE	MAE	MAPE
Axis Bank (AXBK)	0.0055	0.0044	1.8975
HCL (HCLT)	0.0141	0.0096	2.5394
Maruti (MRTI)	0.0055	0.0044	1.8975

As Tables 4 and 2 show, a smaller MAPE is reported compared to the one reported by Hiransha et al. [9]. While Hiransha et al. [9] reached the MAPE of 4.88, the proposed CNN-LSTM model reached 1.83.

So, as can be seen in Fig. 12, on a large scale, all the models follow the trend of the stock market, but, in a zoomed view of the predictions, the models are not accurate. We improved the curve accuracy by adding CNN in the LSTM model, which forced the curve to stay softer, following the close price. Therefore, the model with CNN succeeds in extracting features from the data and improving the results.

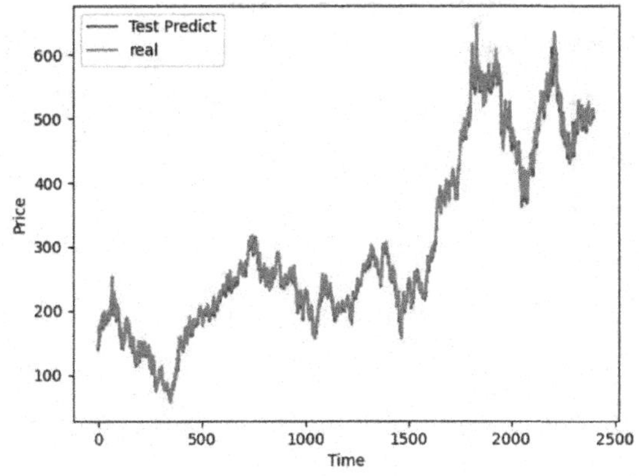

Fig. 9. Predictions made by CNN-LSTM model in Axis Bank dataset. Time in days.

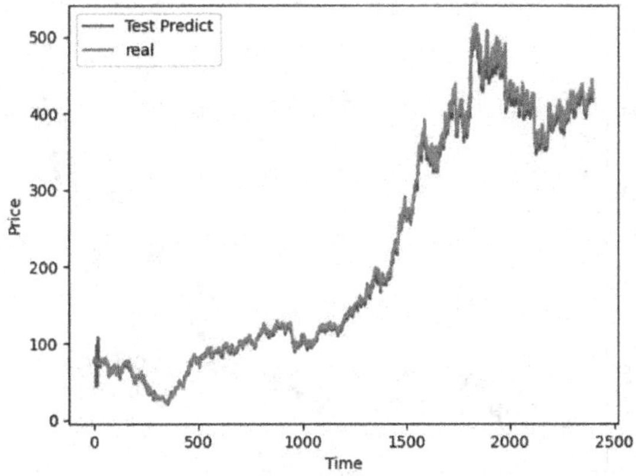

Fig. 10. Predictions made by CNN-LSTM model in HCL Technologies dataset. Time in days.

4.4 Results Discussion

As seen in Table 6, the proposed approach overcomes the baseline model proposed by Hiransha *et al.* [9]. The results show that it will be possible to build a model that can precisely predict the prices in a near future. But, despite the reported results, it is important to observe that the market floats with external factors, which makes the model use quite risky.

As Table 6 shows, the CNN-LSTM model has better results. The intuition of getting smaller errors maybe is the information of more days used by the CNN-LSTM and the data processing made by the convolution layer. Gathering more data with different prices and fluctuations can result in a more general model.

Comparing the much higher error of the SVR to that of the LSTM and CNN-LSTM, it is visible how the use of only one day can be detrimental to

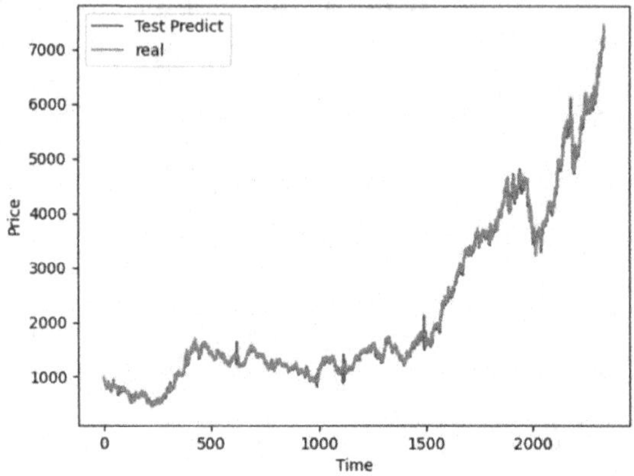

Fig. 11. Predictions made by CNN-LSTM model in Maruti dataset. Time in days.

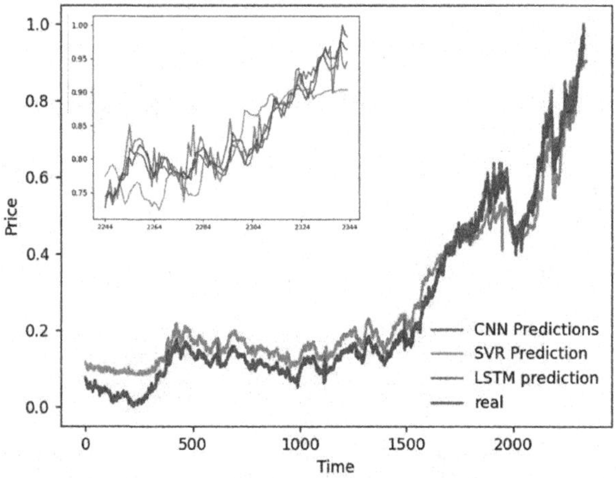

Fig. 12. Prediction made by all models with a zoomed view of the last 100 days predicted in MRTI dataset.

Table 6. MAPE of all models trained. AXBK = Axis Bank; HCLT = HCL; MRTI = Maruti.

Stock	SVR	LSTM	CNN	CNN-LSTM	Hiransha et al. [9]
AXBK	10.84	2.32	2.11	1.89	4.88
HCLT	21.86	2.12	2.54	2.53	4.40
MRTI	22.15	2.59	1.84	1.89	4.05

the algorithm's performance. As only one day is used, long-term behavior is not taken into account, which tends to generate results that are inconsistent with those expected.

5 Conclusion and Future Works

In this work, we analyzed how models work in long and short periods to predict close stock prices, a challenging task. Three different architectures were evaluated: a traditional machine learning algorithm (SVR) and two deep learning algorithms (LSTM and CNN).

As can be seen, adding a CNN layer to the LSTM model increased the performance by 0.75% in the best case. This finding aligns with the primary research question addressed in this study. The results also address the second research question, once the models could generalize to a dataset different from the one a model was trained on.

As the input is a time series, the LSTM model overcomes the SVR one. Another analysis is that the convolutional layers can extract features in tabular data and increase the results when added to the LSTM model.

In future work, two research paths to reduce the error will be explored: (1) an ensemble model, which combines some models responses, like CNN, RNN, LSTM, and transformers, and (2) gathering external information about the companies to predict the close price, such as the response of sentiment analysis about news regarding the companies and other negative or positive aspects of the company.

References

1. Agrawal, M., Shukla, P.K., Nair, R., Nayyar, A., Masud, M.: Stock prediction based on technical indicators using deep learning model. Comput. Mater. Continua **70**(1), 287–304 (2022)
2. Aguirre, A.A.A., Méndez, N.D.D., Medina, R.A.R.: Artificial intelligence applied to investment in variable income through the MACD (moving average convergence/divergence) indicator. J. Econ. Financ. Adm. Sci. (2021)
3. An, Z., Feng, Z.: A stock price forecasting method using autoregressive integrated moving average model and gated recurrent unit network. In: 2021 International Conference on Big Data Analysis and Computer Science (BDACS), pp. 31–34. IEEE (2021)
4. Das, S., Behera, R.K., Rath, S.K., et al.: Real-time sentiment analysis of twitter streaming data for stock prediction. Procedia Comput. Sci. **132**, 956–964 (2018)
5. Freitas, C., Silva, P., Moreira, G., Luz, E.: Rede neural convolucional e lstm para biometria baseada em eeg no modo de identificaçao. In: Anais do XXII Simpósio Brasileiro de Computação Aplicada à Saúde, pp. 256–267. SBC (2022)
6. Gomes, I.d.O.: Estratégias para operações de day trade na B3. Master's thesis, Escola de Economia de São Paulo da Fundação Getulio Vargas (2018)
7. Goodfellow, I., Bengio, Y., Courville, A.: Deep Learning. MIT Press, Cambridge (2016)
8. Hansun, S., Young, J.C.: Predicting lq45 financial sector indices using RNN-LSTM. J. Big Data **8**(1), 1–13 (2021)
9. Hiransha, M., Gopalakrishnan, E.A., Menon, V.K., Soman, K.: NSE stock market prediction using deep-learning models. Procedia Comput. Sci. **132**, 1351–1362 (2018)

10. Huang, Y., Capretz, L.F., Ho, D.: Neural network models for stock selection based on fundamental analysis. In: 2019 IEEE Canadian Conference of Electrical and Computer Engineering (CCECE), pp. 1–4. IEEE (2019)
11. Jayadev, S., Anupama, R., Sebastian, S.: Testing the market efficiency of NSE: a study on random walk theory. Contemp. Res. Financ. 34 (2021)
12. Jiang, W.: Applications of deep learning in stock market prediction: recent progress. Expert Syst. Appl. **184**, 115537 (2021)
13. Jin, Z., Yang, Y., Liu, Y.: Stock closing price prediction based on sentiment analysis and LSTM. Neural Comput. Appl. **32**(13), 9713–9729 (2020)
14. Kurani, A., Doshi, P., Vakharia, A., Shah, M.: A comprehensive comparative study of artificial neural network (ANN) and support vector machines (SVM) on stock forecasting. Ann. Data Sci. 1–26 (2021)
15. LeCun, Y., Bengio, Y., et al.: Convolutional networks for images, speech, and time series. Handb. Brain Theory Neural Netw. **3361**(10), 1995 (1995)
16. Lorena, A.C., De Carvalho, A.C.: Uma introdução às support vector machines. Revista de Informática Teórica e Aplicada **14**(2), 43–67 (2007)
17. Lu, W., Li, J., Wang, J., Qin, L.: A CNN-BiLSTM-am method for stock price prediction. Neural Comput. Appl. **33**(10), 4741–4753 (2021)
18. Pang, X., Zhou, Y., Wang, P., Lin, W., Chang, V.: An innovative neural network approach for stock market prediction. J. Supercomput. **76**(3), 2098–2118 (2020)
19. Rakhra, M., et al.: Crop price prediction using random forest and decision tree regression:-a review. Mater. Today Proc. (2021)
20. Santuci, A., Sbruzzi, E., Araújo-Filho, L., Leles, M.: Evaluation of forex trading strategies based in random forest and support vector machines. IEEE Lat. Am. Trans. **20**(9), 2146–2152 (2022)
21. Singh, R., Srivastava, S.: Stock prediction using deep learning. Multimedia Tools Appl. **76**(18), 18569–18584 (2017)
22. Singh, S., Parmar, K.S., Kumar, J.: Soft computing model coupled with statistical models to estimate future of stock market. Neural Comput. Appl. **33**(13), 7629–7647 (2021)
23. Srivastava, N., Hinton, G., Krizhevsky, A., Sutskever, I., Salakhutdinov, R.: Dropout: a simple way to prevent neural networks from overfitting. J. Mach. Learn. Res. **15**(1), 1929–1958 (2014)
24. Vijh, M., Chandola, D., Tikkiwal, V.A., Kumar, A.: Stock closing price prediction using machine learning techniques. Procedia Comput. Sci. **167**, 599–606 (2020)

Ensemble Architectures and Efficient Fusion Techniques for Convolutional Neural Networks: An Analysis on Resource Optimization Strategies

Cícero L. Costa[1,2](✉), Danielli A. Lima[1], Celia A. Zorzo Barcelos[2], and Bruno A. N. Travençolo[2]

[1] Federal Institute of Triângulo Mineiro (IFTM) Campus Patrocínio, Uberaba, MG, Brazil
{cicero,danielli}@iftm.edu.br

[2] Federal University of Uberlândia (UFU) Campus Santa Mônica, Uberlândia, MG, Brazil
{celiazb,travencolo}@ufu.br

Abstract. The human gastrointestinal tract is prone to various abnormalities, including lethal diseases such as cancer, necessitating better endoscopic performance and standardized screening. Endoscopic scoring systems lack generalizability, emphasizing the need for artificial intelligence-based solutions. Using the HyperKvasir dataset, we employed deep learning, specifically Convolutional Neural Networks, or shortly CNNs, to analyze endoscopic images and videos. Our study focused on improving the classification of gastrointestinal tract diseases by proposing various CNN ensembles and fusion techniques. Through the use of seven CNN models and effective merging techniques, we achieved enhanced performance. Validation involved literature review and experiments. DenseNet-161 influenced the merger process, and integrating ResNet152 and VGG further enhanced effectiveness. Resource analysis included GPU model, RAM usage, and execution time. Results demonstrated comparable performance to the previous model, with F1-score of 0.910 and Matthews correlation coefficient, MCC for short, of 0.902, using 10 GB GPU RAM (compared to 15.8 GB). With 24.7 GB GPU RAM, F1-score of 0.913 and MCC of 0.905 were achieved. These findings advance our understanding of ensemble architectures and fusion techniques.

Keywords: Convolutional neural networks · Gastrointestinal abnormalities · Ensembles performance · Image classification performance · Fusion techniques · Resource consumption

1 Introduction

The human gastrointestinal (GI) tract is susceptible to several abnormal mucosal findings, including life-threatening diseases [2]. GI cancer alone accounts for millions of new cases annually, emphasizing the need for improved endoscopic performance and systematic screening [10]. Gastrointestinal exams and colonoscopy

M. C. Naldi and R. A. C. Bianchi (Eds.): BRACIS 2023, LNAI 14196, pp. 107–121, 2023.
https://doi.org/10.1007/978-3-031-45389-2_8

are essential procedures to investigate the human GI tract [9]. These tests play a vital role in the diagnosis and management of gastrointestinal conditions, contributing to the early detection, treatment and prevention of serious complications [2,5,7]. However, current endoscopic scoring systems lack standardization and are subjective [2,7].

In this context, artificial intelligence (AI)-enabled computer-assisted diagnostic systems, particularly machine learning, have shown promise in healthcare, but the scarcity of medical data impedes progress [2,8]. To solve this, we used a database (DB), called HyperKvasir, a large dataset of gastrointestinal images and videos collected during real exams [2]. The dataset contains over 1.1×10^5 images and 374 videos and represents anatomical landmarks as well as pathological and normal findings [2].

Over the years, machine learning has evolved into deep learning algorithms, relying primarily on the Deep Neural Network (DNN). Convolutional neural networks (CNN), a type of DNN, have emerged as a powerful tool for image analysis and classification, including medical imaging tasks. CNN Ensemble architectures have been widely employed to improve forecast accuracy by combining the outputs of various models. These sets leverage the diversity of individual CNN models to improve overall performance. In addition, fusion techniques are employed to effectively integrate predictions from multiple CNN models [11,16].

In this work, our main objective is to propose a new ensemble architecture and efficient fusion techniques for CNNs in the classification of gastrointestinal tract diseases using the HyperKvasir dataset, aiming to obtain better results than in the literature and to optimize computational resources. To achieve this, we performed a thorough literature review to identify relevant studies on the use of deep learning methods in similar health domains. In addition, we performed several experiments to evaluate the effectiveness of our proposed approach.

2 Theoretical Foundation

Since the emergence of computers, there have been research efforts to make them reproduce biological characteristics; these are known as bioinspired research. Among the bioinspired research, there is one that seeks to simulate the functioning of the brain through artificial neural networks. These networks have undergone many transformations, as reported in the papers [4,11–13,16–18]. This section provides an overview of the HyperKvasir database [2], the dataset utilized in this study. We reviewed the literature on deep learning in digital imaging (DI) and consider Borgli et al.'s general model [2] as a reference for our research. Our objective is to establish a robust foundation by analyzing the dataset and surveying related studies.

2.1 HyperKvasir Dataset

The HyperKvasir dataset[1] is composed of images and videos. The dataset content was collected between 2008 and 2016, in a hospital in Norway. In this work,

[1] Available at: https://datasets.simula.no/hyper-kvasir.

10639 labels available in the dataset are used. The images are separated into 23 classes, in Fig. 1 it is possible to see examples of images contained in the dataset. Class labels are Z-line, Pylorus, Retroflex stomach, Barrett's, Short segment, Esophagitis grade A, Esophagitis grade B-D, Cecum, Retroflex rectum, Terminal ileum, Polyps, Ulcerative colitis grade 0–1, Ulcerative colitis grade 1, Ulcerative colitis grade 1–2, Ulcerative colitis grade 2, Ulcerative colitis grade $2-3$, Ulcerative colitis grade 3, Hemorrhoids, Dyed lifted polyps, Dyed resection margins, BBPS 0–1, BBPS 2–3, Impacted stool. The dataset offers a file (`.csv`) with the division of classes studied by Borgli [2].

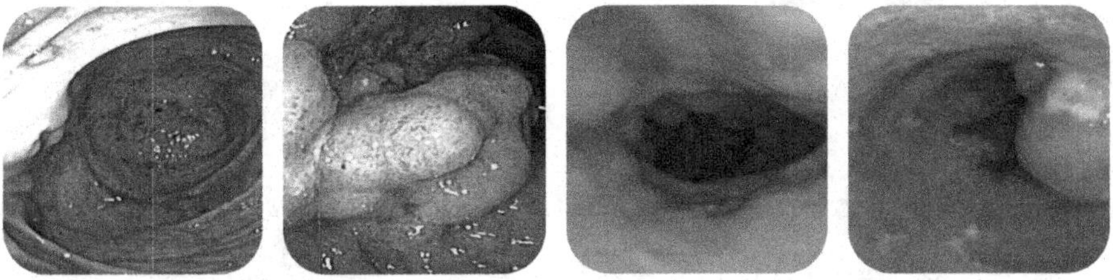

Fig. 1. Example of images present in the gastrointestinal disease image database. These images are not sequential.

2.2 Related Works

In our study, we started by establishing a solid foundation using the reference article [2]. Expanding upon this work, we created a comprehensive graph, as illustrated in Fig. 2, to visually illustrate the interconnectedness of relevant papers in our research field. This graph provides a valuable representation of the network of related literature, with a specific focus on the HyperKvasir image and video dataset for gastrointestinal endoscopy, as discussed by Borgli et al. [2].

Upon analyzing the graph, we identified a total of 41 studies connected to the article [2], resulting in a set of 42 relevant studies for our research. However, we established inclusion criteria, considering only studies published after 2019, that is, after the publication date of the base article. Additionally, we excluded systematic literature reviews or survey studies from our analysis. The 10 remaining studies were evaluated for their degree of similarity to the base article, represented by the similarity index SbP (similar based-paper), ranging from 12% to 100%. The higher the SbP[2] value, the greater the similarity between the article and the base work (previous paper [2]), which is relevant to the results obtained in our research.

To gather related works for our paper, each article was thoroughly reviewed based on the following parameters: `Study` name and year, `Task` performed, CNN `Architecture` used, `Methodology` approach, `Dataset` and (`SbP`) Similarity on the base paper. These parameters were used to assess and categorize the papers,

[2] Available at: https://www.connectedpapers.com/main.

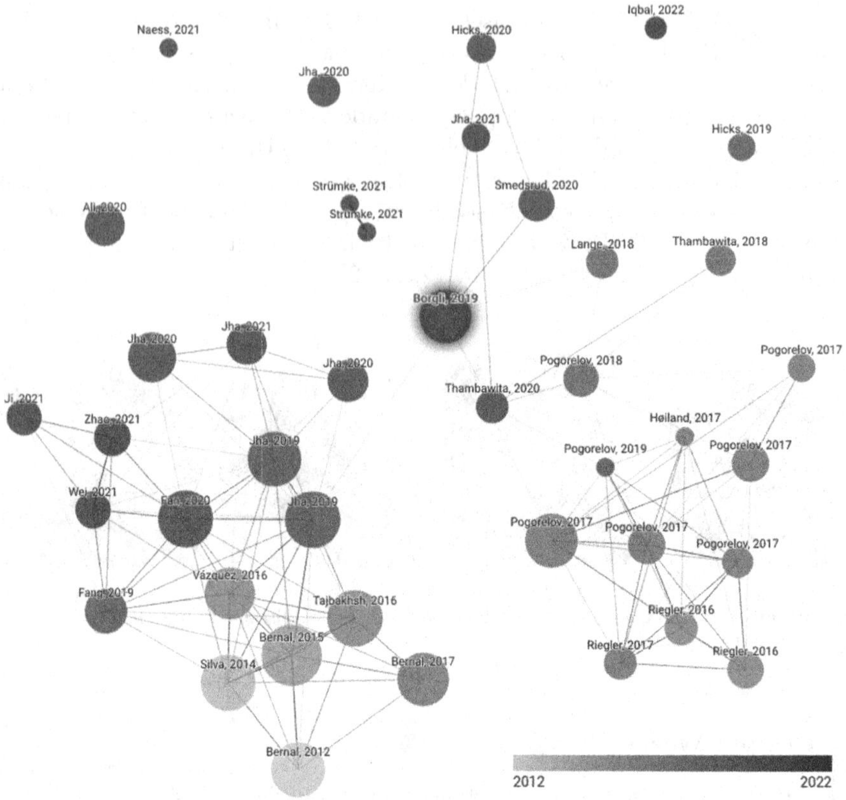

Fig. 2. Graph for connected papers with HyperKvasir image and video dataset for gastrointestinal endoscopy, Borgli et al. [2].

ensuring that they align with the focus and objectives of our research. By analyzing each article based on these criteria, we were able to identify and select relevant works that contribute to our study.

Table 1 shows several studies in the context of gastrointestinal endoscopy. The studies cover a range of tasks such as polyp classification, segmentation, detection, localization, and abnormality identification. Various deep learning architectures, including CNNs like ResNet-152, DenseNet-161, U-Net, Pix2Pix and HGANet are utilized in these studies. Different methodologies and techniques such as Fuzzy C-Means Clustering, ResUNet, Conditional Random Fields (CRF), Test-Time Augmentation (TTA), and Adversarial Training are also employed. Multiple datasets are used for evaluation, including the HyperKvasir and the EAD2019 datasets. The achieved similar based-paper (SbP) rank is used as a performance metric, with higher values indicating better results, which means is most similar to paper [2].

Table 1. State of the art summarization considering GI dataset for medical diagnoses.

Study	Task	Architecture	Methodology	Dataset	SbP
HyperKvasir, a comprehensive multi-class image and video dataset for gastrointestinal endoscopy [2] (2020)	Gastrointestinal polyp Classification	Pre-trained (ResNet-50, ResNet-152, DenseNet-161 Averaged ResNet-152 + DenseNet-161, ResNet-152 + DenseNet-161+MLP	Five different deep CNN were trained and evaluated using standard classification metrics.	HyperKvasir	100
Kvasir-SEG: A Segmented Polyp Dataset [10] (2020)	Gastrointestinal polyp Segmentation	Fuzzy c-mean clustering, ResUNet CNN	Pre-processing steps, FCM algorithm, Data augmentation, ResU-Net implementation details, Qualitative comparison of FCM clustering and ResUNet results	Kvasir-SEG	19.7
The endotect 2020 challenge: evaluation and Comparison of Classification, Segmentation and Inference time for endoscopy [5] (2021)	Gastrointestinal polyp detection and segmentation	CNN (ResNet-152), CNN (Mask Scoring R-CNN, DeepLab V3+)	Automatic segmentation of polyps, Real-time analysis	HyperKvasir	24.8
An Extensive Study on Cross-dataset Bias and Evaluation Metrics Interpretation for Machine Learning Applied to Gastrointestinal tract Abnormality Classification [20] (2020)	Gastrointestinal tract diseases Identification and Segmentation	CNN (ResNet-152, Dense Net-161) and addtional MLP	GF-based approaches, Deep neural networks, Cross-dataset evaluations, Automated identification of GI tract diseases	CVC-12k, CVC-356-plus, CVC-612-plus, 2018 Medico	20.6
Real-Time Polyp Detection, Localization and segmentation in colonoscopy Using Deep Learning [7] (2021)	Gastrointestinal polyp detection, localization and segmentation	CNN (YOLOv4 with Darknet53 backbone), Segmentation networks (Colon SegNet, UNet+ ResNet34 backbone, Deep-Labv3+, PSPNet, HRNet)	Object detection and localization using YOLOv4 with Dark-net53 backbone and Cross-Stage-Partial-Connections (CSP), Semantic segmentation + different UNet, Deep-Labv3+, PSPNet, HRNet	Kvasir-SEG	13.8
Medico Multimedia Task at Media Eval 2020: Automatic Polyp Segmentation [8] (2020)	Gastrointestinal polyp segmentation	CNN, Dice similarity coefficient (DSC) and mean Intersection over Union (mIoU)	Algorithm Speed Efficiency, Framesper-second (FPS) while segment colonoscopic images	Kvasir-Seg	12.8
An objective comparison of detection and segmentation algorithms for artefactsin clinical endoscopy [1] (2020)	Hollow-organs generalization, detection and segmentation	Mask R-CNN, RetinaNet, Cascade R-CNN, DeepLabV3	Transfer learning, ensemble techniques, out-of-sample generalization, 2-training separate batches, 7 prevalent artefact types	EAD2019 (2192 unique video, 475 video frames + mask annotations, addi= tional 195, 122, and 51 videos)	12
A comprehensive study on colorectal polyp segmentation with ResUNet++, conditional random Field and Test-Time Augmentation [9] (2021)	Gastrointestinal polyp Segmentation	CNN ResUNet++	Conditional Random Field (CRF) and TestTime Augmentation (TTA), Dice coefficient (DSC), Intersection over Union (IoU), mean IoU (mIoU), AUC-ROCand data augmentation	Kvasir-SEG, CVC-ClinicDB, CVC-ColonDB, ETIS-Larib Polyp DB, ASU-Mayo Clinic Colonoscopy Video Database, CVC-Video ClinicDB	13.1
Pyramidal segmentation of Medical Images using Adversarial Training [14] (2021)	Gastrointestinal polyp Segmentation and localization	U-Net and Pix2Pix	Learning to segment within several grids, Grid augmentation, Cross-data training and testing	Kvasir-SEG (validation, testing), CVCClinic DB (testing)	12.7
Automated identification of human gastrointestinal tract abnormalities based on deep convolutional neural network with endoscopic images [6] (2022)	Gastrointestinal polyp abnormality identification	DCNN (HGANet)	HGANet with multiple routes, various image resolutions, and several convolutional layers. Pre-processing involves cropping, downsampling and removal of undesired artifacts. Augmentation techniques are applied to balance the classes.	Kvasir-Capsule	12.4

These studies provided valuable insights and advancements in leveraging deep learning and clustering techniques for gastrointestinal endoscopy. They contribute to the development of automated systems for polyp classification, segmentation, and abnormality detection, which can improve the efficiency and accuracy of medical diagnoses. Besides that, these papers served as the foundation for our study, as they utilized various CNNs for different approaches to GI problems.

2.3 Background Model

In Borgli et al. [2], the authors proposed a classification model represented in Fig. 3. The model is composed of pre-trained CNNs, DenseNet161 and ResNet152. Each CNN model is a function, M, composed of a set of subfunctions (convolution, pooling, batch normalization, softmax, optimize, etc.) which, in this case, given an input image $\overrightarrow{\chi}$, a learning rate value η and the number of epochs e, returns an output $\overrightarrow{P}$, according to the function:

$$\overrightarrow{P}_i = M(\overrightarrow{\chi}, \eta, e) \tag{1}$$

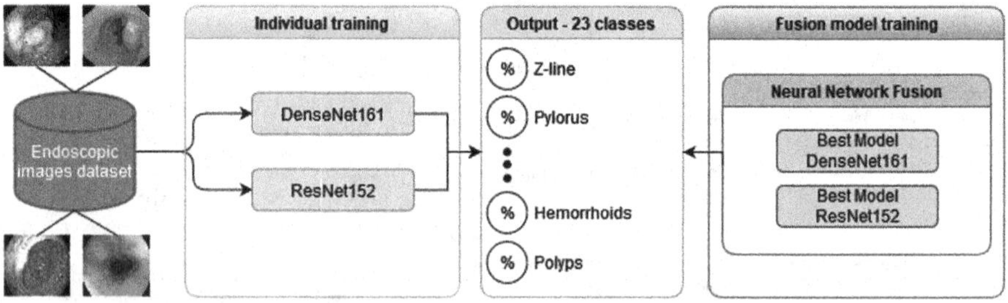

Fig. 3. Previous model [2] for CNN models training and fusion process.

The output, $\overrightarrow{P}$, is a probability vector that indicates the probability that $\overrightarrow{v}_i$ belongs to one of the classes of the problem. The vector $\overrightarrow{P} = [c_1, c_2, c_3, \cdots, c_{23}]$, where $c \in C$ for each GI class and $|C| = 23$. Given a dataset with 10639 gastrointestinal images, $\overrightarrow{\chi}$, divided into two splits, one set with 5315 images and another set with 5324 images. The model proposed in [2] alternates sets for training and validation. The final response of the model is the average of the results of the two splits. For each split, the authors trained the CNNs models M_1 and M_2, respectively, DenseNet161 and ResNet152, using $\eta = 0.001$, $e = 50$ and optimize SGD, stochastic gradient descent; M_1 and M_2 generated the responses $\overrightarrow{P}_1$ and $\overrightarrow{P}_2$, respectively. After training, the best weight set $\overrightarrow{w}$ for each model was found and the best-trained models M_1^b and M_2^b are saved. Using the trained models, a model M^ν is created:

$$\overrightarrow{P}_i^\nu = M^\nu(\overrightarrow{\chi}, \eta, e, M_1^b, M_2^b), \tag{2}$$

since $\overrightarrow{P}_i^\nu = \frac{(\overrightarrow{P}_1^b + \overrightarrow{P}_2^b)}{2}$, $\overrightarrow{P}_1^b$ and $\overrightarrow{P}_2^b$ are output of M_1^b and M_2^b, respectively. For each split, the CNN model M^ν was trained using $\eta = 0.001$ and $e = 50$.

3 Proposal

In this paper, we have two proposals. The first proposal sought to overcome the literature results using fewer GPU resources than the second proposal, which is an expansion of what the authors propose in [2]. The second proposal sought to verify whether other CNN fusions, even using more GPU resources, outperformed the literature results. The CNNs models used are accessible through the Pytorch framework.

3.1 Fusion and Ensemble Processes

Our first proposal was to individually train a set of CNNs models $\{M_1, M_2, M_3, ..., M_n\}$ and get their respective answers $\{\overrightarrow{P}_1, \overrightarrow{P}_2, \cdots, \overrightarrow{P}_n\}$, for $n = 7$. The CNNs were DenseNet121, DenseNet161, DenseNet201, EfficientNet_b0, MobileNetV2, ResNet152 e VGG16. Each model was trained adopting the following values for $\eta = \{0.0001, 0.0003, 0.0005, 0.001, 0.003, 0.005\}$, $e = 50$ and optimize SGD, as shown in Fig. 4. After training, each model was tested and the responses were fused. Each network could participate or not in the fusion. Fusions occur between trained models with the same learning rate, the total of fusions were $2^n \times 6$.

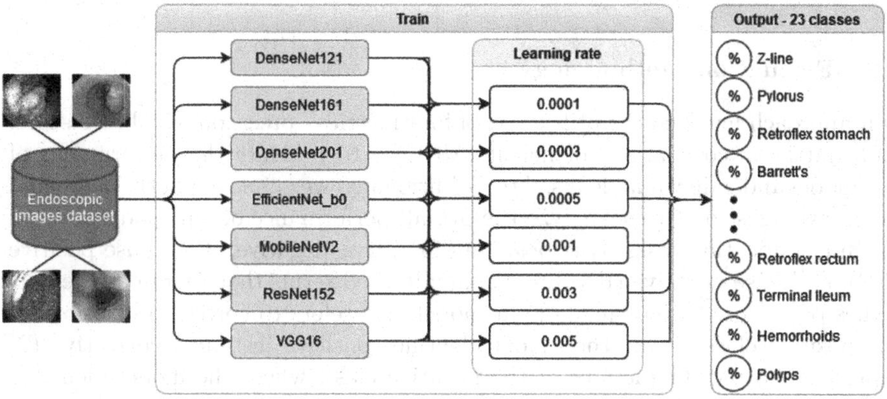

Fig. 4. Our proposal for ensemble architecture for training and fusion process.

For this proposal, two fusion alternatives were analyzed, by average and by voting. In average fusion, each trained model, M_i, generates an output $\overrightarrow{P}_i$ and the fusion is given by $\overrightarrow{P}_\tau = \frac{1}{n}\sum_{i=1}^{n} \overrightarrow{P}_i$, where n is the number of models of CNNs and $\overrightarrow{P}_\tau$ is the average output of the models that make up the fusion, according to Fig. 5(a).

In fusion by voting, considering the values in Fig. 5(a), each network that makes up the fusion votes in the class that receives the highest percentage of

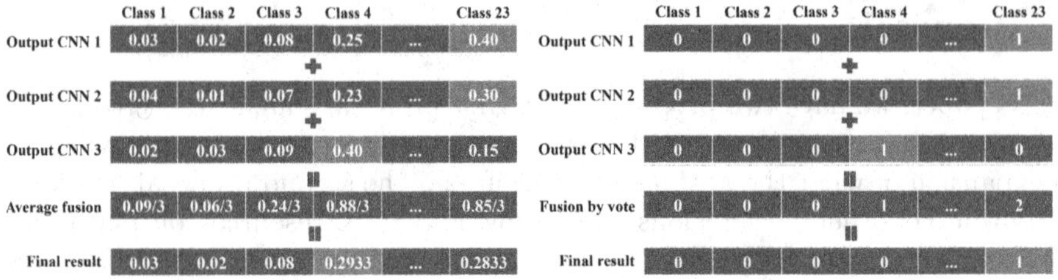

(a) Fusion by mean scheme. (b) Fusion by voting scheme.

Fig. 5. Different fusion schemes for combining models.

probability, as shown in Fig. 5(b). In case of a tie, the first tiebreaker considered the number of times the class was in first and second place in the voting. If the tie remains, among the classes that met the first tiebreaker criterion, the one with the highest percentage of probability is chosen.

In our second proposal, we created models with different fusions of pre-trained CNNs, so M^{ν} was the composition of best models $\{M_1^b, M_2^b, \cdots, M_n^b\}$, for different values of $2 \leq n \leq 7$. The results of the proposals are presented in Sect. 4. In the first proposal, with trained models, it is possible to perform fusions between models without having to train them again. In the second proposal, each new fusion generates a new model that needs to be trained.

3.2 Evaluation Methodology

Our approach involved the utilization of four metrics - precision, recall, F1-score, and Matthews correlation coefficient (MCC) – to evaluate the performance of our model and gain valuable insights. Additionally, we employed both macro and micro averages to further analyze the overall performance of our model [15].

In classification tasks, true positive (TP), true negative (TN), false positive (FP), and false negative (FN) are commonly used terms that represent the outcomes of the predictions made by a model. TP refers to correct predictions of the positive class, where the model identifies positive instances correctly. TN represents correct predictions of the negative class, where the model identifies negative instances correctly. FP refers to incorrect predictions of the positive class, where the model identifies negative instances as positive. FN represents incorrect predictions of the negative class, where the model identifies positive instances as negative [15].

Matthews correlation coefficient, which is a measure of the quality of binary (two-class) classification models. It takes into account TP, TN, FP and FN to provide a balanced assessment of the model's performance, as shown in Eq. 3. The MCC ranges from $[-1, +1]$, where a value of $(+1)$ indicates a perfect classification, (0) indicates a random classification, and (-1) indicates a completely wrong classification. MCC values closer to $(+1)$ indicate better performance of the classification model [3].

$$MCC = \frac{TP \times TN - FP \times FN}{\sqrt{(TP+FP)(TP+FN)(TN+FP)(TN+FN)}} \quad (3)$$

Micro-average is a method of aggregating the performance metrics across all classes in a multi-class classification problem. The formulas for micro-average precision (miP) (Eq. 4), recall (miR) (Eq. 5), and F1-score ($miF1$) (Eq. 6) are as follows [19]:

$$miP = \frac{TP}{TP+FP} \quad (4)$$

$$miR = \frac{TP}{TP+FN} \quad (5)$$

$$miF1 = \frac{2 \times (miP \times miR)}{(miP + miR)} \quad (6)$$

Macro-average, on the other hand, calculates the performance metrics for each class individually and then takes the average across all classes. The formulas for macro-average precision (maP) (Eq. 7), recall (maR) (Eq. 8), and F1-score ($maF1$) (Eq. 9) are as follows [19]:

$$maP = \frac{\sum_i^n (Precision_i)}{n} \quad (7)$$

$$maR = \frac{\sum_i^n (Recall_i)}{n} \quad (8)$$

$$maF1 = \frac{\sum_i^n (F1\_score_i)}{n} \quad (9)$$

where precision$_i$, recall$_i$, and F1-score$_i$ represent the precision, recall, and F1-score of class i, and n is the total number of classes. By using micro and macro averaging, we can gain insights into the overall performance of the classification model, considering both the individual class performance and the overall performance across all classes.

4 Results and Discussion

In this section, we present a comprehensive analysis of various aspects related to convolutional neural networks (CNNs) and their fusion configurations. Firstly, we discuss the analysis for individual CNNs. Next, we delve into the evaluation of fusion configurations and CNN performance metrics. Furthermore, we investigate the fusion with optimal training CNN models. Lastly, we conduct a comprehensive analysis of resource consumption for the CNN models. By examining these four aspects, we gain a comprehensive understanding of the individual and fused CNN models, their performance metrics, optimal training configurations, and resource requirements. This knowledge enables us to make informed decisions and design more effective and resource-efficient CNN-based systems.

4.1 Analysis for Individual Convolutional Neural Networks

In this section, we evaluate the individual performance of the CNN models and their effectiveness in tackling the given task. This analysis provides insights into the strengths and weaknesses of each individual model. Table 2 present differents CNNs and configurations, such as learning rate (LR), in addition, performance metrics, including Matthews correlation coefficient (MCC), precision, recall, and F1-score. Each row represents a different CNN model, denoted by M_1 to M_7. Analyzing the results, it can be observed that different CNN models achieve varying levels of performance across the evaluated metrics. Among the models, M_2 (DenseNet161) stands out with the highest precision, recall, F1-score, and MCC values. On the other hand, M_5 (MobileNetV2) and M_7 (VGG16) exhibit slightly lower performance in terms of precision, recall, F1-score, and MCC.

Table 2. Individual convolutional neural networks results.

ID	CNN Models	LR	Macro Average			Micro Average			MCC
			Precision	Recall	F1-Score	Precision	Recall	F1-Score	
M_1	DenseNet121	0.003	0.614874	0.600373	0.598622	0.892939	0.892939	0.892939	0.883851
M_2	DenseNet161	0.003	0.619034	0.601613	0.604546	**0.902527**	**0.902527**	**0.902527**	**0.894189**
M_3	DenseNet201	0.003	0.619914	0.596281	0.599139	0.897172	0.897172	0.897172	0.888382
M_4	EfficientNet_b0	0.005	0.595228	0.607846	0.595481	0.890215	0.890215	0.890215	0.880982
M_5	MobileNetV2	0.003	0.599915	0.597632	0.592839	0.88561	0.88561	0.88561	0.876007
M_6	ResNet152	0.005	**0.625207**	**0.60677**	**0.609434**	0.900744	0.900744	0.900744	0.892271
M_7	VGG16	0.003	0.589345	0.593563	0.587409	0.884576	0.884576	0.884576	0.874854

4.2 Analysis of Fusion Configurations and CNN Performance Metrics

In this section, we explore different fusion techniques and assess their impact on the overall performance of the system. Table 3 presents the results obtained from the fusion of multiple convolutional neural networks (CNNs), ensembles, using the average method. Each row in the table represents a different fusion scenario, denoted by F_i, where multiple CNN models (M_1 to M_7) are combined, where (`Y`) indicates the presence of a model in the fusion configuration, while (`N`) denotes the absence of that model. The table also includes the LR used for each fusion scenario. The evaluation metrics used to assess the performance of the fusion approach are precision, recall, F1-score, and MCC. Analyzing the results, it can be seen that different ensemble architectures and their fusion responses generate varying levels of performance in the evaluated metrics.

Among the merger scenarios, F_3 displays the highest F1-score value for macro-average. This scenario combines specific CNN models (M_2, M_4, M_6 and M_7) and achieves remarkable performance in correctly classifying positive and negative instances.

Table 3. Results obtained with fusion of CNN by average.

F_i	M_1	M_2	M_3	M_4	M_5	M_6	M_7	LR	Macro Average			Micro Average			MCC
									Precision	Recall	F1-Score	Precision	Recall	F1-Score	
1	N	Y	N	Y	N	Y	Y	0.003	0.63117	0.60845	0.60839	**0.91014**	**0.91014**	**0.91014**	**0.90247**
2	Y	Y	Y	Y	N	Y	Y	0.003	0.64021	0.61205	0.61434	0.91014	0.91014	0.91014	0.90247
3	N	Y	N	Y	N	Y	Y	0.005	0.63945	0.61798	**0.61869**	0.90995	0.90995	0.90995	0.90229
4	Y	Y	Y	Y	Y	Y	Y	0.003	0.63344	0.61381	0.61497	0.90995	0.90995	0.90995	0.90227
5	Y	Y	N	Y	N	Y	Y	0.003	0.63598	0.61137	0.61184	0.90995	0.90995	0.90995	0.90227
6	Y	Y	Y	N	N	Y	Y	0.003	0.63994	0.61046	0.61387	0.90967	0.90967	0.90967	0.90194
7	Y	Y	N	Y	N	Y	Y	0.005	0.62651	0.61649	0.61539	0.90967	0.90967	0.90967	0.90197
8	Y	Y	Y	N	Y	Y	Y	0.003	0.63522	0.61048	0.61225	0.90958	0.90958	0.90958	0.90185
9	N	Y	N	N	N	Y	N	0.003	0.63471	0.60612	0.61022	0.90742	0.90742	0.90742	0.89949

These scenarios show the effectiveness of ensemble methods to improve classification accuracy. Fusion F_1 stands out, achieving relatively high values of accuracy, recall, F1-score and MCC, for micro average. This suggests that the combination of models M_2, M_4, M_6 and M_7 with an LR of $\eta = 0.003$ leads to successful predictions with high accuracy and completeness. Considering the CNN models that appeared more frequently in the fusion experiments, the models M_2, M_6 and M_7 were used in a greater number of experiments. This suggests that these models have a greater impact on the overall performance of the ensemble architectures. Table 4 shows the results of the CNN merger using a voting mechanism where each model in the ensemble makes an independent prediction, and the final prediction is based on the highest number of votes.

Table 4. Results obtained with CNN fusion by vote.

F_i	M_1	M_2	M_3	M_4	M_5	M_6	M_7	LR	Macro Average			Micro Average			MCC
									Precision	Recall	F1-Score	Precision	Recall	F1-Score	
1	N	Y	Y	Y	N	Y	Y	0.003	0.63590	0.60927	0.61024	0.90892	0.90892	0.90892	0.90114
2	Y	Y	Y	Y	N	N	Y	0.003	**0.64163**	**0.61383**	**0.61582**	**0.90892**	**0.90892**	**0.90892**	**0.90115**
3	Y	Y	N	Y	N	Y	Y	0.003	0.63075	0.61247	0.61249	0.90882	0.90882	0.90882	0.90106
4	Y	Y	Y	N	N	Y	Y	0.003	0.64030	0.61237	0.61572	0.90882	0.90882	0.90882	0.90103
5	N	Y	Y	Y	Y	Y	Y	0.003	0.62637	0.61044	0.61016	0.90845	0.90845	0.90845	0.90064
6	Y	Y	Y	Y	N	Y	N	0.003	0.62882	0.60789	0.60804	0.90835	0.90835	0.90835	0.90052
7	Y	Y	Y	Y	Y	Y	Y	0.003	0.62904	0.60964	0.60973	0.90817	0.90817	0.90817	0.90033
8	N	Y	N	N	N	Y	Y	0.003	0.63761	0.61121	0.61459	0.90807	0.90807	0.90807	0.90022
9	N	Y	N	N	N	Y	N	0.005	0.63084	0.61210	0.61472	0.90385	0.90385	0.90385	0.89564

Upon analyzing the results, it is evident that the performance of the fusion models varies depending on the specific combination of CNN models used. F_2 stands out as it achieves the highest values in terms of F1-score for macro (0.61582), F1-score for micro average (0.90892) and MCC (0.90115). This combination includes models M_1, M_2, M_3, M_4, and M_7, indicating that these models contribute significantly to the overall performance of the fusion model.

4.3 Fusion with Optimal Training CNN Models

In this section, we explore the integration of specific CNN models to further enhance the system's performance and achieve superior results. Table 5 presents the results of the fusion of trained CNN models using different combinations. In our experiments, in F_4^b, considering the amount of CNNs that make up the fusion, we had the best performance compared to other combinations, considering both micro-average, F1-score (0.913) and MCC (0.905). Overall, the analysis of the fusion results indicates that the combinations involving M_2^b (DenseNet161), M_4^b (EfficientNet_b0), M_6^b (ResNet152), and M_7^b (VGG16) generally lead to better performance, with higher F1-scores and MCC values. The presence of M_1^b and M_5^b does not contribute significantly to the overall performance improvement.

Table 5. Fusion with trained CNN models.

F_i^b	M_1^b	M_2^b	M_3^b	M_4^b	M_5^b	M_6^b	M_7^b	LR	Macro Average			Micro Average			MCC
									Precision	Recall	F1-Score	Precision	Recall	F1-Score	
1*	N	Y	N	N	N	Y	N	0.001	0.633	0.615	0.617	0.910	0.910	0.910	0.902
2	N	Y	N	N	N	N	Y	0.003	0.634	0.617	0.620	0.908	0.908	0.908	0.900
3	N	Y	N	N	N	Y	Y	0.003	0.634	0.621	0.625	0.912	0.912	0.912	0.905
4	N	Y	N	Y	N	Y	Y	0.003	0.633	0.621	0.623	**0.913**	**0.913**	**0.913**	**0.905**
5	N	Y	Y	Y	N	Y	Y	0.003	0.630	0.621	0.623	0.912	0.912	0.912	0.905
6	Y	Y	Y	Y	N	N	Y	0.003	0.627	0.616	0.618	0.911	0.911	0.911	0.903
7	Y	Y	Y	Y	N	Y	Y	0.003	0.629	0.619	0.621	0.913	0.913	0.913	0.905

4.4 Resource Consumption Analysis for CNN Models

In this section, we present Table 6, which provides details such as the CNN model name, GPU model used for execution, GPU RAM capacity, execution time in minutes, and the number of parameters for each model. To measure the execution time and GPU consumption, the `timeit` module and the `psutil` library were used, respectively. All network models, M_1 to M_7, utilize the Tesla V100-SXM2-16 GB GPU model. Additionally, the F_1^{b*} model [2] employs the Tesla V100-SMX2-16 GB, while fusion models F_2^b to F_7^b, for Table 5 and Table 6, utilize the Nvidia A100-SXM-40 GB GPU model.

These data allow us to analyze the computational cost associated with achieving the results presented in Table 3, Table 4 and Table 5.

In Table 3, the F_1 result, including M_2, M_4, M_6, and M_7, achieved the highest F1-score (0.91014) with the least number of models used for micro-average. The models were executed individually on the GPU, resulting in a total GPU consumption equal to the highest consumption among the individual models, which is 10 GB for M_2. Thus, the proposed ensemble F_1 with CNN model averaging has a maximum GPU consumption of 10 GB.

In Table 4, using the technique of fusion by vote, the ensemble F_2 achieved the highest F1-score of 0.90892 for micro and F1-score of 0.61582 for macro. The set F_2 consisted of models M_1, M_2, M_3, M_4 and M_7. The GPU consumption for the set F_1 corresponds to that of the model M_2, which is 10 GB.

Table 6. Resource consumption for network models.

Individual Models					Fusion Models				
CNN models	GPU model	RAM GPU (GB)	Execution Time (m)	Parameters	CNN models	GPU model	RAM GPU (GB)	Execution Time (m)	Parameters
M_1	tesla v100-sxm2-16gb	6.2	92.0	6977431	F_1^{b*}	tesla v100-sxm2-16gb	15.8	113.2	84713742
M_2	tesla v100-sxm2-16gb	10.0	94.9	26522807	F_2^b	nvidia a100-sxm-40gb	15.7	55.5	160877582
M_3	tesla v100-sxm2-16gb	8.9	89.9	18137111	F_3^b	nvidia a100-sxm-40gb	22.7	82.9	219068517
M_4	tesla v100-sxm2-16gb	4.6	80.9	4037011	F_4^b	nvidia a100-sxm-40gb	24.7	98.5	223105528
M_5	tesla v100-sxm2-16gb	4.2	86.2	2253335	F_5^b	nvidia a100-sxm-40gb	31.0	89.6	241242639
M_6	tesla v100-sxm2-16gb	8.1	101.3	58190935	F_6^b	nvidia a100-sxm-40gb	28.9	110.3	190029135
M_7	tesla v100-sxm2-16gb	7.0	109.8	134354775	F_7^b	nvidia a100-sxm-40gb	35.9	124.5	248220070

In Fusion with optimal training CNN models, as shown in Table 5, for F_1^{b*}, the F1-score is 0.910, which matches our proposal in Table 3. The approach in [2], F_1^{b*}, requires 15.8 GB of GPU, as the best models M_2^b and M_6^b are trained together. Building upon the combination of models proposed in [2], we introduce ensembles F_2^b to F_7^b, with F_4^b achieving the best result. Figure 6 depicts a bubble chart illustrating that we have achieved comparable results (indicated by the blue and red bubbles) when compared to the fusion model F_1^{b*} in Table 5 (green bubble), as presented in [2]. Our proposal F_1 in Table 3 attained the same results while utilizing 10 GB of GPU, which is 36.7% less than the consumption of [2] with 15.8 GB GPU. The purple bubbles demonstrate that our ensemble architectures using Fusion with optimal training CNN models obtain better results than [2], albeit at a higher GPU cost.

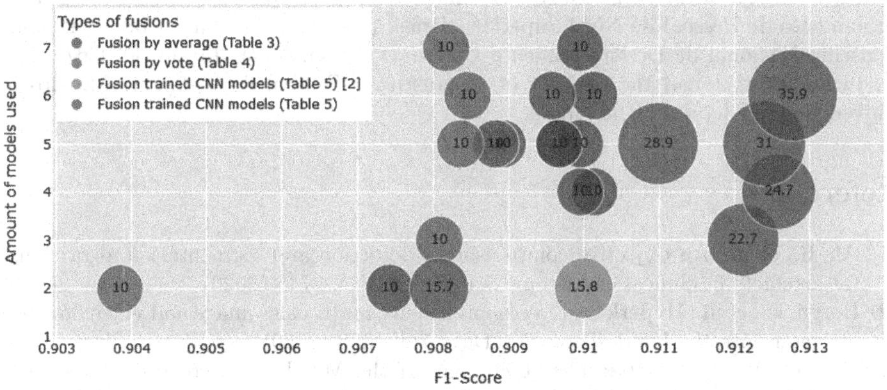

Fig. 6. Bubble chart for resource consumption analysis and comparative results. The diameter of the bubbles is associated with GPU consumption. (Color figure online)

5 Conclusions

In conclusion, to aid in the diagnosis of diseases of the human gastrointestinal tract, through image classification, our objective was to propose ensemble architectures and efficient fusion techniques for CNNs, aiming to obtain better results than in the literature and to optimize computational resources.

A literature review and extensive experiments were conducted to validate the effectiveness of the proposed approach. The findings of this study highlight the significant contribution of DenseNet161 and ResNet152 to the fusion process in all experiments. Furthermore, our findings demonstrate a similar level of performance compared to the previous model, as indicated by an F1-score of 0.910 and MCC of 0.902. Remarkably, our approach achieves this performance using just 10 GB of GPU RAM, in contrast to the previous model's requirement of 15.8 GB.

We were able to achieve a minimally higher F1 value of 0.913 and MCC of 0.905, but using 24.7 GB of GPU RAM. These findings contribute to our understanding of individual model performance, fusion techniques, and resource utilization, paving the way for the design and implementation of more efficient and effective deep learning-based systems in endoscopic scoring.

The main challenge was accessing sufficient computational resources. Future research should explore model compression techniques to reduce computational requirements. Additionally, further experiments and analysis are needed to identify optimal model combinations and refine the fusion process. Evaluating fusion alternatives on different datasets is crucial for assessing performance and generalizability.

Acknowledgements. This study was financed in part by Coordenação de Aperfeiçoamento de Pessoal de Nível Superior - Brasil (CAPES) - Finance Code 001* and Conselho Nacional de Desenvolvimento Científico e Tecnológico (grant 306436/2022-1). In addition, it had the support of the Instituto Federal do Triângulo Mineiro e Universidade Federal de Uberlândia.

References

1. Ali, S., et al.: An objective comparison of detection and segmentation algorithms for artefacts in clinical endoscopy. Sci. Rep. **10**(1), 2748 (2020)
2. Borgli, H., et al.: Hyperkvasir, a comprehensive multi-class image and video dataset for gastrointestinal endoscopy. Sci. Data **7**(1), 283 (2020)
3. Chicco, D., Jurman, G.: The advantages of the Matthews correlation coefficient (MCC) over f1 score and accuracy in binary classification evaluation. BMC Genom. **21**, 1–13 (2020)
4. Fukushima, K.: Neocognitron: a self-organizing neural network model for a mechanism of pattern recognition unaffected by shift in position. Biol. Cybern. **36**(4), 193–202 (1980)
5. Hicks, S.A., Jha, D., Thambawita, V., Halvorsen, P., Hammer, H.L., Riegler, M.A.: The EndoTect 2020 challenge: evaluation and comparison of classification, segmentation and inference time for endoscopy. In: Del Bimbo, A., et al. (eds.) ICPR 2021.

LNCS, vol. 12668, pp. 263–274. Springer, Cham (2021). https://doi.org/10.1007/978-3-030-68793-9_18
6. Iqbal, I., Walayat, K., Kakar, M.U., Ma, J.: Automated identification of human gastrointestinal tract abnormalities based on deep convolutional neural network with endoscopic images. Intell. Syst. Appl. **16**, 200149 (2022)
7. Jha, D., et al.: Real-time polyp detection, localization and segmentation in colonoscopy using deep learning. IEEE Access **9**, 40496–40510 (2021)
8. Jha, D., et al.: Medico multimedia task at mediaeval 2020: automatic polyp segmentation. arXiv preprint arXiv:2012.15244 (2020)
9. Jha, D., et al.: A comprehensive study on colorectal polyp segmentation with resunet++, conditional random field and test-time augmentation. IEEE J. Biomed. Health Inform. **25**(6), 2029–2040 (2021)
10. Jha, D., et al.: Kvasir-SEG: a segmented polyp dataset. In: Ro, Y.M., et al. (eds.) MMM 2020. LNCS, vol. 11962, pp. 451–462. Springer, Cham (2020). https://doi.org/10.1007/978-3-030-37734-2_37
11. Krizhevsky, A., Sutskever, I., Hinton, G.E.: Imagenet classification with deep convolutional neural networks. In: Advances in Neural Information Processing Systems, vol. 25, pp. 1097–1105 (2012)
12. LeCun, Y., Bengio, Y., Hinton, G.: Deep learning. Nature **521**(7553), 436–444 (2015)
13. LeCun, Y., et al.: Backpropagation applied to handwritten zip code recognition. Neural Comput. **1**(4), 541–551 (1989)
14. Naess, E., Thambawita, V., Hicks, S.A., Riegler, M.A., Halvorsen, P.: Pyramidal segmentation of medical images using adversarial training. In: Proceedings of the 2021 Workshop on Intelligent Cross-Data Analysis and Retrieval, pp. 33–38 (2021)
15. Sarkar, D., Bali, R., Sharma, T.: Practical Machine Learning with Python (2018). https://doi.org/10.1007/978-1-4842-3207-1
16. Sze, V., Chen, Y.H., Yang, T.J., Emer, J.S.: Efficient processing of deep neural networks: a tutorial and survey. Proc. IEEE **105**(12), 2295–2329 (2017)
17. Sze, V., Chen, Y.H., Yang, T.J., Emer, J.S.: Efficient processing of deep neural networks. Synth. Lect. Comput. Archit. **15**(2), 1–341 (2020)
18. Szegedy, C., Ioffe, S., Vanhoucke, V., Alemi, A.A.: Inception-v4, inception-resnet and the impact of residual connections on learning. In: Thirty-First AAAI Conference on Artificial Intelligence (2017)
19. Takahashi, K., Yamamoto, K., Kuchiba, A., Koyama, T.: Confidence interval for micro-averaged f 1 and macro-averaged f 1 scores. Appl. Intell. **52**(5), 4961–4972 (2022)
20. Thambawita, V., et al.: An extensive study on cross-dataset bias and evaluation metrics interpretation for machine learning applied to gastrointestinal tract abnormality classification. ACM Trans. Comput. Healthc. **1**(3), 1–29 (2020)

Deep Learning Applications

Dog Face Recognition Using Deep Features Embeddings

João P. B. Andrade[1(✉)], Leonardo F. Costa[1], Lucas S. Fernandes[1], Paulo A. L. Rego[1], and José G. R. Maia[2]

[1] MDCC, Universidade Federal do Ceará, Campus do Pici - Bloco 910, Fortaleza, CE, Brazil
jpandrade@alu.ufc.br

[2] Instituto UFC Virtual - Universidade Federal do Ceará, Campus do Pici - Bloco 901, Fortaleza, CE, Brazil
gilvan@virtual.ufc.br

Abstract. Over 470 million dogs are kept as pets around the world. Dogs are owned at an average number of 1.6% per household. The US has the most dog pets, where about 68% of households own at least one pet. Lost and missing dogs are a severe source of suffering and problems for their families. So, this paper addresses the problem of facial dog identification. This technology can benefit many applications, such as handling the missing pet problem, granting pets access to their houses, more intelligent zoonosis control, pet health care, and tracking stray pets. We evaluate a Residual Convolutional Neural Network, specifically ResNet-34, for facial identification in dogs. We tested in DogFaceNet and Flickr-dog datasets with and without two face preprocessing techniques: a central crop and an aligned facial extraction. Experimental results show promising results surpassing the state-of-the-art: 97.6% and 82.8% accuracies for DogFaceNet and Flickr-dog, respectively. Moreover, we also provide recall metrics for the best models.

Keywords: Residual Neural Networks · Deep Features · Facial Identification · Dogs

1 Introduction

Effective identification and tracking of pets is a valuable technology for modern society. For example, many countries have more dogs than children, according to recent official data releases by [15]. Stray animals, in general, could pose a serious hazard to other species and human health in urban areas, especially to children. On the other hand, missing pets is a frequent problem whose conditions of occurrence are minimal. Since dogs are autonomous animals, it only takes a little carelessness to make them run away from home, especially when there are festive events with fireworks. On top of that, the human ability to discern animals by their traits is reasonably limited; consequently, it is difficult to identify and track animals effectively with the naked eye [4].

Supported by organization FUNCAP.

M. C. Naldi and R. A. C. Bianchi (Eds.): BRACIS 2023, LNAI 14196, pp. 125–139, 2023.
https://doi.org/10.1007/978-3-031-45389-2_9

In [7], the authors sought to understand the process of coping with grief experienced by people who experienced the loss of their pets. They concluded that these processes have characteristics similar to those present in mourning processes arising from the loss of significant people. That said, canine recognition can be an attractive alternative for smart cities to deal with complex problems related to stray or lost animals, access control for pets, dog catching, zoonosis control, and others.

As evidence of the interest in identifying dogs, some work seeks to identify dogs using the dog's nose print, an even more challenging problem with the images further reduced to just the dog's nose. [13] reaches AUC of 88.81% with a model with the dual global descriptor, which can exploit the multi-scaled features of image working in a dataset with 6000 dogs with 20,000 photographs of their nose prints. A recent study, [23], moved by the problem that in Sri Lanka, the authors' country, a lost dog being found is an infrequent occurrence, presents a model of Convolutional Neural Network (CNN) for classification and recognition of images for facial recognition of dogs. The CNN used consists of a customized model VGG-16, obtaining about 94% accuracy in the dog recognition task.

It is noteworthy that facial recognition is one of the least invasive options and is less subject to operational problems when the animal does not cooperate with the use of devices or collars [15]. Of course, in the case of dogs, the effectiveness of this approach may vary if the animal is not cared for similarly after it is lost (hair and other characteristics may change, for example). There are algorithms in the literature that deal with this problem, but several of them are arguably overly complex or may present a still limited performance for such a challenging but significant problem. Thus, the present work seeks to consolidate results in this area by investigating the use of deep feature embedding vectors for dog face recognition, focusing on the facial identification task.

These are the main contributions of this paper:

- We used ResNet-34, a residual CNN for the face identification task in dogs, on two datasets in which each dataset was evaluated in its standard form and applied two preprocessing techniques for each dataset. We evaluate the use of this computation-intensive method, its benefits, and whether they are justifiable, taking its accuracy into account.
- We compare three dog face preprocessing methods using two public datasets [15,16] found in similar works. While most papers only provide performance analysis considering accuracy, this paper also includes recall.
- We obtained state-of-the-art performance using a fast approach that outperforms previous methods, which usually resort to more complex, computationally expensive models.

The remainder of this paper is organized as follows. In Sect. 2, we presented the related works. The datasets that address the problem of facial identification in dogs are described in Sect. 3. In Sect. 4, we described the approach presented in this work. In Sect. 5, we present the methodology for the experimental evaluation of our approach. In Sect. 6, we presented the results. Moreover, finally, in Sect. 7, we presented final remarks and future works.

2 Related Work

It is worth mentioning that most of the works found in the literature are related to the task of dog breed classification. This is a challenging problem due to the diversity of the existing native breeds and the breeds resulting from crossbreeding and genetic modifications, where many of these are visually similar to each other.

[14] presented a method for fine-grained image classification based on using key regions during the feature extraction process to improve the performance of the classification. This method was evaluated on the dog breed classification problem using a dataset of images downloaded from Google, Image-net, and Flickr, which comprises 133 dog breeds and $8,351$ images. The proposed method achieved a recognition rate of 67% and used conventional tools such as SIFT and color histograms of the animal's face region, excluding everything else in the image.

[17] used Dense-SIFT to extract features over the input image. These features are split and combined before feeding a CNN classifier. This method resulted in 94.2% accuracy for 121 dog breeds, using a database obtained by crawl image data methods on the web. Similarly, [4] created a (closed) dataset for dog breed and wild wolf classification. These authors compared different CNN architectures using transfer learning and obtained 92.48% accuracy.

[19] proposed another method based on deep learning, using supervised clustering based on a multi-part-convolutional neural network and expectation-maximization. The method was tested on many datasets, including the Snapshot Serengeti dataset [20], which contains approximately $78,000$ labeled images of animals in Africa, obtaining an accuracy of 98.4%, and the authors reinforce the idea that using CNNs with supervised training works very well in extracting features from images in such a type of problem.

Other works relate to using different techniques to track and find lost pets, whereas many approaches resort to installing GPS and other kinds of sensors. [21] proposed Bokk Meow, a mobile application whose goal is to help users locate their animals using GPS once the animals might have this type of sensor installed on their collars. The authors argue that their application contributes to lesser stray/lost pets. However, Bokk Meow demands an active Wi-Fi connection, and its tracking features do not work appropriately in dense urban building areas.

[1] proposed a methodology for noseprint-based dog recognition using a deep, Siamese convolutional network called DNNet. This CNN is composed of feature extraction and (expensive) self-attention modules. Their method seems promising, presenting an accuracy above 98.97% and resorts to affordable equipment for image acquisition. However, such an approach sounds hard to automate. Moreover, human operators should handle dogs physically, which may cause risks or discomfort to the operator and the dog when there is no collaboration from the animal.

The work of [22] presents a smartphone application for dog face detection. The system uses the YOLO deep convolutional neural network to simultaneously predict face bounding boxes and class confidences. The Doggie Smile app helps to take pictures of the dogs as they face the camera. The proposed mobile app can simultaneously assess the gaze directions of three dogs in a scene more than 13 times per second, measured on the iPhone XR. The reported average accuracy of the dog face detection system is 92%.

3 Dog Face Recognition Datasets and Methods

There are datasets for many tasks related to fine-grained dog image classification, such as Columbia Dogs and Stanford Dogs, that are used in [9,14], respectively. Those datasets were used in previous works for augmenting data and for fine-tuning pre-trained models before the dog recognition task was carried out. For example, [16] also resorted to these datasets to train and evaluate fog face detectors and aligners.

3.1 Dog Face Recognition Datasets

Table 1 summarizes the datasets known in the literature for the face identification task in dogs, their main characteristics, and which ones we used in our experiments. Our only motivation for choosing these datasets was that they were public and freely available for download.

Table 1. Datasets found in the literature for the task of facial identification in dogs.

Dataset	Number of Indentities	Number of Images	Is public?
Pet2Net [2]	16	213	no
Snoopybook [15]	18	251	no
Flickr-dog [15]	**42**	**374**	**yes**
Weerasekara et al. [23]	320	2500	no
DogFaceNet [16]	**1393**	**8363**	**yes**

The *DogFaceNet* dataset [16], contains 8,363 images from 1,393 dog identities. Each dog has at least two JPEG images with 224×224 RGB pixels. Despite the previous works of [15,16] claim that DogFaceNet is the first public dataset for dog face recognition and verification.

Examples of the DogFaceNet[1] are shown at the top of Fig. 1, with 3 identities with 5 images each. It is essential to highlight that the images were cropped and aligned beforehand. However, this process introduces black borders in the images. Moreover, there are many samples where the dog appears in profile or their face is partially occluded.

The Flickr-dog dataset [15] is 22.3 times smaller than DogFaceNet, containing only 374 images of dogs from two breeds: pugs and huskies. The dataset includes 21 identities for each dog breed, totaling 42 for individual dogs and thus 5 images

[1] https://github.com/GuillaumeMougeot/DogFaceNet.

per identity. This means that Flickr-dog provides 33.1 times fewer identities than DogFaceNet, so the task of face recognition is also related to the challenge of discerning from multiple dogs of the same breed.

Each image in the dataset has 250×250 RGB pixels containing a horizontally aligned face of a dog. Also, the images are normalized such that all images are cropped, (loosely) aligned, and resized. Five sample images are shown at the bottom of Fig. 1 for three individuals: lighting conditions vary significantly, there are artifacts, and also black borders are evident.

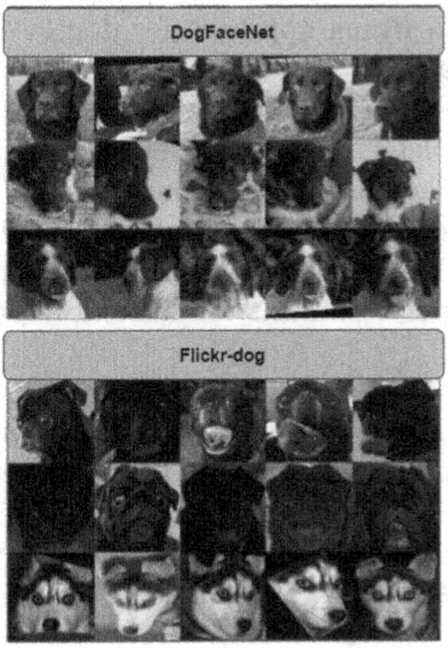

Fig. 1. Public dog face recognition datasets: DogFaceNet (top) and Flickr-dog (bottom). Many faces are misaligned and/or appear at challenging poses in both datasets.

3.2 Dog Face Recognition Methods

In the work of [23], a VGG16-based [18] CNN is trained to classify dogs in 5 categories using transfer learning. Then, 128-dimensional embeddings are extracted from this CNN and used for identification by training an unsupervised model for distance calculation. According to the authors, experimental evaluation resulted in more than 90% accuracy using their dataset where dogs appear in various poses and sizes, but the work is missing evaluation details of how dog faces are processed. Moreover, the dataset is private.

[15] proposed a pioneer work in dog face recognition. The authors evaluated four classic approaches (EigenFaces, FisherFaces, LBPH, and Sparse) against two CNNs, BARK and WOOF, for retrieving lost dogs based on facial features. These CNNs significantly outperformed the other methods, obtaining an accuracy of 89.4% and 81.1% for WOOF and BARK, respectively, for the single-breed

Snoopybook dataset, which is not publicly available. Moreover, the accuracy is considerably lower considering the evaluation on the Flickr-dog dataset: 66.9% and 67.6% for WOOF and BARK, respectively.

In [3], in order to identify animals, the authors compared two different CNN architectures (VGG-16 and Inception-V3) originally applied in object detection. These two architectures are applied to the Flickr-dog and DogsVsCats datasets. The accuracy achieved by VGG-16 on Flickr-dog was 95% and 98% on DogsVsCats, while Inception-V3 reached 87% on Flickr-dog and 85% on the DogsVsCats dataset. DogsVsCats dataset is a dataset with a different purpose, which includes images of cats, so it was not considered among the datasets in this work.

[16] proposed a deep learning approach for dog face verification and introduced the DogFaceNet dataset. They evaluated two models based on VGG and ResNet using 224 × 224 RGB images as input. Deep feature embedding vectors were generated using Triplet Loss with soft and hard triplet mining, generated both online and offline. These authors obtained a verification accuracy of 86%, considered poor compared to the typical performance of similar approaches for human face recognition. They stated that errors arose from frequent occlusions in the test dataset, dog's pose, similarity of different dogs of the same breed, and light exposure.

[12] proposed the use of soft biometrics for improving identification. Based on a Fast R-CNN, their method performs a coarse-to-fine identification, i.e., starting from the breed filtering, then carrying out the identification. Their results suggest that breed information improved accuracy for different dog breeds, and overall accuracy is 76.53% for the Flickr-dog using an Xception back-end [5]. The authors improved this result to 84.94% by using a likelihood-adjusted decision network.

[2] proposed and evaluated a deep learning-based approach to pet detection and recognition, named Pet2NetID, based on a dataset provided by Pet2Network, a social network for pets and their owners. They achieved 94.59% accuracy by combining transfer learning and object detection approaches with Inception V3 and SSD Inception V2 on the provided Pet2Net dataset, and 77.19% was achieved when training with the Pet2Net dataset on the Flickr-dog dataset. The approach proposes to work with wild images without any preprocessing, including other objects besides the pets themselves, and the identification uses information from the whole animal, not only the face.

In general, as seen in Table 2, these works use deep learning methods without preprocessing experimentation (e.g., face detection) and do not provide recall results, only accuracy. We observed this and included it in our experiments, described in the following sections, developing a unique approach to the problem with a more robust evaluation than the related works in this section.

Table 2. Comparison between related works.

Related Work	Method	Use Recall?	Private dataset?
Weerasekara et al. [23]	CNN	No	Yes
Moreira et al. [15]	CNN	No	Yes
Capone et al. [3]	CNN	No	No
Batic & Culibrk [2]	CNN	No	Yes
Proposed Method *	**CNN**	**Yes**	**No**

4 Proposed Dog Face Recognition Method

Since many works in the literature already resorted to transferring learning and fine-tuning existing models, our investigation prioritized following a different strategy, so we trained models from scratch. As stated before, the incoming images are subjected to optional preprocessing, which can occur in two ways. However, only cropping is allowed in this step, so no further filtering or image enhancement is applied in favor of an honest comparison against other methods.

4.1 Preprocessing and Training Pipeline

The pipeline of evaluation using the methods is illustrated in Fig. 2. We took three different preprocessing paths and evaluated them separately, these being: **(A)** the dataset in its pure form; **(B)** a version where each image was cropped in the center, which we call Central Crop (CC); and **(C)** another path using Facial Detection (FD). The use of each technique is detailed below.

A centralized crop is applied to assume the images are satisfactorily aligned, i.e., the dog's eyes lie near the center of the image, but assuming there is a border whose existence is questionable for facial recognition purposes. The CC preprocessing algorithm extracts the central pixels from a given dimension, thus reducing the input image size.

FD includes alignment and performs as follows. First, the dog face is detected using a CNN-based object detector using dlib's [10] Max-Margin Object Detection (MMOD) loss layer that implements the object detection scheme as described in [11]. Despite using a concise representation, i.e., this CNN's weights require only about 700 KB. However, MMOD inference is the most time-consuming operation in this pipeline. Second, six facial landmarks are then estimated using an ensemble of regression trees [8] with cascade $depth = 20$ and $tree\ depth = 5$. Finally, these points (shown in Fig. 3) are converted into another layout before they are used to extract aligned dog faces with $padding = 0.2$ since mapping this 5-point into an aligned box is less prone to errors. Such conversion is based on the proportions of the eye and nose landmarks.

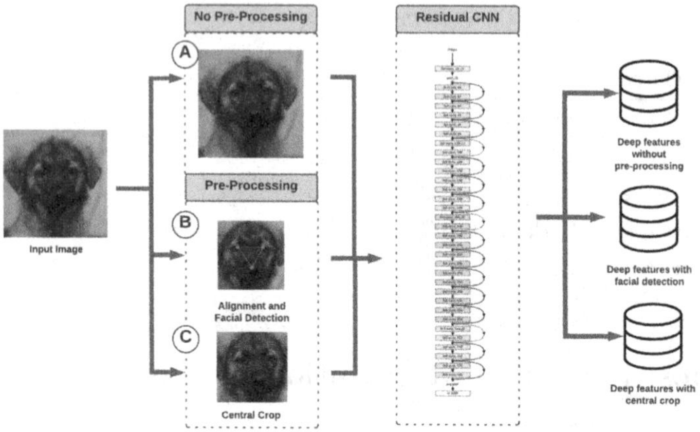

Fig. 2. In order to study the impact of using different preprocessing techniques, we performed tests without preprocessing (A), with alignment and face detection preprocessing (B), and with central crop (C) before using Residual CNN for feature generation.

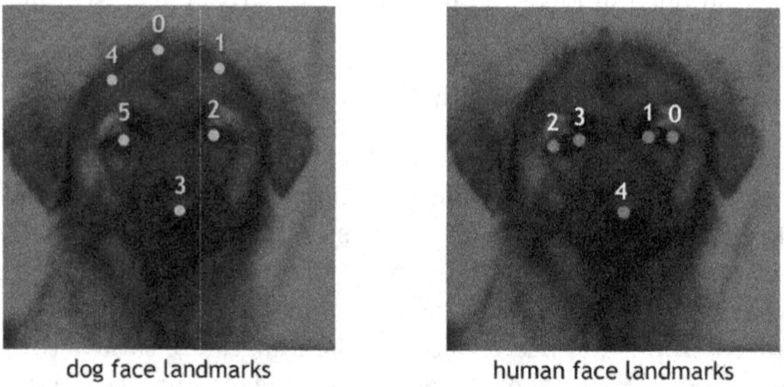

Fig. 3. Dog face landmark detection returns 6 points (left), which are converted to the 5-point layout (right) before alignment and cropping.

4.2 Preprocessing the Datasets

Once we have clarified the two preprocessing techniques, in this subsection, we detail the generation of datasets' versions with each preprocessing and the details that occurred in each technique and dataset.

DogFaceNet CC results from a central crop of 180×180 pixels of each image in DogFaceNet, so the total individuals and images are kept. *DogFaceNet FD* corresponds to the DogFaceNet dataset after face and alignment. We then used the same $80 - 20$ split used in the original DogFaceNet and *DogFaceNet CC*, but some images are missing due to face detection errors. This resulted in 1,108 identities for training, corresponding to 5,978 images. There are 276 identities for testing, corresponding to 1,694 images. This process lost 9 identities corresponding to 0.65% of individuals (ids 44, 67, 287, 449, 494, 948, 1131, 1274, and

1310). On the other hand, the 691 images lost correspond to 8.26% of the images from the original dataset. It is worth highlighting that missing images cannot be included in the training-test split.

Flickr-dog CC corresponds to the Flickr-dog dataset after a central crop for the 200 × 200 inner pixels. In turn, *Flickr-dog FD* corresponds to Flickr-dog after face detection and alignment. This process resulted in 41 of the 42 identities, totaling 218 images. This process lost one identity corresponding to 2.38% of individuals since the face of "Wilco" could not be detected in any of its images (first row of Fig. 1). On the other hand, 156 images lost correspond to 41.71% of the images from the original dataset. Such a fact is quite a loss in terms of images. However, it is essential to remember that all the 218 remaining images will be used for model evaluation.

In Fig. 4, there are some examples of removed images where the algorithm could not detect the face and examples of dogs that lost all images and consequently lost their identity for both datasets. The final numbers for the datasets are described in Table 3.

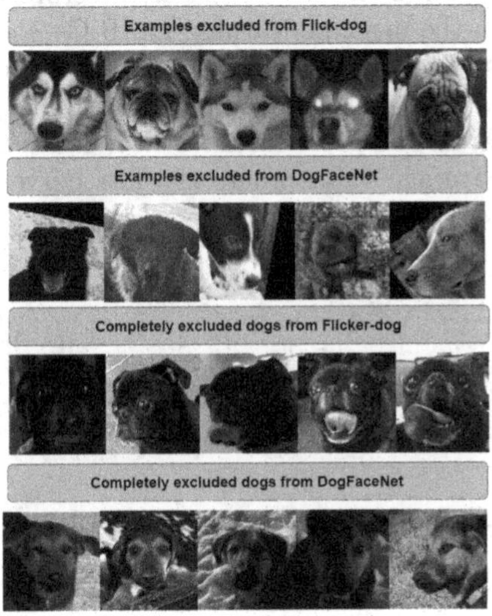

Fig. 4. Examples of images removed from datasets because the algorithm could not detect the dog's face

5 Experimental Evaluation

In order to evaluate the use of the residual network in the dataset and each of the generated versions, we performed the identification experiments evaluating different metrics of interest in this context. We carried out the experiments on a computer running Linux operating system (Ubuntu 20.04.2 LTS), equipped with

Table 3. Number of identities and images of each of the dog face datasets.

Dataset	Number of Indentities	Number of Images
Flickr-dog	42	374
Flickr-dog CC	42	374
Flickr-dog FD	41	218
DogFaceNet	1,393	8,363
DogFaceNet CC	1,393	8,363
DogFaceNet FD	1,393	7,473

an Intel(R) Core(TM) i7-9700K CPU @ 3.60 GHz, 32 GB RAM, a 1 TB SDD, and an NVIDIA GeForce RTX 2080ti graphics card.

The ResNet-34 convolutional neural network was used for the experiments, as it is a good trade-off between performance and computational cost. It is a residual neural network with about 21, 282, 000 parameters, requiring about 4 GFLOPS for activation. Figure 5 shows the ResNet-34 architecture. The implementation of ResNet-34 present in the machine learning DLIB C++ library was used [10].

The split training and testing procedures were explained individually in specific topics for each dataset. We defined that all training sessions would have 80, 000 steps, to standardize all procedures for each dataset. We chose this number because it is the average value where the loss curve is already relatively stabilized.

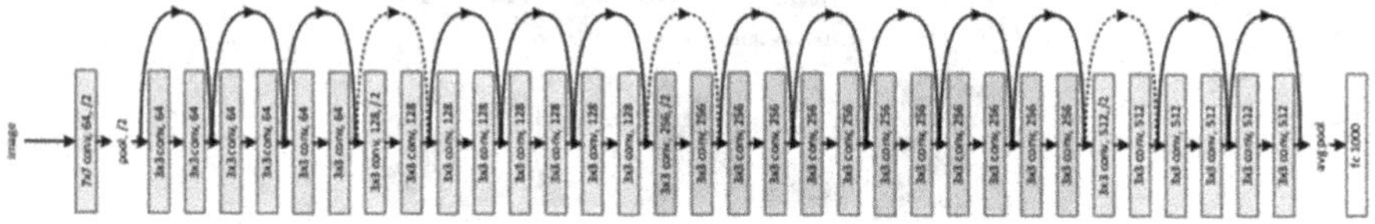

Fig. 5. ResNet34 Architecture (Adapted from [6]).

5.1 Performance Metrics

The metrics used in this work are accuracy and recall. Both are defined in terms of *true positives (tp)*, *false positives (fp)*, *true negative (tn)* and *false negative (fn)* samples. Accuracy is defined in Eq. 1 and recall in Eq. 2.

$$Accuracy = \frac{tp + tn}{tp + tn + fp + fn} \tag{1}$$

$$Recall = \frac{tp}{tp + fn} \tag{2}$$

Precision and F1-score sound virtually meaningless for our experiments since the evaluation considers an upper triangular matrix where most elements in that matrix represent different individuals. In other words, we would be roughly counting the main diagonal versus the rest of the matrix in the limit, so these performance metrics are influenced mainly by the model's ability to discard dissimilar individuals.

5.2 Preparation of the Datasets

Notice that the baseline work [16] adopts a $90-10$ split, i.e., only 697 images (8.33% of the dataset) are used for testing. To better generalize the model used here, we adopted the $80-20$ split, making the problem even more challenging by having fewer images for training and providing a more extensive test set. For the DogFaceNet case, 5,978 images for training in each case (CC and FD), excluding all the FD algorithm images, did not detect a face, and 1,794 test images for original DogFacneNet, 1,704 images in CC, and 1,520 in DogFaceNet FD.

The experiment performed with Flickr-dog proceeded as follows. Since this is a small-scale dataset, the method heavily depends on some data augmentation strategy or external data.

The results are inferior without training with external data and only with data augmentation: an accuracy limited to 76.55% and recall limited to 65.68%, also performed using the $80-20$ split.

We decided to use the model trained with the different versions of the DogFaceNet dataset (original, CC, and FD) and test on a range corresponding to 20% of the Flickr-dog dataset, which was also submitted and generated versions with CC and FD preprocessing. For this case, the original with 65 images, Flickr-dog CC with 69, and FD with 45 images.

6 Results

Preliminarily, we demonstrated the loss for each of the training variations of the DogFaceNet dataset in Fig. 6 and noticed that the curves are very similar, reaching very close values at the end of the training. The curve of the FD training stabilized faster, showing that it would not require pervasive training for this case, unlike the other two pieces of training that stabilized closer to the end of the training.

Still regarding the training, in Fig. 7, we showed through the embeddings extracted from each dataset that ResNet-34 produced an effective separation between classes, where classes are the "same dog" and "different dogs", with Euclidean distance and relative frequency as dimensions. It is also possible to see in these distinction curves that the overlap area of the CC is visually smaller than the other two pieces of training.

The results are presented in Table 4, containing each dataset variation, accuracy, and recall of the method. The best result in the DogFaceNet dataset is

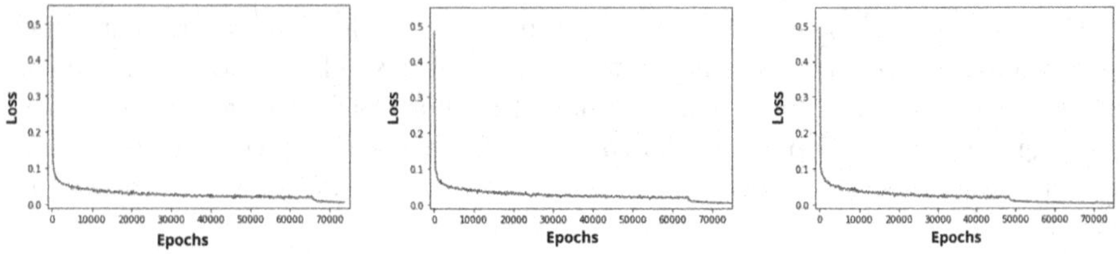

(a) Original. (b) Facial detection. (c) Central crop.

Fig. 6. Loss curves for DogFaceNet.

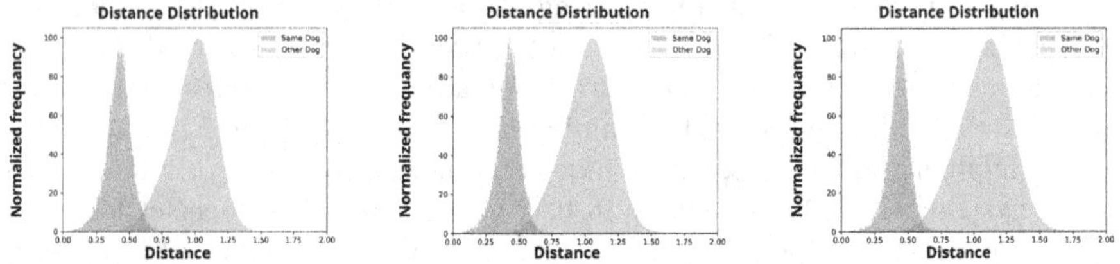

(a) Original. (b) Face detection. (c) Central crop.

Fig. 7. Distance distribution curves for DogFaceNet.

98.43% of accuracy with DogFaceNet with facial detection for training and testing. In the central cropped dataset version, the method achieves 85.64% of recall without preprocessing in training. Furthermore, the best result in the Flickr-dog is 91.53% of accuracy with facial detection in training and testing.

Overall, the facial alignment FD allowed for achieving the best accuracy for both datasets used in this work, which probably makes the network more objective, not paying attention to details that are not part of the dog's face but are in the image.

To compare more fairly, we do not use our results obtained in tests on the FD version with the other works in the literature. We consider that the modification made and the loss of images generate an impact, making the data set different. Using the CC technique with no image loss, we demonstrated better results than all the other works by using FD for training and CC for testing, as shown in Table 5.

7 Conclusion

This work investigated the problem of dog face recognition. We compared three strategies to preprocess the input images: no preprocessing, central crop, and face detection with alignment. We explored the performance of a robust baseline model, ResNet-34, showing that it can be competitive or even outperform the results found in the literature. Moreover, this paper also includes *recall* as an essential performance metric regarding the verification that is missing in most

Table 4. Results of accuracy and recall using ResNet-34 model

Train Dataset	Test Dataset	Accuracy (%)	Recall (%)
DogFaceNet	DogFaceNet	96.8496	74.7830
DogFaceNet CC	DogFaceNet	96.0873	80.5709
DogFaceNet FD	DogFaceNet	93.8303	71.1918
DogFaceNet	DogFaceNet CC	97.2604	**85.6420**
DogFaceNet CC	DogFaceNet CC	97.6937	80.6135
DogFaceNet FD	DogFaceNet CC	98.0256	77.3877
DogFaceNet	DogFaceNet FD	96.3672	76.9618
DogFaceNet CC	DogFaceNet FD	97.6051	80.8224
DogFaceNet FD	DogFaceNet FD	**98.4347**	76.8899
DogFaceNet	Flickr-dog	82.1635	69.9620
DogFaceNet CC	Flickr-dog	72.5916	71.4286
DogFaceNet FD	Flickr-dog	86.5385	55.5556
DogFaceNet	Flickr-dog CC	84.7596	58.9354
DogFaceNet CC	Flickr-dog CC	81.8841	**75.0000**
DogFaceNet FD	Flickr-dog CC	89.4872	73.7374
DogFaceNet	Flickr-dog FD	81.0577	64.2586
DogFaceNet CC	Flickr-dog FD	82.8218	71.4286
DogFaceNet FD	Flickr-dog FD	**91.5385**	74.7475

Table 5. Comparison with other methods

Method	Dataset	Accuracy (%)
Moreira et al. [15]	Flickr-dog	67.60
Batic & Culibrk [2]	Flickr-dog	77.19
Lai et al. [12]	Flickr-dog	84.94
Capone et al. [3]	Flickr-dog	87.00
Proposed method	**Flickr-dog**	**89.48**
Mougeot et al. [16]	DogFaceNet	86.00
Yoon et al. [24]	DogFaceNet	97.33
Proposed method	**DogFaceNet**	**98.02**

related works, especially considering the method's ability not to repudiate the dog.

Experimental results show that face alignment does not play an important role for improving recall in dog face recognition, where the results on datasets with preprocessing were slightly better than the others.

Finally, it is worth emphasizing that the alignment process pays off on Flickr-dog since (1) this strategy improved the state-of-the-art considering both accuracy and recall, and (2) our experiments evaluated the entire Flickr-dog since the model was trained with external data. The latter observation is consistent with the hypothesis found in previous works [23] that augmentation of dog breeds yields better accuracy when the literature focus on classification and lacks a proper analysis for identification.

Future work includes extending the existing dog identification datasets in terms of both new identities and additional annotations, adding support for cat face recognition, investigating effective approaches for animal identification and re-identification, and the assessment of other techniques for the problem of animal biometric recognition.

Acknowledgements. The authors would like to thank The Ceará State Foundation for the Support of Scientific and Technological Development (FUNCAP) for the financial support (6945087/2019).

References

1. Bae, H.B., Pak, D., Lee, S.: Dog nose-print identification using deep neural networks. IEEE Access **9**, 49141–49153 (2021)
2. Batic, D., Culibrk, D.: Identifying individual dogs in social media images. arXiv preprint arXiv:2003.06705 (2020)
3. Capone, V., Figueiredo, C., Valle, E., Andaló, F.: CrowdPet: deep learning applied to the detection of dogs in the wild. Multimedia Tools Appl. **76**(14), 15325–15340 (2017)
4. Chaturvedi, K.: Wolf and dog breed image classification using deep learning techniques. Ph.D. thesis, Dublin, National College of Ireland (2020)
5. Chollet, F.: Xception: deep learning with depthwise separable convolutions. In: Proceedings of the IEEE Conference on Computer Vision and Pattern Recognition, pp. 1251–1258 (2017)
6. He, K., Zhang, X., Ren, S., Sun, J.: Deep residual learning for image recognition. In: Proceedings of the IEEE Conference on Computer Vision and Pattern Recognition, pp. 770–778 (2016)
7. Karasu, S., Alkar, Ö., et al.: A qualitative investigation on the mourning period that occurs after the loss of pets. Veteriner Hekimler Derneği Dergisi/J. Turk. Vet. Med. Soc. **91**(2), 86–97 (2020)
8. Kazemi, V., Sullivan, J.: One millisecond face alignment with an ensemble of regression trees. In: Proceedings of the IEEE Conference on Computer Vision and Pattern Recognition, pp. 1867–1874 (2014)
9. Khosla, A., Jayadevaprakash, N., Yao, B., Li, F.F.: Novel dataset for fine-grained image categorization: Stanford dogs. In: Proceedings of CVPR Workshop on Fine-Grained Visual Categorization (FGVC), vol. 2. Citeseer (2011)
10. King, D.E.: Dlib-ML: a machine learning toolkit. J. Mach. Learn. Res. **10**, 1755–1758 (2009)
11. King, D.E.: Max-margin object detection. CoRR abs/1502.00046 (2015). http://arxiv.org/abs/1502.00046
12. Lai, K., Tu, X., Yanushkevich, S.: Dog identification using soft biometrics and neural networks. In: 2019 International Joint Conference on Neural Networks (IJCNN), pp. 1–8. IEEE (2019)
13. Li, B., Wang, Z., Wu, N., Shi, S., Ma, Q.: Dog nose print matching with dual global descriptor based on contrastive learning. arXiv preprint arXiv:2206.00580 (2022)
14. Liu, J., Kanazawa, A., Jacobs, D., Belhumeur, P.: Dog breed classification using part localization. In: Fitzgibbon, A., Lazebnik, S., Perona, P., Sato, Y., Schmid, C. (eds.) ECCV 2012. LNCS, vol. 7572, pp. 172–185. Springer, Heidelberg (2012). https://doi.org/10.1007/978-3-642-33718-5_13

15. Moreira, T.P., Perez, M.L., de Oliveira Werneck, R., Valle, E.: Where is my puppy? Retrieving lost dogs by facial features. Multimedia Tools Appl. **76**(14), 15325–15340 (2017)
16. Mougeot, G., Li, D., Jia, S.: A deep learning approach for dog face verification and recognition. In: Nayak, A.C., Sharma, A. (eds.) PRICAI 2019. LNCS (LNAI), vol. 11672, pp. 418–430. Springer, Cham (2019). https://doi.org/10.1007/978-3-030-29894-4_34
17. Ouyang, J., He, H., He, Y., Tang, H.: Dog recognition in public places based on convolutional neural network. Int. J. Distrib. Sens. Netw. **15**(5), 1550147719829675 (2019)
18. Simonyan, K., Zisserman, A.: Very deep convolutional networks for large-scale image recognition. arXiv preprint arXiv:1409.1556 (2014)
19. Sundaram, D.M., Loganathan, A.: A new supervised clustering framework using multi discriminative parts and expectation-maximization approach for a fine-grained animal breed classification (SC-MPEM). Neural Process. Lett. **52**(1), 727–766 (2020)
20. Swanson, A., Kosmala, M., Lintott, C., Simpson, R., Smith, A., Packer, C.: Data from: snapshot serengeti, high-frequency annotated camera trap images of 40 mammalian species in an African savanna (2015). https://doi.org/10.5061/dryad.5pt92
21. Tangsripairoj, S., Kittirattanaviwat, P., Koophiran, K., Raksaithong, L.: Bokk meow: a mobile application for finding and tracking pets. In: 2018 15th International Joint Conference on Computer Science and Software Engineering (JCSSE), pp. 1–6. IEEE (2018)
22. Turečková, A., Holík, T., Komínková Oplatková, Z.: Dog face detection using yolo network. In: Mendel. Brno University of Technology (2020)
23. Weerasekara, D., Gamage, M., Kulasooriya, K.: Combined approach of supervised and unsupervised learning for dog face recognition. In: 2021 6th International Conference for Convergence in Technology (I2CT), pp. 1–5. IEEE (2021)
24. Yoon, B., So, H., Rhee, J.: A methodology for utilizing vector space to improve the performance of a dog face identification model. Appl. Sci. **11**(5), 2074 (2021). https://doi.org/10.3390/app11052074

Clinical Oncology Textual Notes Analysis Using Machine Learning and Deep Learning

Diego Pinheiro da Silva[1(✉)], William da Rosa Fröhlich[2], Marco Antonio Schwertner[1], and Sandro José Rigo[1]

[1] Universidade do Vale do Rio dos Sinos, Rio Grande do Sul, Brazil
{diego192,marco.schwertner,rigo}@edu.unisinos.br

[2] Pontifícia Universidade Católica do Rio Grande do Sul, Rio Grande do Sul, Brazil
william.frohlich@edu.pucrs.br

Abstract. Advances in textual classification can foster quality in existing clinical systems. Our research explored experimentally text classification methods applied in non-synthetic oncology clinical notes corpora. The experiments were performed in a dataset with 3,308 medical notes. Experiments evaluated the following machine learning and deep learning classification methods: Multilayer Perceptron Neural network, Logistic Regression, Decision Tree classifier, Random Forest classifier, K-nearest neighbors classifier, and Long-Short Term Memory. An experiment evaluated the influence of the corpora preprocessing step on the results, allowing us to identify that the classifier's mean accuracy was leveraged from 26.1% to 86.7% with the per-clinical-event corpus and 93.9% with the per-patient corpus. The best-performing classifier was the Multilayer Perceptron, which achieved 93.90% accuracy, a Macro F1 score of 93.61%, and a Weighted F1 score of 93.99%.

Keywords: Artificial Intelligence · Natural Language Processing · Oncology · Deep Learning · Machine Learning · Healthcare

1 Introduction

Today's world is moving toward changing healthcare from reactive and hospital-centered to preventive and personalized treatment approaches [1]. Therefore, physicians must be updated with these new treatments and patients' health information. Furthermore, decision-making is affected by a broad range of parameters. For instance, different information sources are used to define oncology diagnosis and treatment options [2]. In this context, the information accumulated in a health institution's database can prove valuable to help physicians decide on the appropriate course of action in specific cases [3]. Nevertheless, it is not feasible for health professionals to manually evaluate and analyze these resources due to the enormous effort it will demand, considering the size of these databases [4], and the exponential growth in medical information [5].

Due to the rapid growth of electronic information, most health information about patients is found as unstructured data, i.e., narrative text or free text [3].

M. C. Naldi and R. A. C. Bianchi (Eds.): BRACIS 2023, LNAI 14196, pp. 140–153, 2023.
https://doi.org/10.1007/978-3-031-45389-2_10

Therefore, the unstructured format of the enormous amount of medical data is the primary motive of recent research initiatives aiming to extract information from medical notes [6–9].

Considering these challenges, tools are needed to support healthcare professionals in their activities beyond a healthcare system's restructuring. Clinical Decision Support Systems (CDSS) provide additional assistance, synthesizing and integrating patient-specific information, performing complex evaluations, and presenting the results to clinicians in an adequate time [10]. In addition, CDSS offers assistance to overcome the difficulties in dealing with massive amounts of information [11–13].

This work focuses on the text classification task, covering the steps to create and process a clinical corpus for the experiments and comparing evaluations with machine learning and deep learning approaches. An oncology clinical notes corpus was created, preprocessed, and transformed to be used by machine learning and deep learning methods. This corpus was obtained from a real-world oncology clinic, de-identified to preserve the patient and professional's identification, and is entirely composed of Brazilian Portuguese language texts. In addition, this work implemented several machine learning and deep learning methods for text classification, compared their performance, and evaluated their results.

Related works about text classification and information extraction methods in oncology clinical notes were studied to understand how to better deal with unstructured data. The papers used one or a combination of natural language processing (NLP), machine learning, and deep learning methods. The related works study indicates text classification and information extraction as essential tasks to deal with unstructured data in healthcare. Several related works used domain-specific corpora. However, none used a corpus with non-synthetic medical notes from the oncology healthcare-specific domain in the Brazilian Portuguese language. Furthermore, most papers used machine learning or deep learning methods to consider the text classification task. Just a few of them presented a comparison between several machine learning and deep learning methods.

Therefore, the main objective of this work consists of applying text classification approaches to support healthcare professionals regarding diagnosis decisions. The survey provided insight into crucial aspects of the area, such as that patients' clinical data are often inputted in a free-text format. There was a lack of structured data repositories on this topic. This topic was described as an essential problem in most studied papers considering that CDSS needs data in a structured format. The corpora creation is an essential element in this research and constitutes one of the contributions of this work because it was created from non-synthetic data with specific preprocessing needs. Other contributions presented in this work are the corpus de-identification and enrichment process, described in the following sections. The evaluation and performance comparison of several machine learning and deep learning classifiers also brings contributions from this work.

2 Material and Methods

This section presents the methodological aspects adopted in this work. This work aims to evaluate text classification techniques to support health professionals' needs regarding diagnosis decisions. Therefore, text classification experiments were conducted using machine learning and deep learning approaches. These techniques were applied in Brazilian Portuguese clinical notes corpora obtained from an EHR system in the oncology domain.

Figure 1 shows the general view containing the elements of the proposed approach for this work. This context is derived from real-world observation, considering actual medical clinics, and expresses health professional needs.

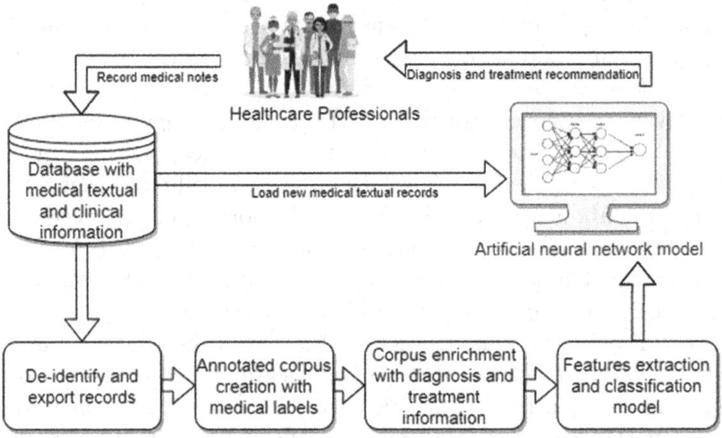

Fig. 1. General view of the approach

The overall process starts with the creation of a record of a medical note by the healthcare professional. This situation generates textual medical records and structured clinical information in the real-world observed cases. In this step, the healthcare professionals use the Oncology EHR system [14] to input their observations about the patient, which are recorded in the system's database. These observations can be composed of free-text and structured data, and both data types are used together to achieve better results.

The upper part of Fig. 1 describes the primary steps for exploring possible answers to the research question in this work. First, developing a flow of operations that starts with the health professional's assistance was evaluated, as identified in the flow "Load new medical textual records" used together with a support service, applying the classification models studied. The application of the models will generate support for determining a response with suggestions for framing and similar contexts, as represented by the "Diagnosis and treatment recommendation" flow that the physician will use. The generation of new information about each patient's clinical event is stored in the Oncology EHR system's database, as indicated in the "Record medical notes" flow.

This initial flow consists of a vision of the future use of the system by physicians and healthcare professionals. The other items in Fig. 1 were studied and experimented.

The first necessary step to generate the corpora and to use the proposed approach in the future was to anonymize and export these records. First, a de-identification process was applied to anonymize the data, preventing the personal identification of each patient and professional. After de-identification, the records were exported, creating the corpus used in text classification approaches. In this step, the textual data was exported along with some structured clinical information.

The corpus was annotated, taking advantage of free-text data labels generated by the medical staff when using the EHR system and stored in the database. Therefore, this corpus could be considered for some of the classification approaches tasks, such as the training tasks.

After the corpus creation and annotation, it was enriched with some structured clinical information. As mentioned previously, the EHR system stores the data as a free-text type or structured type. Some of these structured data are the diagnosis and treatment information used to enrich the corpus.

The final step involves all the preprocessing steps necessary to train the evaluated classification algorithms. Before training the artificial neural network model, the corpus was preprocessed as described in Sect. 2. In this step, the corpus annotations and enriched data were assessed. When necessary, textual features were extracted. After that, the corpus and its features were used as input to train the following Artificial Neural Network (ANN) models: Multilayer Perceptron (MLP) and Long short-term memory (LSTM). Part of the corpus was reserved for testing to evaluate each ANN model's performance.

We developed experiments combining approaches such as corpus format, new corpus with updated information, document or paragraph processing levels, and different machine learning and deep learning classifiers (as described in Sect. 3). The experiments' results are evaluated in Sect. 3.3.

As a result of the work overviewed above, creating a corpus with non-synthetic oncological medical notes and implementing a de-identification and enrichment process of the corpus are highlighted. In addition, the evaluation and performance comparison of a machine learning and a deep learning classifier are also highlighted as contributions of this work.

In the following sections, each component or step involved in the general approach of Fig. 1 is detailed.

The Machine Learning and Deep Learning architectures will be described, along with some parametrization aspects, complementary information on the feature engineering, and the general view of the planned experiments set.

The additional corpora preprocessing, the feature extraction details, and the model specifications are described in the following subsections.

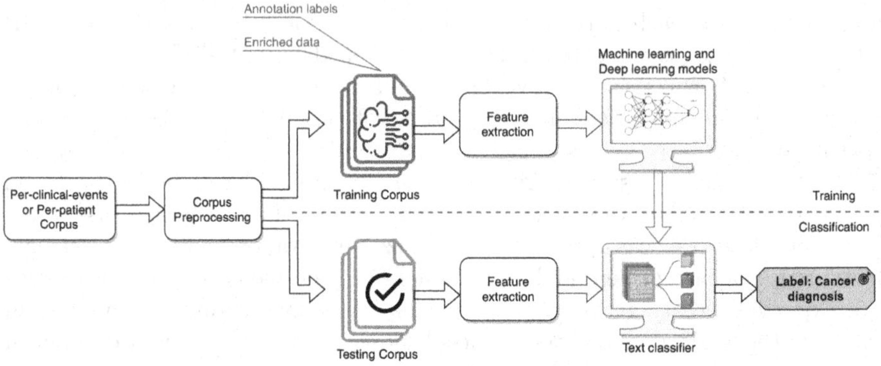

Fig. 2. Overview of the model applied in this research

Corpus Preprocessing. After the corpora creation process, it was necessary to preprocess them to normalize the text. The same preprocessing was performed for both per-clinical-event and per-patient corpora types. The text analysis and preprocessing techniques applied are described below.

C	D	E	F
SENTENCE	SETENCE_CATEGORY	ARGUMENT_2	ARGUMENT_2_CATEGORY
Exame físico normal	Ectoscopia / Pele e anexos	C50.9 - Mama	Diagnostico
Continua Anastrozol, Eligard e Zometa			
Solicito Ex lab e TCs de reestadiamento.	Conduta	C50.9 - Mama	Diagnostico
# Carcinoma ductal invasivo de mama D / Triplo negativo / Pré menopausa			
# 4x AC + 12 T até 23/09/2015			
# RXT 25/11/2015			
# Seguimento 5 meses > Recidiva óssea (CO : lesão em oitavo arco costal D)			
# 17 x Xeloda e Zometa			
# SVCS pelo PC - TEP - Stent com Dr. Luis Otávio em GYN	Histórico Oncológico	C50.9 - Mama	Diagnostico
Melhora dos sintomas de SVCS após retirada de PC e stent			
HB 13,8 L 3600 Plaq 120 mil Cr 0,9	Sintomas	C50.9 - Mama	Diagnostico
Boa tolerância ao Tratamento	Toxicidade	C50.9 - Mama	Diagnostico
Hematoma em tórax a E	Ectoscopia / Pele, anexos/ Mamas	C50.9 - Mama	Diagnostico
Impalpáveis	Linfonodos	C50.9 - Mama	Diagnostico
Ritmo cardíaco regular em dois tempos, BNF s/sopro	Cardiovascular	C50.9 - Mama	Diagnostico
MV + ARA	Respiratório	C50.9 - Mama	Diagnostico
Flácido, indolor, sem visceromegalias	Abdominal	C50.9 - Mama	Diagnostico
sem edema	Membros	C50.9 - Mama	Diagnostico
Ca de mama E T2N1M0 Triplo negativo EIIA R Osso			
em tratamento de primeira linha	Impressão	C50.9 - Mama	Diagnostico
Libero ciclo 18 Xeloda e Zometa	Conduta	C50.9 - Mama	Diagnostico
# Colectomia E por obstrução intestinal _Adenocarcinoma G2, invasão perir			
# 2x Roswell Park	Histórico Oncológico	C18.9 - Cólon	Diagnostico
Insônia	Sintomas	C18.9 - Cólon	Diagnostico
Boa tolerância a Qt	Toxicidade	C18.9 - Cólon	Diagnostico

Fig. 3. A sample of a corpus

Before the text preprocessing, it was possible to observe that a small group of diagnoses concentrates on the most frequent occurrences. Hence the diagnoses with less than 50 occurrences were joined into a single group called "Outros" ("Others"). Furthermore, to evaluate the neural network's performance according to the dataset sparsity, a new version of the dataset was created with the 12 most occurring diagnoses.

The following tasks were performed:

- Tokenization: split the text into tokens that correspond to words;
- Stop-words filtering: removal of most common words in the Brazilian Portuguese language, punctuation, and special characters;
- Case folding: conversion all words to lowercase.

An additional manual analysis was done on the corpus in the per-clinical-events corpora. The text was assessed to understand how it could be transformed to improve ANN algorithms. Repeated medical notes were removed from several annotation labels. These notes could weaken the representation strength of the corpus.

A complementary experiment was performed to evaluate how this step leveraged the results of the classifiers. As described in Sect. 3.1, a significant improvement was achieved by applying this step.

Feature Extraction. The text from medical notes must be transformed into a structure that the classifiers could use. For that reason, the Bag-of-Words (BoW) method was used, which is a representation that turns arbitrary text into fixed-length vectors by counting how many times each word appears. This representation is useful to be used by the classifier algorithms.

This work used the medical notes bag-of-words (BoW) to extract the features to be used in machine learning and deep learning training. In the per-clinical-event corpora, the BoW was generated for each medical note. Likewise, the per-patient corpora were generated for each patient with all their medical notes.

Before creating the medical notes BoW, the text was normalized as described in the preprocessing Sect. 2. This preprocessing step aims to reduce the number of useless words, special characters, and punctuations, which would not make a difference in the classifier model's training. It also helped to reduce the computational effort to create the BoW.

The BoW applied to the medical notes resulted in a sparse representation, i.e., the vector sequence of numbers representing each word contained too many zeros. The Principal Component Analysis (PCA) technique was applied to reduce the data sparsity. The PCA technique converts a set of observations of possibly correlated features into values of linearly uncorrelated features. The PCA with 500 features was used.

Machine Learning and Deep Learning Architectures. This work applied machine learning and deep learning methods for the text classification task, comparing their results. Several machine learning classification algorithms were applied to evaluate which one had the best performance. Furthermore, an LSTM deep learning algorithm was also applied to compare traditional machine learning and a deep learning recurrent neural network.

The following Machine Learning algorithms were evaluated: Multilayer Perceptron (MLP) neural network [15]; Logistic Regression [16]; Decision Tree classifier [17]; Random Forest classifier [18]; K-Nearest Neighbors (KNN) classifier [19]. Furthermore, a Long-short Term Memory (LSTM) deep learning experiment was also performed. The machine learning classification and the deep learning architecture used are described, regarding their algorithms and their theoretical background, in the following papers [20–22].

The datasets were divided into two groups: one with 80% of the data to train the models and 20% of the data to test the models. The data were shuffled to keep the categories' proportion, and then they were divided as aforementioned.

The machine learning algorithms were implemented using scikit-learn[1], and the deep learning LSTM was performed using Keras library[2], both were implemented in Python. In the first set of experiments, seven tests were performed, with the following architecture details: an MLP with one hidden layer with 500 neurons; an MLP with two hidden layers with 800 and 500 neurons; a Logistic Regression classifier; a Decision Tree with a maximum of twenty levels and three samples by leaf; a Random Forest with a maximum of twenty levels and three samples by leaf; an Extra Trees with a maximum of twenty levels and three samples by leaf; a KNN classifier with a unitary K.

The second set of experiments was performed this time with the 12 most occurring diagnoses in the dataset. For this experiment, the best-performing machine learning was selected to compare with an LSTM deep learning recurrent neural network. The machine learning algorithm had the following architecture: an MLP with two hidden layers with 800 and 500 neurons. The deep learning algorithm had the following architecture: an LSTM with the library Keras built on top of Tensorflow in a python implementation, in which the parametrization used was composed of Batch size 128, the Dropout rate of 0.2, validation split of 0.2, Optimizer with adam, Loss measure was categorical cross-entropy. Also, to prevent overfitting, EarlyStopping was used. Standard values for the parameters were used in these experiments, as described in the literature. All the described models were evaluated using standard metrics indicated in the literature, such as accuracy and macro and weighted F1 scores.

3 Experiments and Results

Several machine learning classifiers and a deep learning recurrent neural network were applied in this work's experiments. The main objective of these experiments was to address the main research question and identify a possible workflow to use the dataset and text classification algorithms to evaluate potential support for healthcare professionals.

The dataset used was composed of an arrangement of the available options, using the per-clinical-event and the per-patient versions. The complete dataset with several machine learning text classification algorithms was used in the first step. Both datasets (per clinical event and per-patient) were used in this experiment. In the second step, a new experiment was carried out involving the per-patient dataset and the algorithms MLP and LSTM. The per-patient dataset was chosen in the second step because all patient's clinical notes were joined

[1] https://scikit-learn.org/.
[2] https://keras.io/.

into a single record, which would perform better considering the LSTM's ability to process entire sequences of data. All methods were executed on a partition, then repartitioned and reran all methods to get the mean Accuracy.

Therefore, two main experiments were performed: a) Machine learning - several machine learning classifiers have been experimented with and their performance compared (described in Sect. 3.1); b) Deep learning - an experiment with a deep learning recurrent neural network was performed (described in Sect. 3.2).

To measure performance in the experiments, we use different metrics in this study: Accuracy, macro, and weighted F1 score [20]. The predicted output as True Positive (TP) indicates text classified as correct, True Negative (TN) when classified incorrectly. False Positive (FP) if a text correctly indicates not belonging to the class. Similarly, a False Negative (FN) is the text is classified incorrectly.

The Accuracy describes the overall performance of the classifier and mathematical Accuracy expressed as below in Eq. (1):

$$\text{Accuracy} = \frac{TP+TN}{TP+TN+FP+FN} \tag{1}$$

The F1-score It is a harmonic measure between precision and sensitivity, expressed in the Eq. (2):

$$\text{F1-score} = \frac{2*TP}{2*TP+FN+FP} \tag{2}$$

The Macro F1-score is defined as the mean of class-wise/label-wise F1-scores, in Eq. (3), where i is the class/label index and N the number of classes/labels:

$$\text{Macro F1-score} = \frac{1}{N}\sum_{i=0}^{N} \text{F1-score}_i \tag{3}$$

In weighted-average F1-score, we weight the F1-score of each class by the number of samples from that class, such as in Eq. (4), where i is the class index and N the number of classes and S the number of elements in the class.

$$\text{Weighted F1-score} = \frac{1}{N}\sum_{i=0}^{N} \text{F1-score}_i * S_i \tag{4}$$

3.1 Machine Learning Experiments

The per-clinical-event corpus (described in Sect. 2) of the smallest clinic in the preprocessed and raw versions were used to perform the classifiers. This is the clinic database and contains 3.308 clinical notes, and 397 distinct patients. The preprocessed dataset was used first with the machine learning classifiers, described in Sect. 2.

To perform the experiments, the dataset was randomly divided into two parts: 80% for training and 20% for testing. A shuffle method was used to generate these two parts, and a different set of training and testing datasets were created each time it was performed. Hence different classifiers' metrics were obtained, but it always kept the performance order.

Table 1. Machine learning classifiers experiments results.

Method	Mean Accuracy	Macro F1 score	Weight F1 score
MLP 1	84.89%	84.21%	84.99%
MLP 2	87.62%	87.44%	87.70%
Logistic regression	84.89%	82.75%	84.75%
Decision tree	71.86%	63.95%	71.98%
Random forest	80.23%	76.09%	79.53%
Extra trees	78.46%	76.71%	78.03%
KNN classifier	85.05%	83.93%	85.20%

The mean accuracy, Macro F1 score, and Weighted F1 score of each classifier are presented in Table 1. These experiments were performed to evaluate which machine learning classifier had the best performance. According to Table 1, the MLP 2 classifier achieved the best accuracy, Macro F1, and Weighted F1 scores. These results are evaluated in Sect. 3.3.

An additional experiment evaluated how the dataset's structure and preprocessing step (Sect. 2) leveraged the classifiers' performance. This experiment used the same dataset from the Clinic with the smalest dataset. The preprocessed and raw versions of the per-clinical-event dataset, plus the preprocessed per-patient dataset, were used with the best performance classifier. According to Table 1, the MLP 2 classifier had the best performance and was used in this experiment.

Table 2 presents the mean accuracy of the MLP 2 classifier with preprocessed and raw versions of the per-clinical-event dataset, plus the preprocessed per-patient dataset.

Table 2. Comparison of the MLP 2 classifier's performance with the per-clinical-event dataset in raw and preprocessed versions, plus the per-patient preprocessed dataset.

Dataset	Mean accuracy
Per-clinical-event raw dataset	26.1%
Per-clinical-event preprocessed dataset	86.7%
Per-patient preprocessed dataset	93.9%

These results are evaluated in Sect. 3.3, exploring the improvements obtained with the integration of the clinical events in a more complete view of the patient history. As it can be observed in Table 2, the mean accuracy improves with more complete patient data.

3.2 Deep Learning Experiments

In this set of experiments, the following machine learning and deep learning classifiers were tested:

- The machine learning classifier that best performed (the MLP 2), according to Sect. 3.1;
- An LSTM (Long-short term memory) deep learning recurrent neural network.

The per-patient corpus (described in Sect. 2) of the smallest clinic in the preprocessed version was used to perform the classifiers. The per-patient corpus was chosen because all patient's clinical notes were joined into a single registry, which would perform better considering the LSTM ability to process entire sequences of data.

Table 3. MLP 2 and LSTM classifiers performed with its performance.

Method	Mean Accuracy	Macro F1 score	Weighted F1 score
MLP 2	93.90%	93.61%	93.99%
LSTM	84.81%	84.57%	84.93%

Table 3 presents the mean accuracy, Macro F1 score and Weighted F1 score of the MLP 2 and the LSTM classifiers. These results are evaluated in Sect. 3.3.

3.3 Results Evaluation

Several experiments were performed to understand the behavior of the selected machine learning and deep learning classifiers with the corpora created and preprocessed (Sect. 2).

First, a set of experiments with seven machine learning classifiers were performed with the per-clinical-event corpus, according to Sect. 3.1. Considering the mean accuracy, Macro F1, and Weighted F1 scores, the classifier that best performed was the MLP 2, as seen in Fig. 4.

Preparing the corpus to be used in machine learning and deep learning classifiers was an important step. An experiment was performed with the MLP 2 classifier, the preprocessed and raw versions of the per-clinical-event dataset, and the preprocessed per-patient dataset (as described in Sect. 2). The performance of each is presented in Table 2. This table shows a significant improvement in the MLP 2 classifier performance with the preprocessing of the dataset (Sect. 2), in both per-clinical-event and per-patient datasets. Furthermore, the per-patient corpus performed better than the per-clinical-events corpus. For that reason, the next experiments used the per-patient corpus.

After the evaluation of the machine learning classifier that best performed, the MLP 2 classifier was selected. A new experiment was performed, to compare the MLP 2 classifier with an LSTM recurrence neural network, using the preprocessed per-patient corpus (as described in Sect. 3.2). Figure 5 shows that the MLP 2 performed better than the LSTM classifier, even though the latter is a more recent neural network. This result can be associated with the fact that this experiment used the smallest per-patient corpus, and deep learning algorithms perform better on large datasets.

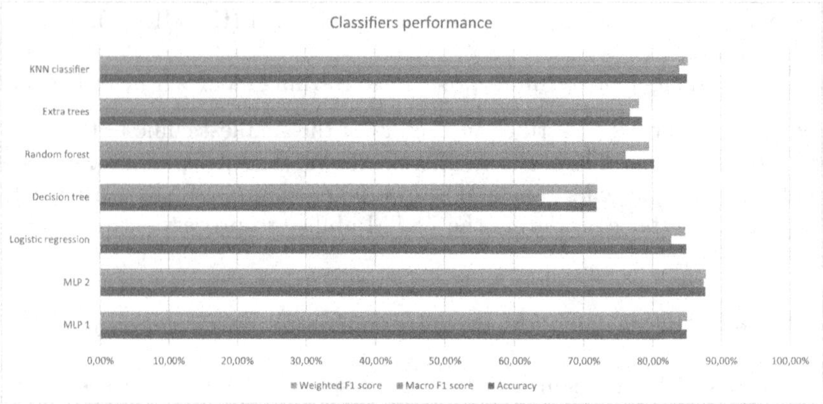

Fig. 4. Machine learning classifiers' performance chart.

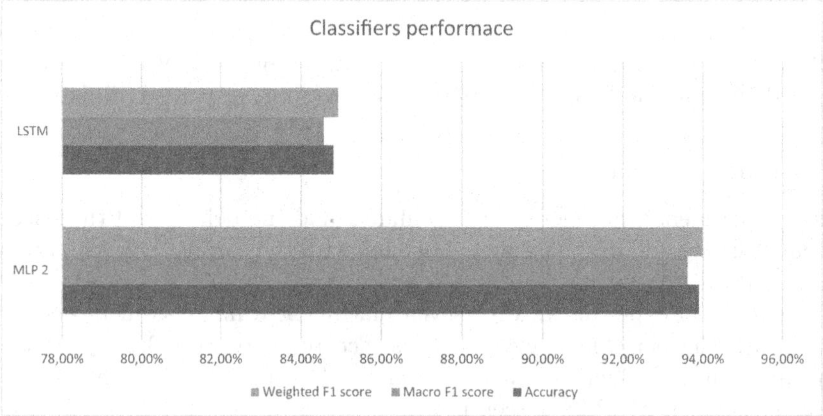

Fig. 5. The performance chart of the best machine learning classifier and deep learning classifier.

4 Conclusion

With a large amount of information generated daily in healthcare, it is unsuitable for humans to process it manually. Furthermore, much of this information is recorded as unstructured data, transforming it into a hard task. It is necessary to develop tools to help healthcare professionals to deal with it, automate the classification and extraction of information from medical notes, and enable this information to be processed by machines.

This work created the corpus based on non-synthetic oncological medical notes from an Oncology EHR system database. This corpus creation process allowed the identification of various preprocessing steps (as described in Sect. 2), and specific term treatment to improve overall results. The experiments achieved good accuracy, especially MLP machine learning and LSTM deep learning methods, showing possibilities of using these resources for medical notes text classification. Therefore, this set of tasks provided strong support indicating resources and processes to text classification in the specific context of health professional support.

Artificial Intelligence, special machine learning, and deep learning algorithms have been widely applied in several industries, sometimes surpassing human accuracy. In the healthcare industry, several processes can be improved by AI, leveraging healthcare professionals. In the oncology area, diagnosis and treatment decision-making is one of these complex processes that AI algorithms can aid.

Considering this work's development, the following list of future steps is suggested:

- Create larger corpus with the medical notes from several oncology clinic databases and conduct new experiments;
- Enhance the corpus preprocessing step by removing low-frequency words, spell checker, replacing acronyms and abbreviations by its standard word;
- Create a domain-specific word embedding from the corpora and apply it with the classifiers. Another option is to use general-purpose word embeddings and fine-tune it with the corpora;
- Improve the enrichment process with more structured data available from the Oncology EHR system, such as prescribed medications, patient's problems, and allergies;
- Tune the implemented classifiers and try different versions, such as a Bidirectional LSTM (Bi-LSTM);
- Integrate the implemented classifiers with the Oncology EHR system to obtain feedback from the healthcare professionals about their accuracy, and suggest the diagnosis based on the patient's clinical history.

References

1. Sabra, S., Alobaidi, M., Malik, K.M., Sabeeh, V.: Performance evaluation for semantic-based risk factors extraction from clinical narratives. In: IEEE 8th Annual Computing and Communication Workshop and Conference (CCWC). Las Vegas, NV, USA, vol. 2018, pp. 695–701 (2018). https://doi.org/10.1109/CCWC.2018.8301742
2. Glatzer, M., Panje, C.M., Sirén, C., Cihoric, N., Putora, P.M.: Decision making criteria in oncology. Oncology **98**(6), 370–378 (2020). Epub 2018 Sep 18. PMID: 30227426. https://doi.org/10.1159/000492272
3. Reyes-Ortiz, J.A., González-Beltrán, B.A., Gallardo-López, L.: Clinical decision support systems: a survey of NLP-based approaches from unstructured data. In: 26th International Workshop on Database and Expert Systems Applications (DEXA). Valencia, Spain vol. 2015, pp. 163–167 (2015). https://doi.org/10.1109/DEXA.2015.47

4. Alemzadeh, H., Devarakonda, M.: An NLP-based cognitive system for disease status identification in electronic health records. In: 2017 IEEE EMBS International Conference on Biomedical & Health Informatics (BHI), Orlando, FL, USA, pp. 89–92 (2017). https://doi.org/10.1109/BHI.2017.7897212
5. Meskó, B.: The guide to the future of medicine: technology and the human touch. In: Webicina KFT (2014)
6. Zhang, R., Ma, S., Shanahan, L., Munroe, J., Horn, S., Speedie, S.: Automatic methods to extract New York heart association classification from clinical notes. In: 2017 IEEE International Conference on Bioinformatics and Biomedicine (BIBM), Kansas City, MO, USA, pp. 1296–1299 (2017). https://doi.org/10.1109/BIBM.2017.8217848
7. Chen, X., Xie, H., Wang, F., et al.: A bibliometric analysis of natural language processing in medical research. BMC Med. Inform. Decis. Mak. **18**(Suppl 1), 14 (2018). https://doi.org/10.1186/s12911-018-0594-x
8. Shickel, B., et al.: Deep EHR: a survey of recent advances in deep learning techniques for electronic health record (EHR) analysis. IEEE J. Biomed. Health Inform. **22**(5), 1589–1604 (2017)
9. Kreimeyer, K., et al.: Natural language processing systems for capturing and standardizing unstructured clinical information: a systematic review. J. Biomed. Inform. **73**, 14–29 (2017). Epub 2017 Jul 17. PMID: 28729030; PMCID: PMC6864736. https://doi.org/10.1016/j.jbi.2017.07.012
10. Hunt, D.L., Haynes, R.B., Hanna, S.E., Smith, K.: Effects of computer-based clinical decision support systems on physician performance and patient outcomes: a systematic review. JAMA **280**(15), 1339–1346 (1998). PMID: 9794315. https://doi.org/10.1001/jama.280.15.1339
11. Bucur, A., van Leeuwen, J., Cirstea, T.C., Graf, N.: Clinical decision support framework for validation of multiscale models and personalization of treatment in oncology. In: 13th IEEE International Conference on BioInformatics and BioEngineering, Chania, Greece, pp. 1–4 (2013). https://doi.org/10.1109/BIBE.2013.6701695
12. Polpinij, J.: The cancerology ontology: designed to support the search of evidence-based oncology from biomedical literatures. In: 24th International Symposium on Computer-Based Medical Systems (CBMS). Bristol, UK, pp. 1–6 (2011). https://doi.org/10.1109/CBMS.2011.5999168
13. Wang, Y., et al.: Clinical information extraction applications: a literature review. J. Biomed. Inform. **77**, 34–49 (2018). ISSN 1532-0464. https://doi.org/10.1016/j.jbi.2017.11.011
14. InterProcess: InterProcess Gemed Oncology - Oncological management system (2019). www.interprocess.com.br/en/gemed-oncology/
15. Naraei, P., Abhari, A., Sadeghian, A.: Application of multilayer perceptron neural networks and support vector machines in classification of healthcare data. In: Future Technologies Conference (FTC). San Francisco, CA, USA, vol. 2016, pp. 848–852 (2016). https://doi.org/10.1109/FTC.2016.7821702
16. Lemon, S.C., Roy, J., Clark, M.A., Friedmann, P.D., Rakowski, W.: Classification and regression tree analysis in public health: methodological review and comparison with logistic regression. Ann. Behav. Med. **26**(3), 172–181 (2003). PMID: 14644693. https://doi.org/10.1207/S15324796ABM2603_02
17. Lavanya, D., Rani, K.U.: Ensemble decision tree classifier for breast cancer data. Int. J. Inf. Technol. Convergence Serv., **2**(1), 17 (2012)

18. DuBrava, S., et al.: Using random forest models to identify correlates of a diabetic peripheral neuropathy diagnosis from electronic health record data. Pain Med. **18**(1), 107–115 (2017). PMID: 27252307. https://doi.org/10.1093/pm/pnw096
19. Tayeb, S., et al.: Toward predicting medical conditions using K-nearest neighbors. In: 2017 IEEE International Conference on Big Data (Big Data), Boston, MA, USA, pp. 3897–3903 (2017). https://doi.org/10.1109/BigData.2017.8258395
20. ul Haq, A., Li, J.P., Memon, M.H., Nazir, S., Sun, R.: A hybrid intelligent system framework for the prediction of heart disease using machine learning algorithms. Hindawi Mobile Inf. Syst. (2018). https://doi.org/10.1155/2018/3860146
21. Haq, A.U., et al.: Intelligent machine learning approach for effective recognition of diabetes in E-Healthcare using clinical data. Sensors **20**, 2649 (2020). https://doi.org/10.3390/s20092649
22. Tai, K.S., Socher, R., Manning, C.D.: Language processing, improved semantic representations from tree structured long short-term memory networks. In: Proceedings of the 53rd Annual Meeting of the Association for Computational Linguistics and the 7th International Joint Conference on Natural Language Processing of the Asian Federation of Natural, pp. 1556–1566 (2015)

EfficientDeepLab for Automated Trachea Segmentation on Medical Images

Arthur Guilherme Santos Fernandes(✉), Geraldo Braz Junior, João Otávio Bandeira Diniz, Aristófanes Correa Silva, and Caio Eduardo Falcõ Matos

Applied Computing Group, Federal University of Maranhão, Av. Portugueses 1996, São Luís, MA, Brazil
{arthurgsf,geraldo,joao.bandeira,ari,caioefalcao}@nca.ufma.br

Abstract. Segmentation of Organs at Risk is a fundamental step during radiotherapy planning for cancer treatment. Its goal is to preserve healthy tissue around the tumor and ensure that the most radiation strikes only cancer cells. Physicians do this job manually, which can be slow and error-prone. Thus, automatic segmentation methodologies can speed up organ delimiting during radiotherapy planning. This work designs a method, EfficientDeepLab, a convolutional neural network architecture trained on CT scans for trachea segmentation, and obtained an 88.6% dice score.

Keywords: Deep Learning · Segmentation · Organs at Risk · Radiotherapy · Convolutional Neural Networks

1 Introduction

Radiotherapy is one type of cancer treatment consisting of blasting high radiation doses into unhealthy cells to destroy or shrink tumors [2]. Identifying the tumor location and the surrounding organs is the first step in radiotherapy planning and helps diminish the ionizing radiation effects on healthy tissue. We call these healthy organs that surround a tumor Organs at Risk **(OaR)** [8,18]. A physician does the manual segmentation of the OaR using images obtained by Computed Tomography **(CT)**.

CT is an imaging exam that allows doctors to evaluate the body's soft and hard tissue health. An x-ray emitting machine spins around the patient while sensors at the other end capture the radiation that trespasses the patient and convert it to a digital image [1,14]. The result is a 3D image that can be sliced down to 2D for better visualization [10]. Figure 1 shows examples of CT slices.

Accurate manual segmentation of OaR is crucial because it provides the exact location of healthy tissue near the tumor so that specialists can prevent unnecessary X-ray exposition and minimize the side effects of radiation (inflammation, fibrosis, ulceration, and even organ failure [21]). However, it depends on the physician's expertise and can be a slow and error-prone process. Smaller

M. C. Naldi and R. A. C. Bianchi (Eds.): BRACIS 2023, LNAI 14196, pp. 154–166, 2023.
https://doi.org/10.1007/978-3-031-45389-2_11

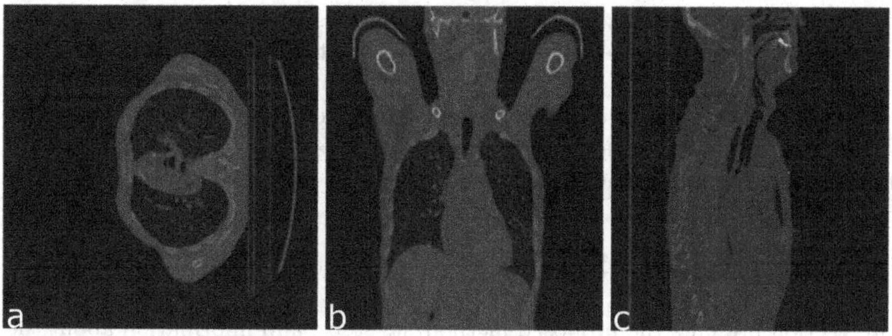

Fig. 1. 2D slices from a CT test. a) Axial View; b) Coronal View; c) Sagital View.

organs like the trachea are hard to segment due to size, texture similarity with surrounding tissue, and lack of detail richness. Because of that, recent research focused on the automatic segmentation of Organs at Risk.

Based on the importance of trachea segmentation for radiotherapy planning and current literature on automatic medical image segmentation, this work proposes utilizing image pre-processing techniques and proposes an Efficient-DeepLab convolutional network architecture for trachea segmentation from computed tomography images. General contributions are: (a) pixel intensity windowing for better contrast on trachea images, and (b) a method that merges EfficientNet and DeepLabV3+ architectures to perform trachea segmentation, increasing the convolutions field-of-view by using atrous convolutions, leading to more detailed segmentation, considerably less network parameters and efficient utilization of computational resources.

2 Related Work

Some works develop methods for trachea segmentation using fully convolutional networks or combining transformers with these networks.

Driven by recent advances in transformer architecture for image segmentation, [24] used it alongside U-Net to perform segmentation on various OaRs, including the trachea.

Some methods use the 3D nature of tomography to extract contextual information about the CT scan, like in [22] and [23].

Other methods combine U-Net and another convolutional network architecture to perform OaR segmentation: [9] used pyramid fusion module to merge features from different scales on the U-Net and [25] applied Generative Adversarial Networks and U-Net together to segment organs from CT images.

There are also some efforts to build public datasets for OaR segmentation. In [17], researchers gathered CT scans from 60 patients diagnosed with lung cancer to train a U-NET with 2D input. Of those 60 CT scans, 40 are publicly available, composing the SEGTHOR dataset.

The literature presents efforts based on methods such as U-NET, but without necessarily worrying about the efficiency of networks in terms of trainable parameters. Starting from this point, we propose a method where the network backbone is built using EfficientNet and network decoder/encoder with DeepLab. The aim is to build a model with few trainable parameters, easily applicable in small datasets, and presenting consistent results.

3 Materials and Methods

This section presents the proposed method, its steps, and information about the image dataset used for training. Figure 2 illustrates the general pipeline of this work, beginning from the image acquisition, passing to the preprocessing work, EfficientDeepLab model training, and segmentation mask output.

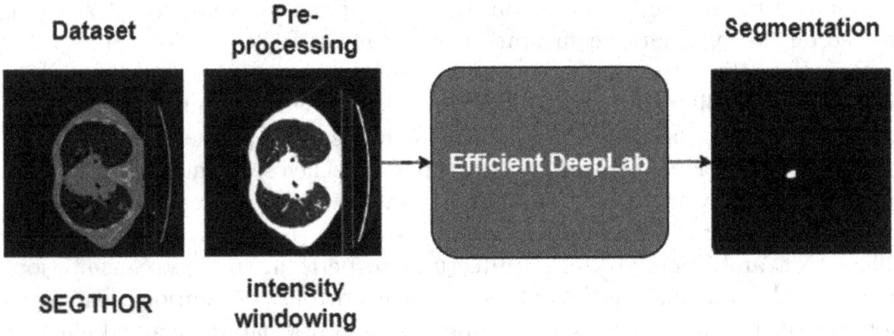

Fig. 2. General workflow of the method.

3.1 Dataset

The images in the experiment come from the SEGTHOR database [17]. There are 40 publicly available 3D CT volumes from patients affected by lung cancer. The volumes were sliced along the longitudinal body axis to generate 7420 2D images. From the total, 1987 shows trachea and the other 5433 don't. This distribution generates an unbalanced dataset, so, for training purposes, 1987 images without trachea were randomly selected, resulting in equally distributed slices.

The data also includes a physician-made manual segmentation for ground truth. Figure 3 shows a trachea region slice and the correspondent demarcation.

The pixel intensity values on a CT image range from -1000 to 1000, according to the Hounsfield scale [11], which is a way to represent tissue density in radiography images (the denser the tissue, the brighter the image). Table 1 summarizes the Hounsfield scale.

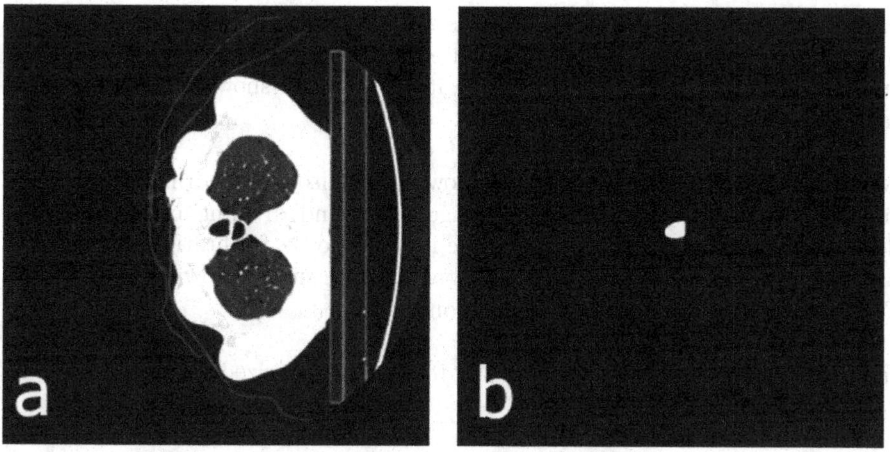

Fig. 3. a) Axial view of trachea slice; b) Manual segmentation made by radiology professional.

Table 1. Hounsfield intensity values by kind of biological tissue.

Substance	Hounsfield Units (HU)
Air	−1000
Fat	−100
water	0
Cerebrospinal Fluid	15
White Matter	20–30
Gray Matter	36–46
Muscle	50
Blood	56–76
Bone	1000

Pre-processing: For better contrast, we applied an intensity windowing method, limiting the pixel values to an interval, thus discarding unwanted information, after this process, a normalization followed by image resizing were applied. In the training phase, a random data augmentation technique was applied.

Intensity Windowing: This technique is employed within the radiology domain to improve CT imaging. Its usage varies depending on the application, allowing for enhanced visualization of particular body tissues. Let α and β represent the lower and upper boundaries of the interval $[\alpha, \beta]$, respectively. The intensity windowing technique involves clipping the pixel values outside this interval to the closest boundary value while letting pixels within the interval remain unchanged. These intervals are called windows, and each range emphasizes different kinds of tissues from the body. The trachea is a hollow tubular organ, so the chosen intensity interval was [−1000, 60] because it enhances the contrast between the trachea muscle walls and the air inside it [6]. Figure 4 shows different window ranges applied to the same image.

Normalization: After the intensity windowing, we also apply normalization, subtracting the pixel values from the image mean and dividing by its standard deviation. This way, the data tend to be more uniform, facilitating the learning process for the model. Equation 1 describes the intensity windowing process, and Eqs. 2 to 4 detail the image normalization method.

Resizing: After the normalization, the images were resized from 512×512 to 256×256.

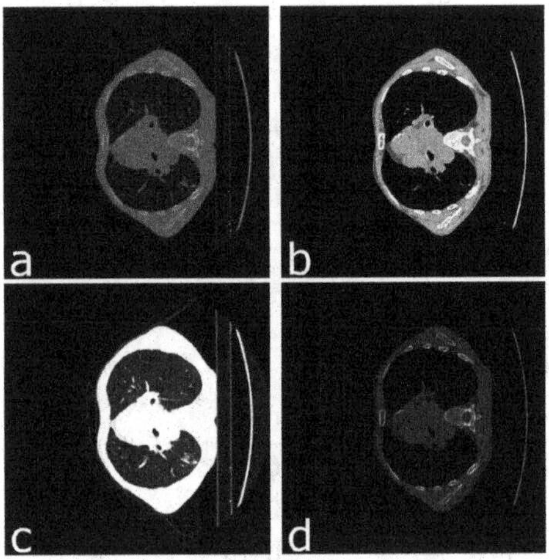

Fig. 4. Examples of different intensity windowing preprocessing applied to CT images.

$$f(x) = \begin{cases} -1000, & \text{if } x < -1000 \\ 60, & \text{if } x > 60 \\ x, & \text{if } x \in [-1000, 60] \end{cases} \tag{1}$$

$$\mu = \frac{\sum_i^n I_i}{n} \tag{2}$$

$$\sigma = \sqrt{\frac{\sum_i^n (I_i - \mu)^2}{N}} \tag{3}$$

$$f_{norm}(I) = \frac{I - \mu}{\sigma} \tag{4}$$

Data Augmentation: This process is executed to insert variance into the training data and make the model more robust and generalist. A random rotation is set to each image, ranging from −45 to 45° with a chance of 50%.

3.2 Model Architecture

For semantic segmentation of the trachea area, we apply a neural network architecture based on DeepLabV3+ [5] combined with EfficientNet [20] for feature extraction. Figure 5 illustrates the encoder/decoder model architecture overview.

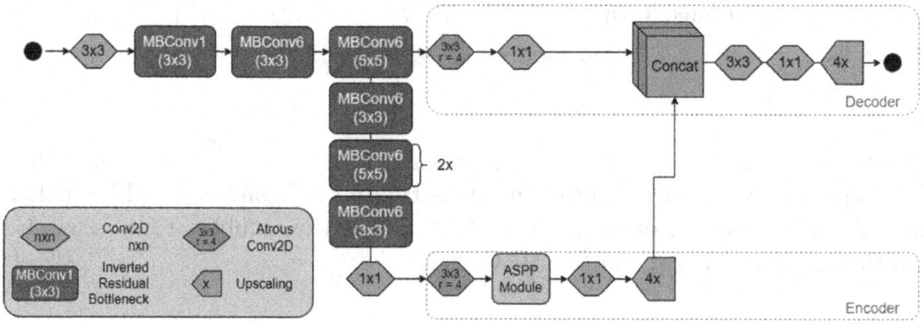

Fig. 5. Architectural overview of EfficientDeeplab.

The main DeepLabV3+ innovations are the atrous convolutions (also called dilatable convolutions) and atrous spatial pyramid pooling [12]. This way, we are capable of merging features from multiple scales. In the original DeepLabV3+, the author uses ResNet [13] architecture as a feature extraction backbone, aggregating intermediate layer feature maps, summing them, and interpolating back to the input resolution, thus obtaining a segmentation output mask. In this work, we modified the feature extraction backbone to use EfficientNet, a scalable, robust, and compact deep learning model that surpasses ResNet in recent benchmarks on ImageNet [7] dataset.

EfficientNet: Built with efficient resource usage in mind, it consists of a family of networks derived from a baseline model obtained with Neural Architecture Search. The baseline network is called the b0, the other models range from b1 to b7, and the differences are the input resolution, the number of channels, and layers in each stage, such that the larger the number on the network's name, the bigger the model (in terms of parameters). The b0 model's main structure is summarized in Table 2.

The main building blocks of the EfficientNet are the MBConv blocks, introduced in MobileNetV2 [19]. They are an inverted residual bottleneck, expanding the number of channels in the intermediate layers and lately squeezing them to their original size. Figure 6 illustrates the structure of the MBConv block.

Table 2. EfficientNet B0 architecture [20].

Stage	Operator	Resolution	#Channels	#Layers
1	Conv 3×3	224×224	32	1
2	MBConv1 3×3	112×112	16	1
3	MBConv6 3×3	112×112	24	2
4	MBConv6 5×5	56×56	40	2
5	MBConv6 3×3	28×28	80	3
6	MBConv6 5×5	14×14	112	3
7	MBConv6 5×5	14×14	192	4
8	MBConv6 3×3	7×7	320	1
9	Conv 1×1 & Pooling & FC	7×7	1280	1

The number that follows the name in MBConv1 and MBConv6 is a multiplier that indicates the number of filters in the leftmost 1×1 convolutional layer. For example, if the number of input channels is 4, then the number of filters will be 4 in the MBConv1 and 24 in the MBConv6.

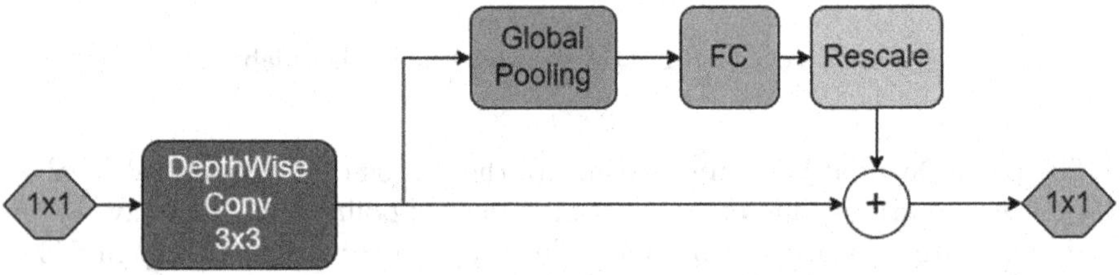

Fig. 6. Schematic diagram of MBConv.

Two outputs from the EfficientNet model are used: the last convolutional layer activations and the intermediate features from the third convolutional block. The first one feeds the encoder part, while the latter feeds the decoder part.

Atrous Convolution: Introduced by DeepLabV1 [3], it is a convolution with dilatable kernels, and its goal is to achieve a larger field of view on feature extraction. The filter is filled with zeros according to a chosen dilation rate r. Figure 7 exemplifies the dilation rate effect over the convolution kernel. Notice that the area covered by the filter goes from 3×3 to 5×5, but the number of parameters is still the same. Varying the dilation rate along the network convolutions helps the model to capture features on different scales. DeeplabV3+ also implements an atrous version of the Depthwise Separable Convolution, where the Depthwise convolution also has a dilation in the kernel.

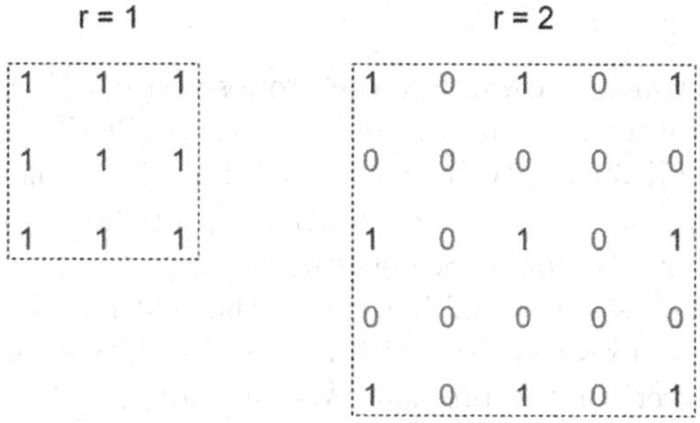

Fig. 7. Atrous Convolution kernel with different dilation rates.

Atrous Spatial Pyramid Pooling: Inspired by the Spatial Pyramid Pooling (SPP) proposed in [12], the authors of DeepLab introduced the atrous version of it in [4]. The method encodes features from different resolutions into fixed-size feature maps and concatenates them together to feed the rest of the network, as illustrated in Fig. 8.

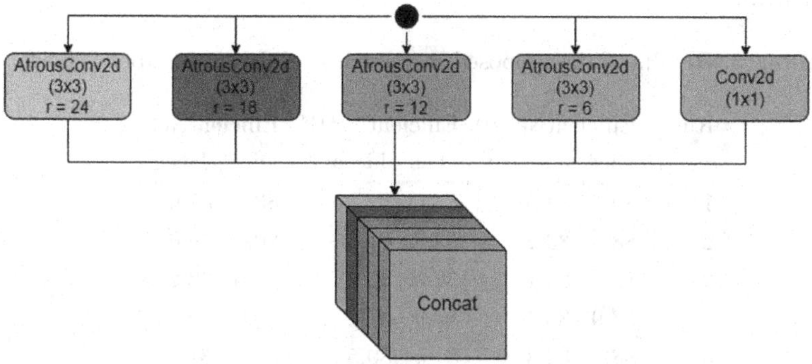

Fig. 8. Architectural overview of Atrous Spatial Pyramid Pooling Module.

Encoder/Decoder: Following the recipe of the most successful methods for semantic image segmentation, this work uses the encoder/decoder structure for extracting features and restoring them to the original resolution. The EfficientNet model serves as the feature extraction backbone alongside the ASPP module, and both compose the encoder part of the model. The decoder inputs are the features from EfficientNet's third layer. They are further convoluted by one atrous separable convolution and one 1×1 convolution layer. The result is concatenated with the encoder output to get convolved and upsampled to the original image resolution, outputting an image segmentation mask.

3.3 Results

We conducted experiments to validate the proposed method. The training environment was a computer with Intel Core ™ i5-10400F CPU @ 2.90 GHz and one Nvidia ® GPU RTX-3060 12 GB of memory. For the implementation of the model, we used Python programming language and the libraries: Tensorflow, Keras, and Tensorflow Advanced Segmentation Models [15].

For a better understanding of how large the feature extraction backbone model needs to be, we experimented with EffcientNet b0, b1, and b2, estimating which one suits better for the problem. We did not use the models above b2 because the dataset is not big enough to supply those models.

The pre-processed SEGTHOR 2D slices were split 70% for training and 30% for the test, per patient. It is important to emphasize that the results use at least 30% fewer data than the works on the SEGTHOR dataset due to the unavailability of 20 test case patients from the original 60 patients on the dataset.

The training was executed for 100 epochs and used ADAM (Adaptive Moment Estimation) to update the network weights [16]. The loss function used is Binary Cross-Entropy Loss + Dice Loss. The training process was repeated 5 times for each model, following the holdout validation method. Table 3 summarizes the dice and Jaccard score for the different EffcientNet backbones experimented with.

Table 3. Results of the proposed method with different backbone choice.

Run	EfficientNetB0		EfficientNetB1		EfficientNetB2	
	Dice	Jaccard	Dice	Jaccard	Dice	Jaccard
1	89.7	84.9	84.9	82.5	83.9	81.6
2	88.5	83.2	83.7	81.3	80.1	79.6
3	88.2	84.1	82.0	80.9	81.6	80.2
4	87.9	83.7	82.9	81.1	79.9	79.0
5	88.5	83.3	81.5	80.5	82.2	81.3
Mean	88.6	83.8	83.0	81.3	81.6	80.3
Std	0.7	0.7	1.4	0.7	1.6	1.1

The best-performing model reached 88.6% Dice score for trachea segmentation with 0.7 Std, demonstrating that the model could generalize with stable metrics. This result was found with EfficientNet b0, which is also a lighter version, indicating that the dataset is not sufficiently extensive for training deeper networks. Notice that EfficientNet b1 and b2 obtained a higher std and at the same time lower dice score.

3.4 Discussion

Some regions of the trachea are more difficult to segment. It happens because the organ varies in shape according to the region depth in the body. At some depths, the organ may also split into two tubes, as depicted in Fig. 9. Here we can see the original image on the left side, the model prediction in the middle, and the manual segmentation on the right.

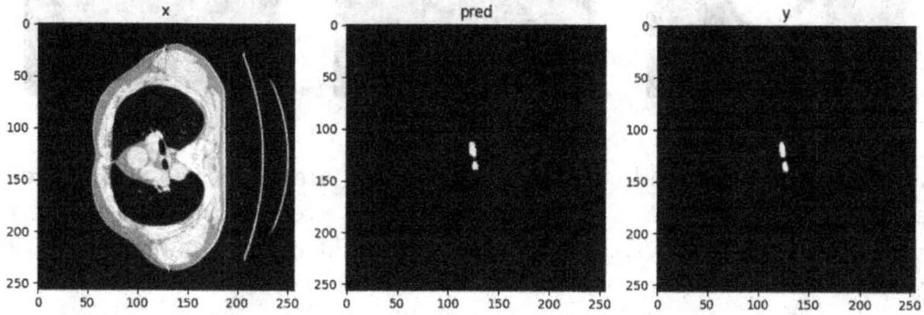

Fig. 9. Study case: 2D slice where the trachea is found splitted into two tubes. The image in the middle is the prediction made by the model. The image on the right is the ground truth, provided by a physician.

The network performs well for the central portions of the trachea, as illustrated in Fig. 10, where EfficientDeeplab was capable of correctly identifying the shape and location of the organ.

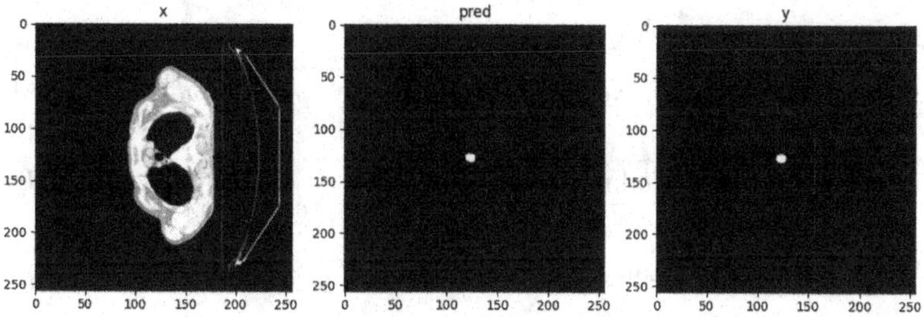

Fig. 10. Study case: prediction of a central portion of the trachea. The image in the middle is the prediction made by the model. The image on the right is the ground truth, provided by a physician

However, for the peripheral slices of the organ and the images without trachea presence, it's difficult for the model to yield the correct output. Some image regions contain similar textural features and shapes as the trachea, so the network outputs false positives, as shown in Fig. 11. Notice that there is no trachea presence in this CT slice, however, there is another body structure with a similar shape and texture that tricks the model into marking this region as the trachea. Correctly segmenting these kind of slices is the main limitation of the model, but it can be overcome by adding post-processing to select only the largest 3D object, considering prediction results over the entire CT scan.

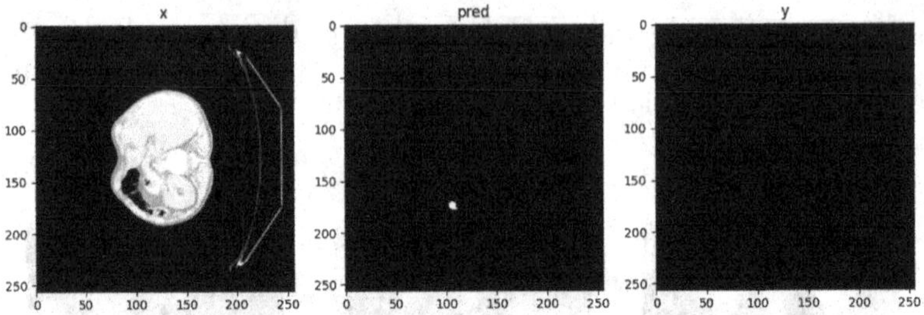

Fig. 11. Study case: the method's main limitations are the slices where there's no presence of trachea. The image in the middle is the prediction made by the model, which detected a false positive. The image on the right is the ground truth, provided by a physician

Table 4 compares this work with some methods that also segment trachea. For the analysis, we need to clarify that our experiment did five random hold-out sampling to avoid biased results based on the dataset organization and to guarantee repetition. The related works did not do this procedure. Also, the other works have access to the complete dataset, which is an advantage for training deep-learning models.

Table 4. Comparison with related works.

Paper	Method	Dataset	Dice Score (%)	#Params
[22]	MultiRes 3D U-Net	SEGTHOR	92.17	120M
[23]	U-Net 2.5D	SEGTHOR	92.56	23M
[9]	U-Net/SPP Module	SEGTHOR	89.00	26M
[25]	U-Net/ GAN	SEGTHOR	88.30	~
This Work	EfficientDeepLab	SEGTHOR (−30%)	88.6 ± 0.7	6M

Despite not being the best-performing model in the literature, it still produces excellent results, close to other works, but with 20x less parameters than [22], which is a considerably smaller number of parameters, being a compact yet robust neural network that consumes less computational resources than similar models. This is achieved using the EfficientNet architecture, a very compact and scalable model that rethinks network scale with resource constraints in mind.

4 Conclusion

This work proposed an automated method for trachea segmentation on CT images using EfficientDeepLab neural network architecture. The obtained results are promising and competitive, and its main contributions are the pre-processing based on intensity windowing, the use of dilated convolutions (atrous), and the atrous version of spatial pyramid pooling, leading to better feature extraction over different scales of resolution, and the efficient utilization of computational resource by adopting the EfficientNet architecture as the feature extraction backbone, which delivers a model with only 6 million parameters, considerably smaller than the methods proposed in the current literature for trachea segmentation with deep learning.

There is space for future improvements, such as adding new pre-processing techniques, automatically estimating a intensity window, aggregating 2.5D information from the volume into the model, adding post-processing to the output, and removing false positives. This way, the methodology can get more robust and precise, contributing to faster radiotherapy planning and consequently improving the quality of life for cancer patients.

Acknowledgments. This work was carried out with the support of the Coordination for the Improvement of Higher Education Personnel - Brazil (CAPES) - Financing Code 001, Maranhão Research Support Foundation (FAPEMA), National Council for Scientific and Technological Development (CNPq) and Brazilian Company of Hospital Services (Ebserh) Brazil (Proc. 409593/2021-4).

References

1. Amaro, E.J., Yamashita, H.: Aspectos básicos de tomografia computadorizada e ressonância magnética. Braz. J. Psychiatry **23**, 2–3 (2001)
2. Baskar, R., Lee, K.A., Yeo, R., Yeoh, K.W.: Cancer and radiation therapy: current advances and future directions. Int. J. Med. Sci. **9**(3), 193 (2012)
3. Chen, L.C., Papandreou, G., Kokkinos, I., Murphy, K., Yuille, A.L.: Semantic image segmentation with deep convolutional nets and fully connected CRFs (2016)
4. Chen, L.C., Papandreou, G., Kokkinos, I., Murphy, K., Yuille, A.L.: DeepLab: semantic image segmentation with deep convolutional nets, atrous convolution, and fully connected CRFs. IEEE Trans. Pattern Anal. Mach. Intell. **40**(4), 834–848 (2018). https://doi.org/10.1109/tpami.2017.2699184
5. Chen, L.C., Zhu, Y., Papaändreou, G., Schroff, F., Adam, H.: Encoder-decoder with atrous separable convolution for semantic image segmentation (2018)
6. Consídera, D.P., et al.: A tomografia computadorizada de alta resolução na avaliação da toxicidade pulmonar por amiodarona. Radiologia Brasileira **39**, 113–118 (2006)
7. Deng, J., Dong, W., Socher, R., Li, L.J., Li, K., Fei-Fei, L.: Imagenet: a large-scale hierarchical image database. In: 2009 IEEE Conference on Computer Vision and Pattern Recognition, pp. 248–255 (2009). https://doi.org/10.1109/CVPR.2009.5206848
8. Diniz, J., Ferreira, J., Silva, G., Quintanilha, D., Silva, A., Paiva, A.: Segmentação de coração em tomografias computadorizadas utilizando atlas probabilístico e redes neurais convolucionais. In: Anais do XXI Simpósio Brasileiro de Computação Aplicada á Saúde, pp. 83–94. SBC, Porto Alegre, RS, Brasil (2021). https://doi.org/10.5753/sbcas.2021.16055. https://sol.sbc.org.br/index.php/sbcas/article/view/16055

9. Feng, S., et al.: CPFNet: context pyramid fusion network for medical image segmentation. IEEE Trans. Med. Imaging **39**(10), 3008–3018 (2020). https://doi.org/10.1109/TMI.2020.2983721
10. Gupta, T., Narayan, C.A.: Image-guided radiation therapy: physician's perspectives. J. Med. Phys./Assoc. Med. Physicists India **37**(4), 174 (2012)
11. Hansen, P.C., Jørgensen, J., Lionheart, W.R.: Computed tomography: algorithms, insight, and just enough theory. SIAM (2021)
12. He, K., Zhang, X., Ren, S., Sun, J.: Spatial pyramid pooling in deep convolutional networks for visual recognition. In: Fleet, D., Pajdla, T., Schiele, B., Tuytelaars, T. (eds.) ECCV 2014. LNCS, vol. 8691, pp. 346–361. Springer, Cham (2014). https://doi.org/10.1007/978-3-319-10578-9_23
13. He, K., Zhang, X., Ren, S., Sun, J.: Deep residual learning for image recognition (2015)
14. Kalender, W.A.: X-ray computed tomography. Phys. Med. Biol. **51**(13), R29 (2006)
15. Kezmann, J.M.: Tensorflow advanced segmentation models (2020). https://github.com/JanMarcelKezmann/TensorFlow-Advanced-Segmentation-Models
16. Kingma, D.P., Ba, J.: Adam: a method for stochastic optimization (2017)
17. Lambert, Z., Petitjean, C., Dubray, B., Ruan, S.: Segthor: segmentation of thoracic organs at risk in CT images (2019). https://doi.org/10.48550/ARXIV.1912.05950. arxiv.org/abs/1912.05950
18. Noël, G., Antoni, D., Barillot, I., Chauvet, B.: Délinéation des organes á risque et contraintes dosimétriques. Cancer/Radiothérapie **20**, S36–S60 (2016). https://doi.org/10.1016/j.canrad.2016.07.032. https://www.sciencedirect.com/science/article/pii/S1278321816301676. recorad: Recommandations pour la pratique de la radiothérapie externe et de la curiethérapie
19. Sandler, M., Howard, A., Zhu, M., Zhmoginov, A., Chen, L.C.: Mobilenetv 2: inverted residuals and linear bottlenecks (2019)
20. Tan, M., Le, Q.: EfficientNet: rethinking model scaling for convolutional neural networks. In: Chaudhuri, K., Salakhutdinov, R. (eds.) Proceedings of the 36th International Conference on Machine Learning. Proceedings of Machine Learning Research, vol. 97, pp. 6105–6114. PMLR (2019). https://proceedings.mlr.press/v97/tan19a.html
21. Tekatli, H., et al.: Normal tissue complication probability modeling of pulmonary toxicity after stereotactic and hypofractionated radiation therapy for central lung tumors. Int. J. Radiat. Oncol. Biol. Phys. **100**(3), 738–747 (2018)
22. Wang, Q., et al.: 3D enhanced multi-scale network for thoracic organs segmentation. SegTHOR@ ISBI **3**(1), 1–5 (2019)
23. Wang, S., et al.: Conquering data variations in resolution: a slice-aware multi-branch decoder network. IEEE Trans. Med. Imaging **39**(12), 4174–4185 (2020). https://doi.org/10.1109/TMI.2020.3014433
24. Yan, X., Tang, H., Sun, S., Ma, H., Kong, D., Xie, X.: After-UNet: axial fusion transformer UNet for medical image segmentation. In: 2022 IEEE/CVF Winter Conference on Applications of Computer Vision (WACV), Los Alamitos, CA, USA, pp. 3270–3280. IEEE Computer Society (2022). https://doi.org/10.1109/WACV51458.2022.00333. https://doi.ieeecomputersociety.org/10.1109/WACV51458.2022.00333
25. Zhao, W., Chen, H., Lu, Y.: W-net: A network structure for automatic segmentation of organs at risk in thorax computed tomography. In: Proceedings of the 2020 2nd International Conference on Intelligent Medicine and Image Processing, IMIP 2020, pp. 66–69. Association for Computing Machinery, New York (2020). https://doi.org/10.1145/3399637.3399642

Multi-label Classification of Pathologies in Chest Radiograph Images Using DenseNet

Alison Corrêa Mendes(✉), Alexandre César Pinto Pessoa, and Anselmo Cardoso de Paiva

Núcleo de Computacão Aplicada, Universidade Federal do Maranhão (UFMA) Caixa Postal, 65.085-580, São Luís, MA, Brazil
{alison.mendes,alexandre.pessoa,paiva}@nca.ufma.br

Abstract. Chest radiography exams are still one of the main methods for detecting and diagnosing certain thoracic pathologies. This study evaluates the performance of a DenseNet in a multi-label classification task on radiography images, using focal loss as the loss function to address the class imbalance problem. For the experiments, 14 different types of findings were considered. Satisfactory results were obtained using the area under the ROC curve (AUC-ROC) as the metric, where the average performance across all classes was 0.861.

Keywords: Radiography · DenseNet · Focal Loss

1 Introduction

Chest radiography is one of the most commonly used medical examinations for detection and diagnosis of thoracic pathologies. In Brazil, this examination is one of the oldest, cheapest, and most widely available [15]. However, there is a need for methods to assist radiology professionals since the analysis of thoracic abnormalities through radiographic images is subjective and can vary depending on the experience and perception of the physician [11].

Artificial neural networks (ANNs) are computational systems inspired by the functioning of the human brain, where hundreds of billions of interconnected neurons process information in parallel [16]. Deep neural networks (DNNs) are ANNs that have multiple hidden layers between their input and output [12]. These networks have been widely used and have achieved good results in the classification of medical images, especially convolutional neural networks (CNNs), a class of ANN commonly applied to image analysis. CheXNet, for instance, has 121 layers and was trained using a dataset of 100,000 frontal chest radiographs, and outperformed the performance of four radiologists [13].

Deep Learning-based techniques are widely employed in image classification tasks, due to their ability to automatically learn discriminative features from raw data [9]. One of the advantages of deep learning over other techniques is that it minimizes the need for preprocessing, segmentation, and feature extraction [14].

M. C. Naldi and R. A. C. Bianchi (Eds.): BRACIS 2023, LNAI 14196, pp. 167–180, 2023.
https://doi.org/10.1007/978-3-031-45389-2_12

The application of hyperparameter optimization methods in deep neural networks to analyze the network configuration mechanisms by capturing specific problem characteristics, enhances the discriminative capability of the produced techniques. This work evaluated the performance of a dense convolutional neural network (DenseNet) using the Focal Loss function in a multi-label classification task on radiographic images.

2 Related Works

The search for methods to aid in the detection and diagnosis of pathologies from X-ray images has attracted the attention of many researchers worldwide, with several published works presenting promising results. However, obtaining a highly accurate and fully automated CNN-based method for medical diagnosis is still a challenging task.

Wang et al. [17] presented the ChestX-ray8, a dataset with 108,948 chest X-ray images from 32,717 unique patients, where each image can have multiple labels. In this study, a quantitative benchmarking of the performance of 4 pre-trained models in the classification of the 8 thoracic pathologies in this dataset was conducted, with the ResNet-50-based model [6] achieving the best results. Later, this dataset was expanded to ChestX-ray14, including 6 additional common thoracic pathologies.

Yao et al. [18] utilized an architecture that learns at multiple resolutions while generating weakly supervised saliency maps. The study also parameterized the LSE-LBA pooling function (Long-Sum-Exp with learnable Lower-Bounded Adaptation). The average AUC for the 14 labels in the ChestX-ray14 dataset in this work was 0.761.

Bhusal et al. [2] implemented a model using DenseNet and a modified cross-entropy loss to handle class imbalance, using specific weights for each class. The highest area under the ROC curve (AUC) was achieved for the Cardiomegaly finding, with a value of 0.896, and the lowest AUC for the Nodule finding with a score of 0.655. The average AUC for the 14 labels in the ChestX-ray14 dataset in this work was 0.762.

Rajpurkar et al. [13] developed ChesXNet, a 121-layer CNN trained on the ChestX-ray14 dataset. Rajpurkar et al. [13] compared the performance of the ChesXNet with four radiologists using the F1 metric, where the network outperformed the average score of the radiologists.

Zhao et al. [20] proposed a model called AM_DenseNet, which employs a dense connection network with an attention module after each dense block to optimize the feature extraction capability of the model. They used focal loss to address the class imbalance problem. The average AUC for the 14 pathology labels in the ChestX-ray14 dataset in this work was 0.8537.

In this work, the utilization of DenseNet-121 combined with focal loss (FL) addresses the class imbalance in the classification of thoracic pathologies in the ChestX-ray14 and CheXpert datasets. In contrast to some previous approaches, we opted to preserve the original DenseNet-121 architecture and instead performed fine-tuning and optimization of the hyperparameters of the focal loss function. Experiments were conducted comparing the results obtained with FL and cross entropy loss, allowing for the evaluation of the benefits of FL and highlighting its effectiveness in improving the model's performance in the context of this task.

3 Background

3.1 DenseNet

DenseNet is a network that connects each of its layers to every other layer in a feed-forward fashion. A feed-forward neural network (FNN) is an artificial neural network in which the connections between nodes do not form a cycle [19, p. 73]. Unlike conventional convolutional neural networks, which have L layers and L connections (one between each layer and its subsequent direct connections), DenseNet has $\frac{L(L+1)}{2}$ direct connections [7].

Preserving the feed-forward nature of the network, Huang et al. [7] proposes a connectivity pattern between the layers, where the lth layer receives the feature maps from all previous layers, $x_0, ..., x_{l-1}$. Its input can be defined as:

$$X_l = H_l([X_0, X_1, ..., X_{l-1}]),$$

where $[X_0, X_1, ..., X_{l-1}]$ refers to the concatenation of the feature maps produced in layers $0, ..., l-1$. The H_l in this equation represents the multiple inputs into a single tensor. Figure 1 illustrates this connectivity between the layers.

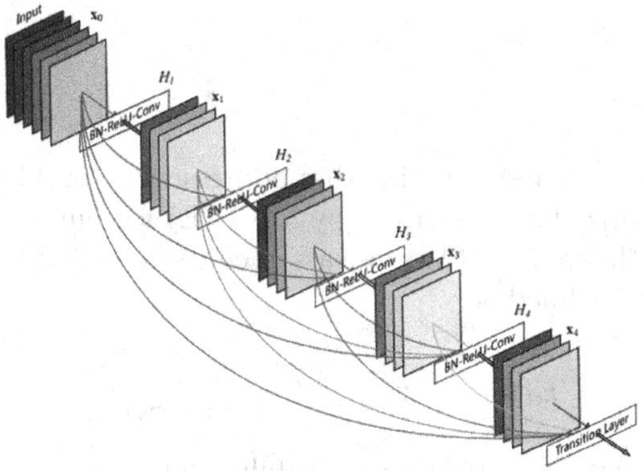

Fig. 1. Example of a Dense Block with 5 layers. This image illustrates how the connections between the layers occur [7].

DenseNet-121. The DenseNet-121 network consists of four dense blocks, where each block corresponds to a certain number of convolutional layers. Dense block 1 is composed of 6 dense layers, dense block 2 consists of 12 dense layers, dense block 3 consists of 24 dense layers, and dense block 4 consists of 16 dense layers. Each convolutional layer consists of batch normalization (BN), rectified linear unit (ReLU) activation function, and convolutional (Conv) operations for feature extraction.

DenseNet-121 uses a transition layer between two adjacent blocks, which consists of a 1×1 convolution, followed by 2×2 average pooling (calculating the average for each patch of the feature map). At the end of the last dense block, there is a pooling layer to which a classifier is attached, as shown in Fig. 2.

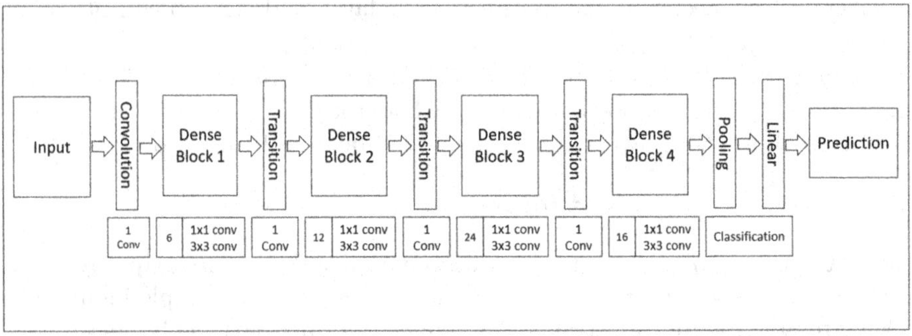

Fig. 2. Structure of DenseNet-121.

3.2 Focal Loss

The Focal Loss (FL) is a loss function that addresses the problem of class imbalance by reformulating the Cross Entropy loss (CE) for binary classification [5], in order to reduce the weight of the loss assigned to well-classified examples. The CE function can be defined as:

$$CE(p, y) = \begin{cases} -\log(p); & \text{if } y = 1, \\ \log(1-p); & \text{otherwise}, \end{cases}$$

where $y \in \{\pm 1\}$, with p being the probability for the class with $y = 1$ and $p \in [0, 1]$. p_t is defined as follows:

$$p_t = \begin{cases} p; & \text{if } y = 1, \\ 1-p; & \text{otherwise}, \end{cases}$$

thus, we have $CE(p, y) = CE(p_t) = -\log(p_t)$.

FL focuses the training on a sparse set of challenging examples, preventing a large number of easy negative examples from overwhelming the model during training [10]. It proposes the addition of a modulating factor $(1 - p_t)^\gamma$ to the cross-entropy loss, with an adjustable focus parameter $y \geq 0$. Formally, the Focal Loss is defined as follows:

$$FL(p_t) = -(1 - p_t)^\gamma \log(p_t).$$

A variant of this equation, called α-balanced can be used. This variant yields slightly improved precision compared to the non α-balanced form [10], and is defined as:

$$FL(\rho_t) = -\alpha_t(1 - \rho_t)^\gamma \log(\rho_t),$$

it can also be observed that when $\gamma = 0$ and $\alpha_t = 1$, FL is equal to CE.

4 Materials and Methods

4.1 Dataset

This work utilized two datasets: ChestX-ray14 [17] and ChesXpert [8]. These datasets employ a multi-label approach, which means that each image may contain multiple pathologies. This section provides a description of each dataset used in this work, including information about the number of images, labels, as well as the origin of the data.

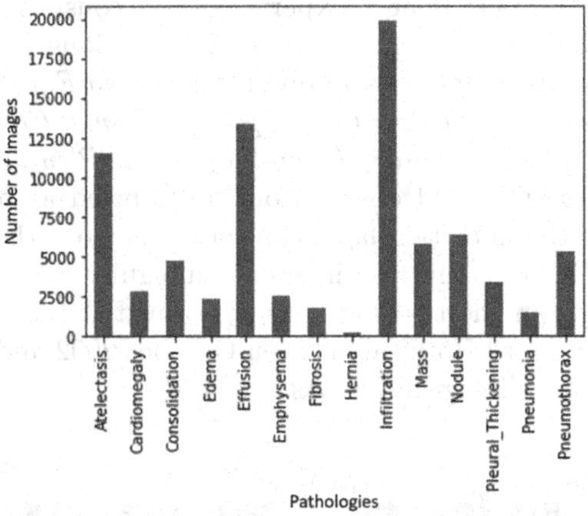

Fig. 3. Number of images for each pathology in the ChestX-ray14 dataset.

ChestX-Ray14. The ChestX-ray14 dataset contains 112,120 radiograph images from 30,805 unique patients with image labels for 14 thoracic pathologies: *Atelectasis*, *Cardiomegaly*, *Consolidation*, *Edema*, *Effusion*, *Emphysema*, *Fibrosis*, *Hernia*, *Infiltration*, *Mass*, *Nodule*, *Pleural Thickening*, *Pneumonia*, and *Pneumothorax*. This dataset is an extension of the 8 common pathology patterns listed in the ChestX-ray8 dataset [17], which are represented in Fig. 4. This dataset was extracted from clinical data in the PACS (Picture Archiving and Communication Systems) at the NIH Clinical Center (National Institutes of Health Clinical Center).

The images have a size of 1024 × 1024 pixels and are in PNG format. Natural Language Processing (NLP) techniques were used to extract the image labels, and the accuracy for the extracted labels was over 90% [17]. Figure 3 shows the number of images for each pathology present in the dataset, highlighting a significant class imbalance.

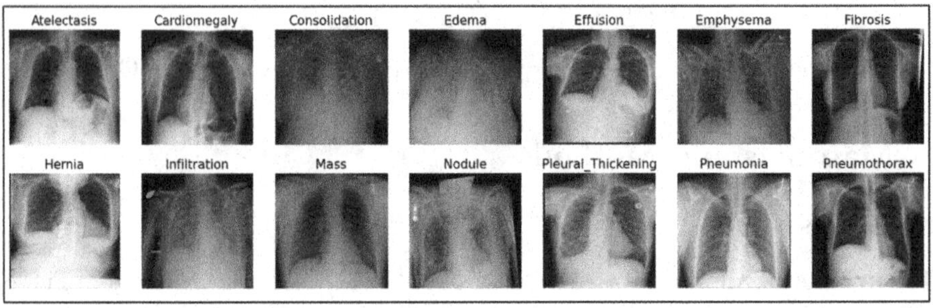

Fig. 4. Example of 14 of the pathologies present in ChestX-ray14 dataset with their respective labels.

CheXpert. The CheXpert (Chest eXpert) dataset consists of 224,316 frontal and lateral chest radiographic images from 65,240 unique patients. Similar to ChestX-ray14, this dataset includes 14 observations: *No Finding*, *Enlarged Cardiom.*, *Cardiomegaly*, *Lung Lesion*, *Lung Opacity*, *Edema*, *Consolidation*, *Pneumonia*, *Atelectasis*, *Pneumothorax*, *Pleural Effusion*, *Pleural Other*, *Fracture*, and *Support Devices.* These 14 observations are labeled as positive, negative, or uncertain, where the uncertain label can represent both the uncertainty of a radiologist in the diagnosis and the inherent ambiguity in the report [8]. This data was extracted from chest radiographs performed at the inpatient and outpatient centers of Stanford Hospital between October 2002 and July 2017, along with their associated radiological reports.

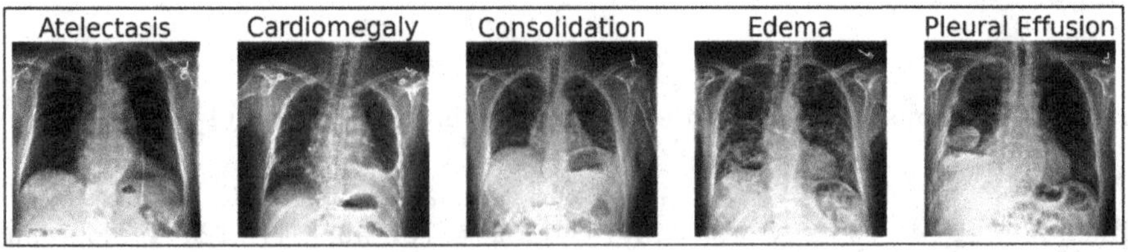

Fig. 5. Example of 5 of the pathologies present in CheXpert dataset with their respective labels.

4.2 Preprocessing

For the ChestX-ray14 dataset, the pathology information for each image was preprocessed by converting it into an array of size 14, with each position representing each pathology. In this array, '0' indicates the absence of the corresponding pathology and '1' indicates the presence of the pathology. Then, the images in the dataset were preprocessed.

The images were divided into two subsets: a training set and a validation set. For each pathology, approximately 80% of the images were separated for training, and 20% were set aside for testing. This division was performed based on the unique patient identifier, ensuring that images from the same patient do not appear in both the training and testing sets. During the image preprocessing process, to standardize the pixel levels between 0 and 1, all intensities were divided by 255 (the maximum possible value). The feature maps were standardized by dividing each value by the standard deviation of the map, assuming a mean of 0, resulting in a standard deviation of 1 for the sample. Additionally, these images were resized to a resolution of 224×224 pixels.

For the CheXpert dataset, information on present pathologies was also preprocessed. The uncertainty labels in this dataset, annotated as '−1', were handled using the 'U-Ones' approach presented in [8], where these uncertainty labels were converted from '−1' to '1'. Additionally, blank labels were converted to '0'. Five pathologies were selected from this dataset based on their predominance in the validation set and their presence in the ChestX-ray14 dataset (see Fig. 6).

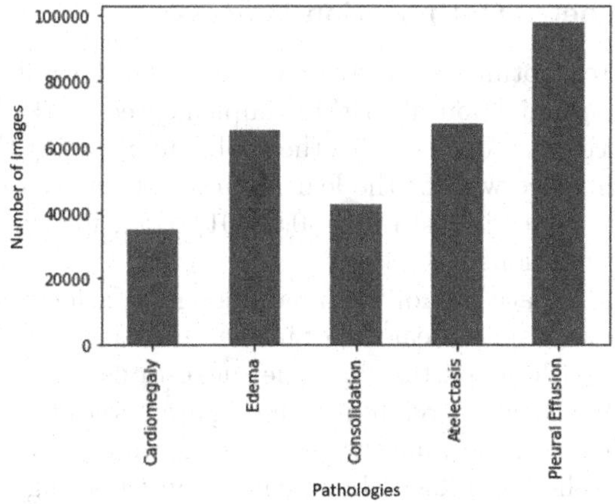

Fig. 6. Distribution of the selected classes in the CheXpert dataset.

The pathology information in CheXpert was converted into a vector of size 5, where '0' indicates the absence of pathology and '1' indicates the presence of pathology. The images in this dataset were preprocessed in the same way as the ChestX-ray14 dataset. Like ChestX-ray14, these images were also resized to a size of 224×224 pixels.

4.3 Training Configuration

This work utilized a transfer learning approach known as fine-tuning, where the weights of a pretrained model are preserved (frozen) in some layers and adjusted (trained) in the remaining layers.

In this work, a DenseNet-121 network implemented in [3] was used. It was pretrained with weights from Imagenet [4]. The top classification layers were excluded. On top of this network, a global average pooling layer and the final sigmoid activation layer for classification were stacked.

Some layers of the pretrained model were frozen to keep their weights constant during training. Then, the model was compiled with the appropriate loss functions for each experiment and an Adam optimizer with a learning rate defined in the hyperparameter optimization process [1].

After that, the model underwent training and testing stages using the preprocessed datasets. The model was trained for 20 epochs with a batch size of 16, using an early stop callback to monitor the loss function. The minimum delta was set to 0.001, which defines the minimum change in the monitored value to qualify as an improvement, and the patience was set to 3, representing the number of epochs without improvement after which the training will be stopped.

Then, the fine-tuning stage was performed. The initial layers of the model were unfrozen, and the model was retrained with a lower learning rate. From there, the trained network is used to predict the labels of the images in the validation set of each dataset used in this work.

4.4 Hyperparameter Optimization Process

The hyperparameter optimization stage utilized the Tree-Structured Parzen Estimator (TPE) optimization algorithm, implemented in [1]. Three hyperparameter search spaces were defined with the goal of maximizing the AUC (ROC).

The first search space was for the learning rate of the network. This search space involved real values in the range [0.00001, 0.01], following a logarithmic pattern in value suggestions.

The second search space was for the α parameter of the focal loss function. For the values in this search space, real values in the range [0.1, 1.0] were considered, also following a logarithmic pattern in value suggestions.

The third search space was related to the γ parameter of the focal loss function. In this search space, the values were set in the range [1.0, 5.0], and like the other two cases, it followed a logarithmic pattern in value suggestions.

In each optimization process, 15 trials were executed. At the end of these processes, the data from this study were saved in a CSV file, including their respective hyperparameters and AUC values.

5 Results and Discussion

In the first experiment, the proposed model was used to predict the 14 pathology labels of the ChestX-ray14 dataset. For this initial experiment, the binary cross-entropy loss (BCE) function, implemented in [3]. The Adam optimizer with a learning rate of 0.00001 was employed.

It was observed that the model handled the *Hernia* class better than the other classes, with an AUC value of 0.926. On the other hand, the *Pneumonia* class had the lowest AUC value of 0.647. In this experiment, the model achieved an AUC above 0.900 in three out of the fourteen classes, while five classes had AUC values above 0.800, and another five were above 0.700. Only for the *Pneumonia* class, the model achieved an AUC value below 0.700.

A new experiment was conducted using the ChestX-ray14 dataset. This time the focal loss (FL) was used as the loss function, with α set to 0.173 and γ set to 2.89. The Adam optimizer with a learning rate of 0.00001 was employed. Once again, the *Hernia* class demonstrated the best performance among the other classes, achieving an AUC of 0.986. On the other hand, the *Pneumonia* class had the lowest performance with an AUC of 0.749. Notably, eight out of the fourteen categories achieved AUC values above 0.850, while only three categories had AUC values below 0.800, as depicted in Fig. 7.

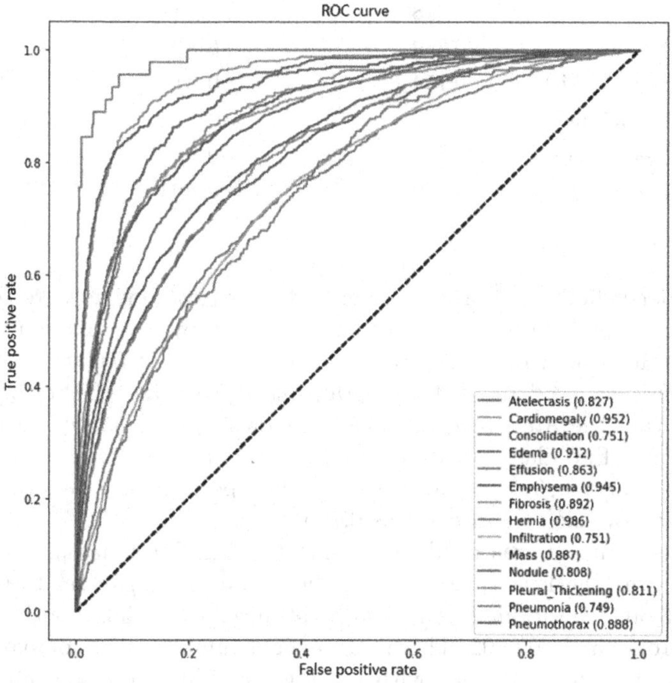

Fig. 7. Performance of the model with Focal Loss using AUC metric for the ChestX-ray14 dataset.

The results of the two experiments are compared in Table 1. It is evident that the model using FL as the loss function achieved superior performance for all pathologies, leading to a higher average AUC across the 14 classes in the FL experiment.

Table 1. Comparison of performance between the two experiments conducted on the ChestX-ray14 dataset.

Pathology	DenseNet-121 + BCE	DenseNet-121 + FL
Atelectasis	0.781	**0.827**
Cardiomegaly	0.908	**0.952**
Consolidation	0.707	**0.751**
Edema	0.859	**0.912**
Effusion	0.831	**0.863**
Emphysema	0.906	**0.945**
Fibrosis	0.823	**0.892**
Hernia	0.926	**0.986**
Infiltration	0.709	**0.751**
Mass	0.818	**0.887**
Nodule	0.750	**0.808**
Pleural Thickening	0.754	**0.811**
Pneumonia	0.647	**0.749**
Pneumothorax	0.825	**0.888**
Average	0.803	**0.861**

To further validate the proposed method, new experiments were conducted using the CheXpert dataset. The binary cross-entropy loss function was used, along with the Adam optimizer with a learning rate of 0.0002. The *Pleural Effusion* class achieved the highest performance, with an AUC of 0.874. It is also noting that the *Atelectasis* class had the lowest performance, with an AUC value of 0.698. Two classes had achieved results above 0.850, two classes had AUC values above 0.700, and only one class had performance below 0.700. The average AUC for this experiment was 0.790.

A new experiment was conducted using the CheXpert dataset. This time, FL was used as the loss function, with the parameter *alpha* set to 0.363 and *gamma* to 3.303. Additionally, the Adam optimizer was utilized with a learning rate of approximately 0.0002. There was a slight improvement compared to the previous experiment. The class with the highest AUC score was *Edema* with an AUC of 0.909 as depicted in Fig. 8. Two classes achieved AUC values above 0.900, and all classes had a performance above 0.700. The average AUC for this experiment was 0.820.

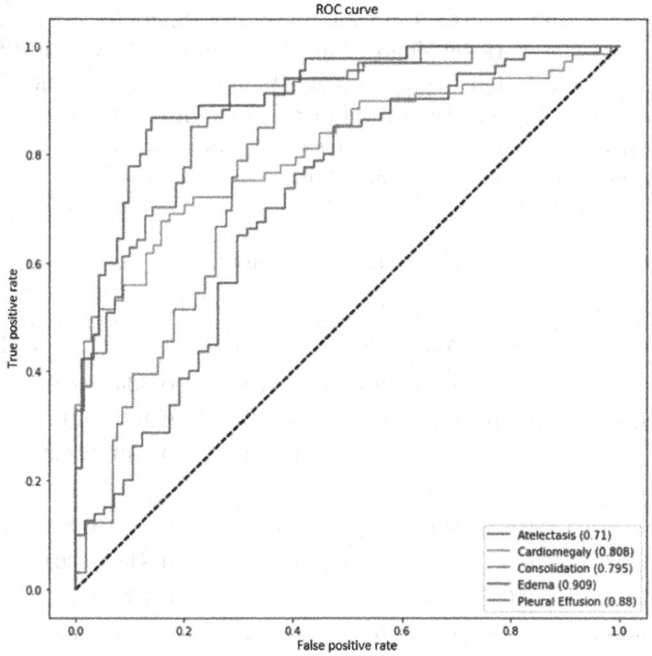

Fig. 8. Model performance with FL using the AUC metric for the CheXpert database.

Table 2 compares the results of the experiments using CheXpert. Similar to the experiments using ChestX-ray14, it was observed that the model utilizing FL as the loss function achieved the best performance for all studied pathologies. Thus, the average for the 5 classes was higher in the experiment using FL.

Table 2. Comparison of performance between the two experiments conducted with CheXpert.

Pathology	DenseNet-121 + BCE	DenseNet-121 + FL
Atelectasis	0.698	**0.710**
Cardiomegaly	0.773	**0.808**
Consolidation	0.742	**0.795**
Edema	0.865	**0.909**
Pleural Effusion	0.874	**0.880**
Average	0.790	**0.820**

We compared the results of our most promising experiment with the results of other studies that also used deep learning techniques to classify thoracic pathologies in the ChestX-ray14 dataset. The results shown in Table 3 indicate that the proposed model obtained a superior performance in terms of the AUC metric for 6 labels. Note that, for the *Emfisem* class, the difference between the proposed model and the work by Zhao et al. [20] is minimal, with a difference of approximately 0.002, similar to the class *Atelectasis* with a difference of 0.007. It is important to mention that the distribution of training sets may vary between different studies. To ensure a fair comparison, it would be ideal to reproduce the experiments using the exact same distribution of datasets.

Table 3. Performance of the proposed model compared to the literature using the AUC metric.

Pathology	Wang	Yao	Bhusal	Rajpurkar	Zhao	Proposed Model
Atelectasis	0.700	0.733	0.79	0.820	**0.834**	0.827
Cardiomegaly	0.814	0.856	0.89	0.904	0.913	**0.952**
Effusion	0.736	0.806	0.76	0.883	**0.889**	0.863
Infiltration	0.613	0.673	0.65	0.720	0.730	**0.751**
Mass	0.693	0.777	0.82	0.861	0.875	**0.887**
Nodule	0.669	0.724	0.65	0.776	**0.818**	0.808
Pneumonia	0.658	0.684	0.68	0.763	**0.769**	0.749
Pneumothorax	0.799	0.805	0.81	0.893	**0.911**	0.888
Consolidation	0.703	0.711	0.75	0.793	**0.818**	0.751
Edema	0.805	0.806	0.86	0.893	0.901	**0.912**
Empysema	0.833	0.842	0.79	0.926	**0.947**	0.945
Fibrosis	0.786	0.743	0.72	0.804	0.819	**0.892**
Pleural Thickening	0.684	0.724	0.71	0.813	**0.836**	0.811
Hernia	0.872	0.775	0.77	0.938	0.885	**0.986**
Average	0.740	0.761	0.762	0.841	0.853	**0.861**

In the experiments conducted using the ChestX-ray14 dataset, the impact of using focal loss on the overall network performance can be observed, resulting in an average performance gain of 5.5% points. However, for pathologies with fewer positive samples, the performance was notably better. In these cases, there was a greater performance improvement for the labels *Pneumonia* (10.2% points), *Mass* (6.9% points), *Fibrosis* (6.9% points), and *Hernia* (6.0% points), which are among the pathologies with fewer samples in the dataset. On the other hand, labels that have more positive samples in the dataset showed a performance gain of less than 5.0% points, such as *Effusion* (3.2% points), *Infiltration* (4.2% points), and *Atelectasis* (4.6% points).

A similar behavior can also be observed in the experiments conducted using CheXpert, which demonstrated an average performance gain of 3.0% points. Labels with a higher number of positive samples exhibited the lowest performance improvement, such as the *Pleural Effusion* label (0.6% points). Conversely, the label with the fewest positive samples in the dataset showed the highest performance gain, as seen in the case of the *Consolidation* label (5.3% points).

6 Conclusion

This work evaluated the performance of a DenseNet for the classification of 14 thoracic pathologies in a multi-label approach. Experiments were conducted with the proposed model using binary cross-entropy loss and also using Focal Loss to address the class imbalance issues present in the datasets.

The proposed model was validated using the ChestX-ray14 and CheXpert datasets. The results of the model using cross-entropy and Focal Loss were compared, and in these experiments, Focal Loss outperformed cross-entropy in all cases, as it effectively addressed the class imbalance problem present in both datasets.

The most promising experiment was conducted using Focal Loss to predict the 14 pathologies in the ChestX-ray14 dataset, achieving an average AUC score of 0.861, which was the highest among the researched works and outperformed the other methods in six out of the fourteen classes.

One of the major challenges of this work was the significant class imbalance in the utilized datasets. As future work, we intend to explore the structure of DenseNet further to improve the results, as well as apply this approach to computed tomography images.

It is important to emphasize that although the model produced satisfactory results, its use does not replace the diagnosis of a radiology professional. However, it is also important to highlight that the results generated by the proposed model can provide valuable references for these professionals.

This work was carried out with the support of the Coordination for the Improvement of Higher Education Personnel - Brazil (CAPES) - Financing Code 001, Maranhão Research Support Foundation (FAPEMA), National Council for Scientific and Technological Development (CNPq) and Brazilian Company of Hospital Services (Ebserh) Brazil (Proc. 409593/2021-4).

References

1. Akiba, T., Sano, S., Yanase, T., Ohta, T., Koyama, M.: Optuna: a next-generation hyperparameter optimization framework. In: Proceedings of the 25th ACM SIGKDD International Conference on Knowledge Discovery & Data Mining, pp. 2623–2631 (2019)
2. Bhusal, D., Panday, D., Prasad, S.: Multi-label classification of thoracic diseases using dense convolutional network on chest radiographs. arXiv preprint arXiv:2202.03583 (2022)
3. Chollet, F., et al.: Keras (2015). https://github.com/fchollet/keras
4. Deng, J., Dong, W., Socher, R., Li, L.J., Li, K., Fei-Fei, L.: Imagenet: a large-scale hierarchical image database. In: 2009 IEEE Conference on Computer Vision and Pattern Recognition, pp. 248–255. IEEE (2009)

5. Good, I.J.: Rational decisions. In: Kotz, S., Johnson, N.L. (eds.) Breakthroughs in Statistics, pp. 365–377. Springer, New York (1992). https://doi.org/10.1007/978-1-4612-0919-5_24
6. He, K., Zhang, X., Ren, S., Sun, J.: Deep residual learning for image recognition. In: Proceedings of the IEEE Conference on Computer Vision and Pattern Recognition, pp. 770–778 (2016)
7. Huang, G., Liu, Z., Van Der Maaten, L., Weinberger, K.Q.: Densely connected convolutional networks. In: Proceedings of the IEEE Conference on Computer Vision and Pattern Recognition, pp. 4700–4708 (2017)
8. Irvin, J., et al.: Chexpert: a large chest radiograph dataset with uncertainty labels and expert comparison. In: Proceedings of the AAAI Conference on Artificial Intelligence, vol. 33, pp. 590–597 (2019)
9. LeCun, Y., Bengio, Y., Hinton, G.: Deep learning. Nature **521**(7553), 436–444 (2015)
10. Lin, T.Y., Goyal, P., Girshick, R., He, K., Dollár, P.: Focal loss for dense object detection. In: Proceedings of the IEEE International Conference on Computer Vision, pp. 2980–2988 (2017)
11. Marcos, L., Bichinho, G.L., Panizzi, E.A., Storino, K.K.G., Pinto, D.C.: Classificação da doença pulmonar obstrutiva crônica pela radiografia do tórax. Radiol. Bras. **46**, 327–332 (2013)
12. Qian, Y., Fan, Y., Hu, W., Soong, F.K.: On the training aspects of deep neural network (DNN) for parametric TTS synthesis. In: 2014 IEEE International Conference on Acoustics, Speech and Signal Processing (ICASSP), pp. 3829–3833. IEEE (2014)
13. Rajpurkar, P., et al.: Chexnet: radiologist-level pneumonia detection on chest X-rays with deep learning. arXiv preprint arXiv:1711.05225 (2017)
14. Santos, M.K., Ferreira, J.R., Wada, D.T., Tenório, A.P.M., Barbosa, M.H.N., Marques, P.M.D.A.: Inteligência artificial, aprendizado de máquina, diagnóstico auxiliado por computador e radiômica: avanços da imagem rumo à medicina de precisão. Radiol. Bras. **52**, 387–396 (2019)
15. Wada, D.T., Rodrigues, J.A.H., Santos, M.K.: Aspectos técnicos e roteiro de análise da radiografia de tórax. Medicina **52**(Suppl. 1), 5–15 (2019)
16. Wang, S.C.: Artificial neural network. In: Wang, S.C. (ed.) Interdisciplinary Computing in Java Programming, pp. 81–100. Springer, Boston (2003). https://doi.org/10.1007/978-1-4615-0377-4_5
17. Wang, X., Peng, Y., Lu, L., Lu, Z., Bagheri, M., Summers, R.: Chestx-ray8: hospital-scale chest X-ray database and benchmarks on weakly-supervised classification and localization of common thorax diseases. In: 2017 IEEE Conference on Computer Vision and Pattern Recognition (CVPR), pp. 3462–3471 (2017)
18. Yao, L., Poblenz, E., Dagunts, D., Covington, B., Bernard, D., Lyman, K.: Learning to diagnose from scratch by exploiting dependencies among labels. arXiv preprint arXiv:1710.10501 (2017)
19. Zell, A.: Simulation neuronaler netze, vol. 1. Addison-Wesley Bonn (1994)
20. Zhao, J., Li, M., Shi, W., Miao, Y., Jiang, Z., Ji, B.: A deep learning method for classification of chest X-ray images. In: Journal of Physics: Conference Series, vol. 1848, p. 012030. IOP Publishing (2021)

Does Pre-training on Brain-Related Tasks Results in Better Deep-Learning-Based Brain Age Biomarkers?

Bruno M. Pacheco[1], Victor H. R. de Oliveira[1], Augusto B. F. Antunes[2], Saulo D. S. Pedro[3], Danilo Silva[1](✉), and for the Alzheimer's Disease Neuroimaging Initiative

[1] Federal University of Santa Catarina (UFSC), Florianópolis, SC, Brazil
bruno.m.pacheco@posgrad.ufsc.br, danilo.silva@ufsc.br
[2] Alliar - NEPIA, Belo Horizonte, MG, Brazil
augusto.antunes@alliar.com
[3] 3778 Healthcare, Belo Horizonte, MG, Brazil
saulo.pedro@3778.care

Abstract. Brain age prediction using neuroimaging data has shown great potential as an indicator of overall brain health and successful aging, as well as a disease biomarker. Deep learning models have been established as reliable and efficient brain age estimators, being trained to predict the chronological age of healthy subjects. In this paper, we investigate the impact of a pre-training step on deep learning models for brain age prediction. More precisely, instead of the common approach of pre-training on natural imaging classification, we propose pre-training the models on brain-related tasks, which led to state-of-the-art results in our experiments on ADNI data. Furthermore, we validate the resulting brain age biomarker on images of patients with mild cognitive impairment and Alzheimer's disease. Interestingly, our results indicate that better-performing deep learning models in terms of brain age prediction on healthy patients do not result in more reliable biomarkers.

Keywords: Brain Age · Deep Learning · ADNI · BraTS · MRI · Transfer Learning

1 Introduction

As human lifespan increases, there is a growing need for reliable methods to assess brain health and age-related changes in the brain. Brain age prediction

Data used in preparation of this article were obtained from the Alzheimer's Disease Neuroimaging Initiative (ADNI) database (adni.loni.usc.edu). As such, the investigators within the ADNI contributed to the design and implementation of ADNI and/or provided data but did not participate in analysis or writing of this report. A complete listing of ADNI investigators can be found at: http://adni.loni.usc.edu/wp-content/uploads/how_to_apply/ADNI_Acknowledgement_List.pdf.

M. C. Naldi and R. A. C. Bianchi (Eds.): BRACIS 2023, LNAI 14196, pp. 181–194, 2023.
https://doi.org/10.1007/978-3-031-45389-2_13

is a promising technique that uses neuroimaging data to estimate the apparent age of an individual's brain, which can serve as an indicator of overall brain health and successful aging, as well as a disease biomarker [5,6,11,17,21]. Deep learning models have shown great potential in accurately predicting brain age from magnetic resonance imaging (MRI) data [4,7,17,27,29].

Training deep learning models for brain age prediction shares several challenges with other neuroimaging tasks, in comparison to traditional computer vision, such as the increased GPU memory used from the 3D data and the extensive pre-processing required to account for the variability in the acquisition process. In particular, available neuroimaging datasets are much smaller than existing natural imaging datasets [13,33], and deep learning models are known to be very dependent on sample sizes. Therefore, data-efficient training strategies are crucial to achieve high performance in brain age prediction.

In this paper, we explore the impact of pre-training deep learning models for brain age prediction. Inspired by the learning process of expert neuroradiologists, we apply transfer learning by pre-training our brain age models on a brain-related task. For comparison, we also train models without pre-training and models pre-trained on natural image classification. We investigate the performance gain from pre-training and evaluate the models' brain age prediction as a biomarker for cognitive impairment. More specifically:

- We pre-train deep learning models on the brain tumor segmentation task and compare them to models without pre-training and with pre-training on the ImageNet natural image classification task;
- We test the brain age models using data from the ADNI studies, and show that the models pre-trained on the brain-related task outperform the other models, achieving the state-of-the-art in brain age prediction;
- We evaluate the brain age prediction of all models as a biomarker for different clinical groups (healthy, mild cognitive impairment, and Alzheimer's disease);
- Our experiments suggest that, despite the common practice, better models in terms of brain age prediction of healthy patients do not result in more reliable biomarkers;
- All of our results are reported on a standardized, publicly available dataset, providing an easy comparison with future research.

2 Related Work

Detecting aging features on brain MRI has been an active area of research for many years [10,18,34]. The use of deep learning for brain age prediction has gained considerable attention in recent years [1,8,20,27,29,30]. In this section, we provide a brief overview of related works on brain age prediction from MRIs using deep learning models.

One of the earliest applications of deep learning to brain age prediction was presented by Cole et al. [7]. The authors employed a neural network comprising a convolutional backbone and a fully connected regression head to analyze a dataset of T1-weighted MRI scans from 2001 healthy subjects aged 18 to 90.

The training is performed solely on images of healthy subjects, following the hypothesis that the brain age of healthy subjects is close to their actual age. Their deep learning model outperformed the machine learning approach (Gaussian Process Regressor). The authors also assessed the reliability of the predictions across individuals and acquisition methods.

Jonsson et al. [17] developed a deep learning model for brain age prediction using brain MRI scans from 1264 healthy subjects aged 18 to 75. They explored the impact of training and testing on distinct datasets, finding that the performance of brain age prediction degraded when the target dataset differed from the training dataset.

Bashyam et al. [4] proposed to improve brain age prediction by utilizing a larger dataset of brain MRI scans from 14,468 subjects. The dataset included data acquired from different sites following different protocols, with subjects aged 3 to 95. The authors pre-trained their neural network on the ImageNet dataset, which is an even larger dataset of natural images. They found that models performing well at chronological age prediction might not be the best at providing brain age estimates that correlate to the diagnosis of diseases such as schizophrenia and Alzheimer's.

Peng et al. [27] proposed quality brain age prediction using a lightweight deep learning model. They used a dataset containing 14,503 subjects from the UK Biobank, with ages ranging from 44 to 80. The authors showed that even though larger models perform well on natural image tasks, smaller models can perform equally well and sometimes even better on medical imaging tasks.

Multiple authors have reported brain age performance on MRI data from ADNI [19,20,22,26,29]. To the best of our knowledge, neither has provided means to reproduce the dataset used for testing the models. The studies either used a random, non-disclosed split, or did not provide which images have been selected from the ADNI database. Therefore, a direct comparison is not possible, as we cannot perfectly replicate the evaluation setting. Nonetheless, we highlight that the best performance reported was a mean absolute error of 3.10 years [20]. Further details on the performance of each approach can be seen in Table 1.

Table 1. Performance of brain age prediction methods on MRIs from ADNI. The authors have not disclosed from which phases of the ADNI study the subjects were drawn. Aβ(-) indicates that the subjects have sustained a negative amyloid beta status over 3 years.

	Test set	MAE
Lam et al. [19]	631 CN subjects (10-fold cross validation)	3.96
Ly et al. [22]	51 CN subjects with Aβ(-)	3.7
More et al. [26]	209 CN subjects	6.56
Lee et al. [20]	330 CN subjects	3.10
Poloni et al. [29]	151 CN subjects over 70 years old	3.66

Overall, these studies demonstrate the potential of pre-training for brain age prediction and the need for more efficient training strategies, as acquiring medical imaging data is very laborious.[1] None of them, however, took advantage of models trained for other brain-related tasks, such as brain tumor segmentation. Previous works also lack a standardized dataset on which we could perform a fair comparison, either because they use private datasets or because they do not share which samples were used for training or testing. Therefore, our paper stands out by comparing brain age models pre-trained on brain tumor segmentation to models without pre-training or pre-trained on natural image classification. Furthermore, we experimented on a standardized and publicly available dataset, providing reproducible results.

3 Materials and Methods

3.1 Data

ADNI. The Alzheimer's Disease Neuroimaging Initiative (ADNI) was launched in 2003 with the primary goal of testing whether serial MRI, positron emission tomography (PET), other biological markers, and clinical and neuropsychological assessment can be combined to measure the progression of mild cognitive impairment (MCI) and early Alzheimer's disease (AD) [28]. The ADNI database contains longitudinal data from clinical evaluations, cognitive tests, biological samples, and various types of imaging data, including MRI, functional MRI and PET. For up-to-date information, see https://adni.loni.usc.edu.

The study's data has been collected over several phases: ADNI-1, ADNI-GO, ADNI-2, and ADNI-3. The dataset contains cohorts of individuals with AD, MCI, and healthy controls (Cognitively Normal, CN). Furthermore, all exams underwent quality control for the image quality (e.g., subject motion, anatomic coverage). In this paper, we employed available MPRAGE T1-weighted MRIs from all phases, filtering out images deemed "unusable" by the quality control assessment. The images are available after gradient non-linearity and intensity inhomogeneity correction, when necessary. An overview of the demographics from each dataset can be seen in Table 2. More detailed information on the images from ADNI-1 used in this work can be found in Wyman et al. [35].

BraTS. The brain tumor segmentation challenge (BraTS) [25] provides a dataset of structural brain MRIs along with expert annotations of tumorous regions [2]. For each subject, four MRI scan modalities are available: T1-weighted, contrast-enhanced T1-weighted, T2-weighted, and T2-Flair. All images are available after preprocessing (registration to a common atlas, interpolation to 1 mm^3, and skull-stripping) [3]. In the 2020 edition[2], the dataset contained 369 images from subjects aged 18 to 86 (avg. of 61.2).

[1] In contrast to natural imaging datasets, as medical imaging requires an expensive procedure and legal authorization from each subject.

[2] https://www.med.upenn.edu/cbica/brats2020/data.html.

Table 2. Overview of the subjects from the ADNI database whose images were used in this work. Age is considered at the time of the first visit.

	CN	MCI	AD	Age Range (mean)	Sex
ADNI-1	229	401	188	55-91 (75.28)	342 F/476 M
ADNI-GO	0	142	0	55-88 (71.25)	67 F/75 M
ADNI-2	150	373	109	55-91 (72.61)	324 F/307 M
ADNI-3	456	354	64	51-97 (72.78)	475 F/399 M

3.2 Preprocessing

We applied minimal preprocessing to the ADNI images. Our major goal was to ensure all images would have the same orientation and spatial resolution, and that they would present no skull or non-brain-tissue information, i.e., only brain voxels would be present in the image. Only the brain information must be available in the image, otherwise, the deep learning models could learn to predict the age based on other structures.

We register all ADNI images to the MNI152 template, interpolate to 2 mm^3 resolution, and apply skull-stripping using HD-BET [16]. To feed the preprocessed 3D MRI scans to the 2D brain age models, each volume was sliced through the axial plane. We discard the slices from the top 40 mm and the bottom 35 mm of the scan to exclude slices with little to no brain information. Therefore, we extract 40 images from each 3D MRI scan.

With respect to the images from the BraTS dataset, no further (see Sect. 3.1) preprocessing is performed. We feed all slices of the 3D MRI scan to the deep learning models.

3.3 Deep Learning Models

We use 2D deep convolutional neural networks for brain age prediction. Our proposed model can be divided into a backbone and a head. Intuitively, the backbone is responsible for extracting relevant features from the input image, while the head combines these features into the final prediction. The backbone consists of several convolution filters that reduce the image size. The head is a single unit with linear activation that is fully connected to the backbone's output. Figure 1 illustrates our proposed model, highlighting both the backbone and the head.

Even though the architecture of the head is task-specific, the backbone's architecture depends only on a few characteristics of the input (e.g., number of channels, minimum size). Furthermore, learned features from one task can be useful for another, at least as a starting point, which is known as *transfer learning* [31]. This allows us to reutilize the backbone of models trained for different tasks, that is, we can extract the backbone from a model trained for some task and use it as the backbone for our model designed for brain age prediction.

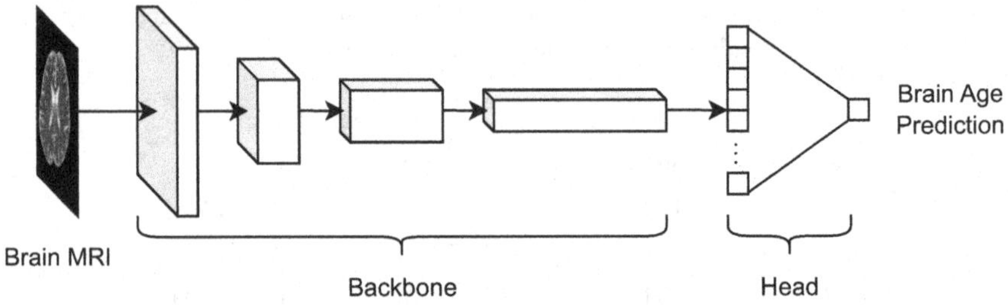

Fig. 1. Illustration of the overall architecture of our brain age model. The backbone is composed of several convolutional operations, extracted from the U-Net or the ResNet architectures. The head is a fully-connected layer applied to the vectorized values of the backbone's output.

We use the backbones from two different architectures: ResNet [12] and U-Net [32]. More specifically, we use the ResNet-50 architecture, available in the torchvision package [24], and the 2D U-Net proposed in [14], which is designed for medical image segmentation. The U-Net is composed of an encoder, a bottleneck, and a decoder. For our backbone, we use the encoder with the bottleneck.

To obtain the brain age prediction of a 3D MRI, we apply the model to the 40 axial slices of the image that contain brain information and take the mean of the outputs [4]. The pipeline of operations can be seen in Fig. 2.

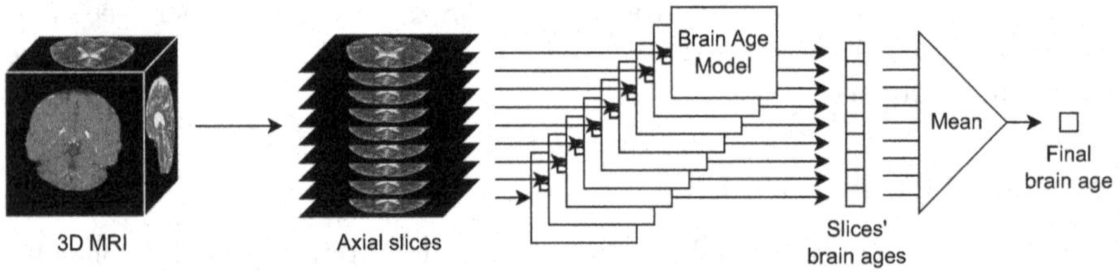

Fig. 2. Brain age prediction of a 3D MRI using a 2D deep learning model.

3.4 Pre-training on Brain Tumor Segmentation

To leverage the knowledge from other brain-related tasks to brain age prediction, we pre-train our backbones in the brain tumor segmentation task. We follow the BraTS challenge setup, with BraTS data, for both U-Net and ResNet backbones. To be able to train a ResNet backbone in a segmentation task, we replace the original head of the ResNet with a U-Net decoder, in a ResUNet architecture [36]. This means that the decoder matches the ResNet backbone with respect to the size of the intermediate feature maps so that the skip connections can be added in the same way as in the original U-Net implementation.

We follow Crimi et al. [15] for training all models on the BraTS data. We first train the models on a random 80/20 split of the BraTS 2020 dataset. The models are evaluated through the Dice score [25], which measures the overlap between the predicted segmentation mask $\hat{Y}$ and the ground truth Y as

$$\text{Dice}(\hat{Y}, Y) = \frac{2|\hat{Y} \cap Y|}{|\hat{Y}| + |Y|}.$$

Based on the models' performance, we fixed the number of epochs to avoid overfitting. Then, the entirety of the BraTS data is used to train the backbones for a fixed number of epochs.

3.5 Training on Brain Age Prediction

To train the deep learning models on brain age prediction, we assume, following previous work, that the brain age of healthy subjects is close to their chronological age [4,7,20]. Thus, we train and evaluate our models solely on the images of subjects belonging to the CN group. To provide easy-to-compare results, we choose to evaluate our models on a standardized dataset. Therefore, we follow the standardized split of the analysis set for ADNI-1 [35], and use the standard test set as our test set, and their training set as our validation set. In other words, we divide the preprocessed ADNI-1 T1-weighted scans from CN subjects following the standardized split to form our validation and test sets. The remaining images (i.e., those from ADNI-GO, ADNI-2, and ADNI-3) compose our training set. Detailed information on the images used in the training set can be found in our code repository[3].

We train all models, regardless of backbone architecture or pre-training, using the Adam optimizer to minimize the mean squared error between the age predicted from each slice and the true age of the CN subjects. The models are first trained on the training set. The performance of these models on the validation set is used for hyperparameter tuning and early stopping. Namely, batch size and learning rate were adjusted, and a moving average[4] of the MAE on the validation set was used to determine the ideal epoch (i.e., the one with the smallest MAE) for stopping the training. The models with the best performance on the validation set are then evaluated on the test set, which is unseen up to then.

3.6 Evaluation

The brain age models were evaluated on the error between the predicted age and the actual age of the CN subjects. More specifically, we use the mean absolute error (MAE) as our standard evaluation metric. We compute the error based on the predicted age of the whole 3D MRI, i.e., after averaging the predictions of all slices as described in Sect. 3.5 and illustrated in Fig. 2.

[3] https://github.com/gama-ufsc/brain-age.

[4] At the end of each epoch, we compute the average over the 5 latest results, including the current one.

Furthermore, we evaluate the capacity of the brain age estimate in differentiating between CN, MCI, and AD patients. For this, we use the brain age delta $\Delta_{BA} = \hat{y}_{BA} - y_{CA}$, which is the difference between the predicted brain age $\hat{y}_{BA}$ and the chronological age y_{CA} of a subject. As the progression toward Alzheimer's diagnosis is associated with aging patterns, it is expected that the Δ_{BA} of a subject in the AD group is greater than that of a subject in the MCI group, and that the latter's Δ_{BA} is still greater than the Δ_{BA} of a subject in the CN group. Therefore, we compute the predicted Δ_{BA} for all images in the three groups (CN, MCI, AD) of the test set and apply a pairwise Mann-Whitney U (MWU) test [9,23]. The MWU test is a nonparametric version of the t-test for independent samples. In our case, the null hypothesis is that the Δ_{BA} from one group is not stochastically greater than the other.

4 Experiments and Results

In our experiments, we evaluate the impact of brain-related pre-training using 3 backbones: the U-Net with random initialization, ResUNet with random initialization, and ResUNet pre-trained on the ImageNet. To improve the reliability of our results, 5 models are trained for each experiment, e.g., 5 U-Net models with random weight initializations are evaluated without brain-related pre-training, against 5 (different) U-Net models with brain-related pre-training. In other words, for each of the 3 backbones, we train 10 models: 5 with no brain-related pre-training, and 5 with brain-related pre-training.

All experiments reported below were performed on a Linux machine with 8 vCPUs, 30 GB of RAM and an Nvidia T4 GPU. Further details regarding the implementation of the experiments and additional results can be seen in code our repository[5].

4.1 Pre-training on BraTS

Following the procedure described in Sect. 3.4, we trained 5 U-Net models and 5 ResUNet models with random initialization on the brain tumor segmentation task. We also used the backbone from ResNets pre-trained on the ImageNet's natural image classification task, therefore, we trained 5 ResUNet models using the ImageNet pre-trained backbone. ImageNet pre-trained models are readily available in the torchvision package.

Using an 80-20 random split, we observed that 30 epochs were enough to achieve peak performance and avoid overfitting for the U-Net models, while 50 epochs were enough for the ResUNet models. The average performance of these models on the random split can be seen in Table 3. We highlight that the models present performance on par with state-of-the-art brain tumor segmentation models [3]. All models were then re-trained (with new initial random weights) on the entirety of the dataset for the same number of epochs.

[5] https://github.com/gama-ufsc/brain-age.

Table 3. Performance of the models on an 80-20 split of the BraTS dataset. Models were trained on 80% of the data and evaluated on 20%. Reported values are the average (and standard deviation) of 5 runs with random initialization, except for the backbone of the ResUNet that uses the backbone of the ResNet pre-trained on ImageNet on all runs.

Model	Dice Score
U-Net	0.8290 ($\sigma = 0.0049$)
ResUNet	0.8112 ($\sigma = 0.0246$)
ResUNet (ImageNet)	0.8129 ($\sigma = 0.0181$)

4.2 Brain Age Prediction

We train 5 models of each combination of backbone and pre-training available. Namely: U-Net backbone with random initialization or pre-trained on BraTS; ResNet backbone with random initialization or pre-trained on BraTS; and ResNet backbone pre-trained on ImageNet or pre-trained on ImageNet and then on BraTS. As described in Sect. 3.5, we use the validation set to define the best set of hyperparameters for each backbone and pre-training combination. More specifically, we used a batch size of 64 images for all models and a learning rate of 10^{-3} for the models with U-Net backbone without pre-training, 10^{-5} for the models with ResNet backbone with ImageNet and BraTS pre-training, and 10^{-4} for all other models. We trained all models for 50 epochs, early-stopping the training when a running average of the MAE on the validation set achieved the smallest value. The average performance of the models on the test set can be seen in Table 4.

Furthermore, after hyperparameter tuning, we re-train the models on the union of the training and the validation sets. This increases the amount of data used for training and increases the similarity between the distribution of the training data and the test data, as both validation and test sets are drawn from the same study (ADNI-1). We defined the training budget for each backbone and pre-training combination as the average of the epochs in which the respective 5 models achieved the early-stopping criterion on the validation set. Note that the test set is not considered in any step of this process, therefore, no data leakage occurs. The average test set performance of the models trained on the training and validation sets can be seen in Table 4, under column "Test MAE (train+validation models)".

In our experiments, pre-training on the brain tumor segmentation task was consistently advantageous. Even though the performance difference was not highly significant at all experiments, the models with the proposed pre-training outperformed their counterpart in all configurations except for the ResNet backbone pre-trained on the ImageNet when trained on the union of the train and validation sets. It is also evident, when comparing the ResNet backbones, that pre-training solely on BraTS was consistently better than pre-training on the ImageNet. Our experiments show that the best brain age predictions are achieved

Table 4. Average performance (over 5 runs) of brain age models. "train+validation models" indicates that the models were trained on the union of train and validation datasets, with hyperparameters defined on previous experiments using the validation set. Values in bold indicate the best performance in their respective set (column).

Backbone	BraTS pre-training	Validation MAE	Test MAE	Test MAE (train+validation models)
U-Net		3.203 ($\sigma = 0.042$)	3.358 ($\sigma = 0.099$)	3.138 ($\sigma = 0.069$)
	X	**3.186** ($\sigma = 0.084$)	**3.284** ($\sigma = 0.071$)	**3.079** ($\sigma = 0.077$)
ResNet-50		3.474 ($\sigma = 0.117$)	3.556 ($\sigma = 0.258$)	3.226 ($\sigma = 0.071$)
	X	3.361 ($\sigma = 0.079$)	3.413 ($\sigma = 0.123$)	3.141 ($\sigma = 0.030$)
ResNet-50 (ImageNet)		3.509 ($\sigma = 0.086$)	3.638 ($\sigma = 0.160$)	3.210 ($\sigma = 0.047$)
	X	3.452 ($\sigma = 0.086$)	3.411 ($\sigma = 0.066$)	3.238 ($\sigma = 0.022$)

by using a U-Net backbone pre-trained on BraTS, even though the ResNet backbones achieved a better brain tumor segmentation performance (see Table 3). Finally, the use of the validation set for training (after hyperparameter tuning) improved the performance of all models, as can be seen in the column "Test MAE (train+validation models)" of Table 4.

4.3 Statistical Analysis of Δ_{BA}

To assess the significance of the resulting brain age indicator, we apply the MWU test to the predicted Δ_{BA} of all models for subjects in the CN, MCI, and AD groups of the validation and the test sets (see Sect. 3.6). The tests on samples between CN and MCI patients and CN and AD patients all pointed to a strong differentiation, with p-values smaller than 0.1% for all models[6], indicating that the Δ_{BA} biomarker is useful to distinguish healthy patients from those with Alzheimer's disease or mild cognitive impairment. The distinction, however, between MCI and AD patients was not as significant, as can be seen in Table 5. Note that the images from MCI and AD patients of the validation set are not used for training or hyperparameter tuning, thus, the validation set can be interpreted as an additional test set in the case of differentiating between MCI and AD.

Even though the models with the U-Net backbones achieved more significant results in the statistical analysis of the biomarker, these results do not allow us to conclude that pre-training (with any of the tasks) had a positive impact, as was observed for brain age prediction. In fact, the exact opposite is observed. Using pre-trained backbones (for both BraTS and ImageNet pre-training) resulted, most of the times, in models that achieved a worse separation between AD and MCI patients, in comparison to their counterparts without pre-training. The

[6] The results of the tests on CN-MCI and CN-AD are available in our code repository https://github.com/gama-ufsc/brain-age.

Table 5. Significance of the MWU test on the distinction between AD and MCI patients using the predicted Δ_{BA} values. We report the average (maximum) p-value over the 5 models trained for each combination of backbone and pre-training. "train+validation models" indicates that the models were trained on the union of train and validation datasets (CN only), with hyperparameters defined on previous experiments using the validation set (CN only). Values in bold indicate high significance (p-value $\leq 5\%$).

Backbone	BraTS pre-training	Validation p-value	Test p-value	Test p-value (train+validation models)
U-Net		**0.011** (**0.017**)	**0.025** (0.053)	0.091 (0.107)
	X	**0.011** (**0.035**)	**0.036** (**0.049**)	0.190 (0.213)
ResNet-50		0.051 (0.066)	**0.046** (0.071)	0.192 (0.245)
	X	0.051 (0.064)	0.051 (0.089)	0.261 (0.298)
ResNet-50 (ImageNet)		**0.044** (0.055)	0.063 (0.112)	0.177 (0.230)
	X	0.095 (0.110)	0.171 (0.225)	0.345 (0.390)

same effect was also observed in the use of the validation set for training, upon which no model achieved a significant distinction between AD and MCI patients.

5 Discussion and Conclusions

In this study, we investigated the transfer learning capacity of a brain-related task to the task of brain age prediction. More specifically, we pre-trained deep learning models for brain age prediction on the task of brain tumor segmentation. In comparison to pre-training on a natural image classification task or performing no pre-training at all, our results suggest that the proposed pre-training may be a better option. The only inconclusive case is on performing both pre-trainings, that is, first pre-training on ImageNet and then on brain tumor segmentation, which yielded mixed results for brain age prediction.

Furthermore, we observed that using the validation set for training the models (after hyperparameter tuning) significantly reduced the brain age prediction error in all scenarios, even though there was only a small increment in the number of subjects in the training set (15% more subjects, as can be verified through Table 2). Using the U-Net backbone pre-trained on BraTS and then trained on the union of train and validation sets showed state-of-the-art results, with MAE values previously unseen on ADNI data (see Table 1). We recall that the validation and test sets are built with images from the ADNI-1 study, while the training set is built from ADNI-GO, ADNI-2, and ADNI-3. As the image acquisition protocols change between the studies, we can assume that there are different characteristics between their images. Therefore, we can expect a distribution shift between the training and the test sets that does not exist between the validation and the test sets. This allows us to conclude that the use of the validation set for training resulted in better models because it decreased the difference between the training and test distributions.

At the same time, by evaluating our models as biomarkers for cognitive impairment levels, we observed results that challenge the standard approach of training deep learning models for brain age. Most of the modifications that improved the brain age predictive performance on healthy subjects (i.e., the chronological age prediction), resulted in models that were less reliable for distinguishing between patients with MCI and AD. A similar behavior was reported by Bashyam et al. [4], in which the models that achieved the lowest MAE did not provide the strongest distinction between healthy subjects and subjects with AD, MCI, Schizophrenia or Depression.

Our results show an inverse relationship between the performance on chronological age prediction (see Table 4) and the reliability of the biomarker for cognitive impairment levels (see Table 5). This is particularly true for the use of the validation set for training brain age models, which significantly reduced the reliability in all scenarios. Given that reducing the distribution shift between training and testing data degraded the performance of the biomarker, we speculate that state-of-the-art models got to a point of overfitting the images of healthy subjects, thus, degrading their performance on images from AD and MCI patients. Therefore, we suggest that the validity of chronological age predictions as a means to develop brain age models is an important investigation topic for the development of the brain age biomarker.

References

1. Armanious, K., et al.: Age-Net: an MRI-based iterative framework for brain biological age estimation. IEEE Trans. Med. Imaging **40**(7), 1778–1791 (2021). https://doi.org/10.1109/TMI.2021.3066857
2. Bakas, S., et al.: Advancing the cancer genome atlas glioma MRI collections with expert segmentation labels and radiomic features. Sci. Data **4**(1), 170117 (2017). https://doi.org/10.1038/sdata.2017.117
3. Bakas, S., et al.: Identifying the Best Machine Learning Algorithms for Brain Tumor Segmentation, Progression Assessment, and Overall Survival Prediction in the BRATS Challenge (2018). https://doi.org/10.48550/ARXIV.1811.02629. Publisher: arXiv Version Number: 3
4. Bashyam, V.M., et al.: MRI signatures of brain age and disease over the lifespan based on a deep brain network and 14,468 individuals worldwide. Brain **143**(7), 2312–2324 (2020). https://doi.org/10.1093/brain/awaa160
5. Cole, J.H., et al.: Brain age predicts mortality. Mol. Psychiatry **23**(5), 1385–1392 (2018). https://doi.org/10.1038/mp.2017.62
6. Cole, J.H., Leech, R., Sharp, D.J., for the Alzheimer's Disease Neuroimaging Initiative: Prediction of brain age suggests accelerated atrophy after traumatic brain injury. Ann. Neurol. **77**(4), 571–581 (2015). https://doi.org/10.1002/ana.24367
7. Cole, J.H., et al.: Predicting brain age with deep learning from raw imaging data results in a reliable and heritable biomarker. Neuroimage **163**, 115–124 (2017). https://doi.org/10.1016/j.neuroimage.2017.07.059

8. Dinsdale, N.K., et al.: Learning patterns of the ageing brain in MRI using deep convolutional networks. Neuroimage **224**, 117401 (2021). https://doi.org/10.1016/j.neuroimage.2020.117401
9. Fay, M.P., Proschan, M.A.: Wilcoxon-Mann-Whitney or t-test? On assumptions for hypothesis tests and multiple interpretations of decision rules. Stat. Surv. **4**, 1–39 (2010). https://doi.org/10.1214/09-SS051
10. Franke, K., Ziegler, G., Klöppel, S., Gaser, C.: Estimating the age of healthy subjects from T1-weighted MRI scans using kernel methods: exploring the influence of various parameters. Neuroimage **50**(3), 883–892 (2010). https://doi.org/10.1016/j.neuroimage.2010.01.005
11. Gaser, C., Franke, K., Klöppel, S., Koutsouleris, N., Sauer, H.: Alzheimer's disease neuroimaging initiative: BrainAGE in mild cognitive impaired patients: predicting the conversion to Alzheimer's disease. PLoS ONE **8**(6), e67346 (2013). https://doi.org/10.1371/journal.pone.0067346
12. He, K., Zhang, X., Ren, S., Sun, J.: Deep residual learning for image recognition. In: 2016 IEEE Conference on Computer Vision and Pattern Recognition (CVPR), pp. 770–778 (2016). https://doi.org/10.1109/CVPR.2016.90
13. Deng, J., Berg, A.C., Li, K., Fei-Fei, L.: What does classifying more than 10,000 image categories tell us? In: Daniilidis, K., Maragos, P., Paragios, N. (eds.) ECCV 2010. LNCS, vol. 6315, pp. 71–84. Springer, Heidelberg (2010). https://doi.org/10.1007/978-3-642-15555-0_6
14. Isensee, F., Jaeger, P.F., Kohl, S.A.A., Petersen, J., Maier-Hein, K.H.: nnU-Net: a self-configuring method for deep learning-based biomedical image segmentation. Nat. Methods **18**(2), 203–211 (2021). https://doi.org/10.1038/s41592-020-01008-z
15. Isensee, F., Jäger, P.F., Full, P.M., Vollmuth, P., Maier-Hein, K.H.: nnU-Net for brain tumor segmentation. In: Crimi, A., Bakas, S. (eds.) BrainLes 2020. LNCS, vol. 12659, pp. 118–132. Springer, Cham (2021). https://doi.org/10.1007/978-3-030-72087-2_11
16. Isensee, F., et al.: Automated brain extraction of multisequence MRI using artificial neural networks. Hum. Brain Mapp. **40**(17), 4952–4964 (2019). https://doi.org/10.1002/hbm.24750
17. Jonsson, B.A., et al.: Brain age prediction using deep learning uncovers associated sequence variants. Nat. Commun. **10**(1), 5409 (2019). https://doi.org/10.1038/s41467-019-13163-9
18. Kondo, C., et al.: An age estimation method using brain local features for T1-weighted images. In: 2015 37th Annual International Conference of the IEEE Engineering in Medicine and Biology Society (EMBC), Milan, pp. 666–669. IEEE (2015). https://doi.org/10.1109/EMBC.2015.7318450
19. Lam, P., Zhu, A.H., Gari, I.B., Jahanshad, N., Thompson, P.M.: 3D grid-attention networks for interpretable age and Alzheimer's disease prediction from structural MRI (2020)
20. Lee, J., et al.: Deep learning-based brain age prediction in normal aging and dementia. Nat. Aging **2**(5), 412–424 (2022). https://doi.org/10.1038/s43587-022-00219-7
21. Liem, F., et al.: Predicting brain-age from multimodal imaging data captures cognitive impairment. Neuroimage **148**, 179–188 (2017). https://doi.org/10.1016/j.neuroimage.2016.11.005
22. Ly, M., et al.: Improving brain age prediction models: incorporation of amyloid status in Alzheimer's disease. Neurobiol. Aging **87**, 44–48 (2020). https://doi.org/10.1016/j.neurobiolaging.2019.11.005

23. Mann, H.B., Whitney, D.R.: On a test of whether one of two random variables is stochastically larger than the other. Ann. Math. Stat. **18**(1), 50–60 (1947). https://doi.org/10.1214/aoms/1177730491
24. Marcel, S., Rodriguez, Y.: Torchvision the machine-vision package of torch. In: Proceedings of the 18th ACM International Conference on Multimedia, Firenze, Italy, pp. 1485–1488. ACM (2010). https://doi.org/10.1145/1873951.1874254
25. Menze, B.H., et al.: The multimodal brain tumor image segmentation benchmark (BRATS). IEEE Trans. Med. Imaging **34**(10), 1993–2024 (2015). https://doi.org/10.1109/TMI.2014.2377694
26. More, S., Antonopoulos, G., Hoffstaedter, F., Caspers, J., Eickhoff, S.B., Patil, K.R.: Brain-age prediction: a systematic comparison of machine learning workflows. Neuroimage **270**, 119947 (2023). https://doi.org/10.1016/j.neuroimage.2023.119947
27. Peng, H., Gong, W., Beckmann, C.F., Vedaldi, A., Smith, S.M.: Accurate brain age prediction with lightweight deep neural networks. Med. Image Anal. **68**, 101871 (2021). https://doi.org/10.1016/j.media.2020.101871
28. Petersen, R.C., et al.: Alzheimer's disease neuroimaging initiative (ADNI): clinical characterization. Neurology **74**(3), 201–209 (2010). https://doi.org/10.1212/WNL.0b013e3181cb3e25
29. Poloni, K.M., Ferrari, R.J.: A deep ensemble hippocampal CNN model for brain age estimation applied to Alzheimer's diagnosis. Expert Syst. Appl. **195**, 116622 (2022). https://doi.org/10.1016/j.eswa.2022.116622
30. Popescu, S.G., Glocker, B., Sharp, D.J., Cole, J.H.: Local brain-age: a U-Net model. Front. Aging Neurosci. **13**, 761954 (2021). https://doi.org/10.3389/fnagi.2021.761954
31. Raghu, M., Zhang, C., Kleinberg, J., Bengio, S.: Transfusion: Understanding Transfer Learning for Medical Imaging. Curran Associates Inc., Red Hook (2019). https://dl.acm.org/doi/10.5555/3454287.3454588
32. Ronneberger, O., Fischer, P., Brox, T.: U-Net: convolutional networks for biomedical image segmentation. In: Navab, N., Hornegger, J., Wells, W.M., Frangi, A.F. (eds.) MICCAI 2015. LNCS, vol. 9351, pp. 234–241. Springer, Cham (2015). https://doi.org/10.1007/978-3-319-24574-4_28
33. Russakovsky, O., et al.: ImageNet large scale visual recognition challenge. Int. J. Comput. Vision **115**(3), 211–252 (2015). https://doi.org/10.1007/s11263-015-0816-y
34. Wang, J., Li, W., Miao, W., Dai, D., Hua, J., He, H.: Age estimation using cortical surface pattern combining thickness with curvatures. Med. Biol. Eng. Comput. **52**(4), 331–341 (2014). https://doi.org/10.1007/s11517-013-1131-9
35. Wyman, B.T., et al.: Standardization of analysis sets for reporting results from ADNI MRI data. Alzheimer's Dement. **9**(3), 332–337 (2013). https://doi.org/10.1016/j.jalz.2012.06.004
36. Zhang, Z., Liu, Q., Wang, Y.: Road extraction by deep residual u-net. IEEE Geosci. Remote Sens. Lett. **15**(5), 749–753 (2018). https://doi.org/10.1109/LGRS.2018.2802944

Reinforcement Learning and GAN

Applying Reinforcement Learning for Multiple Functions in Swarm Intelligence

André A. V. Escorel Ribeiro[1], Rodrigo Cesar Lira[1](✉), Mariana Macedo[2], Hugo Valadares Siqueira[3], and Carmelo Bastos-Filho[1]

[1] University of Pernambuco, Pernambuco, Brazil
{aaver,rcls,carmelofilho}@ecomp.poli.br
[2] University of Toulouse, Occitania, France
mmacedo@biocomplexlab.org
[3] Federal University of Technology, Paraná, Brazil
hugosiqueira@utfpr.edu.br

Abstract. Swarm intelligence (SI) algorithms have become popular due to their self-learning characteristics and adaptability to external changes. They can find reasonable solutions to complex problems without in-depth knowledge. Much of the success of these algorithms comes from balancing the exploration and exploitation tasks. This work evaluates the application and performance of a reinforcement learning approach applied to a well-known swarm intelligence algorithm, Particle Swarm Optimization (PSO). We use the reinforcement learning agent Proximal Policy Optimization (PPO) to dynamically change the swarm communication topology according to the problem. We analyze the PSO's behavior, influenced by the reinforcement learning agent, through methods such as interaction networks and fitness analysis. We show that the RL approach can transfer the knowledge learned from one function to other functions, and that dynamic changes of topology over time makes PSO much more efficient than setting only one specific topology, even when using a Dynamic topology. Our results then suggest that changing topologies might be more efficient than having a Dynamic topology, and that indeed Local and Global topologies have an important role in the best swarm performance. Our results take a step further on explaining the performance of SI and automatizing their use for non-experts.

Keywords: Proximal Policy Optimization · Particle Swarm Optimization · Reinforcement Learning · Swarm Intelligence

1 Introduction

Swarm Intelligence (SI) is a branch of Computational Intelligence in which a collective behavior is exhibited by a group of decentralized and self-organized simple reactive agents interacting with each other and the environment. The

M. C. Naldi and R. A. C. Bianchi (Eds.): BRACIS 2023, LNAI 14196, pp. 197–212, 2023.
https://doi.org/10.1007/978-3-031-45389-2_14

interaction among them generates a collective adaptation to allow them to solve complex problems [7]. The simple reactive agents are represented by positions in the search space with simple historical memories that explore it while exchanging information about their experiences with other agents. Based on the individual's experience and the information received by other members of the swarm, the agents can adapt the behavior in the search space and, with enough time, find and refine reasonable solutions to the presented problems. Swarm-based algorithms emerge as an alternative to classical optimization methods in high-dimensional optimization problems [1]. They are an excellent alternative for optimization problems since they considerably reduce the computational cost and do not require a complete understanding of the problem regarding the characteristics of the search space.

Many SI algorithms are based on animal social behavior metaphors, such as Ant colony optimization (ACO) [2] inspired by the behavior of ant colonies, Particle swarm optimization (PSO) [6] inspired by a flock of birds, Artificial Bee Colony (ABC) [5] inspired by the bee hives. Swarm-based meta-heuristics are applied in several problems, such as nuclear engineering [18] and diagnosing diseases [14], among others. We also find applications to solve tasks related to data science and image and signal processing [13,16,19].

Many efforts have been made to improve the performance of swarm-based algorithms. In the case of PSO, Xu et al. [21], and Wu et al. [20] suggested the application of reinforcement learning due to this method's characteristic of being able to learn which actions are best for a given state. It allows the dynamic modification of the behavior of the PSO through the adjustment of communication topology using a reinforcement learning agent, generating better results for complex problems and increasing the convergence speed. Recently, Lira et al. [8] proposed a self-adaptive metaheuristic that considers the real-time information acquired during execution. For the algorithm to adapt, a reinforcement learning agent collects information and chooses actions that modify the metaheuristic's behavior. Despite the good preliminary results presented by reinforcement learning to create advanced approaches for swarm-based algorithms, it needs to be clarified if behaviors learned in some scenarios could be adapted in other scenarios not experienced by the algorithms.

This paper evaluates the transfer learning capability of a reinforcement learning strategy when different topologies can be selected along the optimization with a PSO algorithm, seeking to understand how the changes influence the various observed metrics and providing information about the learning of reinforcement agents and possible patterns that can be found. We find that RL is able to transfer the knowledge from one function to the other functions, and that changing topologies can be more effective than using a dynamic topology for PSO.

This paper is divided as follows: Sect. 2 briefly describes Particle Swarm Optimization, Proximal Policy Optimization, and Interaction Networks. Section 3 describes the methodology and parameterization for the experiments. Section 4 presents our findings and results, and we finish in Sect. 5 with our conclusions.

2 Background

2.1 Particle Swarm Optimization

In 1995, Kennedy and Eberhart [6] proposed Particle Swarm Optimization (PSO) after observing the social behavior of flocks of animals, such as birds. PSO is one of the metaheuristics of swarm intelligence most known and used in the literature [9]. Particle Swarm Optimization consists of a group of simple agents, called particles, that will be scattered in a search space. While a stopping criterion is not reached, the particles update their velocities and positions at each iteration, keeping the information on the best solutions found by them ($\vec{p}_i$) and the best solutions found by their neighbors ($\vec{n}_i$). This information is used to calculate its next movement within the search space. In each PSO iteration, the velocity (v_i) and position (x_i) information of each particle is updated according to Eqs. 1 and 2. Eventually, with enough iterations, the swarm will likely return a solution approaching the optimum position in the search space.

$$\vec{v}_i(t+1) = \chi\Big\{\vec{v}_i(t) + c_1\epsilon_1[\vec{p}_i(t) - \vec{x}_i(t)] + c_2\epsilon_2[\vec{n}_i(t) - \vec{x}_i(t)]\Big\} \tag{1}$$

$$\vec{x}_i(t+1) = \vec{v}_i(t+1) + \vec{x}_i(t), \tag{2}$$

c_1 and c_2 are the acceleration coefficients, ϵ_1 and ϵ_2 are uniform random numbers, and χ is the constriction factor defined by $\chi = \frac{2}{|2-\varphi-\sqrt{\varphi^2-4\varphi}|}$, where $\varphi = c_1 + c_2$.

The communication topology defines the neighborhood in PSO. It describes the relations among particles, influencing the way the swarm behaves. Global (gbest) and Local (lbest) are two well-known topologies for PSO. Global is a fully connected topology where all particles can communicate with the entire swarm, allowing the best solution found to be shared quickly in the entire swarm. The Global topology causes a quick convergence and may not adequately explore the search space. For the Local topology, each particle is connected to k immediate particles in the swarm. It creates a ring-like communication structure using k=2. The particles have information only from the particles next to them so that the sub-swarms can independently converge on several optimal points. Local topology has slower convergence but allows a better exploration of the search space [7]. However, These two topologies are more suitable for specific problems, which led Oliveira et al. [4] to develop a more balanced topology that operates adaptively. In this approach, stagnant particles look for better particles to communicate. This approach adds a new attribute to the particle, called p_k-$failure$, which is incremented every iteration that the particle does not improve its fitness. If the p_k-$failure$ value exceeds a threshold (p_k-$failure^T$), the particle looks for a new neighbor to communicate. The choice of the neighbor for the particle to communicate with is probabilistic, based on roulette wheel selection, so that particles with better fitness have more chances of being chosen.

2.2 Reinforcement Learning

Reinforcement learning (RL) refers to learning which action should be taken in a situation (i.e., state) to achieve one or more goals [14]. At each iteration, a reinforcement learning agent receives observations of the current state scenario and takes an action from a list of actions allowed in that problem. After the action, the agent receives a reward, which measures whether the action was beneficial. Through trial and error, the agent maps and learns the actions that obtained the best rewards for the observed states, seeking to choose the ones that maximize the accumulative reward.

In reinforcement learning, the agent's strategy to map the relationship between the observed states and the actions that must be taken is called policy. The agent aims to find the best strategy (i.e., policy) within the environment that maximizes the reward function. Therefore, the agent can learn which policy works better for a given environment. Reinforcement learning has been used with optimization meta-heuristics, including PSO [14], seeking to improve the algorithm's convergence speed. For applications in continuous and complex problems where mapping the set of states and actions is difficult, it is possible to use a deep reinforcement learning approach. The name deep reinforcement comes from deep learning because of the use of deep neural networks to map the set of states and actions.

Proximal Policy Optimization (PPO) Schulman et al. [17] proposed the Proximal Policy Optimization (PPO), a policy gradient method more stable, efficient, and more straightforward than other predecessors, as Trust Region Policy Optimization (TRPO). PPO works to improve a policy, performing slight modifications. Its main improvements are using clipped surrogate objective, value function clipping, reward and layer scaling, orthogonal, and Adam learning rate [3]. PPO performed well in multiple benchmark problems for Reinforcement Learning [17].

2.3 Interaction Network

Oliveira et al. [11] proposed the Interaction Network (IN) aiming to understand the swarm dynamics better. IN is a framework that assesses the flow of information generated from the agents' interactions. The Interaction Network captures the exchange of information between agents, seeking to understand how the swarm influence each other. Oliveira et al. [12] have demonstrated that using interaction networks helps to compare, for instance, the balance between exploration and exploitation tasks across algorithms. IN is represented using a graph where each node represents an agent, and the edges represent the interactions between agents. The edges can be modelled in many ways, here, we modelled as shown in Eq. 3. In this network, we do not consider how much one particle influenced each other, but who influenced whom over time that signs the communication topology structure [10].

•

$$I_{t_{i,j}} = \begin{cases} 1 \text{ if } j \in \vec{n}_i \\ 0 \text{ otherwise} \end{cases} \quad (3)$$

IN can be evaluated individually or by the accumulation of successive networks. Iterations can be accumulated using a Time Window (TW) to capture the social interactions among agents in a frequency of iterations. The TW allows an analysis of which agents were neighbors to each other within an interaction interval so that large time windows make the interaction networks show the interactions that are most repeated. Short TWs contain the most recent interactions, while TWs = 1 contain instant interactions.

3 Methodology

We used the reinforcement learning framework for swarm intelligence, created by Lira et al. [8] based on Python programming language and RLLLib[1]. We chose the PSO in our experiments since it is a widely known and deployed swarm intelligence metaheuristics in the literature [9].

The simulation runs in time steps, allowing different swarm configurations. At the beginning of each time step, the RL agent acts, selecting Local, Global, or Dynamic topology. After a preset number of iterations, the RL agents evaluate the reward and the new simulation state before starting the same cycle until the stop criterion is reached. We used this preset number of iterations equal to 10 in this paper. At the end of the training stage, the agent should be able to recommend the best topologies for functions with similar characteristics. We used PPO [17] as the reinforcement learning agent to solve this problem. The RL agent is responsible for learning the topologies that best adapt to the tested functions and modifying the topology in the PSO based on the characteristics learned during execution.

We used simple and widely used functions to evaluate the algorithms in unimodal and multimodal scenarios for simplicity and first validation. Yet, it is challenging to appropriately cover all needed scenarios to evaluate a methodology [15], we argue that these scenarios are well-explored in the literature in regard to performance [9]. Thus, we selected Schwefel 2.21, Sphere, and Schwefel 2.22 unimodal functions, and Rastrigin, Griewank, and Schwefel multimodal functions. The search space chosen for these functions is based on Plevris and Solozano [15] (Table 1). Then, two instances of RL agents were trained for each type of function per scenario. We trained with Rastrigin for multimodal functions, and we used Schwefel 2.22 for unimodal functions. We expect that the RL learns the characteristics needed from each type of function by training only in one example.

A given metaheuristic may perform well on a function with few dimensions and poorly on a function with multiple dimensions. This problem is called the

[1] https://docs.ray.io/en/latest/rllib/index.html.

Table 1. Description of the benchmark functions used in the simulations.

	Function	Search Space	Equation
Unimodal	Sphere	$[-100, 100]$	$\sum_{i=1}^{D} \lvert x_i^2 \rvert$
	Schwefel 2.21	$[-100, 100]$	$max(\lvert x_i \rvert), i \in \{0...D\text{-}1\}$
	Schwefel 2.22	$[-100, 100]$	$\sum_{i=1}^{D} \lvert x_i \rvert + \prod_{i=0}^{D} \lvert x_i \rvert$
Multimodal	Rastrigin	$[-5.12, 5.12]$	$\sum{}^{D}{}_{i=1}(x_i{}^2 - 10cos(2\pi x_i) + 10 \cdot D$
	Griewank	$[-100, 100]$	$\frac{1}{4000} \sum_{i=1}^{D} x^2 - \prod_{i=0}^{D} cos(\frac{x_i}{\sqrt{i}}) + 1$
	Schwefel	$[-500, 500]$	$418.9829 \cdot D - \sum_{i=1}^{D} x_i sin(\sqrt{\lvert x_i \rvert})$

"Curse of Dimensionality" – a well-known problem in data science that refers to the phenomena that arise when analyzing and organizing information in spaces of many dimensions that do not occur when few dimensions are implemented [15]. Due to this problem, two different scenarios were empirically chosen for the number of dimensions and particles, Scenario 1 with 20 particles and 50 dimensions and Scenario 2 with 10 particles and 25 dimensions. In both scenarios, we used 1000 iterations as the stop criterion for the PSO simulations. These values were chosen to guarantee convergence for multidimensional problems [15], and we point out that the number of particles is smaller than usually used in the literature in order to make the problem more complex with a smaller dimensionality. Additionally, in PSO, we used the parameters $c_1 = c_2 = 2.05$, and $p_k\text{-}failure^T = 1$.

For the evaluation, we selected three metrics: (i) fitness, which is an indication of algorithm success; (ii) the distribution of the selected topology among Local, Global, and Dynamic, i.e., which actions the Reinforcement Learning agent recommended the most; (iii) interaction networks (IN) [10], which will be used to observe the accumulated interactions of the agents during a defined time interval, allowing us to analyse the importance of the selected topologies over time. We execute the PSO without reinforcement, seeking to observe how the topologies perform in each chosen function. This topology performance information is used to compare the results obtained with the reinforcement, so the topologies with the best performance in the tested functions might be the majority of the agent's recommendations.

4 Results

We divided our results into three subsections. First, we show the fitness performance of the RL proposal compared with PSO using Global, Local, and Dynamic topologies. Next, we focus on our proposal, evaluating the topologies chosen in each scenario. Finally, we analyze the Interactions Networks in the RL approach.

4.1 Evaluating the Fitness

In Figs. 1 and 2, we present the boxplots of the best fitness values found in 50 simulations in each scenario. We see that RL applied to PSO reached, in general, better performance than using the PSO with a specific topology for six functions.

For Scenario 1 (Fig. 1), using unimodal functions, we see that the RL performed as well as the best communication topology found for each function. RL could have been more efficient for the multimodal functions. In Schwefel function, it reached the worst results. In Scenario 2 (Fig. 2), we see our approach reaching competitive results. However, the results showed again that the training in Rastrigin led RL to achieve bad results in Schwefel.

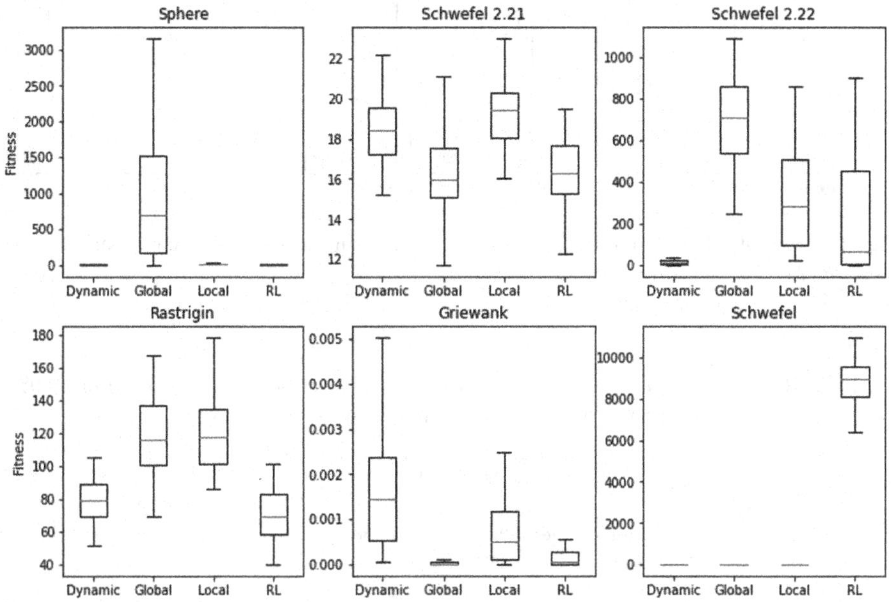

Fig. 1. Boxplot of the best fitness found on 50 simulations of each algorithm in Scenario 1.

We then compared the results using a signal-ranked Wilcoxon test with a confidence rate of 99.9% in Tables 2 and 3. '–' indicates no statistical difference between the solutions, '▲' indicates the RL approach achieved better results than the other algorithm, and '▼' represents that our proposal reached worse results than the algorithm compared. Based on the Wilcoxon test results, we can assure the RL capability for solving different functions, even when we train only one of them with similar characteristics. Only in Schwefel function it did not work well.

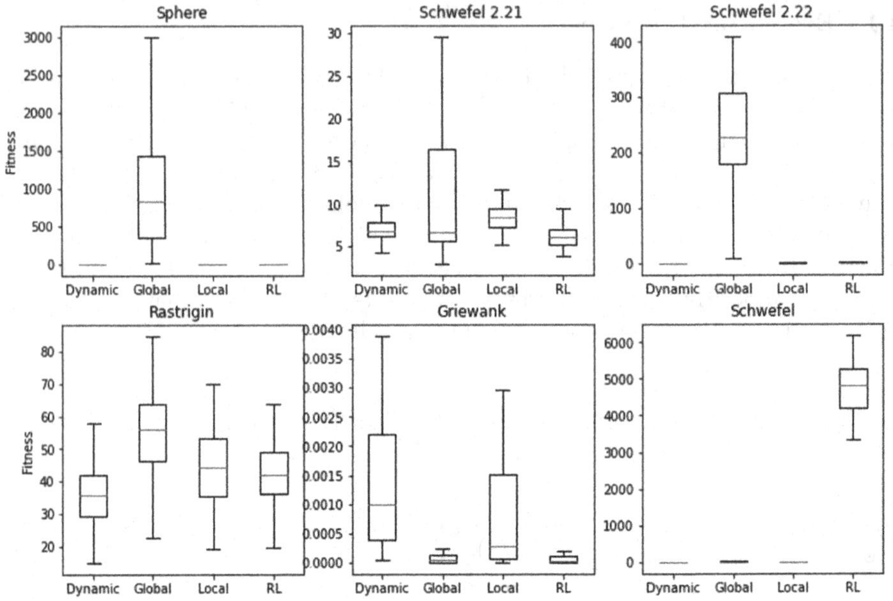

Fig. 2. Boxplot of the best fitness found on 50 simulations of each algorithm in Scenario 2.

Table 2. Results of fitness values and Wilcoxon test with a confidence level of 99.9% comparing the RL with the other algorithms for Scenario 1.

		Scenario 1			
Function		RL	Local	Global	Dynamic
Sphere	Mean Fitness	2.61	15.38	1163.9	2.92
	STD	4.72	9.87	1446.39	6.87
	Wilcoxon		▲	▲	–
Schwefel 2.21	Mean Fitness	16.25	19.25	16.17	18.46
	STD	1.67	1.39	1.83	1.65
	Wilcoxon		▲	–	▲
Schwefel 2.22	Mean Fitness	269.48	312.32	693.32	41.37
	STD	471.29	239.16	225.67	106.0
	Wilcoxon		–	▲	–
Rastrigin	Mean Fitness	71.90	121.60	115.97	80.0
	STD	16.93	23.80	23.97	14.30
	Wilcoxon		▲	▲	–
Griewank	Mean Fitness	0.000349	0.000764	0.040017	0.002407
	STD	0.000714	0.000855	0.279387	0.003118
	Wilcoxon		–	–	▲
Schwefel	Mean Fitness	8902.70	19.25	16.17	18.46
	STD	1031.57	1.39	1.83	1.65
	Wilcoxon		▼	▼	▼

Table 3. Results of fitness values and Wilcoxon test with a confidence level of 99.9% comparing the RL with the other algorithms for Scenario 2.

		Scenario 2			
Function		RL	Local	Global	Dynamic
Sphere	Mean Fitness	9.1e-07	3.5e-03	104.7	2.66e-04
	STD	3.0e-06	4.760e-03	1031.26	9.6e-05
	Wilcoxon		▲	▲	▲
Schwefel 2.21	Mean Fitness	6.28	8.46	12.3	7.05
	STD	1.51	1.71	10.73	1.53
	Wilcoxon		▲	–	–
Schwefel 2.22	Mean Fitness	10.9	58.55	242.55	0.15
	STD	68.82	126.82	114.5	0.44
	Wilcoxon		–	▲	–
Rastrigin	Mean Fitness	43.35	44.48	55.58	36.47
	STD	10.88	12.13	14.57	10.39
	Wilcoxon		–	▲	–
Griewank	Mean Fitness	2.45e-4	0.001318	0.051423	0.001681
	STD	0.000637	2.226e-3	0.2	0.001983
	Wilcoxon		▲	–	▲
Schwefel	Mean Fitness	4731.377060	8.5	12.3	7.05
	STD	733.83	1.71	10.73	1.53
	Wilcoxon		▼	▼	▼

4.2 Evaluation of the Selected Topologies

We now analyse the distribution of the selected topologies by RL over time steps. We expect that even by training on one example of a benchmark function (Schwefel 2.22 or Rastrigin), the RL agent will be able to learn a good policy for solving similar functions.

We plot the percentage of time that a topology was chosen over time step coupled with the best fitness evolution over iteration in Figs. 3 and 4. We can observe that in the first phase, "exploration phase", the Global topology was chosen most of the time, but the "exploitation phase" varies across experiments. The Dynamic topology was most chosen for the "exploitation phase" indicating that the swarm needs more diversity from the connections to improve the fitness. In the "exploration phase", being widely connected is more important than having a diverse set of connections. Therefore, regardless of being unimodal and multimodal functions, it might be true that diversity on the connections is better as the swarm starts to exploit.

The fitness improvement is larger while using the Global topology, but we argue that this is not due to the fact that this topology is more efficient for the swarm. Actually, this might be true because of the easiness of improving

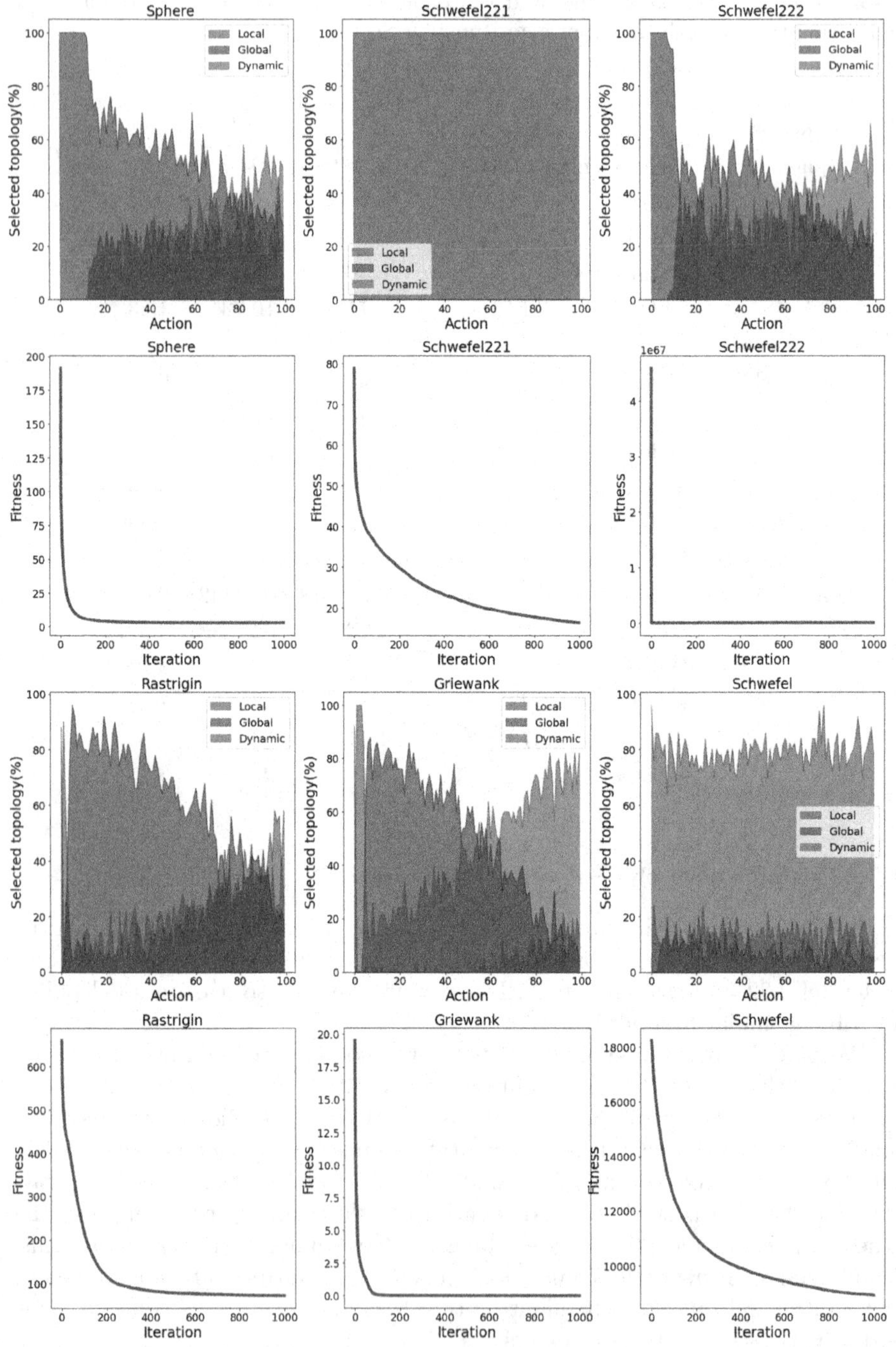

Fig. 3. Percentage of times that a topology was selected by the agent, with its respective fitness evolution on the bottom of each plot for Scenario 1.

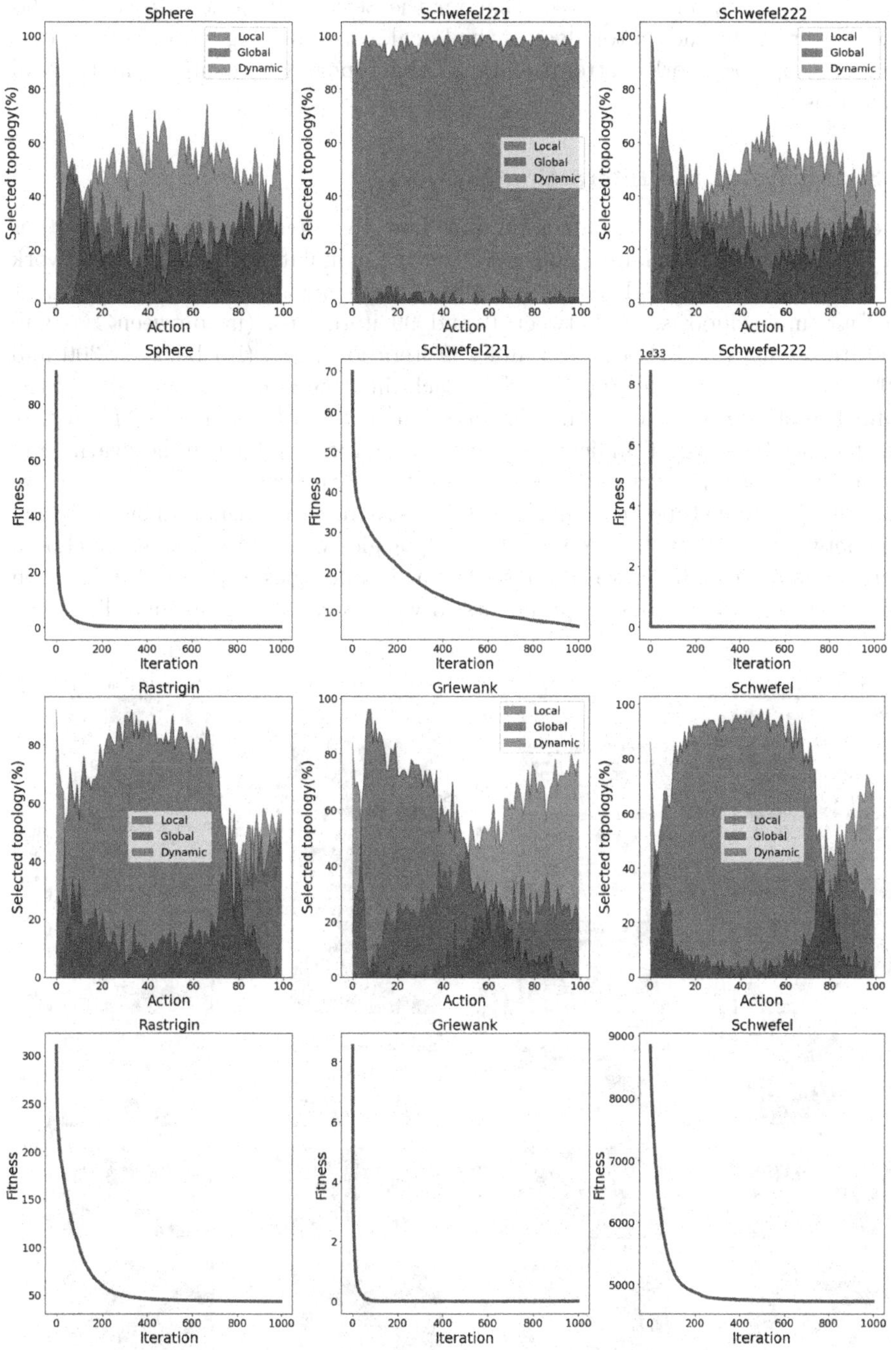

Fig. 4. Percentage of times that a topology was selected by the agent, with its respective fitness evolution on the bottom of each plot for Scenario 2.

in a "exploration phase". We see that for the Schwefel function the swarm did not converge, the chosen topology is the Local, corroborating with the literature that this topology works better than the Global topology for complex multimodal functions.

4.3 Analysing the Interaction Network

We are also interested in understanding how the agents influence each other in their movement over iterations. We use the cumulative Interaction Network (IN) to analyze the social interactions of the best simulation for each experiment in four-time windows: (i) between 0 and 99 iterations, (ii) between 100 and 199 iterations, (iii) between 200 and 299 iterations, and (iv) between 300 and 999 iterations, shown in Figs. 5 and 6. Each line represents the intensity of the influence of one particle on the displacement of the other particles. Therefore, strong lines (yellow-red) indicate particles that strongly influence the swarm, and strong columns represent particles that are strongly influenced by the swarm. We can identify which topology impacted the most across time windows by analyzing the networks. In Sect. 4.2, we see which topologies were more frequently chosen across simulations; here, we can observe which topologies impacted the most on the movement for the best experiments. If we observe strong diagonal lines and

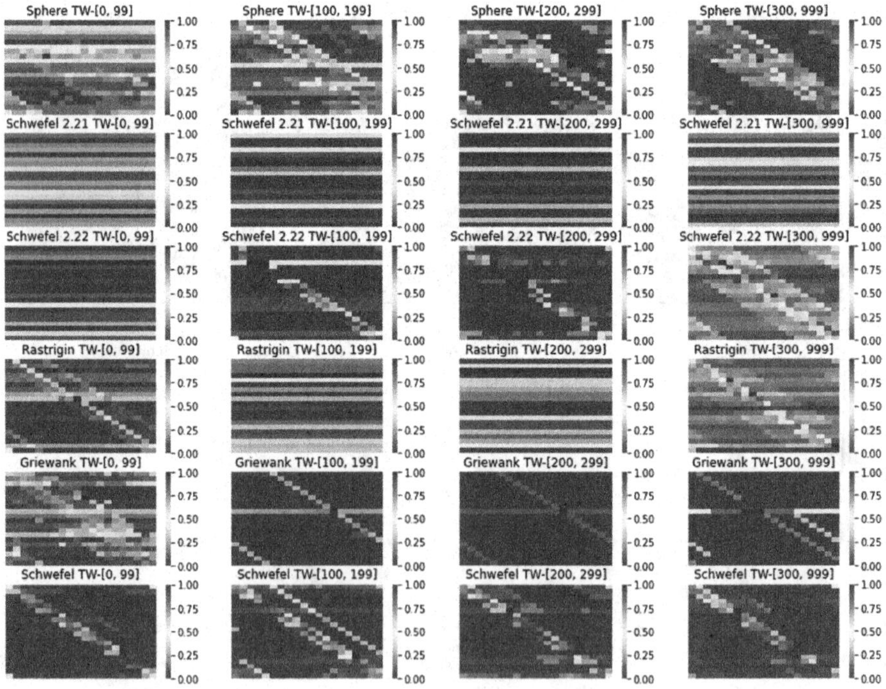

Fig. 5. Interaction Network generated from the simulation with the best fitness for each function of Scenario 1.

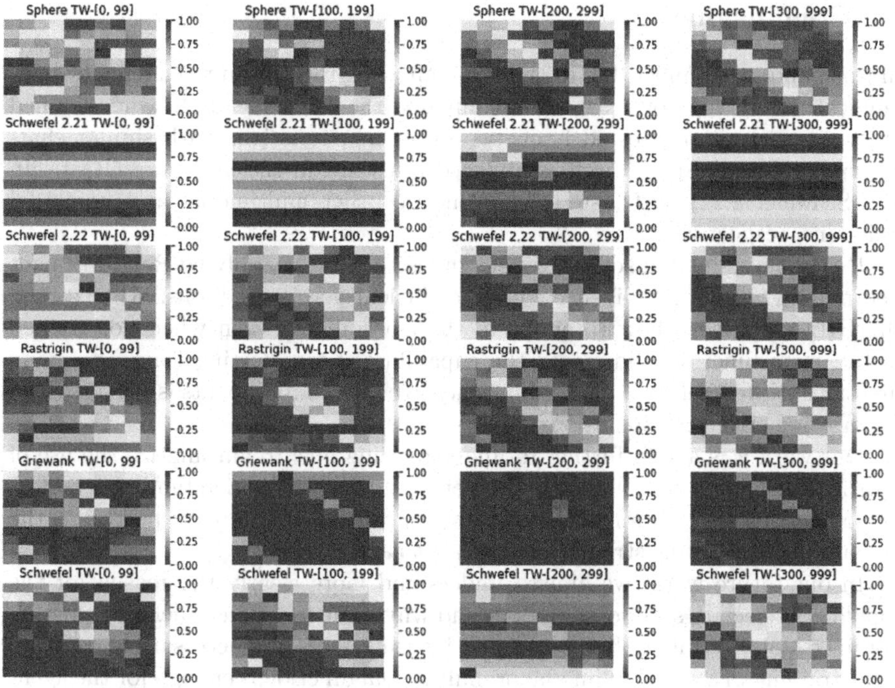

Fig. 6. Interaction Network generated from the simulation with the best fitness for each function of Scenario 2.

random points, the Local, Global, and Dynamic topology substantially affected the displacement, respectively.

We observe that for the best simulations, in Scenario 1 (Figs. 3 and 5), for unimodal functions, the Global topology was more chosen combined with the Dynamic at the end of the simulation which can be observed on the networks. Nevertheless, the Local topology strongly affects the movement from the middle to the end of the simulation (by looking at the diagonals from Sphere and Schwefel 2.22). For the multimodal functions, the Local topology also appears as an essential element for the best simulations, even though it was not the most chosen one for Rastrigin and Griewank functions.

In Scenario 2 (Figs. 4 and 6), we observe some similarities to Scenario 1, but the Dynamic topology is more present on the networks. The importance of the Dynamic topology is in line with its performance, depicted in Table 2. In contrast to the fact that the Global topology was frequently chosen for the multimodal functions, the effect of this topology could have been more substantial than the other topologies.

5 Conclusions

In this paper, we applied RL to the PSO, allowing the swarm to change its communication topology over time. We compared the efficiency of RL when trained on two functions and tested it on two other new functions with similar characteristics. We chose two well-established functions in the literature (Rastrigin and Schwefel 2.22) and tested them on two other multimodal and unimodal functions, respectively.

Using our simulated scenario, we demonstrated that applying Reinforcement Learning in Swarm Intelligence could be efficient across functions. We observed that RL could learn how to adapt to the environment even when not trained in the same function, indicating the capability of transferring learning among functions. Nevertheless, a more comprehensive set of experiments is still essential for drawing stronger conclusions.

Our work was a step further in understanding how to automatize the use of Swarm Intelligence for unknown problems and how to understand the performance and patterns from SI. Swarm Intelligence still requires expertise in the domain, so it is not as straightforward as it can become.

In our future work, we aim to understand more clearly the reason for the topologies selected, seeking to understand why some functions, such as Schwefel, obtained worse results in Reinforcement Learning. It is also necessary to evaluate if the training in a single unimodal or multimodal function is enough for the agent to learn the characteristics presented by the functions. The RL agent may more accurately identify the characteristics of the observed functions using multiple functions with the same characteristics in the training phase.

Acknowledgements. The authors thank the Federal Institute of Pernambuco (IFPE), Brazilian National Council for Scientific and Technological Development (CNPq), processes number 40558/2018-5, 315298/2020-0, and Araucaria Foundation, process number 51497, for their financial support. Mariana Macedo was supported by the Artificial and Natural Intelligence Toulouse Institute (ANITI) - Institut 3iA: ANR-19-PI3A-0004.

References

1. Bansal, J.C., Singh, P.K., Pal, N.R. (eds.): Evolutionary and Swarm Intelligence Algorithms, Studies in Computational Intelligence, vol. 779. Springer International Publishing, Cham (2019). https://doi.org/10.1007/978-3-319-91341-4
2. Dorigo, M., Maniezzo, V., Colorni, A.: Ant system: optimization by a colony of cooperating agents. IEEE Trans. Syst. Man Cybern. Part B (Cybern.) **26**(1), 29–41 (1996). https://doi.org/10.1109/3477.484436
3. Engstrom, L., et al.: Implementation matters in deep policy gradients: a case study on PPO and TRPO (2020)
4. Junior, M.A.C.O., Bastos Filho, C.J.A., Menezes, R.: Using network science to define a dynamic communication topology for particle swarm optimizers. In: Menezes, R., Evsukoff, A., González, M. (eds.) Complex Networks. Studies in Computational Intelligence, vol. 424, pp. 39–47. Springer, Heidelberg (2013). https://doi.org/10.1007/978-3-642-30287-9_5

5. Karaboga, D., et al.: An idea based on honey bee swarm for numerical optimization. Technical report, Technical report-tr06, Erciyes University, Engineering faculty (2005)
6. Kennedy, J., Eberhart, R.: Particle swarm optimization. In: Proceedings of ICNN 1995 - International Conference on Neural Network, vol. 4, pp. 1942–1948 (1995). https://doi.org/10.1109/ICNN.1995.488968
7. Kennedy, J.: Swarm Intelligence, pp. 187–219. Springer, US, Boston, MA (2006)
8. Lira, R.C., Macedo, M., Siqueira, H.V., Bastos-Filho, C.: Integrating reinforcement learning and optimization task: Evaluating an agent to dynamically select PSO communication topology. In: Tan, Y., Shi, Y., Luo, W. (eds.) Advances in Swarm Intelligence. ICSI 2023. LNCS, vol. 13969, pp. 38–48. Springer, Cham (2023).https://doi.org/10.1007/978-3-031-36625-3_4
9. Macedo, M., et al.: Overview on binary optimization using swarm-inspired algorithms. IEEE Access **9**, 149814–149858 (2021). https://doi.org/10.1109/ACCESS.2021.3124710
10. Oliveira, M., Bastos-Filho, C.J.A., Menezes, R.: Towards a network-based approach to analyze particle swarm optimizers. In: 2014 IEEE Symposium on Swarm Intelligence, pp. 1–8 (2014). https://doi.org/10.1109/SIS.2014.7011791
11. Oliveira, M., Bastos-Filho, C.J.A., Menezes, R.: Using network science to assess particle swarm optimizers. Soc. Netw. Anal. Min. **5**(1), 3 (2015). https://doi.org/10.1007/s13278-015-0245-5
12. Oliveira, M., Pinheiro, D., Andrade, B., Bastos-Filho, C., Menezes, R.: Communication diversity in particle swarm optimizers. In: Dorigo, M., et al. (eds.) ANTS 2016. LNCS, vol. 9882, pp. 77–88. Springer, Cham (2016). https://doi.org/10.1007/978-3-319-44427-7_7
13. Parpinelli, R.S., Lopes, H.S.: New inspirations in swarm intelligence: a survey. Int. J. Bio-Inspired Comput. **3(1)**, 1–16 (2011). https://doi.org/10.1504/IJBIC.2011.038700
14. Pervaiz, S., Ul-Qayyum, Z., Bangyal, W.H., Gao, L., Ahmad, J.: A systematic literature review on particle swarm optimization techniques for medical diseases detection. Comput. Math. Methods Med. **2021**, 1–10 (2021). https://doi.org/10.1155/2021/5990999
15. Plevris, V., Solorzano, G.: A collection of 30 multidimensional functions for global optimization benchmarking. Data **7**(4), 46 (2022). https://doi.org/10.3390/data7040046
16. Poli, R., Kennedy, J., Blackwell, T.: Particle swarm optimization: an overview. Swarm Intell. **1**, 33–57 (2007)
17. Schulman, J., Wolski, F., Dhariwal, P., Radford, A., Klimov, O.: Proximal policy optimization algorithms (2017). https://doi.org/10.48550/ARXIV.1707.06347
18. da Silveira Câmara Augusto, J.P., dos Santos Nicolau, A., Schirru, R.: PSO with dynamic topology and random keys method applied to nuclear reactor reload. Progr. Nucl. Energy. **83**, 191–196 (2015). https://doi.org/10.1016/j.pnucene.2015.03.009
19. Wauters, T., Verbeeck, K., De Causmaecker, P., Vanden Berghe, G.: Boosting metaheuristic search using reinforcement learning. In: Talbi, EG. (eds.) Hybrid Metaheuristics. Studies in Computational Intelligence, vol 434, pp. 432–452. Springer, Heidelberg (2013). https://doi.org/10.1007/978-3-642-30671-6_17

20. Wu, D., Wang, G.G.: Employing reinforcement learning to enhance particle swarm optimization methods. Eng. Optim. **54**(2), 329–348 (2022). https://doi.org/10.1080/0305215X.2020.1867120
21. Xu, Y., Pi, D.: A reinforcement learning-based communication topology in particle swarm optimization. Neural Comput. Appl. **32**(14), 10007–10032 (2020). https://doi.org/10.1007/s00521-019-04527-9

Deep Reinforcement Learning for Voltage Control in Power Systems

Mauricio W. Barg[1,2], Barbara S. Rodrigues[1], Gabriela T. Justino[1](✉), Kleyton Pontes Cotta[1], Hugo R. V. Portuita[1], Flávio L. Loução Jr.[1], Iran Pereira Abreu[3], and Antônio Carlos Pigossi Jr.[3]

[1] Research, Development and Innovation Dept (RD&I), Radix Engineering and Software Development, Rio de Janeiro, RJ, Brazil
{pdi,barbara.rodrigues}@radixeng.com.br
https://www.radixeng.com/

[2] Informatics Department, PUC-RJ, Rio de Janeiro, RJ, Brazil

[3] Real Time Operation, ISA-CTEEP, Jundiaí, SP, Brazil

Abstract. The great complexity and size of electrical power systems makes their operation and control a challenging task. Maintaining a stable voltage profile to assure the security and stability of the system is one of many tasks that must be conducted daily by power system operators and its automatic control equipment. This work proposes a deep reinforcement learning framework for controlling the equipment responsible for keeping the voltages across the system buses within their limits. More specifically, a smart agent that is capable of deciding the best course of action in order to keep the system's voltages within a specified range while taking into account system's conditions is proposed. Besides the traditional deep reinforcement learning approach, three novel reinforcement learning variations named windowed, ensemble and windowed ensemble Q-Learning, which alter the agent's learning process for voltage control, are presented and tested on IEEE 13, 37 and 123 bus systems, simulated on OpenDSS.

Keywords: Voltage Control · Deep Reinforcement Learning · Ensemble Learning

1 Introduction

Electric Power Systems (EPS) are complex systems with both physical and digital resources that must work together in order to deliver electricity safely and efficiently to all kinds of customers be them industrial, commercial or residential [17]. Over the last decades the demand for energy is constantly increasing as people wish for better quality of life [22]. With this growth, maintaining the quality of service (QoS) and system safety becomes harder and new tools and resources must be included into the process of operating and controlling the system.

Mainly, the operation of EPSs is executed by several trained professionals called system operators (SO) who must constantly observe the system's condition

M. C. Naldi and R. A. C. Bianchi (Eds.): BRACIS 2023, LNAI 14196, pp. 213–227, 2023.
https://doi.org/10.1007/978-3-031-45389-2_15

and conduct different tasks such as dispatch of generators, frequency control, fault mitigation, among others [29]. One of these tasks known as voltage control regards keeping voltage on all system buses between certain limits which are usually defined by regulatory associations and take into account the systems' correct functioning. As the demand fluctuates through the day, the voltages on system's buses also change and operators must maneuver and switch several equipment in order to keep it under control.

As most other tasks, voltage control must be executed by the SOs in real time who usually must react in a short time frame as the system's resources may be at stake. Voltage can also be controlled by automatic control equipment that react to voltage fluctuations according to some embedded logic using control mechanisms that monitor the voltage at certain system's points and issues control commands that keep it in between the predefined limits [3].

Due to the need for fast reaction, operators usually do not have time for detailed analysis of their actions. They rely on their experience and on actions that have worked in the past, but this may not be optimal because the system varies due to various unknown factors. Automatic control equipment also has fixed logic that does not consider the constantly changing dynamics of the system.

Over the years several different techniques have been proposed in order to address the voltage control problem. As it is formally known, *volt-var* control has seen applications of a plethora of different methods, such as classical optimization [24,25], meta-heuristics [20,21], classic control [1,2], neural networks [8], fuzzy logic [15,16], etc. Also, it has been successfully applied both to distribution and transmission systems.

There are many different approaches to voltage control since it can be seen as both a planning and a real-time operation problem [6,7,13,14] [13]. Online control uses different techniques and methodologies [11,12,23,32], from classical control to reinforcement learning. Usually, equipment such as shunt capacitor banks and OLTC transformers are controlled, but other digital technologies and distributed generation resources are also used. Recent studies have increasingly explored the use of reinforcement learning in energy systems, with different training techniques and architectures [30], Q-Learning [4,28], and deep reinforcement learning [5,31]. Bus voltages and branch power flows are used as state representations.

Most voltage control works focus on offline scenarios and use a variety of methods, mainly capacitor banks and transformer taps. When controlled online, adaptations are necessary, and reinforcement learning is a well-suited technique.

This work proposes a framework that uses reinforcement learning to control the system's voltage in real-time, taking into account equipment constraints and different system parameters. Three new methodologies that consider system complexity are also proposed. The methodologies are tested on three IEEE distribution circuits with 13, 37, and 123 buses.

2 Background

In this section, we will provide a brief description of the voltage control problem and discuss the application of reinforcement learning as a solution to the problem.

2.1 Voltage Control

The power system experiences fluctuations in consumption and generation, causing voltage to vary and potentially violate safety limits set by authorities. Power system operators have to make quick decisions to keep voltage within limits by maneuvering equipment or shedding loads, but this can sacrifice assertiveness. Due to the complexity of the power system, operators rely on previous experiences when controlling voltage, causing some equipment to be overused.

However, operators are not fully responsible for controlling voltages. There are equipment that have automatic control mechanisms and monitor voltages at specific points in the system, but these usually rely on simple feedbacks and do not always provide the best control decisions. Furthermore, as the system changes along the day, the parameter configuration of these controls would be better if adjusted, which most of the time, doesn't happen.

With this in mind, the tool proposed in this work aims to be an automatic system operator that is capable of replacing both human operators and equipment control by deciding and taking the best course of action in order to keep voltages inside its limits.

In later sections, the voltage control problem will be described further with more technical details.

2.2 Deep Reinforcement Learning

According to [27], reinforcement learning is a machine learning paradigm based on interactions between an agent and an environment. The agent interacts with the environment through actions and observations, evaluating the effect of its actions on the environment and seeking the actions that lead to the most desired outcomes. Essential elements for reinforcement learning include a state-action map, a policy to guide actions, a reward signal to evaluate the effect of actions on the environment, and intrinsic values for the states of the environment that represent the benefit of being in those states. By repeating the process of interaction, observation, and adaptation, it is expected that the agent will learn the best actions in an activity or process (Fig. 1).

In environments where the number of states are small, this method of exploring and learning is very effective. However, as the number of states grow in an environment, experiencing all states becomes unfeasible. In this context, the deep reinforcement learning methodology arises [9,10,18,19]. This variation allows the agent to generalize experiences for states it has not seen before by using a deep neural network

In further sections, the application of reinforcement learning to the voltage control problem will be discussed.

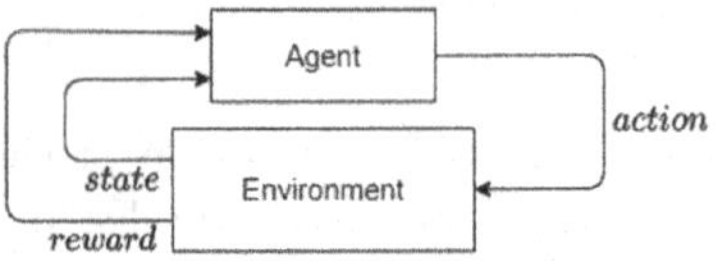

Fig. 1. Reinforcement learning process [27].

3 Problem Definition

As indirectly explained on previous sections:

> Voltage Control is the act of **operating** and **configuring** different **control equipment** in the **right moments** in order to keep **bus' voltages** within **specified limits**, which are determined by physical and commercial aspects.

More formally, the problem can be written as shown in Eq. 1.

$$\begin{array}{r} \min \sum_{t=0}^{T} \sum_{b \in B} distance(V_{bt}, \overline{V}) \\ \text{subject to} \\ V_{-} \leq V_{bt} \leq V_{+}\ (a) \\ V_{bt} = f(P, Q, [e_{0t}, e_{1t}, ..., e_{nt}])\ (b) \\ [e_{il}] \leq [e_{it}] \leq [e_{iu}], i = 0...n\ (c) \\ [-1] \leq [e_{it+1} - e_{it}] \leq [+1], i = 0...n\ (d) \end{array} \quad (1)$$

where t represents the current time step and T the maximum possible time step and n is the total number of equipment. In the real world this time would be continuous as voltage is controlled all the time. Although in order to simulate it, the operation period must be discretized. A certain system bus is represented by b and B is the set of all system buses. V_{bt} is the voltage at bus b on time t and $\overline{V}$ is the voltage target. For the constraints, in constraint (a), called the *voltage limit constraint*, V_{-} and V_{+} represent the lower and upper voltage limits across all buses and in constraint (b), V_{bt} is represented as a function of the system's active (P) and reactive (Q) powers and all equipment set-points at time t, which is how the voltage is essentially controlled (by changing equipment set-points). In constraint (c), also known as the *equipment set-point constraint*, since control equipment have different set-points restrictions, e_{il} and e_{iu} represent the minimum and maximum set-points that an equipment e_i can have at a certain time (t). Finally, constraint (d) called the *maximum set-point change*, represents how much the set-point of an equipment can change from one time step to another. This is to account for delays that an equipment may have in order to change its set-point.

It is important to note that the data needed for the optimization (i.e. the voltage on each bus) is obtained online and in real time. That means that V_{bt} can only be obtained at time step t and not before that. Therefore, the optimization process is conducted at each time step, as shown in Fig. 2.

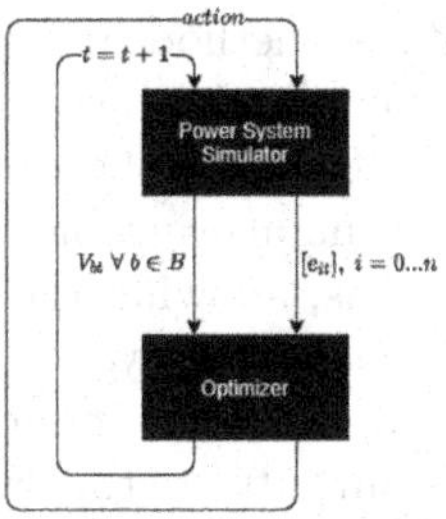

Fig. 2. The optimization process flow.

In order to control the voltage, the existing equipment in the network must be manipulated. In electrical networks, there are many kinds of equipment that are capable of regulating voltage: capacitors, reactors, transformers, synchronous condensers, FACTS equipment, etc. Although there is no clear advantage of using one equipment over another, different actions may have different short and long term impacts on the system.

Mainly, voltage can be controlled by two means: automatic control equipment, which monitor the network and adjust themselves accordingly and by network operators, who choose a certain action (based on experience and/or intuition) which they think is best for the current scenario. The automatic control equipment usually have fixed control logic which is certainly not suitable for every scenario. Furthermore, because there's usually no time to conduct detailed studies, operators tend to rely purely on their own knowledge and past experiences increasing the bias towards certain actions and overusing individual equipment.

4 Proposed Solution

Because power systems are composed by multiple discrete and continuous variables, the amount of possible configurations it can attain is *quasi-infinite*, rendering the classical tabular reinforcement learning not feasible. Therefore, a deep reinforcement learning technique is developed to deal with this characteristic.

Besides the usual deep reinforcement learning (DRL) approach, three different methodologies are proposed. These propositions involve slight modifications to the reinforcement learning technique that intend to adapt the procedure to specific characteristics of power system operation.

Many different kinds of reinforcement learning solutions exist depending on the type of problem that is being solved. For the voltage control problem, Q-Learning is a well suited method. Besides being very simple to implement, it is a model-free method, meaning that the model of the environment does not need to be fully known (which would be very hard for power systems). Furthermore, it can account for the effect that an action may have on a future state, that is not immediately achieved after taking said action.

Although there are many suitable methods to perform reinforcement learning, most of them have limitations, including Q-Learning, due to the need to create a finite map of state to actions. Discretizing the state can be infeasible or lead to inaccuracies if there are many continuous variables involved. Deep reinforcement learning overcomes these limitations, allowing the agent to generalize to states never seen before through a deep neural network.

In deep Q-Learning, the neural network's inputs are every variable that represents the system's state and its outputs are the "worth" or also called Q-Values for each action. During the training process, as the agent explores the environment, the neural network is trained with information from the system's state, the actions taken by the agent and their effects. During operation, the system's state is simply fed as an input to the network and the action with the highest Q-Value is chosen.

Over the years, many advancements have been made in deep Q-Learning. In this work, two of them are considered: double Q-Learning and experience replay. Both these techniques are better described in [10]. In addition, two new different modifications are proposed to the deep Q-Learning framework. These modifications do not modify how the learning process works, bur rather change how the techniques are applied to the problem at hand.

The first technique will be named *windowed* Q-Learning. In problems where episodes are too long, the agent may take longer to learn optimal actions for the entire length of the episode. Also, by being long, the episode may have very distinct behaviors on its course. That is, in a certain problem with a constant number of steps per episode, steps i to j may be very different in behavior from steps m to n. The proposed windowed Q-Learning methodology attempts to address this characteristics by dividing the episode into windows Fig. 3(a) and for each window a different agent is trained. That way, each agent only sees states relative to a certain interval and can learn the specificities of each window better when training. During operation, depending on which window the episode falls into, a different agent is consulted for making decisions.

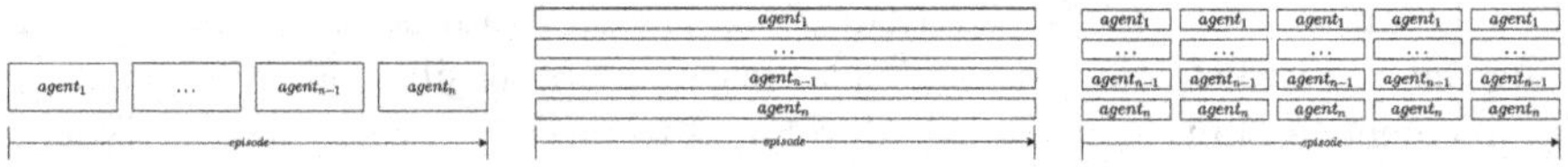

(a) Windowed. (b) Ensemble (c) Windowed Ensemble.

Fig. 3. Proposals Q-Learning.

The second technique will be called *ensemble* Q-Learning and takes inspiration from the way that humans learn. Because when subjected to different experiences people learn things differently, this technique proposes that multiple agents are trained for the same problem Fig. 3(b). These agents can either be equal and rely on the variability of the environment in order to experience things differently or can perceive the environment differently for example by having different reward functions. When operating, every agent is consulted when deciding which action to take. The decision can be made by following several criteria such as averaging every agent's value for the actions and taking the highest average (Eq. 2) or by taking the action with the maximum value over every agent (Eq. 3).

$$a = \operatorname{argmax} \begin{pmatrix} \operatorname{avg}(\phi_{11}, \phi_{12}, \phi_{1j}) \\ \operatorname{avg}(\phi_{21}, \phi_{22}, \phi_{2j}) \\ \operatorname{avg}(\phi_{i1}, \phi_{i2}, \phi_{ij}) \end{pmatrix} \tag{2}$$

$$a = \operatorname{argmax}(\max \phi_{ij}) \; \forall \; i, j \tag{3}$$

where ϕ_{ij} is the value given to action i by agent j.

The third and final technique is a combination of the other two and is called windowed ensemble Q-Learning. In addition to splitting the episode into windows, for each window multiple agents are trained Fig. 3(c). This allows for the combination of both methods' advantages. One disadvantage may be increased training times. During operation, the actions are also chosen by combining both methods: when the episode moment falls into a certain window, all agents trained for that window are consulted when choosing the action.

4.1 Voltage Control as a DRL Problem

In this section, the voltage control problem in an electrical system is modeled. Episodes have a fixed length of 1440, corresponding to each minute in a 24-hour period, and the system loads are updated at each step. The agent has the option to execute actions in the system, such as increasing or decreasing the stage of a capacitor bank, increasing or decreasing a transformer tap, or doing nothing. States are determined by the total active and reactive loads of the system, the current minute of the simulation, and the states of the transformers and capacitor banks. The state representation has a direct impact on the agent's learning, and rewards are defined to lead the voltage at all system buses closer to a certain target and keep them within certain operational limits. The rewards are defined following Algorithm 1. All learning parameters are detailed in the Case Study section.

5 Case Study

In this chapter the proposed techniques will be tested on simulated power systems. The utilized systems are the 13, 37 and 123 bus IEEE test circuits [26]. For each system, four models were trained: a pure reinforcement learning model, a windowed model, a ensemble model and a windowed ensemble model. Additionally, for comparison effect the system was run without any form of control and also with the control capabilities available at the simulation software (OpenDSS).

For every system, the parameters used during training were very similar with the exception of the number of episodes which was 100 for both the 13 and 123

Algorithm 1: Rewards

Initialize $r = 0$ **if** *action a was the opposite of an action taken in the last 30 minutes* **then**
  $r = r - v_1$
for *each bus* **do**
  if *voltage got closer to the target* **then**
    $r = r + v_2$
  else if *voltage got further from the target* **then**
    $r = r - v_3$
  else if *the action chosen was different from "do nothing"* **then**
    $r = r - v_4$
  else if *voltage is at* $\pm 1\%$ *from the target* **then**
    $r = r + v_5$
  else if *voltage violates upper or lower limits* **then**
    $r = r - v_6$
  else
    $r = r + v_7$

bus systems and 200 for the 37 bus system. For the double deep Q-Learning with experience replay, the parameters are as follows: the used memory size N was 1024, the minimum number of samples to start training was 512 and the batch size, 256. The discount factor γ was 0.9. For epsilon, instead of keeping it static, a decay technique was used. That way, epsilon decays linearly with the episodes from 1.0 to a minimum of 0.3.

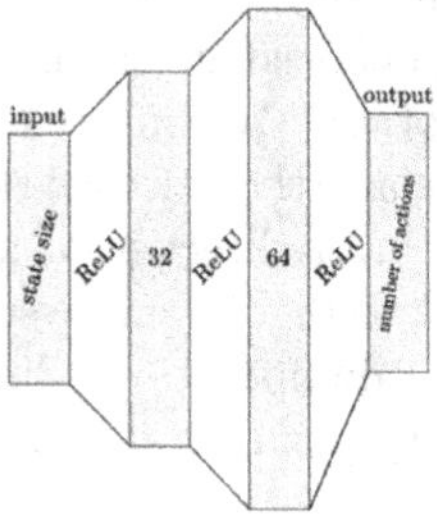

Fig. 4. Neural Network Structure.

The neural network structure is shown on Fig. 4 and its learning rate α was 0.001. The chosen architecture is quite simple when considering its size and activation functions. This is due to the nature of deep reinforcement learning which does not require complex structures in order to predict the action values and also because of the tested systems' sizes. Its input and output layer sizes depends on the system's characteristics, that is the size of its state and the number of available actions, both of which depends on the number of capacitors and tap changing transformers that can be controlled. This data is shown on Table 1.

Table 1. Test Circuits' State and Action Space Sizes.

	13-bus	37-bus	123-bus
State Size	10	7	15
nº of Actions	15	9	25

Finally, for the rewards as shown in Algorithm 1, the values used ($v_1 = -1$, $v_2 = 0.7$, $v_3 = -0.8$, $v_4 = -0.8$, $v_5 = 1$, $v_6 = -1$ and $v_7 = 0.2$). The upper and lower voltage limits considered are 1.05 and 0.92 p.u. while the target is 1.0 p.u. An exception is the ensemble approach where two alternative agents with more rigorous limits are used. The first alternative agent uses 1.03 and 0.95 p.u. for the upper and lower limits while the second uses 1.02 and 0.98 p.u. Also, for the ensemble agents the "vote" on the best action is conducted by following Eq. 2.

6 Results and Discussion

After the training procedure, the agents were tested on the same systems on different days. The days were simulated by randomly choosing a load profile and introducing some random noise into the loads. As stated in the previous section, the agents' performance is compared to the systems' without any form of control, with the control present on the system itself and also with a "pure" deep reinforcement learning approach as proposed by [9,10,18,19] and implemented, with slight variations, by [4,5,28,30,31].

In order to obtain the results, a total of 50 days for each technique and system were simulated. A few metrics are used to show the achieved results: the total number of violations, that is, the number of times the voltage surpassed the limits of 1.05 and 0.92 p.u., the maximum and minimum voltage achieved across the 50 days, the average real power loss across the 50 days and for the reinforcement learning methods, the number of actions the agents took (Tables 2a, 2b and 2c).

For the number of actions, the values are only available for the reinforcement learning techniques, since for the no control no actions are taken and for the system control method, it is not possible to obtain this value from the simulation. Furthermore, a separate day is simulated in order to closely demonstrate the results and the average system voltage across the day is shown, along with the distribution of the voltages during the day (Fig. 5 and 6) and a diagram showing the voltage during the day at each system bus (Fig. 7, 8 and 9).

Regarding the training process, all agents were satisfactorily trained. The training times were varied but the *windowed* methodology has shown a reduced training time even when compared to the traditional *reinforcement learning* approach. The results indicate that the proposed techniques are capable of controlling the voltage on power systems.

Table 2. System Results.

(a) 13-bus.

	Violations	Max Voltage	Min Voltage	Avg Losses (kW)	Avg Actions
No Control	0	1.049	0.925	115294.8	not applicable
System Control	68201	1.066	0.940	114300.1	not applicable
Reinforcement Learning	3552	1.052	0.912	117684.6	3.5
Windowed	0	1.050	0.926	116669.5	3.6
Ensemble	21789	1.062	0.934	115054.9	6.4
Windowed Ensemble	33434	1.059	0.944	114875.3	11.3

(b) 37-bus.

	Violations	Max Voltage	Min Voltage	Avg Losses (kW)	Avg Actions
No Control	66	1.034	0.919	154368.0	not applicable
System Control	326	1.031	0.919	154992.1	not applicable
Reinforcement Learning	31026	1.045	0.899	155482.1	29.0
Windowed	0	1.042	0.935	155700.8	24.5
Ensemble	0	1.042	0.932	154746.1	23.1
Windowed Ensemble	0	1.040	0.942	158196.7	20.3

(c) 123-bus.

	Violations	Max Voltage	Min Voltage	Avg Losses (kW)	Avg Actions
No Control	0	1.048	0.969	97619.3	not applicable
System Control	0	1.050	0.976	97532.7	not applicable
Reinforcement Learning	0	1.037	0.958	109314.9	4.8
Windowed	0	1.037	0.961	109851.9	7.5
Ensemble	0	1.034	0.958	97665.4	2
Windowed Ensemble	105	1.051	0.959	112275.1	7.4

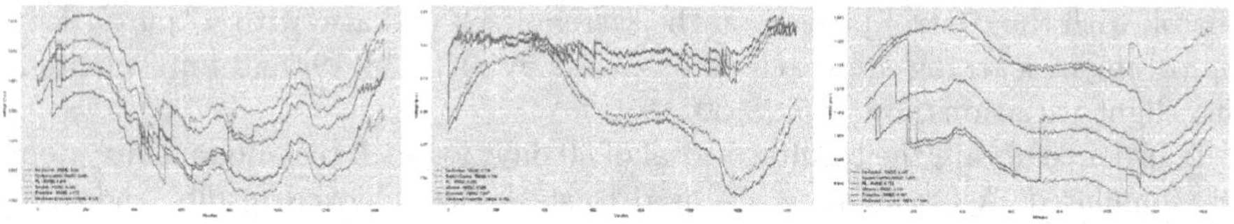

(a) IEEE 13-bus. (b) IEEE 37-bus. (c) IEEE 123-bus.

Fig. 5. Average system voltage during the day results.

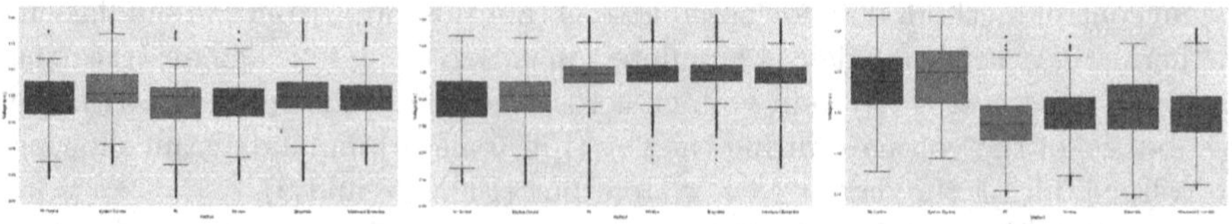

(a) IEEE 13-bus. (b) IEEE 37-bus. (c) IEEE 123-bus.

Fig. 6. Voltage distribution during the day results.

For the 13-bus system, because it's already a fairly balanced and small system, the results were marginal. Nevertheless, the *windowed* methodology was capable of removing all voltage violations when compared to the control already present on the system while keeping the number of required actions low and not affecting the real power losses significantly. Most importantly, Fig. 5(a), 6(a), and 7 show also that the system's voltage profile was improved with the voltages getting closer to the target.

For the 37-bus system, while the pure *reinforcement learning* approach actually increased the number of voltage violations, the other three proposed methodologies completely eliminated them. The improvements are clear on the results shown on Fig. 5(b), 6(b), and 8. It is possible to see that the voltages are much closer to the target of 1 p.u. Also, the voltage profile in general is much closer to ±1% of the target green areas on Fig. 8. In this case, the *windowed ensemble* methodology has executed the task in the least number of actions. For all the methodologies, the average losses were kept fairly close to its original values.

Finally, for the 123-bus system, while there were no violations on the system both with and without control, the proposed methodologies were capable of significantly improving its voltage profile. For this system, the average losses for three of the four methodologies were increased by a significant margin. The *ensemble* technique, has shown a better performance overall, controlling the voltage satisfactorily while also keeping the losses at a better acceptable value and the number of actions at a minimum.

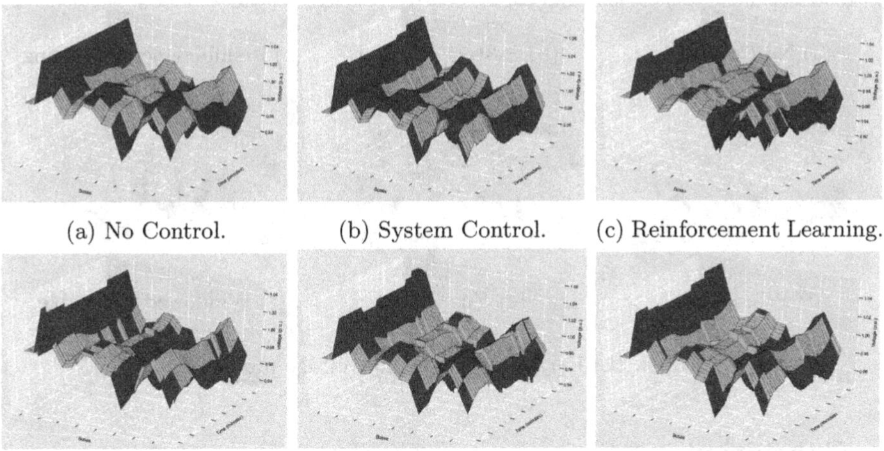

(a) No Control. (b) System Control. (c) Reinforcement Learning.

(d) Windowed. (e) Ensemble. (f) Windowed Ensemble.

Fig. 7. IEEE 13-bus single-day voltage profile.

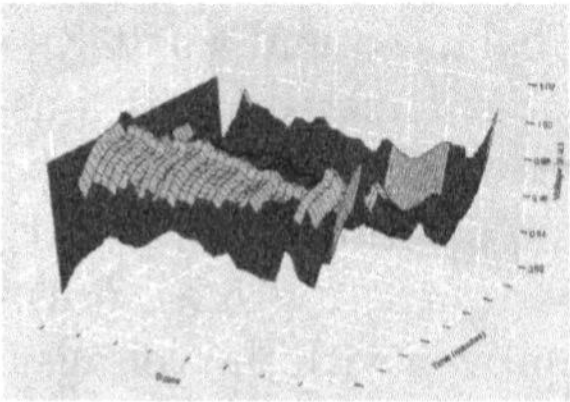

(a) No Control.

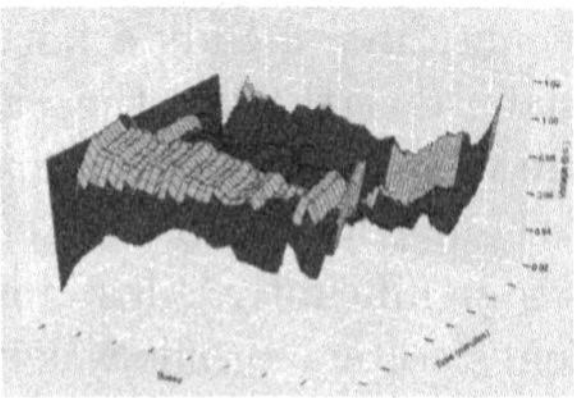

(b) System Control.

(c) Reinforcement Learning.

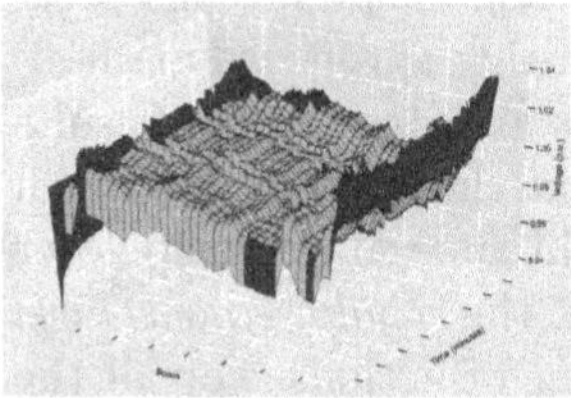

(d) Windowed.

(e) Ensemble.

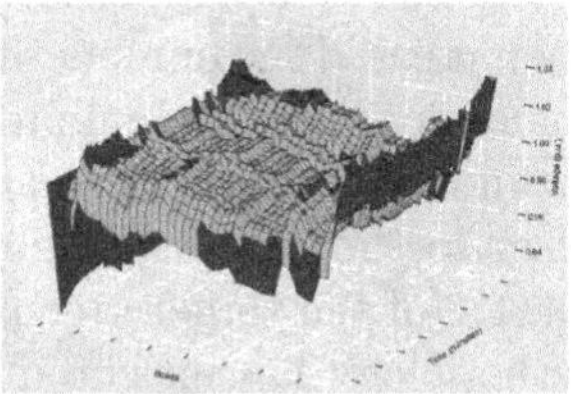

(f) Windowed Ensemble.

Fig. 8. IEEE 37-bus single-day voltage profile.

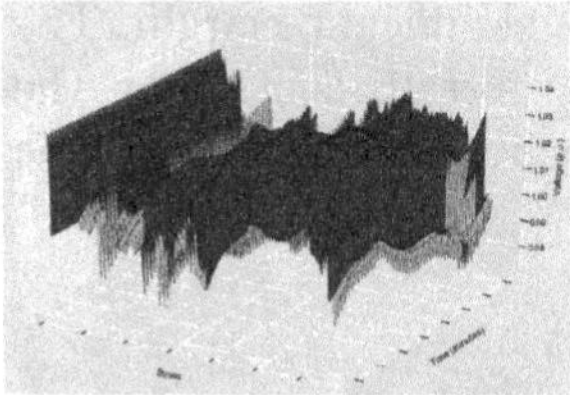

(a) No Control.

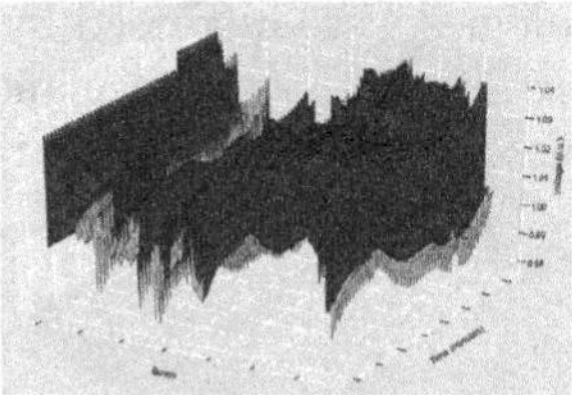

(b) System Control.

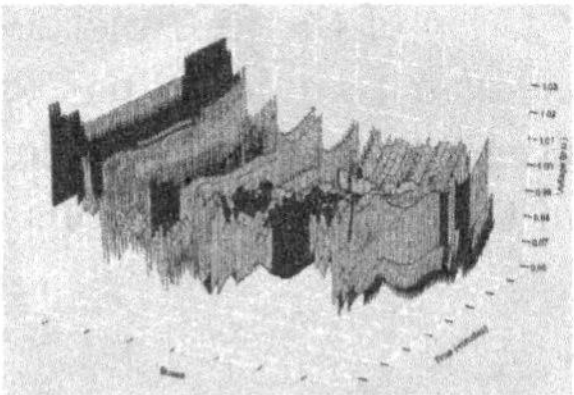

(c) Reinforcement Learning.

(d) Windowed.

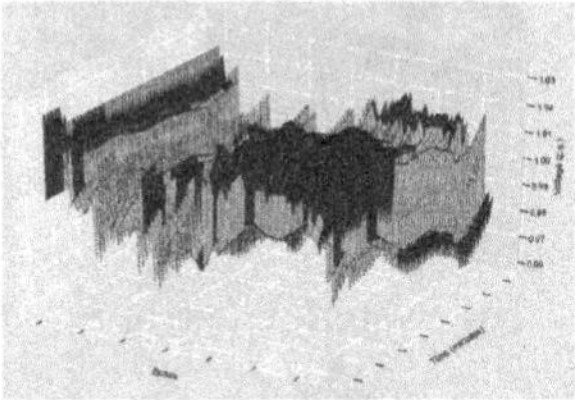

(e) Ensemble.

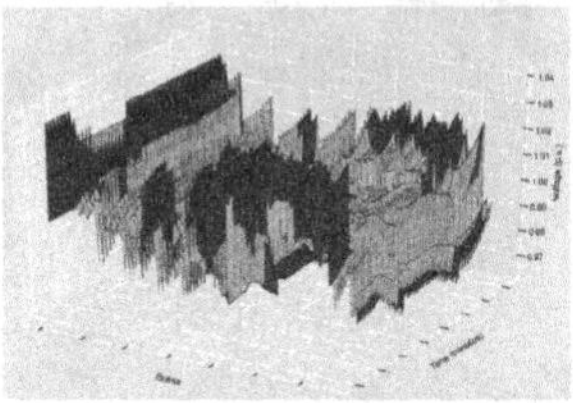

(f) Windowed Ensemble.

Fig. 9. IEEE 123-bus single-day voltage profile.

7 Conclusion

Reinforcement Learning has been around for some time and has shown great results across many different scientific and real-world problems. When combined with the power of deep neural networks, deep Q-Learning can tackle a plethora of problems. In this work, a deep Q-Learning methodology was proposed to solve the voltage control problem on electrical power systems. Besides the regular reinforcement learning approach, three other novel methodologies were proposed with the goal of improving the technique's performance on this specific problem. The results show that the application of the techniques was successful and has shown great value when compared to the traditional DRL approach and also with the systems' own control. The trained intelligent agents are capable of controlling the system voltage in a completely autonomous way while keeping the number of actions taken low and having little effect on the real power losses.

Acknowledgments. The authors would like to thank Companhia de Transmissão de Energia Elétrica Paulista (ISA-CTEEP) for the financial support to Project PD-00068-0044-2019 - intelligent real-time decision support tool for transmission operations centers, developed under the Research and Development program of the National Electric Energy Agency (ANEEL R&D), which the engineering company carried out Radix Engenharia e Software S/A, Rio de Janeiro, Brazil.

References

1. Baran, M.: Ming-Yung Hsu: Volt/VAr control at distribution substations. IEEE Trans. Power Syst. **14**(1), 312–318 (1999). https://doi.org/10.1109/59.744549. https://ieeexplore.ieee.org/document/744549/
2. Borozan, V., Baran, M., Novosel, D.: Integrated volt/VAr control in distribution systems. In: 2001 IEEE Power Engineering Society Winter Meeting. Conference Proceedings (Cat. No. 01CH37194), Columbus, OH, USA, vol. 3, pp. 1485–1490. IEEE (2001). https://doi.org/10.1109/PESW.2001.917328. http://ieeexplore.ieee.org/document/917328/
3. Corsi, S.: Voltage Control and Protection in Electrical Power Systems: From System Components to Wide-Area Control. Springer, London (2015). https://doi.org/10.1007/978-1-4471-6636-8
4. Custódio, G., Ochoa, L., Trindade, F., Alpcan, T.: Using Q-learning for OLTC voltage regulation in PV-rich distribution networks (2020)
5. Diao, R., Wang, Z., Shi, D., Chang, Q., Duan, J., Zhang, X.: Autonomous voltage control for grid operation using deep reinforcement learning (2019). http://arxiv.org/abs/1904.10597
6. Franco, J.F., Rider, M.J., Lavorato, M., Romero, R.: A mixed-integer LP model for the optimal allocation of voltage regulators and capacitors in radial distribution systems. Int. J. Electr. Power Energy Syst. **48**, 123–130 (2013). https://doi.org/10.1016/j.ijepes.2012.11.027. https://linkinghub.elsevier.com/retrieve/pii/S0142061512006801
7. Gallego, R., Monticelli, A., Romero, R.: Optimal capacitor placement in radial distribution networks. IEEE Trans. Power Syst. **16**(4), 630–637 (2001). https://doi.org/10.1109/59.962407. https://ieeexplore.ieee.org/document/962407/
8. Gu, Z., Rizy, D.: Neural networks for combined control of capacitor banks and voltage regulators in distribution systems. IEEE Trans. Power Deliv. **11**(4), 1921–1928 (1996). https://doi.org/10.1109/61.544277. https://ieeexplore.ieee.org/document/544277/
9. van Hasselt, H., Guez, A., Silver, D.: Deep reinforcement learning with double Q-learning (2016). http://arxiv.org/abs/1509.06461
10. Hessel, M., et al.: Rainbow: combining improvements in deep reinforcement learning. In: Proceedings of the AAAI Conference on Artificial Intelligence, vol. 32 (2018)

11. Homaee, O., Zakariazadeh, A., Jadid, S.: Real-time voltage control algorithm with switched capacitors in smart distribution system in presence of renewable generations. Int. J. Electr. Power Energy Syst. **54**, 187–197 (2014). https://doi.org/10.1016/j.ijepes.2013.07.010. http://linkinghub.elsevier.com/retrieve/pii/S0142061513003074
12. Hu, Z., Wang, X., Chen, H., Taylor, G.: Volt/Var control in distribution systems using a time-interval based approach. IEE Proc.-Gener. Transm. Distrib. **150**(5), 548 (2003). https://doi.org/10.1049/ip-gtd:20030562
13. Khanabadi, M., Ghasemi, H., Doostizadeh, M.: Optimal transmission switching considering voltage security and N-1 contingency analysis. IEEE Trans. Power Syst. **28**(1), 542–550 (2013). https://doi.org/10.1109/TPWRS.2012.2207464. https://ieeexplore.ieee.org/document/6253283/
14. Levitin, G., Kalyuzhny, A., Shenkman, A., Chertkov, M.: Optimal capacitor allocation in distribution systems using a genetic algorithm and a fast energy loss computation technique. IEEE Trans. Power Deliv. **15**(2), 623–628 (2000). https://doi.org/10.1109/61.852995. https://ieeexplore.ieee.org/document/852995/
15. Liang, R.H., Chen, Y.K., Chen, Y.T.: Volt/Var control in a distribution system by a fuzzy optimization approach. Int. J. Electr. Power Energy Syst. **33**(2), 278–287 (2011). https://doi.org/10.1016/j.ijepes.2010.08.023. https://linkinghub.elsevier.com/retrieve/pii/S0142061510001596
16. Liu, Y., Zhang, P., Qiu, X.: Optimal volt/var control in distribution systems. Int. J. Electr. Power Energy Syst. **24**(4), 271–276 (2002). https://doi.org/10.1016/S0142-0615(01)00032-1. https://linkinghub.elsevier.com/retrieve/pii/S0142061501000321
17. Meier, A.V.: Electric Power Systems: A Conceptual Introduction. Wiley survival guides in engineering and science, IEEE Press: Wiley-Interscience, Hoboken (2006). oCLC: ocm62616191
18. Mnih, V., et al.: Playing atari with deep reinforcement learning. arXiv preprint arXiv:1312.5602 (2013)
19. Mnih, V., et al.: Human-level control through deep reinforcement learning. Nature **518**(7540), 529–533 (2015)
20. Niknam, T.: A new approach based on ant colony optimization for daily Volt/Var control in distribution networks considering distributed generators. Energy Convers. Manag. **49**(12), 3417–3424 (2008). https://doi.org/10.1016/j.enconman.2008.08.015. https://linkinghub.elsevier.com/retrieve/pii/S0196890408003087
21. Niknam, T., Firouzi, B.B., Ostadi, A.: A new fuzzy adaptive particle swarm optimization for daily Volt/Var control in distribution networks considering distributed generators. Appl. Energy **87**(6), 1919–1928 (2010). https://doi.org/10.1016/j.apenergy.2010.01.003. https://linkinghub.elsevier.com/retrieve/pii/S030626191000005X
22. Rashid, M.H. (ed.): Electric renewable energy systems. Elsevier/AP, Academic Press is an imprint of Elsevier, Amsterdam (2016)
23. Ribeiro, L.C., Schumann Minami, J.P.O., Bonatto, B.D., Ribeiro, P.F., de Souza, A.C.Z.: Voltage control simulations in distribution systems with high penetration of PVs using the OpenDSS. In: 2018 Simposio Brasileiro de Sistemas Eletricos (SBSE), pp. 1–6. IEEE. https://doi.org/10.1109/SBSE.2018.8395639. https://ieeexplore.ieee.org/document/8395639/
24. Roytelman, I., Wee, B., Lugtu, R.: Volt/var control algorithm for modern distribution management system. IEEE Trans. Power Syst. **10**(3), 1454–1460 (1995). https://doi.org/10.1109/59.466504. https://ieeexplore.ieee.org/document/466504/

25. Saric, A.T., Stankovic, A.M.: A robust algorithm for Volt/Var control. In: 2009 IEEE/PES Power Systems Conference and Exposition, Seattle, WA, USA, pp. 1–8. IEEE (2009). https://doi.org/10.1109/PSCE.2009.4840211. https://ieeexplore.ieee.org/document/4840211/
26. Schneider, K.P., et al.: Analytic considerations and design basis for the IEEE distribution test feeders. IEEE Trans. Power Syst. **33**(3), 3181–3188 (2018). https://doi.org/10.1109/TPWRS.2017.2760011. https://ieeexplore.ieee.org/document/8063903/
27. Sutton, R.S., Barto, A.G.: Reinforcement Learning: An Introduction. Adaptive Computation and Machine Learning Series, 2nd edn. The MIT Press, Cambridge (2018)
28. Vlachogiannis, J., Hatziargyriou, N.: Reinforcement learning for reactive power control. IEEE Trans. Power Syst. **19**(3), 1317–1325 (2004). https://doi.org/10.1109/TPWRS.2004.831259. https://ieeexplore.ieee.org/document/1318666/
29. Wood, A.J., Wollenberg, B.F., Sheblé, G.B.: Power Generation, Operation, and Control, 3rd edn. Hoboken, Wiley-IEEE (2013)
30. Xu, H., Domínguez-García, A.D., Sauer, P.W.: Optimal tap setting of voltage regulation transformers using batch reinforcement learning. IEEE Trans. Power Syst. **35**(3), 1990–2001 (2019). https://doi.org/10.1109/TPWRS.2019.2948132. https://arxiv.org/abs/1807.10997
31. Yang, Q., Wang, G., Sadeghi, A., Giannakis, G.B., Sun, J.: Two-Timescale Voltage Control in Distribution Grids Using Deep Reinforcement Learning. arXiv:1904.09374 (2019)
32. Zhang, W., Liu, Y.: Multi-objective reactive power and voltage control based on fuzzy optimization strategy and fuzzy adaptive particle swarm. IEEE Trans. Power Syst. **30**(9), 525–532 (2019). https://doi.org/10.1016/j.ijepes.2008.04.005. https://linkinghub.elsevier.com/retrieve/pii/S0142061508000380

Performance Analysis of Generative Adversarial Networks and Diffusion Models for Face Aging

Bruno Kemmer(✉), Rodolfo Simões, Victor Ivamoto, and Clodoaldo Lima

University of São Paulo, São Paulo, SP, Brazil
{bruno.kemmer,simoesrodolfo,vivamoto,c.lima}@usp.br

Abstract. Computational face aging enables predicting a person's future appearance using algorithms, with the goal that the output age is close to the expected age and that the individual's characteristics are maintained. In this work, we evaluate the performance of four generative models on facial aging. Two models are based on generative adversarial networks (GANs), HRFAE, and SAM, and the other two are based on diffusion models, Pix2pix-zero and Instruct-pix2pix. The first two were explicitly trained to generate an aged version of the original person, and the others have a zero-shot generation; in other words, they are generic models that perform different tasks, including facial aging. Since diffusion models have been gaining attention because of their diversity and high-quality image generation, comparing their results against models specifically designed for the task using meaningful metrics is essential. Therefore, we compared these models using the FFHQ Aging database and with the metrics: Mean absolute error (MAE) of the predicted age, Fréchet inception distance (FID), and the cosine similarity of the FaceNet's embeddings.

Keywords: Face Aging · Generative Adversarial Networks · Diffusion Models

1 Introduction

Computational methods of facial aging aim to generate an aged face while maintaining individual characteristics. However, several factors can cause aging: excess sun exposure, smoking, a polluted environment, stress, and genetic factors. In addition, surgical and non-surgical aesthetic procedures can mitigate the effects of time, as well as the use of cosmetic substances at the time of image capture. For these reasons, facial aging is a complex and non-deterministic process.

Recently, this theme has been the subject of many publications [1,5,8,38], as it can help in an automated way in the search for missing people; in the

V. Ivamoto—Partially supported by the Coordenação de Aperfeiçoamento de Pessoal de Nível Superior – Brasil (CAPES) – Finance Code 001.

M. C. Naldi and R. A. C. Bianchi (Eds.): BRACIS 2023, LNAI 14196, pp. 228–242, 2023.
https://doi.org/10.1007/978-3-031-45389-2_16

identification of criminals, and also for entertainment purposes. Furthermore, its use is also present in biometric tasks (identification of individuals based on physical or behavioral characteristics), as it makes it possible to reduce the distance between the individual characteristics present at the time of training and the current state of the faces, especially if the training has been done with old images [14,18].

Many recent publications have used generative models to perform facial aging, obtaining realistic results. Initially, adversarial autoencoders were used, and later, generative adversarial models (GANs) and, recently, diffusion models [36,40].

This study aims to evaluate four generative models on the facial aging task. In particular, compare if zero-shot diffusion models can generate aged images on par with GANs explicitly trained for the face aging task. To achieve that, we analyze the outputs of two state of art models based on GANs, HRFAE, and SAM, with two diffusion models, Pix2pix-zero and Instruct-pix2pix. Using the following metrics: mean average error (MAE) of the predicted age (measuring age regression), Fréchet inception distance, FID (checking for image reality), and the cosine similarity of the embedding of a pre-trained face recognition network, FaceNet (gauging the adherence to the individual characteristics).

The article is organized as follows. Section 2 aims to show related works that have a degree of similarity with the study. Section 3 presents a brief review of GANs and Diffusion models, a summary of the main aspects of each model used in this comparison, a few methods used inside the models, the FFHQ Aging image database, and evaluation metrics. Section 4 details the methodology used to compare the techniques. Section 5 presents the conclusion.

2 Related Work

In the literature, many works compare the aging effects on human faces using generative models, in particular, employing generative adversarial networks, GANs, and a fewer number applying diffusion models on the task. However, we couldn't find a work that compares these two techniques using the same dataset.

The review [8] presents the main models that perform facial aging using deep networks, the evolution of the number of publications, a taxonomy of existing techniques, and the most important databases with faces and proper metadata. Also, in the review, the authors evaluate three aging models for high-definition images recently published: HRFAE [38], LIFE [23], and SAM [1]. In the comparison, they use external photos of 4 young people, between 24 and 32 years old, and present how the models behave when trying to age the photos for three age groups: 65, 50–69, and 60. To perform a quantitative analysis, the authors estimate the age of the original photo using a pre-trained model ArcFace [4] and each generated photo. In addition, the FID metric is used to quantify the perception of the reality of the images. Finally, they measured the cosine distance between the result of the original image and the aged one in the last layer of the pre-trained network.

Moreover, in work [17], the authors review the literature and compare four aging models based on GANs: CAAE [39], IPCGAN [35] and RCRIIT [9]. In the comparison, the authors used the same set of images from three aging databases: FG-NET, UTKFaces, and CACD.

Additionally, in HRFAE [38], the authors compared their model with IPC-GAN [35] and S2GAN [37] using the FFHQ Aging database.

In SAM [1], the authors compare their model with LIFE [23] and HRFAE [38], stating that at the time of publication, they were considered state-of-the-art works. The database used to evaluate the results is CelebA-HQ, as it contains celebrities' faces and ages. As in [8], the authors used ArcFace to calculate the cosine similarity of each pair of images. The study also presents a qualitative analysis of the images generated for each work. In their analysis, the authors generated 80 images and chose the image in which the ArcFace model had the closest prediction of the desired one. The SAM and HRFAE models can generate faces for a predefined age. However, for comparison purposes, the same protocol was used. Finally, the authors used the ArcFace model during training and employed the Microsoft Azure Face API to access identity maintenance. A survey was also carried out in which humans evaluated photos of the same individual and answered which images they preferred, using the desired age and the quality of the generated image as metrics.

3 Material and Methods

3.1 Generative Adversarial Networks

Generative adversarial networks, GANs [7], are composed of two neural networks: the generator, which is a network to generate examples close to real data by learning the distribution of the training data, and the discriminator, a classification network that aims to separate the generated data from the real ones. These two networks compete with each other during the training stage, and the generator tries to produce examples so close to the real ones as to deceive the discriminator, which tries to improve the detection of the generated images. Over time architecture improvements have been introduced to improve the image quality and mitigate training problems.

HRFAE. In High-Resolution Face Age Editing, HRFAE [38], the authors used an encoder-decoder architecture to perform age editing of photos in high resolution (1024×1024). The G generator consists of an E encoder and a D decoder. The model receives the input image x_0, which passes through the encoder generating two copies of a latent vector $E(x_0)$. The pre-trained DEX age estimator [29] is used to determine the age of the input image α_0. This age will be encoded through a binary encoding module with a sigmoid activation function. Its decoder D has two tasks: it receives the latent vector $E(z)$ and produces an image as similar as possible to the input image $G(x_0, \alpha_0)$, and also makes the aged image $G(x_0, \alpha_1)$ realistic and close to the desired age.

To achieve this goal, its cost function has three components, as shown in Eq. 1:

1. Adversarial Loss (L_{GAN}) uses PatchGAN [15] with the objective function of LSGAN [21].
2. Age Classification Loss (L_{class}) in which the pre-trained DEX model is used to estimate the age obtained with a categorical cross-entropy loss function.
3. Reconstruction Loss (L_{recon}) monitors the model's ability to reconstruct the original image at the initial age. $L_{recon} = \|G(x_0, \alpha_0) - x_0)\|_1$

$$\lambda = \lambda_{recon} L_{recon} + \lambda_{class} L_{class} + L_{GAN}, \tag{1}$$

where λ_{recon} and λ_{class} are hyper-parameters that weigh between maintaining the identity (reconstruction of the original image at the initial age) and the aging effect (classification at the correct age).

In training performed by the authors, they used the Flickr-Faces-HQ, FFHQ, and high-definition image base [16]. It presents fewer images of older people, so they used the StyleGAN network to generate 300,000 synthetic images. They thus obtained a base of images balanced in the criteria of the ages present in it. The authors only used synthetic images in age groups with insufficient real images. Ultimately, they obtained 47,990 images ranging in age from 20 to 69.

In this work (HRFAE), the authors published, along with their results, metadata of the images containing pose, age, and gender prediction, and the segmentation of the regions of the faces present in FFHQ, calling it FFHQ Aging. This is the database used in the present work.

SAM. In Style-Based Age Manipulation, SAM [1], the authors developed an architecture that enables facial aging or rejuvenation using a real image x and a desired age α_t as input. To achieve this, they perform an image-to-image translation. The first step is to find the best latent vectors representing x in the w^* space of an unconditional GAN (StyleGAN) that can reconstruct the original face, to this task, they employed a previously trained coder (pSp) [26]. Its output, a series of style vectors, makes it possible to reconstruct the original image when passing through the *StyleGAN* generator.

A second encoder, E_{age}, is trained to capture the difference (residual) between the reconstructed image obtained by the first encoder (pSp) and the aged image. In this second encoder, a pre-trained DEX network [29] was used to guide the training toward the desired age, and a pre-trained network for face recognition ArcFace [4] was employed to maintain the individual characteristics of the original image. These networks were not changed during the training of the E_{age} encoder, keeping their parameters fixed.

The outputs of both encoders are summed and become the latent input vector that StyleGAN uses. Additionally, because it is an image-to-image translation, the model uses a cyclic loss to reconstruct the original image after a cycle consistency pass.

The cost functions used were:

- **Pixel-to-Pixel Similarity** $\mathcal{L}_2(x_{age}) = \|x - SAM(x_{age})\|_2$
- **Perceptual Similarity Loss**: $\mathcal{L}_{LPIPS}(x_{age}) = \|F(x) - F(SAM(x_{age}))\|_2$.
- **Regularization Loss**: This regularization causes the style vectors to be close to the average of the latent vectors. The authors identified that its use improves image quality by removing unwanted artifacts in the images produced.
- **Identity Loss**: Difference in cosine similarities between the output and input image, weighted by the number of years between the images. If the difference is many years, there is expected to be a loss of identity, $\mathcal{L}_{ID}$.
- **Age Loss**: To verify the quality of aging/rejuvenation in the generated image, a pre-trained DEX network was used, $\mathcal{L}_{age} = \|\alpha_t - DEX(SAM(x_{age}))\|_2$.

3.2 Diffusion Models

Diffusion models aim to destroy the data distribution structure slowly and systematically in a way that enables learning a reverse diffusion process, which recreates the original structure of the data, generating a very flexible and computationally tractable generative model of the data [33].

In the work Denoising Probabilistic Diffusion Models (DDPM) [12], the authors achieved good results when trying to predict the noise $\mathcal{N}(\mu, \sigma^2)$ but fixing σ^2. The noise was also added linearly in the forward step, using a linear schedule. Moreover, the reconstruction was done using a U-Net architecture with attention blocks.

Later, some enhancements were proposed [22], adding the noise by cosine scheduling in the forward step instead of a linear scheduler, as they noted that this approach improved learning by destroying the image signal more slowly. The neural network that reconstructs the data began to learn the parameters in σ^2, increased the depth of the layers and decreased their number, increased the number of attention layers and the number of heads of attention, and other improvements.

In Denoising Diffusion Implicit Models, DDIM [34] showed that using a non-Markovian approach in the forward step of DDPM models made it possible to use only a subset (progressive) of the steps t in a sampling trajectory. Thus, it considerably accelerated the sampling process because, instead of 1,000 noise removal steps used in DDPM models, it was possible to obtain good results with 200 or fewer. Equation 2 shows the addition of noise in the sampling process obtained on the image, x_t, at step t, with Gaussian noise ϵ, unitary variance α_t and the original image x_0. In Eq. 3 is shown the inverse, predicting the noise that will be removed x_t towards x_0, where $\epsilon_t \sim \mathcal{N}(\mathbf{0}, \boldsymbol{I})$.

$$x_t = \sqrt{\alpha_t} x_0 + \sqrt{1-\alpha_t}\epsilon \tag{2}$$

$$\mathbf{x}_{t-1} = \sqrt{\alpha_{t-1}} \underbrace{\left(\frac{\mathbf{x}_t - \sqrt{1-\alpha_t}\epsilon_\theta^{(t)}(\mathbf{x}_t)}{\sqrt{\alpha_t}} \right)}_{\text{“}x_0\text{ predicted”}} \underbrace{+ \sqrt{1-\alpha_{t-1}-\sigma_t^2} \cdot \epsilon_\theta^{(t)}(\mathbf{x}_t)}_{\text{“Direction towards }x_t\text{”}} + \sigma_t \epsilon_t \tag{3}$$

Latent Diffusion Models. Latent diffusion models, LDM [27] gained notoriety for being able to generate realistic text-guided images. With the availability of the pre-trained latent diffusion model Stable Diffusion, trained with high-resolution images present in the LAION database [32] quickly gained prominence due to its open source code and high-quality text-to-image generation. One of the work's main innovations was using a dimension reduction technique, a variational autoencoder model, and performing the diffusion process in this reduced domain. Therefore, this allowed the use of self-attention modules, as these increase complexity quadratically based on the input data. Additionally, during the model training, the energy consumed and processing time decreased considerably compared to previously proposed architectures.

The model was divided into two phases:

1. In the compression phase, they used a variational autoencoder model to learn the perceptual domain of the images. Thus, the diffusion model does not try to add and remove noise from the raw input image, usually in high dimension, but rather from a latent vector (intermediate layer of the autoencoder used), considerably reducing the computational complexity, as it reduces the input size by eight times. This variational autoencoder uses a perceptual loss and has an opposing objective function based on image segments, thus ensuring consistency in each segment.
2. In the generative learning phase, a U-Net [28] structure was used with cross-attention mechanisms conditioned to different input forms (text and image) via a domain-specific encoder.

One technique used is classifier-free guidance [13], in which two text representations are concatenated, an empty array, and the input sentence. Two latent vectors are also concatenated, which will have their noise removed. Therefore, the output will have two components, one in which the input text is oriented $\hat{x}_{cond\_text}$ and another without $\hat{x}_{incond}$. A model parameter h_1 controls how much the component linked to the input text affects the final image, as seen in Eq. 4. Larger values of h_1 force the generated image to be more faithful to the input text, and smaller values give more freedom in the final image.

$$\hat{x} = \hat{x}_{incond} + h_1 * (\hat{x}_{text\_cond} - \hat{x}_{incond}) \quad (4)$$

Image Editing Using Diffusion Models. There are a few ways to perform image editing with diffusion models. Three of them are:

- Methods that use DDIM inversion to find the latent vector that best reconstructs the original image and perform editing in the forward step of noise removal. These methods do not modify the parameters of the diffusion models used. Pix2pix-zero [24] adopts this approach.
- Fine-tuning the weights of pre-trained diffusion models to fit the examples to be edited, as in DreamBooth [30] and Textual Inversion [6].
- Train a diffusion model to do image editing, as in Production-Ready Face Re-Aging for Visual Effects [40] and in Instruct-pix2pix [2].

DDIM Inversion. DDIM Inversion technique [34] is a way of editing real images and consists of finding the latent variable x_T, which, when traversing the deterministic sampling path, will produce a realistic approximation of the original image x_0. Once this is done, it becomes possible to edit just some parts of the coding that conditions the image generation, also being able to replace words while maintaining the characteristics of the original image. Another way of editing forms involves using a mask to edit only a region of interest.

CLIP. In Contrastive Language-Image Pre-Training, CLIP [25], a neural network was trained on 400 million image pairs in the form of an image and its subtitles obtained from the internet.

This network was proposed to be used without the need to be retrained for specific tasks since it aligns in the same dimensional space both the representation of the image (after passing through an encoding model), I_n, and the text (also after going through a process of tokenizing, encoding, and padding to have the same dimension as the image), T_n, since both have the same dimension and the result matrix is the cross product of both $I_n \cdot T_n$. Its objective function tries to maximize the similarity of matching pairs and minimize the similarity of unrelated pairs. The representation obtained by the network has been used in multiple image-related tasks since it aligns the images with text context.

Pix2pix-Zero. In Zero-shot Image-to-Image Translation [24], the authors propose a method for editing real images that preserve the characteristics of the original images. Following the following steps:

1. Perform a DDIM inversion to obtain the original latent vector that best represents this real image in the model used[1].
2. Find an editing direction, they used the GPT-3 text generator template [3] to generate sentences with a source term (e.g. dog) and a target term (e.g. cat). These sentences pass through a CLIP model to obtain the representations in that domain. The subtraction of the average of these representations will, theoretically, be the editing direction of the images.
3. Get a caption (the most text adherent to the image) so that when editing occurs, it happens in the word that best represents the changed term. As the cross-attention modules end up generating masks relating the words to the image, when editing uses this information, it maintains what is not being changed. The authors used the BLIP model [20] to generate the caption in the work.
4. They performed the editing through cross-attention modules. To do so, they reconstructed the image without applying any editing, just using the input text to obtain the cross-attention modules for each step t. Once this was done, they added the editing direction and calculated the gradient loss in relation to the input x_t. This caused the edit to focus on the region represented by that word.

[1] The authors used Stable diffusion 1.4.

Instruct-Pix2pix. In Instruct-pix2pix [2], a method was shown to train a model that can follow human editing instructions on images. The method receives an input image and a text with the instruction and performs editing in the forward step.

This is possible because the authors used the GPT-3 [3] template to generate editing instructions and captions for the original and edited images. After that, the authors used the pre-trained network Stable diffusion [27] to generate pairs of images referring to the created captions, producing a base of more than 450,000 examples. With these examples, a new diffusion model was trained to generate edited images given an input image and editing instructions.

The authors highlighted that text-to-image diffusion models (such as Stable Diffusion) could generate drastically different images for slightly modified texts. To mitigate this problem, the authors used the technique presented in work prompt-to-prompt [10] in which the weights in the cross-attention modules are calculated relating the words with regions of the image, so editing is restricted to the regions related to the words being edited. Furthermore, this model has a parameter ρ[2] that makes it possible to control the similarity between the two images. To do this automatically, 100 examples $\rho \sim \mathcal{U}(0.1; 0.9)$ were sampled and filtered based on a distance metric in the CLIP representation space.

Another important point was that the authors used two parameters in the classifier free guidance: one that controls how much the image corresponds to the input image c_{image}, and the other interferes with how much the instruction should be followed $c_{instruction}$.

3.3 FFHQ Aging Image Database

Flickr-Faces-HQ (FFHQ) [16] is a database of high-quality human faces intended to be a benchmark for GANs. Consists of 70,000 high-resolution 1024×1024 images obtained from the platform Flickr, chosen by their permissive sharing permissions, and pre-processed (cut and aligned) using the dlib library [19]. The authors comment that this base of images contains much more variations in terms of ethnicity and the background of the photos compared to CelebA-HQ, and that it also contains several accessories, such as glasses and sunglasses, hats, etc.

In LIFE [23], the authors published, along with their results, additional metadata containing pose, prediction of age and gender, and the segmentation of the regions of the faces present in the FFHQ database, which they called FFHQ Aging. It is worth noting that the age ranges present in the FFHQ-Aging metadata were estimates obtained by the authors using the Appen platform that, in addition to the age range, returns a confidence interval of the age prediction. To reduce the error that incorrect estimates may bring, in the comparison presented in this work, only images with a confidence interval equal to 100% were chosen.

[2] Ratio of diffusion steps with cross-attention weights.

3.4 Metrics

Fréchet Inception Distance (FID). [11] is currently one of the most used GAN evaluation methods and uses the previously trained classifier with InceptionNet architecture. The last layer of the network (fully connected) is removed, resulting in its output of a vector with dimension 2048, representing the image of the attributes detected by this network. Statistics of these representations, both real and synthetic (generated) images, are used during the FID calculation.

Low FID values mean that the distributions are close, which is what you want to happen when comparing the distribution of the generated images and the real ones.

Cosine Similarity of Face Embeddings. One way that multiple articles use to quantify identity maintenance in the aged-generated image is to use the embeddings of a pre-trained face recognition network [1,23,35] since it was optimized to find features in the faces that make each individual unique.

Therefore, calculating the cosine similarity of the original image embedding and the aged version enable quantifying how close they are in the representation space that captures the identity characteristics.

4 Experiments

These experiments aim to compare the performance of conditioned diffusion models with GANs trained specifically for the face aging task, verifying whether the aged image outputs would be of equivalent or superior quality.

The image base used was the FFHQ Aging, in which the photos have an estimated age in already defined groups. In the experiments, it was selected age ranges close to those used by the authors in their works.

In each group, 50 images of unique individuals were drawn for both women and men. An interval of 20 years was used between the age ranges, except for the last one, which has a difference of 60 years, to verify an extreme case.

In this work, we measured the mean predicted age per group and the mean absolute error (MAE), comparing the predicted age with the one estimated by the DEX model in the initial image.

The similarity of cosines was also calculated between the representations obtained when using the pre-trained network FaceNet [31]. FaceNet has a performance similar to ArcFace [4], but since SAM uses ArcFace in its cost function, the comparison is fairer using another pre-trained network.

In the experiments carried out with the model Instruct-pix2pix changing the value of how much the aging instruction will be followed $c_{instruction}$, it was observed that if the parameter has the value 0, the image is not aged as the instruction is not followed, which is expected. The error drops considerably by increasing the value of $c_{instruction}$ to 10. However, the images are no longer close in the FaceNet representation. Given these results, the parameter $c_{instruction}$ of the Instruct-pix2pix model was used with the value 3.

It is important to emphasize that the results were analyzed by separating the input images between men and women. However, as no significant differences were noted in the results, the data were unified.

Fig. 1. Examples of images in which the original image reconstruction step using the DDIM inversion technique. Some did not obtain good results, which could compromise the quality of the generated aged images.

Table 1. Model comparison between the aging images.

		Experiments				
		10–30	**20–40**	**30–50**	**40–70**	**10–70**
HRFAE	**Predicted age**	18,9	24,9	47,8	43,9	25,4
	MAE	11,1	15,1	3,9	26,1	44,6
	Cosine Sim.	0,85	0,91	0,88	0,94	0,90
	FID	28,32	22,16	21,11	20,26	22,02
SAM	**Predicted age**	26,3	36,3	47,6	61,4	59,1
	MAE	**4,3**	**4,4**	**3,7**	**8,6**	**10,9**
	Cosine Sim.	0,77	0,75	0,71	0,69	0,49
	FID	114,21	109,87	121,06	107,24	135,25
Pix2pix-zero	**Predicted age**	23,2	27,9	35,6	45,4	31,3
	MAE	6,8	12,1	14,4	24,6	38,7
	Cosine Sim.	0,62	0,82	0,79	0,81	0,52
	FID	102,28	84,04	82,75	96,31	104,35
Instruct-pix2pix	**Predicted age**	19,8	28,7	39,4	51,1	21,2
	MAE	11,2	11,3	10,6	18,9	48,8
	Cosine Sim.	**0,93**	**0,94**	**0,93**	**0,92**	**0,92**
	FID	45,43	38,89	40,74	42,52	52,47

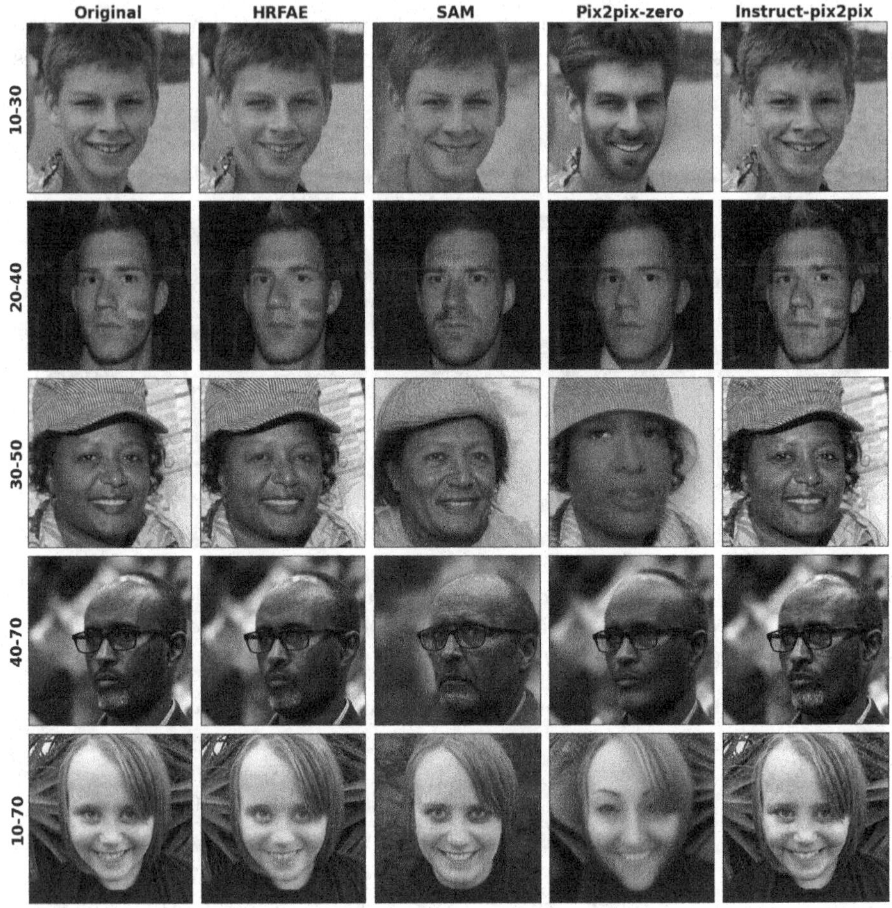

Fig. 2. Comparison of results in facial aging tasks in each model. The first column, Original, is the input image with the initially estimated age. The four images to its right are the outputs of the models trying to get the image aged with the expected age. For example, 10–30 is the task of aging a face with approximately 10 years to 30 years.

In the results presented by Table 1, it is possible to notice that SAM manages to obtain the lowest error averages, possibly due to having an age estimator in its cost function. However, the similarity between the aged image and the original is smaller than the results of Instruct-pix2pix. The high similarity between the images obtained by HRFAE can be explained because the minor changes performed by the network. Therefore, the images are not properly aged, as seen in the error and low FID values.

In the Pix2pix-zero model, two factors can affect the maintenance of the characteristics of individuals: one of them is the fact that the model uses editing directions obtained through a single average direction. This can cause the model

Fig. 3. Comparison of results in facial aging tasks in each model. The first column, Original, is the input image with the estimated initial age, and the four images on the right are the results of the models trying to get the aged image with the expected age.

to age all images unconditionally, which can be seen by the low similarity and high FID values. Another factor is that it needs to reconstruct the original image using the DDIM inversion technique; a poor reconstruction can compromise the final aged image. Figure 1 shows examples of when this occurs.

Observing Figs. 2 and 3, the SAM model is the only one that achieves reasonable performance when aging from approximately ten years to older ages. As it is a growth phase, the face has many changes, and the other models failed to capture these structural changes. The Pix2pix-zero model had blur results on some images, which is expected when using an average edit direction to get from one domain to the next. Instruct-pix2pix maintained the identity of the individuals and achieved considerable aging. However, in all cases, the skin texture degraded, which is common over the years (especially in cases where there is a lot of sunlight or smoking), but it is not guaranteed that aging will be like this. The SAM model seems to have more variability in how it ages faces.

5 Conclusion

This study analyzed the performance of four models in the facial aging task: two models that used GANs in their architecture and two conditional diffusion models that enabled image editing. Of these, two models stood out in this analysis, SAM, a model based on a pre-trained GAN (StyleGAN2) and specialized in the task of facial aging and rejuvenation, having components of its cost function specific to guarantee identity and age change. The second model to be highlighted was Instruct-pix2pix, a generic conditional diffusion model, which only receives instruction in the form of text, being able to edit the images in different ways in a zero-shot fashion. In the study, it was only sent an instruction to age the face to a desired age, and still got realistic aged images.

Currently, specialist GANs still demonstrate superior performance compared to generic diffusion models, such as those used in the experiments. This demonstrates a great capacity in the diffusion model since Instruct-pix2pix obtained considerable results even though it was not specifically trained for this task. Furthermore, diffusion models have gained notoriety recently, with many works presenting new techniques in recent years, so rapid advances in their architectures are expected.

References

1. Alaluf, Y., Patashnik, O., Cohen-Or, D.: Only a matter of style: Age transformation using a style-based regression model. ACM Trans. Graph. **40**(4), 1–12 (2021)
2. Brooks, T., Holynski, A., Efros, A.A.: Instructpix2pix: learning to follow image editing instructions (2022)
3. Brown, T.B., et al.: Language models are few-shot learners. In: Proceedings of the 34th International Conference on Neural Information Processing Systems. NIPS'20, Curran Associates Inc., Red Hook, NY, USA (2020)
4. Deng, J., Guo, J., Xue, N., Zafeiriou, S.: ArcFace: additive angular margin loss for deep face recognition. In: 2019 IEEE/CVF Conference on Computer Vision and Pattern Recognition (CVPR), pp. 4685–4694 (2019)
5. Despois, J., Flament, F., Perrot, M.: AgingMapGAN (AMGAN): high-resolution controllable face aging with spatially-aware conditional GANs. In: Bartoli, A., Fusiello, A. (eds.) ECCV 2020. LNCS, vol. 12537, pp. 613–628. Springer, Cham (2020). https://doi.org/10.1007/978-3-030-67070-2_37
6. Gal, R., et al.: An image is worth one word: personalizing text-to-image generation using textual inversion (2022)
7. Goodfellow, I., et al.: Generative adversarial nets, p. 9 (2014)
8. Grimmer, M., Ramachandra, R., Busch, C.: Deep face age progression: a survey. IEEE Access **9**, 83376–83393 (2021)

9. Heljakka, A., Solin, A., Kannala, J.: Recursive chaining of reversible image-to-image translators for face aging. In: Blanc-Talon, J., Helbert, D., Philips, W., Popescu, D., Scheunders, P. (eds.) ACIVS 2018. LNCS, vol. 11182, pp. 309–320. Springer, Cham (2018). https://doi.org/10.1007/978-3-030-01449-0_26
10. Hertz, A., Mokady, R., Tenenbaum, J., Aberman, K., Pritch, Y., Cohen-Or, D.: Prompt-to-prompt image editing with cross attention control (2022)
11. Heusel, M., Ramsauer, H., Unterthiner, T., Nessler, B., Hochreiter, S.: GANs trained by a two time-scale update rule converge to a local Nash equilibrium. In: Proceedings of the 31st International Conference on Neural Information Processing Systems, NIPS'17, pp. 6629–6640. Curran Associates Inc., Red Hook, NY, USA (2017)
12. Ho, J., Jain, A., Abbeel, P.: Denoising diffusion probabilistic models. In: Larochelle, H., Ranzato, M., Hadsell, R., Balcan, M., Lin, H. (eds.) Advances in Neural Information Processing Systems, vol. 33, pp. 6840–6851. Curran Associates, Inc. (2020)
13. Ho, J., Salimans, T.: Classifier-free diffusion guidance (2022)
14. Huang, Y., Hu, H.: A parallel architecture of age adversarial convolutional neural network for cross-age face recognition. IEEE Trans. Circuits Syst. Video Technol. **31**(1), 148–159 (2021)
15. Isola, P., Zhu, J.Y., Zhou, T., Efros, A.A.: Image-to-image translation with conditional adversarial networks. In: Proceedings of the IEEE Conference on Computer Vision and Pattern Recognition (CVPR) (2017)
16. Karras, T., Laine, S., Aila, T.: A style-based generator architecture for generative adversarial networks. arXiv:1812.04948 [cs, stat] (2019)
17. Kemmer, B., Simões, R., Lima, C.: Face aging using generative adversarial networks. In: Razavi-Far, R., Ruiz-Garcia, A., Palade, V., Schmidhuber, J., et al. (eds.) Generative Adversarial Learning: Architectures and Applications. Intelligent Systems Reference Library, vol. 217. Springer, Cham (2022). https://doi.org/10.1007/978-3-030-91390-8_7
18. Khanna, A., Thakur, A., Tewari, A., Bhat, A.: Cross-age face verification using face aging, pp. 94–99 (2020)
19. King, D.E.: Dlib-ml: a machine learning toolkit. J. Mach. Learn. Res. **10**, 1755–1758 (2009)
20. Li, J., Li, D., Xiong, C., Hoi, S.: BLIP: bootstrapping language-image pre-training for unified vision-language understanding and generation (2022)
21. Mao, X., Li, Q., Xie, H., Lau, R.Y.K., Wang, Z., Smolley, S.P.: Least squares generative adversarial networks. In: 2017 IEEE International Conference on Computer Vision (ICCV), pp. 2813–2821 (2017)
22. Nichol, A., Dhariwal, P.: Improved denoising diffusion probabilistic models (2021)
23. Or-El, R., Sengupta, S., Fried, O., Shechtman, E., Kemelmacher-Shlizerman, I.: Lifespan age transformation synthesis. In: Vedaldi, A., Bischof, H., Brox, T., Frahm, J.-M. (eds.) ECCV 2020. LNCS, vol. 12351, pp. 739–755. Springer, Cham (2020). https://doi.org/10.1007/978-3-030-58539-6_44
24. Parmar, G., Singh, K.K., Zhang, R., Li, Y., Lu, J., Zhu, J.Y.: Zero-shot image-to-image translation (2023)
25. Radford, A., et al.: Learning transferable visual models from natural language supervision (2021)
26. Richardson, E., et al.: Encoding in style: a StyleGAN encoder for image-to-image translation. In: Proceedings of the IEEE/CVF Conference on Computer Vision and Pattern Recognition (CVPR), pp. 2287–2296 (2021)
27. Rombach, R., Blattmann, A., Lorenz, D., Esser, P., Ommer, B.: High-resolution image synthesis with latent diffusion models (2021)

28. Ronneberger, O., Fischer, P., Brox, T.: U-net: Convolutional networks for biomedical image segmentation. In: Navab, N., Hornegger, J., Wells, W.M., Frangi, A.F. (eds.) Medical Image Computing and Computer-Assisted Intervention - MICCAI 2015. Lecture Notes in Computer Science(), vol. 9351, pp. 234–241. Springer, Cham (2015). https://doi.org/10.1007/978-3-319-24574-4_28
29. Rothe, R., Timofte, R., Van Gool, L.: DEX: deep expectation of apparent age from a single image. In: 2015 IEEE International Conference on Computer Vision Workshop (ICCVW), pp. 252–257 (2015)
30. Ruiz, N., Li, Y., Jampani, V., Pritch, Y., Rubinstein, M., Aberman, K.: DreamBooth: fine tuning text-to-image diffusion models for subject-driven generation (2022)
31. Schroff, F., Kalenichenko, D., Philbin, J.: FaceNet: a unified embedding for face recognition and clustering. In: 2015 IEEE Conference on Computer Vision and Pattern Recognition (CVPR), pp. 815–823 (2015)
32. Schuhmann, C., et al.: LAION-400M: open dataset of clip-filtered 400 million image-text pairs (2021)
33. Sohl-Dickstein, J., Weiss, E., Maheswaranathan, N., Ganguli, S.: Deep unsupervised learning using nonequilibrium thermodynamics. In: Bach, F., Blei, D. (eds.) Proceedings of the 32nd International Conference on Machine Learning. Proceedings of Machine Learning Research, vol. 37, pp. 2256–2265. PMLR, Lille, France (2015)
34. Song, J., Meng, C., Ermon, S.: Denoising diffusion implicit models. arXiv:2010.02502 (2020)
35. Wang, Z., Tang, X., Luo, W., Gao, S.: Face aging with identity-preserved conditional generative adversarial networks, pp. 7939–7947 (2018)
36. Wolleb, J., Sandkühler, R., Bieder, F., Cattin, P.C.: The swiss army knife for image-to-image translation: multi-task diffusion models (2022)
37. Yang, H., Huang, D., Wang, Y., Jain, A.: Learning face age progression: a pyramid architecture of GANs, pp. 31–39 (2018)
38. Yao, X., Puy, G., Newson, A., Gousseau, Y., Hellier, P.: High resolution face age editing. CoRR abs/2005.04410 (2020)
39. Zhang, Z., Song, Y., Qi, H.: Age progression/regression by conditional adversarial autoencoder, vol. 2017-January, pp. 4352–4360 (2017)
40. Zoss, G., Chandran, P., Sifakis, E., Gross, M., Gotardo, P., Bradley, D.: Production-ready face re-aging for visual effects. ACM Trans. Graph. **41**(6), 1–15 (2022)

Occluded Face In-painting Using Generative Adversarial Networks—A Review

Victor Ivamoto, Rodolfo Simões, Bruno Kemmer, and Clodoaldo Lima(✉)

University of São Paulo, São Paulo, SP 03828-000, Brazil
{vivamoto,simoesrodolfo,bruno.kemmer,c.lima}@usp.br

Abstract. Face image de-occlusion and inpainting is a challenging problem in computer vision with several practical uses and is employed in many image preprocessing applications. The impressive results achieved by generative adversarial networks in image processing increased the attention of the scientific community in recent years around facial de-occlusion and inpainting. Recent network architecture developments are the two-stage networks using coarse to fine approach, landmarks, semantic segmentation map, and edge maps that guide the inpainting process. Moreover, improved convolutions enlarge the receptive field and filter the values passed to the next layer, and attention layers create relationships between local and distant information. This article presents a brief review of recent developments in GAN-based techniques for de-occlusion and inpainting of face images. In addition, it describes and analyzes network architectures and building blocks. Finally, we identify current limitations and propose directions for future research.

Keywords: face de-occlusion · facial inpainting · GAN · generative adversarial network · biometry · review

1 Introduction

Facial de-occlusion and inpainting are special instances of image inpainting. They are used in the restoration of damaged images [26], removal of unwanted content [26] and data augmentation [8]. Moreover, face de-occlusion and inpainting are important preprocessing steps in many computer vision tasks with numerous applications. This is because occlusions break the entire structure of the face and hide the identity of the subject, resulting in performance degradation in many

This study was financed in part by the Coordenação de Aperfeiçoamento de Pessoal de Nível Superior - Brasil (CAPES) - Finance Code 001.

Supplementary Information The online version contains supplementary material available at https://doi.org/10.1007/978-3-031-45389-2_17.

M. C. Naldi and R. A. C. Bianchi (Eds.): BRACIS 2023, LNAI 14196, pp. 243–258, 2023.
https://doi.org/10.1007/978-3-031-45389-2_17

applications [37]. Occlusions degrade the performance of face parsing [30,32], object and face detection [21] and facial expression analysis [32]. Furthermore, occlusions hide landmarks used in face alignment and frontalization [2,41].

Face inpainting also poses several challenges that are difficult to overcome. First, the human face carries biometric information unique to each subject, revealing identity, age, sex, emotions, ethnicity, and even culture and religion. This biometric information must be preserved in the restored image. Second, there are many plausible solutions for filling the missing holes in an image, where the ground-truth is just one option. For example, given a face covered with a surgical mask, the mouth may be smiling in the ground-truth image whereas it is closed after reconstruction. Third, the set of possible solutions is restricted by the overall content, as the restoration must preserve the subject's skin and hair texture, facial symmetry, structure and expression, along with variations of illumination and pose. Fourth, occlusions can appear anywhere in the image and may be of any shape and size. Large occlusions covering both sides off the face are more difficult to restore than small ones covering just one side. Fifth, unique facial marks such as makeup, tattoos, scars, stains, wrinkles, and accessories are difficult to recover with no reference image. Finally, the restored area must be visually consistent with the neighboring region, creating an imperceptible transition between them [1,38,44].

Researchers developed new methods to overcome these challenges and improve the quality of the image. The most prominent methods are GAN-based networks that are able to reconstruct an image with photo-realism. Modifications in the GAN architecture with the inclusion of new building blocks, network elements and loss functions address specific facial inpainting issues. This review[1] summarizes these developments, building a solid foundation for future research. The rest of the article is organized as follows. Section 2 describes network architecture and components and presents methods of training stability. Section. 3 discusses the current limitations found in the literature. Finally, we conclude in Sect. 4.

2 Theoretical Background

The network used for image inpainting consists of a number of components and elements that contribute to the final result. This section discusses the main network structures found in the literature.

2.1 Network Architecture

Generative Adversarial Networks (GAN). Generative Adversarial Networks (GAN) consist of a generator and discriminator networks [12]. The generator creates images from simple random noise, usually following a uniform or spherical Gaussian distribution [13]. The discriminator is a classifier that distinguishes between real and fake images. Both networks play an adversarial game

[1] See protocol at https://github.com/vivamoto/bracis-2023.

in which the generator tries to fool the discriminator by gradually improving the image quality. They are trained in alternation until the discriminator is unable to distinguish the synthetic from real images [12]. Figure 1 shows the GAN architecture.

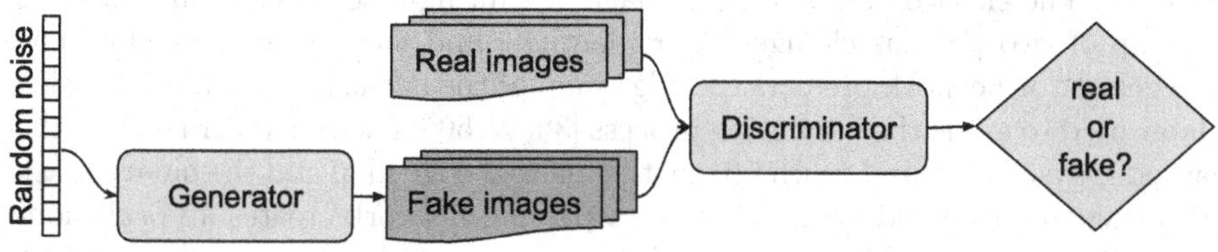

Fig. 1. Original GAN architecture proposed by Goodfellow *et al.* [12]. The generator receives a random noise vector as input and creates fake images. The discriminator is a classifier that evaluates whether the image is real or fake.

In image inpainting and de-occlusion, the generator input is a set of occluded images instead of random noise. After training, only the generator is used to infer new images and the discriminator is removed. Figure 2 illustrates the basic GAN architecture used in image de-occlusion and inpainting. Variations of this architecture found in the literature are described in the next sections.

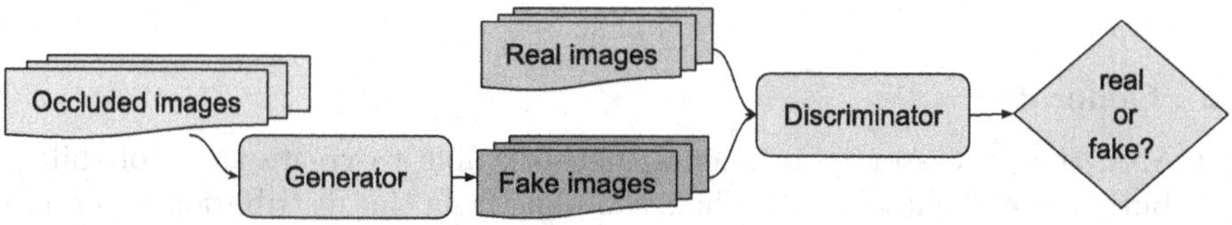

Fig. 2. GAN architecture used in image de-occlusion and inpainting. Instead of random noise, the generator receives occluded images as input and creates occlusion-free images.

Two-Stage Network. Splitting the inpainting process into two or more stages improves the image quality. In this setting, each stage is responsible for a portion of the restoration process. The most common approaches are coarse-to-fine and prior information.

In the coarse-to-fine approach, the first stage creates an initial coarse prediction of the de-occluded image and the second stage takes the result of the first stage as input and refines the prediction [45]. This method gained popularity for its higher performance compared to single-stage networks [4,9,14,44,45]. Figure 3 illustrates a two-stage network with the coarse-to-fine approach.

Prior information such as landmarks, edges, or semantic segmentation maps provides spatial and structural information, guiding the inpainting process. This allows the inpainting network to build the face with realistic structure and facial expressions. In general, the prior network is trained to detect landmarks, edges or semantic segmentation and create the respective maps which are used by the inpainting network to guide the completion process. The landmark map improves the perceptual quality of the image, providing spatial consistency in unaligned faces [38]. The effect of landmarks in image inpainting is so strong that swapping the map of two persons changes their identities and face expressions [43]. The edge generator network predicts the edge map of the occlusion-free image, which is later used to guide the inpainting process [36,42,50]. The generator receives the masked grayscale ground-truth image, the masked edge map and the binary mask indicating the occluded area. Likewise, a parsing network creates an occlusion-free semantic segmentation map of the original occluded image, which guides the de-occlusion process [35,46]. Alternatively, the parsing network can provide semantic regularization, where the semantic segmentation map of the generated image is compared with the ground-truth [11,27].

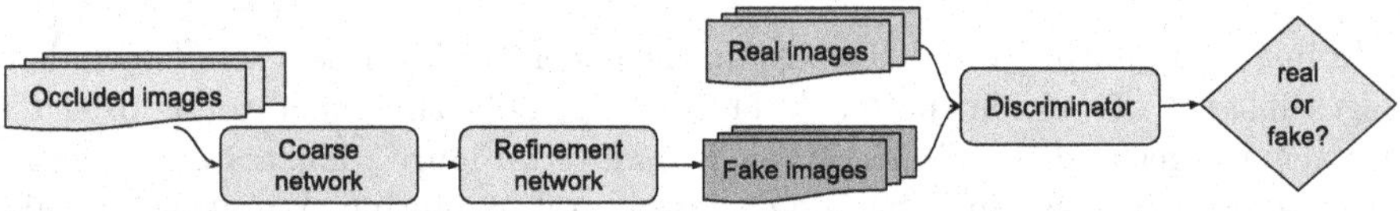

Fig. 3. The two-stage architecture consists of a coarse and a refinement network.

2.2 Generator

In GANs, the generator is any neural network able to create the probability distribution of real data [12,13]. Then, sampling from this distribution generates completely new images. The input to the generator can be a vector of random noise, the incomplete image, semantic segmentation map, edge map, landmarks or binary mask. In face image de-occlusion and inpainting, the generator can be an encoder-decoder, U-net, multi-branch network or any variation.

Encoder-Decoder and U-Net. An encoder-decoder is a generative model trained to reconstruct the input data in an unsupervised way [34]. The network has a symmetric architecture comprised of an encoder and a decoder. The encoder consists of a stack of down-sampling layers that compress the original data into a low dimensional representation. The decoder part contains a series of up-sampling layers that recover the original information. Optionally, a bottleneck layer can be inserted between the encoder and the decoder. This layer converts the encoder's last layer into a vector with similar functionality as the random noise vector in the original GAN.

The encoder-decoder architecture with the bottleneck layer is appropriate for image inpainting with GANs. The encoder converts the occluded image into a vector, and the decoder reconstructs the de-occluded face from this vector.

The U-Net has a similar architecture as the encoder-decoder, where the main difference is the skip connections concatenating each encoder layer with the corresponding symmetrical decoder layer. In the original architecture [40], the U-Net encoder is a series of 3×3 convolutions followed by ReLU and 2×2 max pooling, while the decoder is a series of up sampling layers with 2×2 kernel, a concatenation with the corresponding encoder layer and 3×3 convolutions with ReLU. Figure 4 illustrates the encoder-decoder and U-Net architectures.

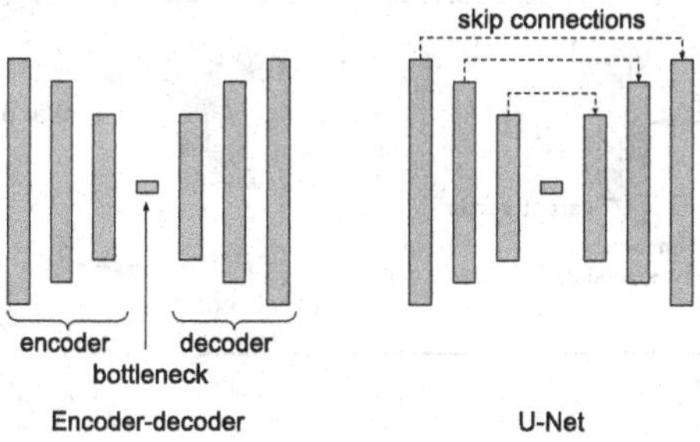

Fig. 4. The encoder-decoder and U-Net architectures consist of an encoder with down sampling layers, a bottleneck layer in the middle and a decoder with up sampling layers. The U-Net has skip-connections concatenating each encoder layer with the corresponding decoder layer. Left: Encoder-decoder. Right: U-Net.

Modified versions of both encoder-decoder and U-Net are commonly used in the generator of GANs used in face de-occlusion and inpainting. The variations include adding dilated convolution [21,25,26,28,29,35,46,47], SE block [21,25,35,47], HDC [10] and self attention blocks [33].

2.3 Discriminator

The discriminator is a classifier that calculates the probability that the image is real rather than synthesized [12]. However, since the inpainted region is a fraction of the entire image, the discriminator is biased towards a generated image being real, resulting in poor inpainting quality. This section describes variations of discriminators that address this issue.

Local and Global Discriminators. The combination of global and local discriminators improves the reconstruction realism and consistency. The global discriminator evaluates the entire image, while the local discriminator judges a small patch around the reconstructed area. The objective function is the sum of the loss functions applied to each discriminator [27]. A less common variation combines the outputs of both discriminators and converts them into a single number representing the probability that the image is real or reconstructed. Specifically, the outputs of both discriminators concatenate and then pass through a fully connected layer. In this setting, the loss is calculated at the combined output [19]. Figure 5 shows an example of an architecture of local and global discriminators with combined outputs. The architecture is a stack of 5×5 convolutions with stride 2 followed by a fully-connected layer that outputs a 1024 vector. The concatenated output of both discriminators passes through a fully-connected layer with sigmoid activation [19].

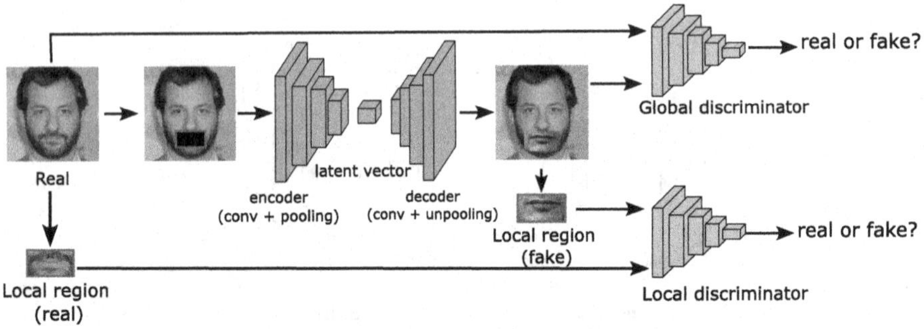

Fig. 5. Network architecture. It consists of one generator and two discriminators. The generator takes the occluded image as input and outputs the occlusion-free image. Two discriminators learn to distinguish the synthesized contents as real and fake. The global discriminator evaluates the entire image, while the local discriminator centers in a small area around the damaged region [19].

PatchGAN. Instead of evaluating the entire image as being real or fake like the standard discriminator, PatchGAN classifies each patch in the input image. This discriminator runs across the image like a convolution and outputs the average of all patches. PatchGAN models high-frequency details, providing texture and style losses[2] [20]. Figure 6 shows the structure of PatchGAN.

SN-PatchGAN. SN-PatchGAN is a fully convolutional spectral-normalized Markovian discriminator. This discriminator computes the loss directly on each point of the last feature map. SN-PatchGAN was designed to inpaint images with regular and irregular shapes of any size and in multiple regions in the image. It provides faster and more stable training, replaces the global and local discriminators and dispenses the perceptual loss [44]. The original discriminator consists of a stack of layers of 5×5 convolutions with stride 2 and spectral normalization. SN-PatchGAN can be interpreted as a 3D classifier, where the loss is applied to each feature element on the feature map of the last layer, as illustrated in Fig. 7.

[2] Python code available at https://github.com/znxlwm/pytorch-pix2pix/blob/3059f2af53324e77089bbcfc31279f01a38c40b8/network.py.

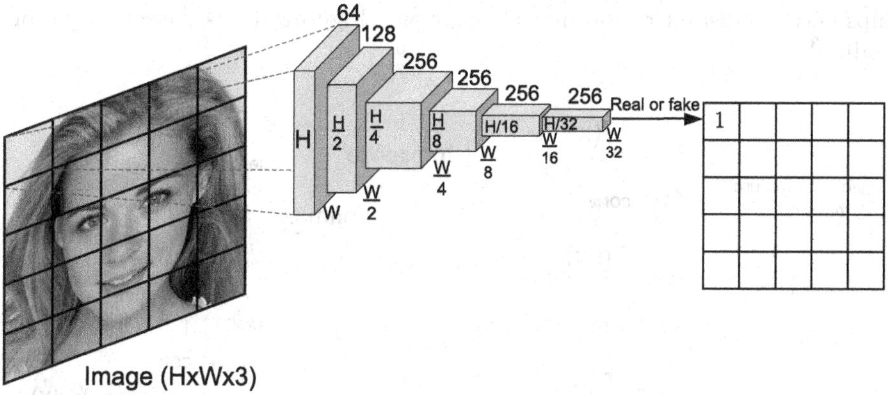

Fig. 6. PatchGAN classifies each patch in the input image as real or fake. This discriminator runs across the image like a convolution and outputs the average of all patches.

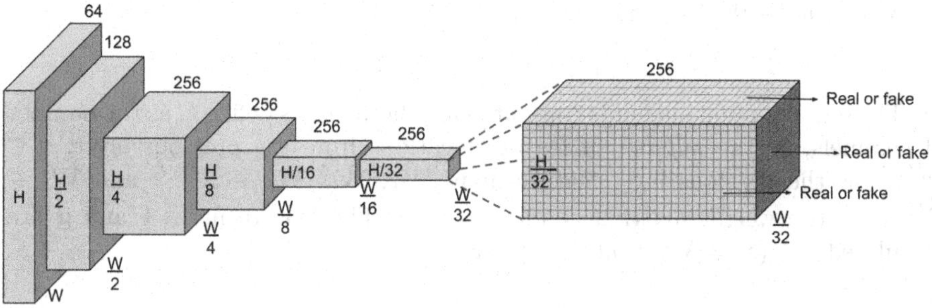

Fig. 7. Fully convolutional spectral-normalized Markovian discriminator (SN-PatchGAN). The discriminator loss is applied in the last feature map, resulting in a 3D classifier [44].

2.4 Building Blocks

In the context of this paper, a block is a group of layers working together that executes a specific task. A block can be inserted between two layers in the generator. This section describes the main building blocks used in GANs, such as self-attention, residual blocks and squeeze and excitation.

Self-Attention. Convolutions process local information limited to the kernel shape and size. When the kernel is inside a hole larger than the kernel size, it captures only invalid pixels, becoming unable to hallucinate meaningful pixels. Therefore, they're not suitable for inpainting regions larger than the kernel size. On the other hand, self-attention is a non-local mechanism that creates relationships between distant regions in the image [48]. Figure 8 shows the self-attention module[3].

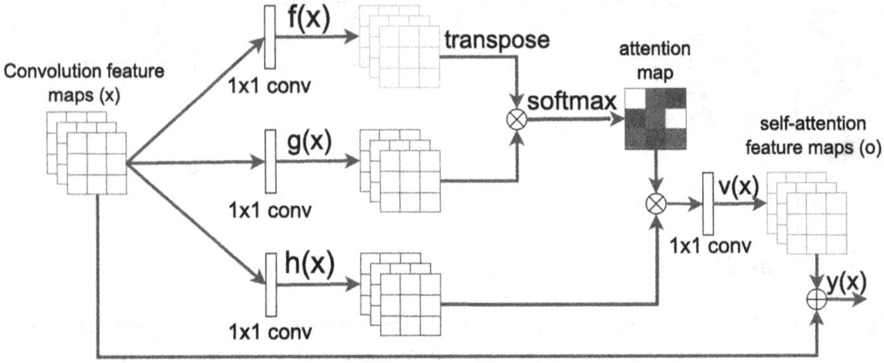

Fig. 8. The self-attention module creates relationships between distant regions in the image. ⊗ denotes the Hadamard product.

The self-attention output is computed as follows. Let's define C as the number of channels, N the number of feature locations from the previous layer, $x \in \mathbb{R}^{C\times N}$ as the previous layer feature map, $\mathbf{W}_f$, $\mathbf{W}_g$, $\mathbf{W}_h \in \mathbb{R}^{\bar{C}\times C}$ and $\mathbf{W}_v \in \mathbb{R}^{C\times \bar{C}}$ as the weight matrices, and $\bar{C} = C/8$. The feature maps $\mathbf{f}$ and $\mathbf{g}$ are calculated as $f(x) = \mathbf{W}_f x$, $g(x) = \mathbf{W}_g x$.

$$\beta_{j,i} = \frac{\exp(s_{ij})}{\sum_{i=1}^{N} \exp(s_{ij})}, \qquad s_{ij} = f(x_i)^\top g(x_j) \tag{1}$$

where the softmax $\beta_{j,i}$ is the probability that the ith location serves the jth region. The output of attention layer $\mathbf{O} = (o_1, ..., o_j, ..., o_N) \in \mathbb{R}^{C\times N}$ is given by:

$$o_j = v\left(\sum_{i=1}^{N} \beta_{j,i} h(x_i)\right), \qquad h(x_i) = \mathbf{W}_h x_i, \qquad v(x_i) = \mathbf{W}_v x_i \tag{2}$$

The final output is given by $y_i = \gamma o_i + x_i$, where γ is a learned parameter.

Residual Block. A residual block (ResBlock) consists of a series of convolutional layers with skip connection, i.e., the input adds to the output as illustrated in Fig. 9.

The residual block avoids gradient dispersion in very deep networks [42] and replaces the standard convolution with dilated convolution [46] or multi-dilated

[3] Python code is available at https://github.com/brain-research/self-attention-gan.

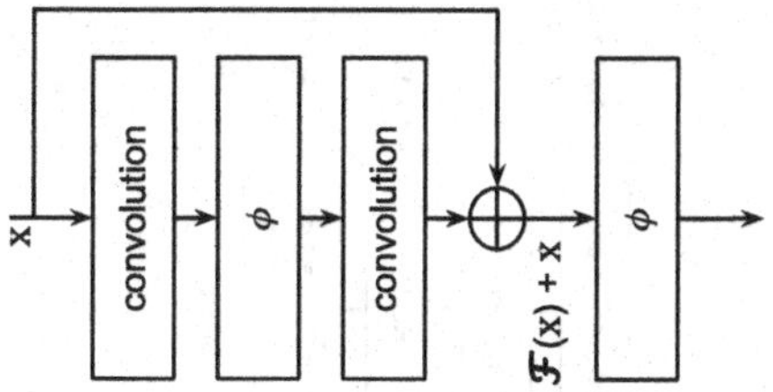

Fig. 9. A residual block consists of two or more convolution layers with skip connection where the input adds to the output. ϕ is the activation function and $\bigoplus$ is element-wise sum.

convolution [26]. Moreover, residual networks are easy to optimize [15], train faster and achieve similar losses compared to non-residual networks [24]. The residual block was originally conceived for image classification [15].

Residual blocks are used in bottleneck layer of encoder-decoder [3,43,46], contraction and expansion sides of U-Net [7,10,26] or as a building block of multi-branch networks [31,32].

Squeeze and Excitation Blocks. The Squeeze-and-Excitation (SE) block models the relationships between channels in the feature maps [18]. The block performs channel-wise feature re-calibration, strengthening meaningful features and weakening worthless ones. SE blocks fit between two layers, achieving higher performance gain at a small computational cost. The squeeze operation uses global average pooling to aggregate each feature map across its spatial dimension, and the excitation operation is a simple gating that produces a collection of weights that are applied to the feature maps. Figure 10 illustrates the architecture of the SE block.

2.5 Training Stability

This section presents two approaches to stabilize the training of GANs. Zhang *et al.* proposed the use of spectral normalization on both the generator and discriminator, as well as employing the two time scale update rule (TTUR) [48].

Two Time Scale Update Rule (TTUR). Using different learning rates for the generator and discriminator in combination with the Adam stochastic optimization improves convergence and stability. In the two time scale update rule (TTUR), the learning rate of the generator is generally lower than the discriminator. Although the TTUR theory ensures convergence, the appropriate learning rates must be empirically found for each network [16]. The learning rates found in the literature for the generator is 1e-4 and for the discriminator are 1e-12 [22,23], 1e-4 [11] and 4e-4 [5,21,48].

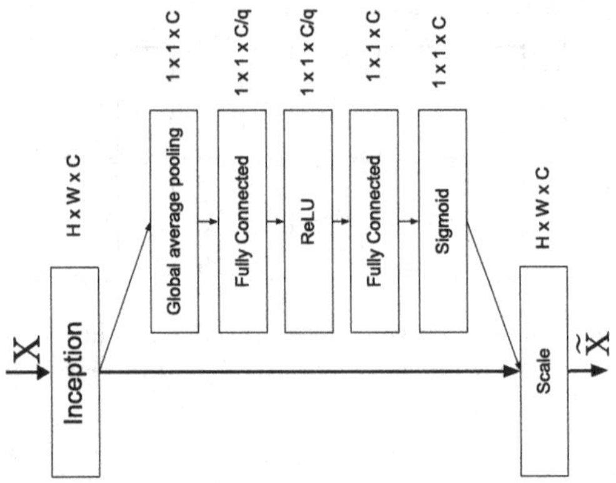

Fig. 10. The squeeze and excitation block scales the feature maps in a given layer according to their importance. The reduction ratio q is a hyperparameter that balances performance and computational complexity. The default is $q = 16$.

Spectral Normalization. Spectral normalization is a weight-normalization technique originally proposed to stabilize the training of the discriminator [39]. Spectral normalization is simple to implement, has low computation cost, and further improves stability when applied in combination with gradient penalty.

Furthermore, when employed in the generator and discriminator, spectral normalization further reduces the discriminator to generator update ratio, decreases the computational cost, and provides more stable training [48]. The spectral normalization is given by Eq. 3:

$$\mathbf{W}_{SN} = \frac{\mathbf{W}}{\eta(\mathbf{W})}, \qquad \eta(\mathbf{W}) = \max_{\|\mathbf{h}\|_2 \leq 1} |\mathbf{W}\mathbf{h}|_2 \tag{3}$$

where $\eta(\mathbf{W})$ is the spectral norm of the matrix $\mathbf{W}$.

3 Limitations

Despite the impressive progress in image de-occlusion and inpainting over the recent years, several challenges remain to be solved. Given the broad extent of current limitations, it's hard to imagine that a single solution will be able to address all situations. This section analyzes key limitations identified during the review and proposes open areas for research.

3.1 Datasets

Model research still requires large amounts of training data, heavily relying on available datasets, and the model generalization capabilities on unseen data largely depends on the trained dataset. An open research area is the development of models, algorithms, and methods resilient to data availability, i.e., models with high generalization capability using few training data.

Moreover, despite the variety in available data, there are few datasets created for face de-occlusion and inpainting. These datasets contain few images compared with other face databases. For this reason, researchers build their own synthetic images based on available public face datasets, usually overlaying an object or a binary mask. This approach may be good for model development, not for inference in real world scenarios which require a large occluded face dataset for testing models.

3.2 Evaluation Metrics

User study measures qualitative attributes that are hard to evaluate with quantitative methods alone. Since researchers employ different methodologies when conducting the qualitative survey, results cannot be compared across published studies. This situation could be avoided if researchers followed a formal protocol describing the survey process. The protocol might use psychophysical similarity measurements already used in the literature, such as Two Alternative Forced Choice (2AFC) and Just Noticeable Differences (JND) used in [49].

Moreover, most quantitative evaluation metrics measure pixel-level statistics that are unable to capture human perception. For historical reasons, they are still widely used for model comparison. The two most used metrics, PSNR and SSIM, carry a simple relationship between them [17]. On the other hand, feature-level metrics capture higher level perceptual quality. LPIPS is the only feature-level metric found in the literature, but it still lags behind human-level perception. More research still needs to be done in an improved version of LPIPS with higher perception level as well as quantitative metrics able to evaluate other qualitative attributes such as effective occlusion removal, naturalness, image realism, consistency, and perception quality.

3.3 Automatic De-occlusion

Most state-of-the-art models require a binary mask with holes in the occluded area. This can be useful for single image restoration and de-occlusion of photographs, but it is unfeasible for videos, real-time de-occlusion and batch processing of several images. Automatic detection and removal of occlusions fails to properly detect occlusions, producing artifacts in the restored region [6]. Moreover, there are still few studies in this area.

3.4 Image Quality

Image quality remains an open problem, for example eyes with different colors, distorted shape of mouth and nose, missing ears, texture discontinuity in the border pixels, artifacts, blur, and bad background filling.

Models using prior information such as landmarks, edges and semantic segmentation maps, or other coarse-to-fine approach, rely on the quality of the predictions of these priors. These predictions have performance degradation on occluded faces, in particular combined with large pose variations, such as top or bottom views and profile.

Large occlusions also degrade performance. The majority of models are trained with 25% of missing region, and a few restore over than 50%. It's specially challenging to remove occlusions covering symmetric parts of the face, for example both eyes, simply because the model doesn't know the color and shape of the eyes. In such cases, the use of prior information helps with the structure of the face, but the texture still remains missing.

3.5 Computational Cost

Current models have high computational cost, restricting the use in edge devices and in real-time applications. Inference time is still very high for real-time applications even with GPU, and the number of parameters may restrict the use in edge devices. Moreover, training a model takes a few days and model design takes a few months. More research is needed to create better algorithms and more efficient methods to reduce these computational costs.

4 Conclusion

This paper reviews GAN-based face image inpainting and de-occlusion studies found in the literature. More specifically, we explored the network architecture and components. The review also analyzed the limitations.

The GAN architecture for image inpainting and two-stage networks were described. Encoder-decoder and U-Net are basic generator architectures that can be combined with other components for additional functionalities. They are also used in single and multi-stage architectures. Local and global discriminators, PatchGAN and SN-PatchGAN improve the GAN's ability to distinguish between real and fake images in local and global levels. Squeeze and excitation blocks perform channel-wise feature re-calibration, weighting the importance of each feature map in a given layer. TTUR, Adam optimizer and spectral normalization accelerate and stabilize GAN training.

Finally, this study discussed the current limitations and challenges found in datasets, evaluation metrics, automatic de-occlusion, image quality and computational cost. Furthermore, we propose insights for future research.

References

1. Bertalmio, M., Sapiro, G., Caselles, V., Ballester, C.: Image inpainting. In: Proceedings of the 27th Annual Conference on Computer Graphics and Interactive Techniques, SIGGRAPH 2000, pp. 417–424. ACM Press/Addison-Wesley Publishing Co., USA (2000)

2. Burgos-Artizzu, X.P., Perona, P., Dollár, P.: Robust face landmark estimation under occlusion. In: 2013 IEEE International Conference on Computer Vision, pp. 1513–1520 (2013)
3. Cai, J., Han, H., Cui, J., Chen, J., Liu, L., Kevin Zhou, S.: Semi-supervised natural face de-occlusion. IEEE Trans. Inf. Forensics Secur. **16**, 1044–1057 (2021)
4. Cao, S., Sakurai, K.: Face completion with pyramid semantic attention and latent codes. In: Proceedings - 2020 8th International Symposium on Computing and Networking, CANDAR 2020, pp. 1–8. Institute of Electrical and Electronics Engineers Inc. (2020)
5. Chen, M., Liu, Z., Ye, L., Wang, Y.: Attentional coarse-and-fine generative adversarial networks for image inpainting. Neurocomputing **405**, 259–269 (2020)
6. Chen, Y.A., Chen, W.C., Wei, C.P., Wang, Y.C.: Occlusion-aware face inpainting via generative adversarial networks. In: Proceedings - International Conference on Image Processing, ICIP, vol. 2017-September, pp. 1202–1206. IEEE Computer Society (2018)
7. Cheung, Y.M., Li, M., Zou, R.: Facial structure guided GAN for identity-preserved face image de-occlusion. In: ICMR 2021 - Proceedings of the 2021 International Conference on Multimedia Retrieval, pp. 46–54. Association for Computing Machinery, Inc. (2021)
8. Din, N., Javed, K., Bae, S., Yi, J.: Effective removal of user-selected foreground object from facial images using a novel GAN-based network. IEEE Access **8**, 109648–109661 (2020)
9. Dong, J., Zhang, L., Zhang, H., Liu, W.: Occlusion-aware GAN for face de-occlusion in the wild. In: Proceedings - IEEE International Conference on Multimedia and Expo, vol. 2020-July. IEEE Computer Society (2020)
10. Fang, Y., Li, Y., Tu, X., Tan, T., Wang, X.: Face completion with hybrid dilated convolution. Signal Process. Image Commun. **80**, 115664 (2020)
11. Ge, S., Li, C., Zhao, S., Zeng, D.: Occluded face recognition in the wild by identity-diversity inpainting. IEEE Trans. Circuits Syst. Video Technol. **30**(10), 3387–3397 (2020)
12. Goodfellow, I.J., et al.: Generative adversarial nets. In: Proceedings of the 27th International Conference on Neural Information Processing Systems, NIPS 2014, vol. 2, pp. 2672–2680. MIT Press, Cambridge (2014)
13. Gulrajani, I., Ahmed, F., Arjovsky, M., Dumoulin, V., Courville, A.: Improved training of Wasserstein GANs. In: Proceedings of the 31st International Conference on Neural Information Processing Systems, NIPS 2017, pp. 5769–5779. Curran Associates Inc., Red Hook (2017)
14. Guo, D., Feng, J., Zhou, B.: Structure-aware image expansion with global attention. In: SIGGRAPH Asia 2019 Technical Briefs, SA 2019, pp. 13–16. Association for Computing Machinery, Inc. (2019)
15. He, K., Zhang, X., Ren, S., Sun, J.: Deep residual learning for image recognition. In: 2016 IEEE Conference on Computer Vision and Pattern Recognition (CVPR), pp. 770–778 (2016)
16. Heusel, M., Ramsauer, H., Unterthiner, T., Nessler, B., Hochreiter, S.: GANs trained by a two time-scale update rule converge to a local nash equilibrium. In: Guyon, I., et al. (eds.) Advances in Neural Information Processing Systems, vol. 30. Curran Associates, Inc. (2017)
17. Horé, A., Ziou, D.: Image quality metrics: PSNR vs. SSIM. In: 2010 20th International Conference on Pattern Recognition, pp. 2366–2369 (2010)
18. Hu, J., Shen, L., Albanie, S., Sun, G., Wu, E.: Squeeze-and-excitation networks. IEEE Trans. Pattern Anal. Mach. Intell. **42**(8), 2011–2023 (2020)

19. Iizuka, S., Simo-Serra, E., Ishikawa, H.: Globally and locally consistent image completion. ACM Trans. Graph. **36**(4), 1–14 (2017)
20. Isola, P., Zhu, J.Y., Zhou, T., Efros, A.A.: Image-to-image translation with conditional adversarial networks. In: 2017 IEEE Conference on Computer Vision and Pattern Recognition (CVPR), pp. 5967–5976 (2017)
21. Jabbar, A., Li, X., Iqbal, M., Malik, A.: FD-stackGAN: face de-occlusion using stacked generative adversarial networks. KSII Trans. Internet Inf. Syst. **15**(7), 2547–2567 (2021)
22. Jam, J., Kendrick, C., Drouard, V., Walker, K., Hsu, G.S., Yap, M.: R-MNet: a perceptual adversarial network for image inpainting. In: Proceedings - 2021 IEEE Winter Conference on Applications of Computer Vision, WACV 2021, pp. 2713–2722. Institute of Electrical and Electronics Engineers Inc. (2021)
23. Jam, J., Kendrick, C., Drouard, V., Walker, K., Hsu, G.S., Yap, M.: Symmetric skip connection Wasserstein GAN for high-resolution facial image inpainting. In: VISIGRAPP 2021 - Proceedings of the 16th International Joint Conference on Computer Vision, Imaging and Computer Graphics Theory and Applications, vol. 4, pp. 35–44. SciTePress (2021)
24. Johnson, J., Alahi, A., Fei-Fei, L.: Perceptual losses for real-time style transfer and super-resolution. In: Leibe, B., Matas, J., Sebe, N., Welling, M. (eds.) ECCV 2016. LNCS, vol. 9906, pp. 694–711. Springer, Cham (2016). https://doi.org/10.1007/978-3-319-46475-6_43
25. Khan, M., Ud Din, N., Bae, S., Yi, J.: Interactive removal of microphone object in facial images. Electronics **8**(10), 1115 (2019)
26. Li, X., Hu, G., Zhu, J., Zuo, W., Wang, M., Zhang, L.: Learning symmetry consistent deep CNNs for face completion. IEEE Trans. Image Process. **29**, 7641–7655 (2020)
27. Li, Y., Liu, S., Yang, J., Yang, M.H.: Generative face completion. In: 2017 IEEE Conference on Computer Vision and Pattern Recognition (CVPR), pp. 5892–5900 (2017)
28. Li, Z., Zhu, H., Cao, L., Jiao, L., Zhong, Y., Ma, A.: Face inpainting via nested generative adversarial networks. IEEE Access **7**, 155462–155471 (2019)
29. Lie, Y., Li, L.: Image inpainting using multi-scale neural network and shift-net. In: Proceedings - 2020 7th International Conference on Information Science and Control Engineering, ICISCE 2020, pp. 704–709. Institute of Electrical and Electronics Engineers Inc. (2020)
30. Lin, J., Yang, H., Chen, D., Zeng, M., Wen, F., Yuan, L.: Face parsing with ROI tanh-warping. In: 2019 IEEE/CVF Conference on Computer Vision and Pattern Recognition (CVPR), pp. 5647–5656 (2019)
31. Liu, J., Jung, C.: Facial image inpainting using multi-level generative network. In: Proceedings - IEEE International Conference on Multimedia and Expo, vol. 2019-July, pp. 1168–1173. IEEE Computer Society (2019)
32. Liu, J., Jung, C.: Facial image inpainting using attention-based multi-level generative network. Neurocomputing **437**, 95–106 (2021)
33. Luo, X., He, X., Qing, L., Chen, X., Liu, L., Xu, Y.: Eyesgan: synthesize human face from human eyes. Neurocomputing **404**, 213–226 (2020)
34. Maggipinto, M., Masiero, C., Beghi, A., Susto, G.A.: A convolutional autoencoder approach for feature extraction in virtual metrology. Procedia Manufacturing **17**, 126–133 (2018). 28th International Conference on Flexible Automation and Intelligent Manufacturing (FAIM2018), 11–14 June 2018, Columbus, OH, USAGlobal Integration of Intelligent Manufacturing and Smart Industry for Good of Humanity

35. Maharjan, R., Ud Din, N., Yi, J.: Image-to-image translation based face de-occlusion. In: Jiang X., F.H. (ed.) Proceedings of SPIE - The International Society for Optical Engineering, vol. 11519. SPIE (2020)
36. Maheshwari, U., Turlapati, V., Kiruthika, U.: Lucid-GAN: an adversarial network for enhanced image inpainting. In: CIVEMSA 2021 - IEEE International Conference on Computational Intelligence and Virtual Environments for Measurement Systems and Applications, Proceedings. Institute of Electrical and Electronics Engineers Inc. (2021)
37. Mathai, J., Masi, I., Abdalmageed, W.: Does generative face completion help face recognition? In: 2019 International Conference on Biometrics, ICB 2019. Institute of Electrical and Electronics Engineers Inc. (2019)
38. Maulana, A., Fatichah, C., Suciati, N.: Facial inpainting using generative adversarial network with feature reconstruction and landmark loss to preserve spatial consistency in unaligned face images. Int. J. Intell. Eng. Syst. **13**(6), 219–228 (2020)
39. Miyato, T., Kataoka, T., Koyama, M., Yoshida, Y.: Spectral normalization for generative adversarial networks (2018)
40. Ronneberger, O., Fischer, P., Brox, T.: U-Net: convolutional networks for biomedical image segmentation. In: Navab, N., Hornegger, J., Wells, W.M., Frangi, A.F. (eds.) MICCAI 2015. LNCS, vol. 9351, pp. 234–241. Springer, Cham (2015). https://doi.org/10.1007/978-3-319-24574-4_28
41. Sadiq, M., Shi, D.: Attentive occlusion-adaptive deep network for facial landmark detection. Pattern Recognit. **125**, 108510 (2022)
42. Wang, F., Li, W., Liu, Y., Gong, Y., Gao, Z., Lu, J.: Face inpainting combining structured forest edge information and gated convolution. In: Proceedings - 2021 3rd International Conference on Natural Language Processing, ICNLP 2021, pp. 213–217. Institute of Electrical and Electronics Engineers Inc. (2021)
43. Wu, Y., Singh, V., Kapoor, A.: From image to video face inpainting: spatial-temporal nested GAN (STN-GAN) for usability recovery. In: Proceedings - 2020 IEEE Winter Conference on Applications of Computer Vision, WACV 2020, pp. 2385–2394. Institute of Electrical and Electronics Engineers Inc. (2020)
44. Yu, J., Lin, Z., Yang, J., Shen, X., Lu, X., Huang, T.: Free-form image inpainting with gated convolution. In: Proceedings of the IEEE International Conference on Computer Vision, vol. 2019-October, pp. 4470–4479. Institute of Electrical and Electronics Engineers Inc. (2019)
45. Yu, J., Lin, Z., Yang, J., Shen, X., Lu, X., Huang, T.: Generative image inpainting with contextual attention. In: Proceedings of the IEEE Computer Society Conference on Computer Vision and Pattern Recognition, pp. 5505–5514. IEEE Computer Society (2018)
46. Yu, L., Zhu, D., He, J.: Semantic segmentation guided face inpainting based on SN-PatchGAN. In: Proceedings - 2020 13th International Congress on Image and Signal Processing, BioMedical Engineering and Informatics, CISP-BMEI 2020, pp. 110–115. Institute of Electrical and Electronics Engineers Inc. (2020)
47. Zhang, H., Li, T.: Semantic face image inpainting based on generative adversarial network. In: Proceedings - 2020 35th Youth Academic Annual Conference of Chinese Association of Automation, YAC 2020, pp. 530–535. Institute of Electrical and Electronics Engineers Inc. (2020)
48. Zhang, H., Goodfellow, I., Metaxas, D., Odena, A.: Self-attention generative adversarial networks. In: Chaudhuri, K., Salakhutdinov, R. (eds.) Proceedings of the 36th International Conference on Machine Learning. Proceedings of Machine Learning Research, vol. 97, pp. 7354–7363. PMLR (2019)

49. Zhang, R., Isola, P., Efros, A.A., Shechtman, E., Wang, O.: The unreasonable effectiveness of deep features as a perceptual metric. In: 2018 IEEE/CVF Conference on Computer Vision and Pattern Recognition, pp. 586–595 (2018)
50. Zhu, W., Wang, X., Wu, Y., Zou, G.: A face occlusion removal and privacy protection method for IoT devices based on generative adversarial networks. Wirel. Commun. Mob. Comput. **2021** (2021)

Classification

Classification of Facial Images to Assist in the Diagnosis of Autism Spectrum Disorder: A Study on the Effect of Face Detection and Landmark Identification Algorithms

Gabriel C. Michelassi[1], Henrique S. Bortoletti[1], Tuany D. Pinheiro[1], Thiago Nobayashi[1], Fabio R. D. de Barros[1], Rafael L. Testa[1], Andréia F. Silva[2], Mirian C. Revers[2], Joana Portolese[2], Helio Pedrini[3], Helena Brentani[2], Fatima L. S. Nunes[1], and Ariane Machado-Lima[1](✉)

[1] School of Arts, Sciences and Humanities, University of Sao Paulo, R. Arlindo Béttio, 1000 - Ermelino Matarazzo, Sao Paulo-SP 03828-000, Brazil
ariane.machado@usp.br

[2] Department of Psychiatry, School of Medicine, University of Sao Paulo, R. Dr. Ovídio Pires de Campos, 785, Sao Paulo-SP 05403-903, Brazil

[3] Institute of Computing, University of Campinas, Av. Albert Einstein, 1251 - Cidade Universitária, Campinas-SP 13083-852, Brazil

Abstract. Since facial morphology can be linked to brain developmental problems, studies have been conducted to develop computational systems to assist in the diagnosis of some neurodevelopmental disorders based on facial images. The first steps usually include face detection and landmark identification. Although there are several libraries that implement different algorithms for these tasks, to the best of our knowledge no study has discussed the effect of choosing these ready-to-use implementations on the performance of the final classifier. This paper compares four libraries for facial detection and landmark identification in the context of classification of facial images for computer-aided diagnosis of Autism Spectrum Disorder, where the classifiers achieved 0.92, the highest F1-score. The results indicate that the choice of which facial detection and landmark identification algorithms to use do in fact affect the final classifier performance. It appears that the causes are related to not only the quality of face and landmark identification, but also to the success rate of face detection. This last issue is particularly important when the initial training sample size is modest, which is usually the case in terms of classification of some syndromes or neurodevelopmental disorders based on facial images.

Keywords: Image processing · Face Detection · Landmark Identification · Anthropometry · Autism Spectrum Disorder · Classification

M. C. Naldi and R. A. C. Bianchi (Eds.): BRACIS 2023, LNAI 14196, pp. 261–275, 2023.
https://doi.org/10.1007/978-3-031-45389-2_18

1 Introduction

Anthropometry is the science that studies the measurements of the human body including height, weight and size of body components [11]. The measurements can be done directly, with measuring equipment such as measuring tapes, or indirectly, using radiography images, three-dimensional images captured by stereophotogrammetry or two-dimensional images captured by digital cameras [31].

Several studies have suggested that facial morphology can be linked to brain developmental problems [2,6,9,13,32]. For instance, studies of facial anthropometric measures in individuals with Autism Spectrum Disorder (ASD) and individuals with typical development (TD) have revealed significant differences between these two groups, such as the distance between the pupils, ear format, strabismus, head circumference [2,23,27,29]. These results motivate the use of machine learning approaches to classify ASD versus TD individuals, based on anthropometric facial features, to aid ASD diagnosis.

ASD is a neurodevelopmental, polygenic and multifactorial disorder with evidence of genetic and environmental factors contributing to its etiology [10,14,20, 24]. Individuals with ASD exhibit persistent social communication impairments, and also lack the ability to develop, maintain and understand relationships, and show restricted and repetitive patterns of behavior, interests, and activities. These symptoms appear during childhood and can impair the person's daily life [3,18]. An early and correct diagnosis is crucial for appropriate intervention. The diagnosis is clinical and expensive, since it requires trained health professionals to apply the instruments, mostly questionnaires, and observation routines of the child's behavior [22]. Therefore, to increase accessing the ASD diagnosis, computer-aided diagnosis systems could be used for large-scale screening for posterior analysis by a specialist. In order to decrease the financial cost and facilitate using such systems, the use of two-dimensional facial images captured by digital cameras or smartphones is preferred than, for instance, three-dimensional images captured by stereophotogrammetry devices [2,12,26], since they can be more easily achieved considering the popularization of mobile devices.

In fact, promising results related to neurodevelopmental disorders have been achieved using two-dimensional facial images captured by digital cameras [16,33]. The computational pipeline of analysis usually consists of three steps: 1) face detection and landmark identification, 2) feature extraction and 3) classifier induction. Although there are several libraries that implement different algorithms for the first step, to the best of our knowledge no study has discussed the effect of the choice of these ready-to-use implementations on the performance of the final classifier.

This paper investigates the effect of using four image processing libraries for facial detection and landmark identification on the classification results of facial images for computer-aided diagnosis of ASD. While a face detection algorithm may cause training sample reduction by discarding images when no faces were detected, a landmark detection algorithm may interfere with the correct extraction of features from the anthropometric measurements. Both issues can affect the final classifier and are explored in this paper.

2 Methods

Four image processing libraries were compared for face detection and landmark identification tasks. In order to investigate the impact of these libraries on final classification performance, we performed two experiments to test two hypothesis:

- **H1:** the number of training images with successful face detection, combined with correct face detection and landmark identification, will affect the performance of the final classifier;
- **H2:** using the same training images, applying different detection and landmark identification algorithms affects the performance of the final classifier.

Both hypotheses were tested by performing the five-modules process depicted in Fig. 1. In summary, the images of an initial training set were acquired following the protocol described in Sect. 2.1 and pre-processed as described in Sect. 2.2 (modules 1 and 2). Each preprocessed image was submitted to different face and landmark detection algorithms to identify five landmark sets: one for each image processing library, except for MediaPipe which was used twice (module 3, described in Sect. 2.3). Each landmark set was used to extract a set of geometric features based on the distance between each pair of landmarks (module 4, described in Sect. 2.4), resulting in five different feature datasets. In these five datasets, each instance is the feature vector of an image. Since the face detection algorithms usually are not able to detect the face in all images (module 3), the five datasets have not the same number of instances. Therefore, two experiments were performed: 1) using each dataset exactly as it is; and 2) using the datasets composed of only the feature vectors of the images where the face was detected by *all* face detection algorithms executed in module 3. Therefore, experiment 1 evaluates the effect of the face detection on the results (testing hypothesis H1), whereas experiment 2 evaluates the effect of the landmark identification (in the same faces) on the classifier performance (testing hypothesis H2). These datasets are used separately to train and test several classifiers (module 5, described in Sect. 2.5). Finally, a statistical analysis was performed for hypotheses testing (Sect. 2.6).

2.1 Image Acquisition Protocol

All images were captured with a Nikon Coolpix L120 compact camera using the protocol shown in Fig. 2. In each acquisition, the camera was attached to a tripod fixed at a distance of 50 cm from the chair where the volunteer was accommodated. The chair was placed against the wall, and the volunteer was instructed to remain with his back fully resting on the chair. A white wall was chosen as the photo background, where a red band 50 cm by 5 cm was placed

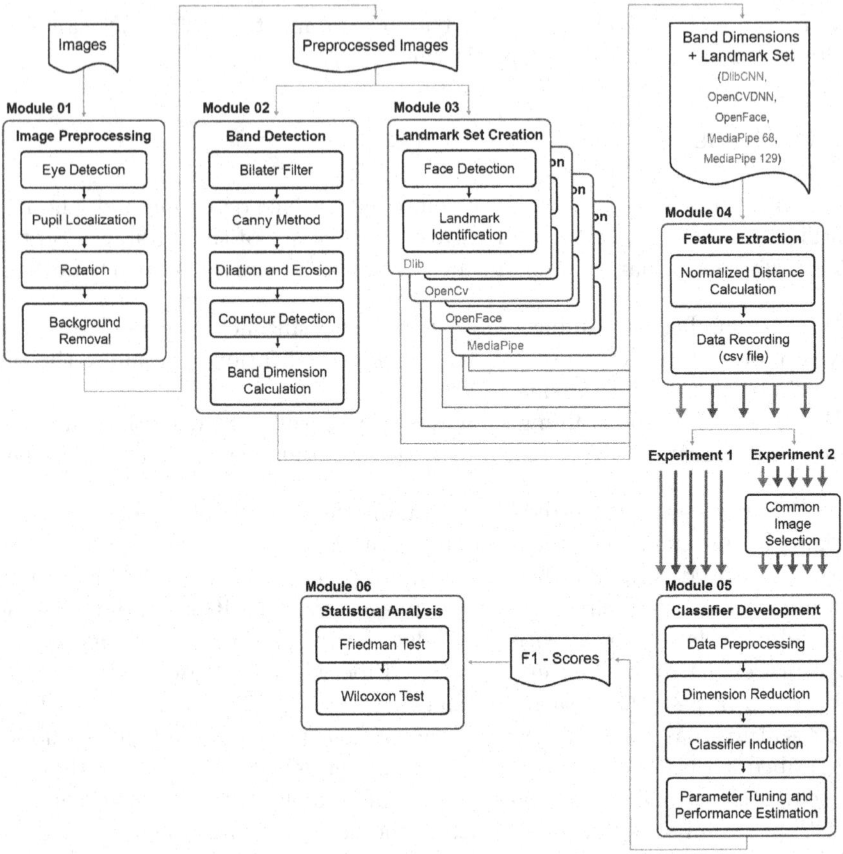

Fig. 1. Overview of the proposed method.

in order to be posteriorly used as a reference for normalizing the images and estimating distortion.

The volunteers were instructed to remain with a facial expression as neutral as possible and five front facial photographs were taken. A video was recorded when neutral expression facial image was not possible or if the face was inclined. Thus, a subsequent video analysis allowed selecting the best frame.

Images were captured from 43 ASD and 74 TD children and adolescents ranging from 5 to 18 years of age. The diagnosis of the ASD participants was confirmed by a reference Hospital the Clinics Hospital of the University of São Paulo and the photographs were taken after the patient's follow-up visit. The TD group consisted of volunteers invited by email. All underwent screening by an infant psychiatrist specialized in ASD to confirm the absence of ASD and/or clinical evidence of genetic syndromes. All participants were authorized by the parents or legal guardians through an informed consent signed at the beginning of each data collection session.

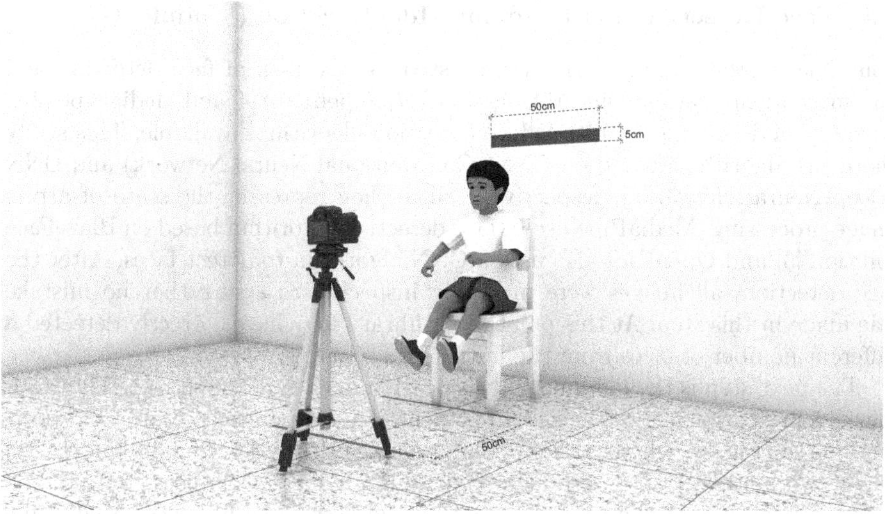

Fig. 2. Protocol used to capture the images, showing the correct position of the individual and the band used as reference for normalization.

This project was approved by the Ethics Board and of the School of Arts, Sciences and Humanities of University of Sao Paulo, protocol number 1.669.832 at August 8th, 2016. All experiments were performed in accordance with the Ethics Board requirements. The informed consent was obtained from all participants and/or their legal guardians.

2.2 Image Preprocessing and Band Detection (Modules 1 and 2)

We acquired more than one frontal image from each volunteer. The best image was selected based on two criteria: facial neutrality and proportion of width and height of the reference band in order to avoid image distortion.

The image preprocessing was performed to rotate and crop all the images. The rotation intends to correct the lateral face inclination and was performed according to the following steps: eye region detection using *Haar cascades*; pupil localization; tracing a line passing over the pupil points; calculation of the angle θ formed by this line and the horizontal axis; image rotation according to θ. Finally, the images were cropped in order to remove the black background resulting from the rotation operation.

The steps for reference band detection were: (i) application of a bilateral filter for image smoothing, (ii) edge detection using the Canny method [8], (iii) application of dilation and erosion operations to correct discontinuities on the edges, (iv) detection of the band contour using the method *findContours* from OpenCV library [30] and (v) calculation of the band width and height.

2.3 Face Detection and Landmark Identification (Module 3)

Four image processing libraries were tested for the task of face detection and landmark identification: Dlib [15], OpenCV [7], OpenFace [4] and MediaPipe [21]. OpenCV and Dlib have multiple face detection algorithms available. This study chose the algorithms based on CNN (Convolutional Neural Network) and DNN (Deep Neural Networks), respectively, since they represent the state of art in image processing. MediaPipe uses a face detection algorithm based on BlazeFace solution [5] and OpenFace also uses a CNN algorithm to detect faces. After the face detection, all images were manually inspected to assure that no mistake was made in this step. At this point, each library may have correctly detected a different number of faces from the initial set of images.

The next step is the landmark identification in each detected face. The landmark identification algorithms of Dlib, OpenCV and OpenFace libraries detect 68 landmarks (Fig. 3), whereas the MediaPipe's algorithm detects 438 landmarks. Thus, two landmark sets were identified using MediaPipe: 1) using only the 68 landmarks closest to the ones detected by the first three libraries (MediaPipe 68); and 2) using the previous 68 landmarks and additional 61 points corresponding to the facial contour (MediaPipe 129) (Fig. 3E). The first landmark set is meant to evaluate the influence of the quality of the identification of the same 68 landmarks on the classifier performance, whereas the second landmark set is meant to evaluate the influence of a higher number of landmarks. Therefore, five landmark sets were obtained and compared in this study (Table 1).

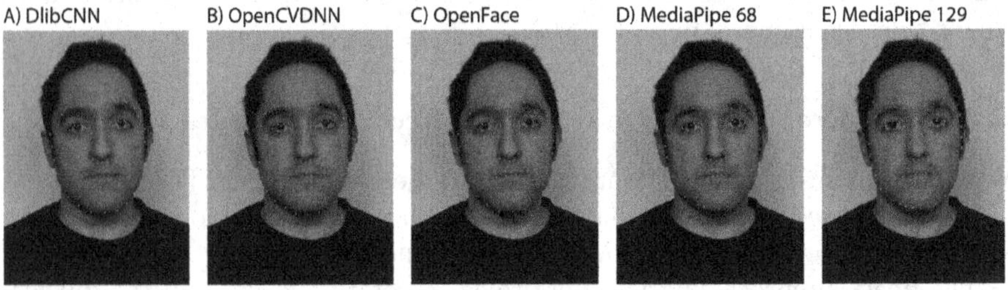

Fig. 3. The 68 landmarks present in DlibCNN, OpenCVDNN, OpenFace, MediaPipe 68 and MediaPipe 129 landmark sets. Source image from [1].

2.4 Feature Extraction and Experiment Split (Module 4)

For each landmark set output from module 3 (Table 1), the Euclidean distance between each pair of landmarks, normalized by the reference band dimensions, was calculated, corresponding to a feature. More specifically, for each pair of landmarks $p(x_1, y_1)$ and $q(x_2, y_2)$, each landmark represented by a coordinate (x, y), the normalized Euclidean distance d_n is calculated according to Eq. 1,

Table 1. Landmark sets compared in this study. Each landmark set was extracted using an image processing library, face detection algorithm, landmark identification algorithm, resulting in a specific number of identified landmarks. The landmark identification algorithm Constrained Local Neural Fields is referred to as CLNF. The landmark identification algorithm Local Binary Fitting is referred to as LBF.

Landmark set	Library	Face detec. alg.	Landmark id. alg.	Nr. of landmarks
OpenFace	OpenCV	CNN	CLNF	68
DlibCNN	Dlib	CNN	Shape Predictor	68
OpenCVDNN	OpenCV	DNN	LBF	68
MediaPipe 68	MediaPipe	BlazeFace	FaceMesh	68
MediaPipe 129	MediaPipe	BlazeFace	FaceMesh	129

where w and h are the reference band width and height (calculated as described in Sect. 2.2), respectively:

$$d_n(p,q) = \sqrt{[(x_1 - x_2)/w]^2 + [(y_1 - y_2)/h]^2} \quad (1)$$

The feature dataset obtained from each landmark set described in Table 1 was used in Experiment 1. In experiment 2, only features from images where the face was correctly detected by all face detection algorithms were used.

2.5 Classifier Development (Module 5)

The last module of the proposed pipeline (Fig. 1) consists in training and evaluating several classifiers varying the procedures for dataset preprocessing, dimension reduction and classifier induction. Here we describe these procedures and how the parameter tuning and performance evaluation were performed.

Dataset Preprocessing. Three preprocessing tasks were applied on the dataset: initial feature filtering, Min-Max normalization and class balancing.

In the initial feature filtering, when two features were highly correlated (Pearson correlation > 0.98), only one was kept.

Min-max normalization was applied on the feature values. For this, after identification of the maximum and minimum values of each feature X in the dataset (X_{max} and X_{min}, respectively), the original value X_i is transformed into a new value $X'_i = (X_i - X_{min})/(X_{max} - X_{min})$.

For class balancing, four approaches were considered: i) undersampling, ii) oversampling with four variants of SMOTE algorithm [25] (Default, SVM, Borderline and KMeans Smote [17]), iii) combination of under and over sampling with SMOTE Tomek, and iv) no balancing.

Dimension Reduction. Dimension reduction methods are extremely important for classification tasks because they eliminate redundant features and can increase classifier performance.

Filter-type feature selection methods tend to have a lower computational cost than wrapper-type methods. In this paper, the implementation of the scikit-learn libraries [28] and skfeature [19] libraries were used and seven different algorithms were applied:

- Principal Component Analysis (PCA), which performs feature extraction by transforming the initial set into a new set containing only the major components;
- Minimum Redundancy Maximum Relevance (mRMR), which performs the selection of features based on two criteria: one is the minimum criterion of redundancy between pairs of features and the other is the maximum relevance through the measurement of information;
- Correlation-based Feature Selection (CFS), which performs the selection of features considering that a suitable subset contains features highly correlated with the target class, but little correlated with each other;
- Fast Correlation-based Filter (FCBF), which performs the selection of features considering the main idea of the CFS method, but calculating the correlation measure based on the entropy concept of information theory;
- ReliefF, which performs the selection of features by means of a relevance ranking of each feature;
- Robust Feature Selection (RFS), which performs the feature selection through sparse dictionary learning;
- Random Forest Select, which performs the feature selection based on the Random Forest classifier.

Classifier Induction (Hyperparameter Tuning and Performance Estimation). Six classifier induction algorithms were used: K-Nearest Neighbors (KNN), Support Vector Machines (SVM), Gaussian Naïve Bayes (NB), Neural Networks (NN), Random Forests (RF) and Linear Discriminant Analysis (LDA).

Hyperparameter tuning of the classifier induction algorithms was performed using a grid search approach to select the hyperparameter values that maximize the classifier f1-score, estimated by stratified 10-fold cross-validation. In each fold of the cross-validation, the synthetic instances produced by the class balancing algorithm were excluded from the test sample. Table 2 shows the hyperparameters and the tested values that were tuned for each classifier. Default parameters were used for Gaussian Naive Bayes classifier.

In summary, all combinations of dataset balancing strategy, dimension reduction and classifier induction algorithm (Table 3) were evaluated, totaling 294 combinations. To measure performance, the F1-score of each classifier resulting from each combination was estimated via stratified 10-fold cross-validation, with no synthetic data in the test folds. In addition, confidence intervals were calculated with 95% confidence.

Table 2. Parameters and values tested for each classifier. *n_features*: the number of features after dimension reduction. The parameter *hidden_layer_sizes* is specified as an ordered pair (x, y) where x is the number of hidden layers and y is the number of neurons per layer.

Classifier	Parameters	Tested values
RF	n_estimators	500, 1000
	min_samples_leaf	1, 2, 5, 10, 15, 20
	max_features	[0.5 * sqrt(*n_features*), 2 * sqrt(*n_features*)]
SVM	kernel	linear, poly, rbf, sigmoid
	C	0.01, 0.1, 1, 5, 10, 50, 100
	gamma	'scale', 'auto', 0.00001, 0.0001, 0.001, 0.01, 0.1, 0.5, 1, 5, 10
	degree	2, 3, 5
NN	hidden_layer_sizes	(1, 100), (1, *n_features*), (3, *n_features*), (1, 50), (3, 50)
	alpha	0.0001, 0.001, 0.01
	activation	'identity', 'logistic', 'tanh', 'relu'
	solver	'lbfgs', 'sgd'
KNN	n_neighbors	1, 2, 3, 5, 10, 20, 25
	weights	uniform, distance
LDA	solvers	'svd', 'lsqr', 'eigen'
	shrinkage	'auto', None

Table 3. Dataset balancing strategies, dimension reduction and classifier induction algorithms used in this study.

Balance Strategies	Dimensionality Reduction	Classifier Induction
No strategy	PCA	SVM
Undersampling	mRMR	KNN
Oversampling SMOTE	CFS	LDA
Oversampling SMOTE Borderline	FCBF	RF
Oversampling SMOTE KMeans	ReliefF	NB
Oversampling SMOTE SVM	RFS	NN
Over- and Under SMOTE Tomek	Random Forest Select	

2.6 Statistical Analysis and Comparison

As described in Sect. 2.5, several classifiers were inducted using different combinations of class balancing algorithms, dimension reduction and classifier induction algorithms (Table 3). Considering the ten folds of the cross-validation, 2940 classifiers were inducted from each dataset (i.e., from each landmark set). Comparing the F1-scores of learned classifiers using each of these combinations in each fold, varying only the landmark set (Table 1), constitutes a test to detect differences in "treatments" across multiple test attempts, where the "treatments" are the use of different libraries for face detection and landmark identification.

Therefore, a table $F_{2940\times5}$ was created where $F[i,j]$ is the F1-score obtained by the classifier inducted using the combination i of class balancing algorithms, dimension reduction, classifier induction algorithm and fold using the landmark set j. As the F1-scores do not present a normal distribution, Friedman's non-parametric test was applied using the F1-scores obtained in experiments 1 and 2 to test the hypotheses H1 and H2, respectively. Considering a critical P-value of 0.05, an inferior P-value means that there is a significant difference between the F1-score obtained using different libraries for face detection and landmark detection.

In addition, in order to investigate if the H1 and H2 test results can be dependent on the classifier induction algorithm, the same test was repeated using only the F1-scores of classifiers inducted by the same algorithms. For these tests a Bonferroni-Holm correction was applied to correct the P-values due to the execution of the six tests.

The Friedman test only evaluates if the average F1-score is significantly different among the five landmark sets. To test if there is a significant difference between each pair of landmark sets, the Wilcoxon test was applied.

3 Results and Discussion

Table 4 (rows corresponding to "Experiment 1") shows the number of images where the face was correctly detected by each library. From an initial set of 117 images (43 images from ASD and 74 images from TD individuals), the face detection algorithms performed differently, detecting the face in 63% to 89% of the images. MediaPipe was the library that detected fewer faces, whereas OpenCV using the DNN algorithm was the library that detected more faces.

Table 4. Number of images represented in each landmark set in the two experiments. Rows corresponding to "Experiment 1" describe the number of faces detected by each library (*: MediaPipe68 and MediaPipe129 are landmark sets based on the same faces extracted by MediaPipe face detection algorithm). Rows corresponding to "Experiment 2" describe the number of faces detected by all libraries.

Experiment	Landmark Set	Library	ASD	TD	No. of detected faces (% of total)
1	OpenFace	OpenCV	37	58	95 (81%)
	DlibCNN	Dlib	39	62	101 (86%)
	OpenCVDNN	OpenCV	38	66	104 (89%)
	MediaPipe 68*	MediaPipe	34	40	74 (63%)
	MediaPipe 129*				
2	OpenFace	OpenCV	33	29	62 (53%)
	DlibCNN	Dlib			
	OpenCVDNN	OpenCV			
	MediaPipe 68	MediaPipe			
	MediaPipe 129	MediaPipe			

Only 62 images (53%) had the face correctly detected by all libraries. These images were used to compose the five landmark sets in Experiment 2 (Table 4 - rows corresponding to "Experiment 2").

Table 5 shows the P-values resulting from the Friedman test performed on all inducted classifiers, as well as for each classifier induction algorithm separately, for both experiments. Considering all learned classifiers (Table 5, row "ALL CLASSIFIERS"), the results show that there is a significant difference between the F1-scores obtained when using different landmark sets (i.e., different libraries for face detection and landmark identification), in both experiments 1 and 2 (P-values < 0.001), allowing that H1 and H2 are accepted. That means that both correct face detection and landmark identification have a significant effect on the final classification. All P-values are also below the critical value of 0.05 considering the test performed for each classifier induction separately (Table 5, rows "SVM" to "LDA"). However, three classifier induction algorithms had their P-values closer to the critical value: SVM in Experiment 1 and Naive Bayes and Random Forests in Experiment 2. This may indicate that different classifier induction algorithms have different susceptibility to data variation obtained using these libraries. However, more experiments should be conduct to investigate this issue.

Table 5. Results of the Friedman tests considering the 2940 generated classifiers per landmark set ("ALL CLASSIFIERS") and per classifier induction algorithm.

	Experiment 1	Experiment 2
ALL CLASSIFIERS	P < .001	P < .001
SVM	0.0133	P < .001
KNN	P < .001	P < .001
NB	P < .001	0.0470
NN	P < .001	P < .001
RF	P < .001	0.0013
LDA	P < .001	P < .001

Table 6 shows the P-values achieved using the Wilcoxon test for each pair of landmark sets in both experiments. In Experiment 1, the results show that the F1-scores achieved using the dlibCNN and OpenCVDNN landmark sets are not significantly different, as well as using MediaPipe 68 and MediaPipe 129. DlibCNN and OpenCVDNN are the landmark sets based on the highest number of images (101 and 104 detected faces, respectively - Table 4) whereas MediaPipe 68 and MediaPipe 129 landmark sets are based on exactly the same images. In Experiment 2, where all landmark sets were extracted from the same 62 images, all P-values were < 0.05, indicating that F1-scores are significantly different between each pair of landmark sets. These results indicate that the number of training images not only affect the classifier performance but also the quality of face and landmark identification. In addition, the significant difference between MediaPipe 68 and MediaPipe 129 only in Experiment 2 may indicate that the additional landmarks in MediaPipe 129 improved the classifier performance when there was a smaller number of training images.

Table 6. Results of the Wilcoxon tests.

Landmark Set		Experiment 1: P-values	Experiment 2: P-values
DlibCNN	OpenCVDNN	0.2896	$P < .001$
DlibCNN	OpenFace	$P < .001$	$P < .001$
DlibCNN	MediaPipe 68	$P < .001$	0.01056
DlibCNN	MediaPipe 129	$P < .001$	$P < .001$
OpenCVDNN	OpenFace	$P < .001$	$P < .001$
OpenCVDNN	MediaPipe 68	$P < .001$	$P < .001$
OpenCVDNN	MediaPipe 129	$P < .001$	$P < .001$
OpenFace	MediaPipe 68	$P < .001$	$P < .001$
OpenFace	MediaPipe 129	$P < .001$	$P < .001$
MediaPipe 68	MediaPipe 129	0.5730	0.0206

Considering the use of these landmark sets to the problem of ASD x TD classification, Fig. 4 shows the highest F1-score (estimated by 10-fold cross-validation) achieved using each landmark set in each experiment. OpenFace, OpenCVDNN and DlibCNN had the best results in Experiment 1, highlighting that OpenFace and DlibCNN presented the narrowest confidence intervals. In fact, in Experiment 1 these three landmark sets are those based on the highest number of detected faces (Table 4). It is noteworthy that, although based on a lower number of images (95 detected faces), the OpenFace landmark set had the highest F1-score (0.9184) with an accuracy of 0.8861, precision of 0.8548 and recall of 1.0. This result was achieved using SMOTE Tomek for class balancing, PCA for dimension reduction and LDA for classifier induction algorithm. The lowest F1-score (0.8436) was achieved using MediaPipe 64, which is consistent with the fact that this landmark set is based on the lowest number of images (74 - Table 4). MediaPipe 129 had a higher F1-score than MediaPipe 64 (0.9029), based on the same images but with additional landmarks. However, it was observed that these two landmark sets are not significantly different in the statistical test using all generated classifiers (Table 6 - Experiment 1). Such relative performance of these landmark sets (DlibCNN, OpenCVDNN and OpenFace presenting higher F1-scores than MediaPipe 64 and MediaPipe 129) was similar in all inducted classifiers (data not shown). In Experiment 2 all F1-scores decreased, which is consistent with the fact that all landmark sets were based on a lower number of images (62). Therefore, it can be seen that not only the correct face detection and reference points are important, but also the number of images used to train a model. The best classifier had a accuracy of 0.88, precision of 0.83, recall of 1.0 and F1-score of 0.8981.

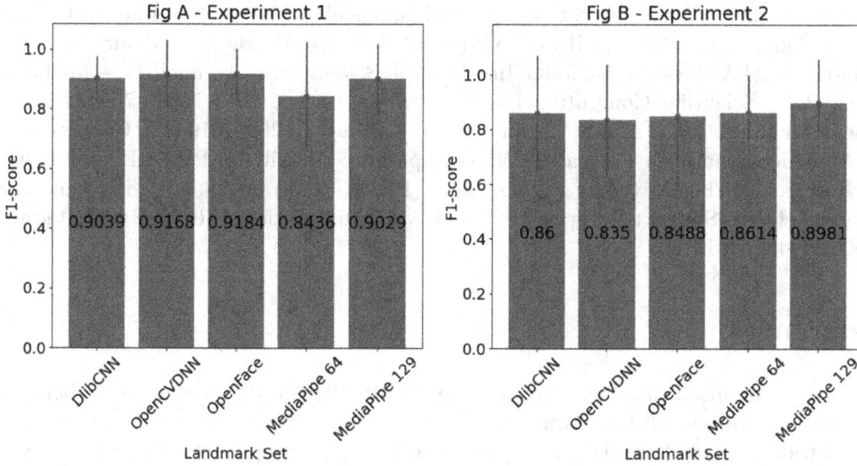

Fig. 4. The highest F1-score obtained using each landmark set in Experiment 1 (Fig A) and Experiment 2 (Fig B). The bar errors correspond to a 95% confidence interval.

4 Conclusion

This paper investigated the effect of employing different image processing libraries on the final result of facial image-based classifiers using ASD x TD classification as case-study. Two hypotheses were tested: that different face detection rates (influence the training sample size), in addition to the landmark identification quality, influence the final classification (H1) and that, using exactly the same images, the face detection and landmark identification per se influence the final classification (H2). The results enabled accepting both hypotheses in this case study.

The analysis of the pairwise differences, as well as of the F1-scores obtained by the different landmark sets in both experiments, indicate that libraries with a higher success rate in face detection (DlibCNN, OpenCVDNN and OpenFace) tend to produce higher F1-score, and thus more suitable.

Finally, considering the problem of ASD x TD classification, after testing several combinations of image processing libraries, dataset balancing, dimension reduction and classifier induction algorithms, this study achieved the F1-score of 0.9184, accuracy of 0.8861, precision of 0.8548 and recall of 1.0, which is important for computer-aided diagnosis. To enhance ASD analysis through face detection, more data from individuals with ASD and controls are crucial.

Acknowledgments. We thank the patients and their families that allowed the execution of this research, as well as the team of the Autism Spectrum Program of the Clinics Hospital (PROTEA-HC). This study was funded by Brazilian National Council of Scientific and Technological Development, (CNPq) (grants 309330/2018-1, 157535/2017-7 and 309030/2019-6) and Scientific and Technological Initiation Program at University of Sao Paulo (PIBIC/PIBIT-CNPq/USP 2020/2021), the São Paulo Research Foundation (FAPESP) - National Institute of Science and Technology - Medicine Assisted by Scientific Computing (INCT-MACC) - grant 2014/50889-7, Sao Paulo Research Foundation (FAPESP) grants #2017/12646-3, #2020/01992-0, Coordenação de Aperfeiçoamento de Pessoal de Nível Superior – Brasil (CAPES), Dean's Office for Research of the University of São Paulo (PRP-USP, grant 18.5.245.86.7) and the National Health Support Program for People with Disabilities (PRONAS/PCD) grant 25000.002484/2017-17.

References

1. Face recognition database (2005). http://cbcl.mit.edu/software-datasets/heisele/facerecognition-database.html
2. Aldridge, K., et al.: Facial phenotypes in subgroups of prepubertal boys with autism spectrum disorders are correlated with clinical phenotypes. Mol. Autism **2**(1), 15 (2011)
3. Association, A.P., et al.: Diagnostic and statistical manual of mental disorders (DSM-5®). American Psychiatric Pub (2013)
4. Baltrusaitis, T., Robinson, P., Morency, L.P.: Constrained local neural fields for robust facial landmark detection in the wild. In: Proceedings of the IEEE International Conference on Computer Vision Workshops, pp. 354–361 (2013)
5. Bazarevsky, V., Kartynnik, Y., Vakunov, A., Raveendran, K., Grundmann, M.: BlazeFace: sub-millisecond neural face detection on mobile GPUs. CoRR abs/1907.05047 (2019)
6. Boehringer, S., et al.: Syndrome identification based on 2D analysis software. Eur. J. Hum. Genet. **14**(10), 1082–1089 (2006)
7. Bradski, G.: Opencv library. Dr. Dobb's Journal of Software Tools (2000)
8. Canny, J.: A computational approach to edge detection. IEEE Trans. Pattern Anal. Mach. Intell. **6**, 679–698 (1986)
9. DeMyer, W., Zeman, W., Palmer, C.G.: The face predicts the brain: diagnostic significance of median facial anomalies for holoprosencephaly (arhinencephaly). Pediatrics **34**(2), 256–263 (1964)
10. Deth, R., Muratore, C., Benzecry, J., Power-Charnitsky, V.A., Waly, M.: How environmental and genetic factors combine to cause autism: a redox/methylation hypothesis. Neurotoxicology **29**(1), 190–201 (2008)
11. Farkas, L.G.: Anthropometry of the Head and Face. Raven Pr (1994)
12. Gilani, S.Z., et al.: Sexually dimorphic facial features vary according to level of autistic-like traits in the general population. J. Neurodev. Disord. **7**(1), 14 (2015)
13. Hammond, P., et al.: 3D analysis of facial morphology. Am. J. Med. Genet. A **126**(4), 339–348 (2004)
14. Johnson, C.P., Myers, S.M.: Identification and evaluation of children with autism spectrum disorders. Pediatrics **120**(5), 1183–1215 (2007)
15. King, D.E.: Dlib-ml: a machine learning toolkit. J. Mach. Learn. Res. **10**, 1755–1758 (2009)

16. Kumov, V., Samorodov, A.: Recognition of genetic diseases based on combined feature extraction from 2D face images. In: 26th Conference of Open Innovations Association (FRUCT). IEEE (2020)
17. Lemaître, G., Nogueira, F., Aridas, C.K.: Imbalanced-learn: a python toolbox to tackle the curse of imbalanced datasets in machine learning. J. Mach. Learn. Res. **18**(17), 1–5 (2017)
18. Levy, S.E., Mandell, D.S., Schultz, R.T.: Autism **374**, 1627–1638 (2009)
19. Li, J., et al.: Feature selection: a data perspective. arXiv:1601.07996 (2016)
20. Lord, C., Cook, E.H., Leventhal, B.L., Amaral, D.G.: Autism spectrum disorders. Neuron **28**(2), 355–363 (2000)
21. Lugaresi, C., et al.: MediaPipe: a framework for perceiving and processing reality. In: Third Workshop on Computer Vision for AR/VR at IEEE Computer Vision and Pattern Recognition (CVPR) 2019 (2019)
22. Mandell, D.S., Novak, M.M., Zubritsky, C.D.: Factors associated with age of diagnosis among children with autism spectrum disorders. Pediatrics **116**(6), 1480–1486 (2005)
23. Miles, J., Hadden, L., Takahashi, T., Hillman, R.: Head circumference is an independent clinical finding associated with autism. Am. J. Med. Genet. **95**(4), 339–350 (2000)
24. Muhle, R., Trentacoste, S.V., Rapin, I.: The genetics of autism. Pediatrics **113**(5), e472–e486 (2004)
25. Chawla, N.V., Bowyer, K.W., Lawrence, O.H., Philip Kegelmeyer, W.: Smote: synthetic minority over-sampling technique. J. Artif. Intell. Res. **16**, 321–357 (2011)
26. Obafemi-Ajayi, T., et al.: Facial structure analysis separates autism spectrum disorders into meaningful clinical subgroups. J. Autism Dev. Disord. **45**(5), 1302–1317 (2015)
27. Ozgen, H., et al.: Morphological features in children with autism spectrum disorders: a matched case-control study. J. Autism Dev. Disord. **41**(1), 23–31 (2011)
28. Pedregosa, F., et al.: Scikit-learn: machine learning in Python. J. Mach. Learn. Res. **12**, 2825–2830 (2011)
29. Rodier, P.M., Bryson, S.E., Welch, J.P.: Minor malformations and physical measurements in autism: data from Nova Scotia. Teratology **55**(5), 319–325 (1997)
30. Suzuki, S.: Topological structural analysis of digitized binary images by border following. Comput. Vis. Graph. Image Process. **30**(1), 32–46 (1985)
31. Weinberg, S.M., Naidoo, S., Govier, D.P., Martin, R.A., Kane, A.A., Marazita, M.L.: Anthropometric precision and accuracy of digital three-dimensional photogrammetry: comparing the genex and 3dMD imaging systems with one another and with direct anthropometry. J. Craniofac. Surg. **17**(3), 477–483 (2006)
32. Zhao, Q., et al.: Digital facial dysmorphology for genetic screening: hierarchical constrained local model using ICA. Med. Image Anal. **18**(5), 699–710 (2014)
33. Zhao, Q., Yao, G., Akhtar, F., Li, J., Pei, Y.: An automated approach to diagnose turner syndrome using ensemble learning methods. IEEE Access **8**, 223335–223345 (2020)

Constructive Machine Learning and Hierarchical Multi-label Classification for Molecules Design

Rodney Renato de Souza Silva and Ricardo Cerri(✉)

Department of Computer Science, Federal University of São Carlos, Rodovia Washington Luis, km 235, São Carlos, SP, Brazil
rodney.souza@unesp.br; cerri@ufscar.br

Abstract. Constructive Machine Learning (CML) is a research field that uses algorithms to generate new instances, similar but not identical to existing ones. It has been widely used to assist the discovery of new drug-like molecules. This is very challenging, given that the search space is discrete, unstructured and enormous. In this work we use CML to learn the intrinsic rules of datasets of molecules to generate novel ones. The chosen CML methods can be divided in two sub groups, text-based and graph oriented. Considering different possibilities to evaluate the methods and the generated molecules, we propose classifying generated molecules in a taxonomy, using a hierarchical multi-label classifier previously trained in a dataset of molecules with known taxonomy information. In this way, it is possible to predict properties and verify the relevance of the generated molecules to existing taxonomies. We also propose a hierarchical diversity measure to compare groups of molecules based on their taxonomy information. The measure showed coherent results and is faster to calculate than the commonly used external diversity measures.

Keywords: *De Novo* Drug Design · Hierarchical Multi-label Classification · Constructive Machine Learning

1 Introduction

One of the objectives of Medicinal Chemistry is to discover new molecules with drug-like characteristics. This is challenging, since searching for molecules with desired chemical properties involves a huge and discrete space. The number of possible candidates is overwhelming and makes impossible to brute force through all possible combinations. About 10^8 substances have been synthesized, whereas the range of potential drug-like molecules is estimated to be between 10^{23} and 10^{60} [14]. Other problem is that the space is unstructured, i.e., adding, removing or changing atoms and bonds of a molecule changes its properties in an inconstant and hard to predict way. Despite improvements in the synthesis and test areas, those procedures are still time consuming and expensive. The discovery process can take 10–15 years with an average cost of US$2.5 billion [9], and one-third of the time and money are spent during the early stages of the process.

M. C. Naldi and R. A. C. Bianchi (Eds.): BRACIS 2023, LNAI 14196, pp. 276–290, 2023.
https://doi.org/10.1007/978-3-031-45389-2_19

Machine Learning (ML) is relevant to manage the volume and complexity of chemical information. A single database (such as PubChem [18]) has more than 96 million chemical structures, with an even larger number of annotations (such as synonyms and properties), targets (drugs with which a molecule interacts), action (how the molecule interacts with its targets), and status within the regulatory-approval process [31]. ML seeks to generalize patterns observed in datasets and predicts unknown characteristics of unseen data. It has been applied in drug design to estimate properties of molecules and predict interactions with biological targets, considerably accelerating drug discovery [6].

Constructive Machine Learning (CML) generates instances similar, but no identical, to known instances. It has shown outstanding results in many areas and is gaining increasing attention [12]. In recent years, CML has been applied to generate new molecules from scratch based on existing ones [24]. Four challenges are observed when applying CML to *de novo* drug design [10]: i) definition of molecules representation; ii) architecture selection; iii) evaluation of approaches for molecules generation and optimization; and iv) design of a reward function, crucial for practical applications of reinforcement learning. Although there is a variety of proposals to overcome these challenges, there are still open problems.

Traditional ML classifiers were developed for binary or multi-class problems. They are not suitable for real-world problems where instances are classified into many classes simultaneously. An example is Hierarchical Multi-label Classification (HMC), where instances are classified into different classes in a taxonomy. These problems require hierarchical multi-label classifiers [30].

Many works validate molecules using classifiers to predict desired molecules properties. However, there are no works applying HMC to classify *de novo* molecules into known properties within a taxonomy. This is important, since we can directly identify general and specific properties, efficiently comparing groups of molecules. Thus, we propose to evaluate our molecules by classifying them into a taxonomy, using a hierarchical multi-label classifier previously trained using molecules with known taxonomy information. To generate molecules, we use the Generative Toolkit for Scientific Discovery (GT4SD) [28], and classify them using a decision tree [30] trained in the ChEBI molecules dataset [8]. For evaluation, we propose a hierarchical distance to compare groups of molecules based on their taxonomy information predicted by the hierarchical decision tree. Our main contributions are: i) the study and comparison of CML models for molecules generation considering their taxonomy; ii) the creation of a library to process the ChEBI dataset for hierarchical classification; iii) the proposal of a new measure to compare groups of molecules based on their taxonomy.

The remainder of this paper is organized as follows. Section 2 presents related works; Sect. 3 reviews fundamental related concepts; Results are presented and discussed in Sect. 4; Finally, Sect. 5 presents conclusions and future works.

2 Related Works

Recurrent Neural Networks (RNN) were applied to learn intrinsic molecules rules [5], comparing distributions of descriptors to verify if they were similar to existing ones. RNNs were also applied to generate molecules from scratch and from scaffolds [15]. Fine tuning was performed in smaller specific datasets with molecules with known biological targets. Principal Component Analysis was applied to the descriptors of the training data and the generated data, showing that the distributions were very close.

Reinforcement Learning (RL) has been extensively investigated due to its potential applications in drug discovery. In a recent study, the combination of RL and Quantitative Structure Activity Relationship (QSAR) was explored as a reward function for identifying active molecules against the Dopamine Receptor Type 2 (DRD2) [24]. The study used RNN and QSAR model to generate structures. The results showed that more than 95% of the structures generated were active against DRD2, demonstrating the potential of RL in drug discovery.

A method called ORGAN combines Generative Adversarial Networks (GANs) with high Quantitative Estimate of Drug-Likeness (QED) [26]. It has three parts: a generator, a discriminator, and QED with Lipinskis rule-of-five as reward function. Variational Autoencoders (VAEs) were also investigated [14]: i) sampling molecules next to known ones; ii) interpolating points from known molecules and converting them back; iii) using a property predictor to search optimal regions, converting points with high values into molecules. The first approach chooses a molecule as reference and samples points around it. The second chooses two molecules as references, smoothly interpolating molecules similar to them. The third approach showed that VAEs trained with the property predictor led to regions where the chosen property is maximized. Adversarial Autoencoders (AAEs) [17] were compared with VAEs, showing that VAEs have a greater coverage. Allowing to trade-off between coverage of the original dataset with generated samples and reconstruction quality, AAEs obtained results similar to VAEs.

Graph Convolutional Policy Network (GCPN) [34] used a generation process guided towards specified desired objectives, restricting the output space based on underlying chemical rules. For goal-directed generation, graph representation, reinforcement learning and adversarial training were used, extended and combined into an unified framework.

A flow-based auto-regressive model (GraphAF) was proposed [27], combining Auto Regressing Flows [25] and graph representation learning. Motifs were also used [21], which are fragments of molecules inferred from data. VAEs were extended to work with graph molecular representation, and generated molecules including motifs, a bond or a atom per iteration. The proposal, called MoLeR (learning to extend molecular scaffolds with structural motif), was argued to be faster than baseline methods. The method uses an off-the-shelf molecular optimization, combining state-of-the-art methods in unconstrained optimization.

From the reviewed works, we generated molecules with the following methods: VAE [14], ORGAN [26], AAE [17], MoLeR [21], GCPN [34] and GraphAF [27].

3 Methodology

This section presents concepts, algorithms and data used in our experiments. Section 3.1 introduces the concepts of molecules properties and Sect. 3.2 presents three molecular representations used in our experiments. The GT4SD library and the Clus system, used to generate and classify molecules, are described in Sects. 3.3 and 3.4. Section 3.5 provides information about the ChEBI taxonomy, used to train our classifier and validate molecules. The next two sections present metrics used to compare datasets of molecules. Section 3.6 describes the internal and external diversity measures, while Sect. 3.7 explains our proposed hierarchical distance. Section 3.8 presents an overview of our methodology.

3.1 Molecules Properties

Each molecule has i) its own structure, i.e., how its atoms are organized; ii) its physical properties (descriptors), such as its weight and solubility; and iii) its biological properties, related to effects in biological targets [4]. Cheminformatics tries to predict molecules physical properties using their chemical structure, and to predict biological targets using known properties [31].

Although we can not say for sure if a molecule has pharmaceutic characteristics, we can observe physicochemical and structural properties within certain ranges. Over 90% of molecules that reach the phase II clinical status have four properties: i) mass lower than 500; ii) $logP$ lower than 5; iii) number of Hydrogen bond donors lower than 5; and iv) number of Hydrogen bond receptors lower than 10. A molecule satisfying these restrictions is a strong candidate to be a drug. Although these are not the only properties for the molecule to be a drug, they have shown to be very influential, and are known as Rule of Five (RO5) (or Lipinski rule) [19]. Even though the RO5 is predictive of oral bioavailability, 16% of oral drugs violate at least one of the criteria and 6% fail on two or more.

An important estimated molecule property is the Quantitative Estimate of Drug-Likeness (QED), which is used to assess molecular target druggability by prioritizing bioactive compounds, directing the tests to most promising molecules [4]. Quantitative Structure Activity Relationship (QSAR) [23] indicates how a chemical compound interacts with biological targets. It can be used as objective function, helping to predict how promising a molecule is to be a drug.

In our experiments, we trained a hierarchical multi-label classifier using molecules extracted from the ChEBI database, having as features the above properties and other descriptors. These same features were extracted from the molecules we generated using CML. These molecules were then classified using the hierarchical classifier. Since molecules with similar functions are close in the feature space [31], we evaluated our generated molecules using similarity measures, comparing groups of molecules generated by different CML algorithms.

3.2 Molecules Representation

The representation of a molecule is a key factor to determine how CML models work and how similarities are evaluated. The Simplified Molecular Input Line

Entry System (SMILES) is a two dimensional representation that uses ASCII symbols: letters for atoms, = and # for bonds, parenthesis for ramifications and numbers for cycles; simple bonds and hydrogen atoms are sub intended. A positive characteristic of SMILES is its ability to capture order and relation of atoms. However, it allows to represent a same molecule in many different ways, which difficulties comparisons. A same molecule can have different similarity scores depending on the chosen atom to start its sequence. Also, a single character perturbation in the representation can lead to changes in the underlying molecular structure or even invalidate it [32]. We used this representation for some CML models in the generation stages, and to store molecules.

Fingerprints uses an array of predefined size, where each position indicates the presence or the absence of a molecular structure. Its positive characteristics are that there is only one possible encoding for each molecule, and it is simple to compare vectors with fixed sizes. However, this representation cannot capture the order neither the relationship between the encoded structures. Therefore, an encoding can be related to numerous molecules, making it impossible to reverse the codification [7]. We used this representation to compare molecules.

The graph representation uses vertices and edges to represent atoms and bonds, providing high coverage of the chemical space, and conveying various chemical features. It is argued that it is the most intuitive and concise representation [33]. We used it in some CML models for the generation process.

3.3 GT4SD Toolkit

The Generative Toolkit for Scientific Discovery (GT4SD) [28] handles molecules, calculates their properties, provides a variety of pretrained models and a framework to train CML models. We used six pretrained models, each of them generating a set of 2000 molecules. They were trained in the ChEMBL [13] dataset and made available by the GT4SD team. The first half of them uses SMILES: Variational Auto Encoder (VAE) [14], Objective Reinforced Generative Adversarial Networks (ORGAN) [26] and Adversarial Auto Encoders (AAE) [17]. The second half process the molecules as graphs: Molecular Scaffolds with Structural Motifs (MoLeR) [21], Graph Convolutional Policy Networks (GCPN) [34] and flow-based autoregressive model for molecular graph generation (GraphAF) [27].

3.4 Clus System

Clus[1] is a decision tree and rule induction system based on Predictive Clustering Trees. It unifies clustering and predictive modeling, and deals with complex prediction tasks such as multi-label classification. Within Clus, we used the hierarchical multi-label classifier Clus-HMC [30] to learn the intrinsic rules of the ChEBI taxonomy. We then used Clus-HMC to predict the taxonomy of our generated molecules based on their descriptors.

[1] https://dtai.cs.kuleuven.be/clus/.

3.5 ChEBI Dataset

The Chemical Entities of Biological Interest (ChEBI) [8] is a dataset that contains information about molecules, such as SMILES representation and descriptors. It uses a Web Ontology Language (OWL) [1] to describe a taxonomy of molecules and their classification. Its information can be preprocessed and represented into a Directed Acyclic Graph (DAG) where each node represents a taxonomy label. It is also possible to prune the DAG, leaving the most important labels and molecules. One can also store the taxonomy information, obtain statistics, and compare groups of molecules.

The ChEBI DAG has three ontologies with nodes belonging to one or more of them, inheriting semantic meaning with their connections. The first ontology classifies molecular entities and chemical substances. The second expresses a molecular role, which is a particular behavior that a material may exhibit. The third expresses atomic particles smaller than an atom, e.g., neutrons and protons.

In order to improve the performance of the hierarchical classifier, we created a library to process and prune the original ChEBI DAG. After pruning, we kept only molecules with role information, since they are more relevant for drug discovery, and only nodes with at least 100 annotated molecules. With this, we ended up excluding the particle graph and many nodes from the structural and role graphs, obtaining a final DAG with 35032 molecules. Table 1 shows the statistics of the pruned DAG. The pruned DAG contains 72 nodes shared between the two ontologies, where 37 are leaves. The table also presents the leaves with most molecules classified for each graph. As the analysis focus on the role information, the most populated leaf by molecules in the role graph has significantly more molecules than the most populated leaf in the structural graph. The hierarchical classifier is trained in the pruned ChEBI DAG and used to predict the taxonomy of the molecules generated by the GT4SD toolkit. Thus, the classifier may infer role or structural information of the generated molecules.

Table 1. Pruned ChEBI taxonomy graph statistics.

	ChEBI Structural graph	ChEBI Role graph
Height	17	12
Number of nodes	278	166
Number of leafs	87	75
Node with most children	CHEBI_29067	CHEBI_33575
Node with most children (quantity)	12	12
Node with most parents	CHEBI_29067	CHEBI_33575
Node with most parents (quantity)	6	6
Leaf with most elements	CHEBI_61379	CHEBI_35610
Leaf with most elements (quantity)	753	1666

3.6 Internal and External Diversity Measures

To compare molecules based on their structural information, the fingerprints representation can be used with the Tanimoto distance (Eq. 1) [2], where a and b are molecules. The internal diversity measures the variety of the molecules in a given set. It is defined in Eq. 2 as the mean of the sum of the Tanimoto distance applied to every pair of molecules in a set, where A is a set of molecules, and a and b are molecules from A. The external diversity is analogous, with the main difference of using two different sets to define the pairs of molecules. It is presented in Eq. 3, with a and b molecules from sets A and B respectively.

The external and internal diversities have a complexity problem. As the sets of molecules grow in size, the time required to calculate the measures grows proportionally to the square of the sizes of the sets [3]. As external diversity is used to compare groups of molecules, we used it as a guideline to propose a faster measure to compare groups of molecules: the hierarchical diversity.

$$T_d(a,b) = 1 - \frac{a \cap b}{a \cup b} \tag{1}$$

$$D_I(A) = \frac{1}{|A|^2} \sum_{(a,b) \in A \times A} T_d(a,b) \tag{2}$$

$$D_E(A,B) = \frac{1}{|A||B|} \sum_{(a,b) \in A \times B} T_d(a,b) \tag{3}$$

3.7 Proposed Hierarchical Diversity Measure

Our proposed diversity uses taxonomies to compare different groups of molecules. A DAG is created for each set of molecules, and one is chosen as reference. Each node from the DAG stores how many molecules are classified in it, and the nodes are ordered constructing the first DAG distribution. The second DAG uses the same labels in the same order as the first DAG to construct its distribution in an analogous way. Finally, the distance of the distributions of molecules is measured using the Wasserstein Distance [29]. It is helpful to compare large groups of molecules since its complexity is way smaller than the external diversity. Equation 4 presents the Wassistein distance between distributions u and v. These are the distributions of the DAGs molecules over their nodes.

$$W_d(u,v) = \inf_{\pi \in \Gamma(u,v)} \int_{\mathbb{R} \times \mathbb{R}} |x - y| \mathrm{d}\pi(x,y) \tag{4}$$

The hierarchical distance is asymmetric. Since we need to choose one of the DAGs as reference, the results may be different depending on the chosen DAG. However, our results in Sect. 4 showed these differences were not significant.

3.8 Methodology Overview

Figure 1 presents an overview of our methodology. The ChEBI dataset, in a OWL format (*ChEBI.OWL*), is preprocessed into a DAG, which is pruned and saved in two files: *ChEBI.arff* and *ChEBI.obj*. Each method from the GT4SD library then generates a *Method.arff* with molecules. Clus-HMC classifier is then trained with *ChEBI.arff* molecules, making predictions for the molecules in the *Method.arff* files. The predictions are stored in a *Method.pred.arff* file, and converted into the *Method.obj* format, which is used along with the *ChEBI.obj* to generate the results discussed in Sect. 4. All our implementations and datasets are freely available[2].

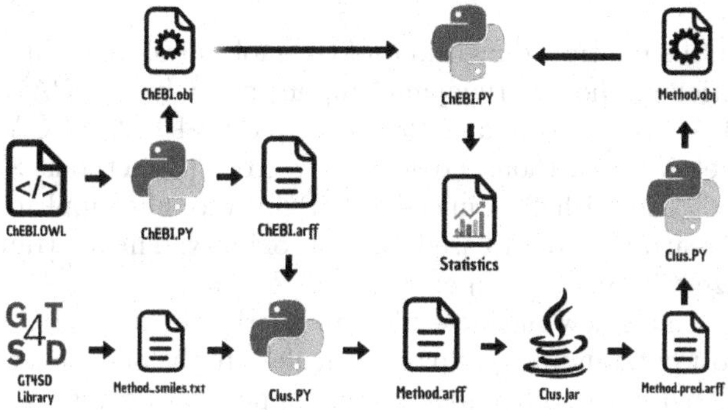

Fig. 1. Illustration of our proposed methodology.

4 Results and Discussions

This section presents an analysis of our generated molecules. For each of the six generators from the GT4SD toolkit, we generated a group of 2 thousand molecules. First, we analyze some of the molecules properties, and visualize molecules using their descriptors space. Next, we present the classification of the molecules by the Clus-HMC classifier. Internal and external diversities are then discussed, together with our proposed hierarchical distance.

For each molecule, 21 descriptors were calculated, including molecular weight, related to the number and type of the molecule atoms, *PlogP*, related to the solubility of the molecule in water, and all properties presented in Sect. 3.1. Table 2 presents information about some of the most important descriptors for each generated set of molecules. The sets generated by the AAE model differs from the other sets, having in general higher mean and variance for the molecular weight and *PlogP* descriptors. We can imply that many molecules in this set violate at least one rule of the RO5 rule. On the other hand, the set generated by GCPN showed the lowest values and variations for these two properties.

[2] https://github.com/yendorr/gt4sd.

Table 2. Statistics from three proprieties of the molecules sets.

	Weight		Plog P		QED	
	Mean	STD	Mean	STD	Mean	STD
AAE	490.55	237.56	6.06	5.79	0.43	0.25
ORGAN	381.42	103.86	3.39	1.75	0.57	0.20
VAE	385.91	106.86	3.43	1.79	0.56	0.21
GCPN	252.56	61.98	2.75	1.25	0.73	0.12
GraphAF	262.61	138.57	2.28	1.62	0.52	0.17
MoLeR	392.27	102.99	3.49	1.88	0.52	0.21

To visualize the multiple dimensions of the molecules, we applied two dimensionality reduction methods: Principal Component Analysis (PCA) [16] and Uniform Manifold Approximation and Projection (UMAP) [22]. PCA was applied in the a dataset with 12 thousand feature vectors (6 methods × 2 thousand molecules), each one with 21 dimensions. PCA was also applied followed by UMAP, which rearranges data based on its proximity. This rearrangement may help to visualize overlapped data and clusters of data.

Figure 2 illustrates how the molecules are distributed in the PCA and UMAP spaces. The most outstanding group of molecules are the ones generated by AAE and GraphAF, while the other four groups appear to be very similar to each other. Although many of the points in the PCA space overlap, we can see where each group is located in relation to each other. In the UMAP space, the points hardly overlap, but it is hard to define a region for each group of molecules.

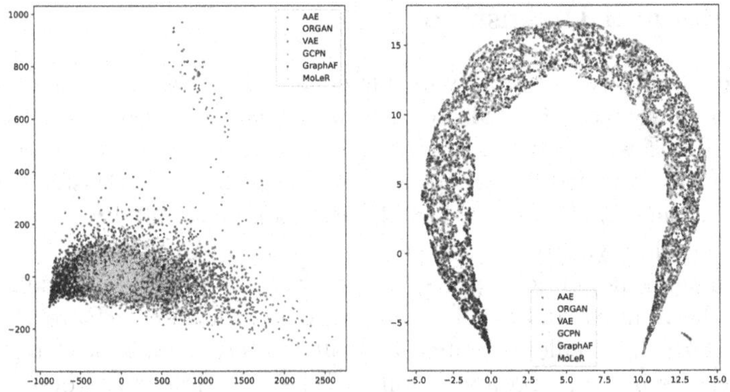

Fig. 2. 2D visualizations of the molecules feature vectors.

Clus-HMC was trained with a 10-fold cross-validation strategy using the 35032 molecules extracted from the ChEBI dataset. We used the Area Under Receiver Operating Characteristic (AUROC) and the Area Under Precision

Recall Curve (AUPRC) to evaluate the classifier. AUROC measures the ability of a classifier to distinguish between positive and negative classes, while AUPRC summarizes the relationship between precision and recall. Table 3 shows the average AUPRC and AUROC values obtained. The values are considerably high, except for the simple average AUPRC which does not consider class imbalance.

Table 3. Hierarchical multi-label classifier results.

Metric name	Metric value
Average AUROC	0.8931
Simple Average AUPRC	0.3541
Weighted Average AUPRC	0.7766

We classified the molecules into classes of the ChEBI taxonomy. Tables 4 and 5 present information about the role and structural DAGs for the molecules generated by each method. Although we can provide the information individually for each molecule, we show the results for groups of molecules in order to have a general view of the properties generated by each method. We can observe that the ORGAN, VAE and MoLeR methods have close statistics, suggesting that their generated molecules are similar to each other. This same analysis can be done with GCPN and GraphAF. The molecules generated by AAE are the ones that differ the most from the other groups of molecules.

Table 4. Role graph information of the ChEBI molecules set.

	AAE	ORGAN	VAE	GCPN	GraphAF	MoLeR
Height	9	10	10	11	10	11
Number of nodes	56	65	72	44	46	69
Number of leafs	16	19	22	11	13	16
Node with most children (quantity)	4	5	5	4	4	5
Node with most parents (quantity)	3	5	5	4	4	5
leaf with most molecules (quantity)	19	8	6	4	4	8

Table 6 shows some classes predicted by the hierarchical classifier for the generated molecules, and the corresponding classification percentages. The classes include biochemical functions, metabolites and subclasses of metabolites, alkaloids and an alkaloid subclass. With the hierarchical classification, we can reduce the number of molecules to search, allowing a more localized and focused search on certain classes or chemical characteristics. Some of the predicted classes have medicinal and pharmacological potential, such as Harmala alkaloids [20], and are present in pharmaceutical drug compounds, such as metabolites [11].

We used internal and external diversity to verify the similarities of the groups of molecules. The main diagonal of Table 7 shows the internal diversity, while

Table 5. Structure graph information of the ChEBI molecules set.

	AAE	ORGAN	VAE	GCPN	GaphAF	MoLeR
Height	14	15	15	15	15	15
Number of nodes	115	142	141	101	111	123
Number of leafs	18	22	22	13	14	17
Node with most children (quantity)	8	9	8	7	7	7
Node with most parents (quantity)	3	5	5	4	4	5
leaf with most molecules (quantity)	6	8	6	4	2	8

Table 6. Percentage of molecules classified in some classes of the ChEBI taxonomy.

	AAE (%)	ORGAN (%)	VAE (%)	GCPN (%)	GraphAF (%)	MoLeR (%)
Biochemical functions	7,45	8,5	8,6	19,15	33,9	6,25
Metabolites	7.4	8.5	8.55	19.15	33.9	6.2
Mammalian metabolite	0.95	0.0	0.1	0.0	0.3	0.1
Human metabolite	0.05	0.0	0.0	0.0	0.05	0.1
Alkaloids	0.8	1.0	0.55	0.45	0.05	0.45
Harmala alkaloids	0.25	0.4	0.15	0.0	0.0	0.15

the external diversity is shown in the other cells. Three main observations can be made: i) the GraphAF and GCPN molecules have the highest internal diversities; ii) the molecules generated by GraphAF and GCPN have the highest external diversities; and iii) the molecules generated by AAE and VAE have the smallest external diversities. These observations are useful to interpret the relations between the sets of molecules, and as a baseline to experiment if our proposed hierarchical distance is also capable to capture these relations.

Table 7. Internal and external diversity between sets of molecules.

	AAE	ORGAN	VAE	GCPN	GaphAF	MoLeR
AAE	0.719	0.727	0.726	0.790	0.843	0.739
ORGAN	0.727	0.728	0.727	0.790	0.842	0.740
VAE	0.726	0.727	0.726	0.790	0.842	0.739
GCPN	0.790	0.790	0.790	0.822	0.869	0.796
GaphAF	0.843	0.842	0.842	0.869	0.895	0.847
MoLeR	0.739	0.740	0.739	0.796	0.847	0.749

Given the difficulty to visualize groups of molecules in Fig. 2, Fig. 3 shows only molecules generated by ORGAN and VAE. They were chosen to validate how our hierarchical distance evaluated two similar sets of molecules. Since the groups are close, a small value is expected. Figure 4 shows the distributions of the same molecules over the DAG nodes. Again, the distributions are very similar, and the hierarchical distance should be small.

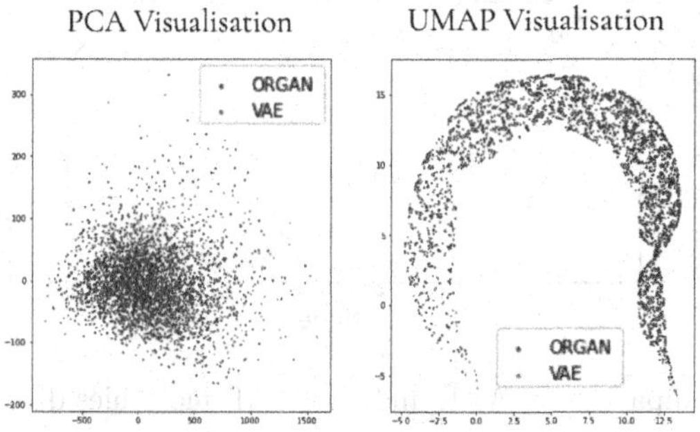

Fig. 3. 2D visualizations of ORGAN and VAE generated molecules.

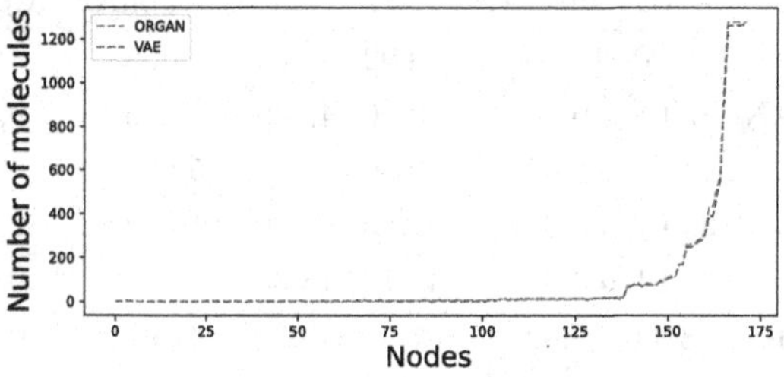

Fig. 4. Comparison of VAE and ORGAN molecules distributions.

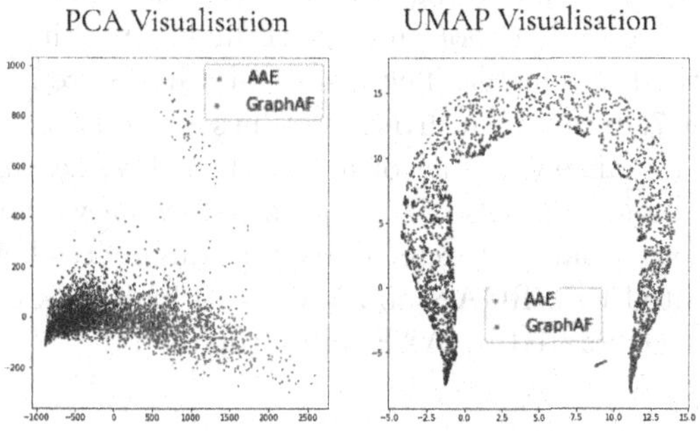

Fig. 5. 2D visualizations of AAE and GraphAF generated molecules.

Figure 5 shows the visualization of the descriptors of the molecules generated by AAE and GraphAF. They were chosen to validate how the hierarchical distance differs two dissimilar sets of molecules. They are not so close to each other, and a high hierarchical distance is expected. Figure 6 shows the distributions of the same molecules over the DAG nodes. We see a sensitive dissonance, which must result in a high hierarchical distance.

Table 8 shows our proposed hierarchical distance values for every pair of sets of molecules. From the results, we can separate the molecules in two clusters. The first one contains molecules generated by GraphAF and GCPN, showing relative

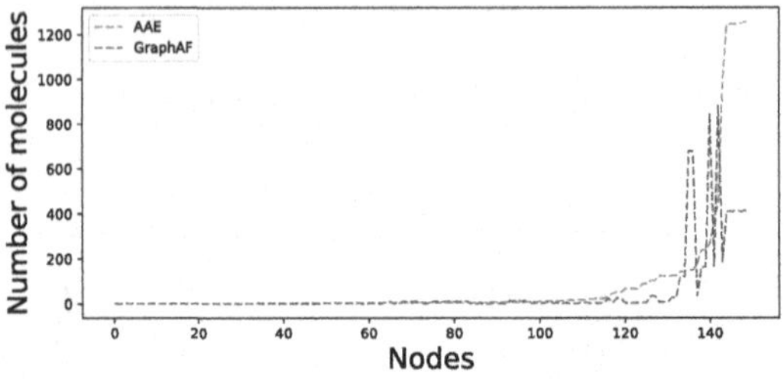

Fig. 6. Comparison of AAE and GraphAF molecules distributions.

Table 8. Hierarchical distance between the molecules sets.

	AAE	ORGAN	VAE	GCPN	GaphAF	MoLeR
AAE	0.00	11.70	9.92	41.62	44.05	8.08
ORGAN	13.08	0.00	3.06	48.96	54.46	9.03
VAE	11.36	3.02	0.00	46.20	52.09	8.24
GCPN	34.07	35.37	32.40	0.00	16.22	32.91
GaphAF	40.81	43.81	40.72	18.35	0.00	40.72
MoLeR	8.44	8.62	7.77	43.08	47.66	0.00

small distance between them and high distances to the other groups. The second cluster is formed by the other molecules, with high similarity among them and high distances to the first cluster. The results of Table 8 are in consonance with the ones from Table 7, except that diversity values are constrained between 0 and 1. The hierarchical distance varies according to the DAG sizes and the dissonance between the sets of molecules. Also, in accordance to what we can see in Figs. 3, 4, 5 and 6, Table 8 shows small hierarchical distance differences between the groups of molecules generated by ORGAN and VAE, and high differences between the groups of molecules generated by AAE and GraphAF.

5 Conclusions and Future Works

In this paper we presented a study on Constructive Machine Learning applied to *de novo* drug-like molecules design. We generated molecules using six different methods from the GT4SD toolkit, and compared them with internal and external diversity measures. We proposed to use a hierarchical multi-label classifier to classify the molecules into classes of a known taxonomy, and compared groups of molecules based on their classification. For this comparison, we proposed a new hierarchical diversity measure that showed to be consistent in comparing groups of molecules based on their taxonomy information.

The ChEBI dataset was used to train a hierarchical multi-label classifier, since it contains taxonomy information from a variety of molecules, and is much smaller than other databases. Our proposal can be used to expand the ChEBI dataset by classifying new molecules in its taxonomy. It also can be used in drug discovery, classifying the generated molecules to narrow down the number of candidates, and selecting the molecules with specific chemical characteristics.

Once we classified groups of molecules into taxonomies, our proposed hierarchical distance obtained results similar to those from external diversity, but in a fraction of the time. External diversity has a limited range between 0 and 1, whereas our hierarchical distance can vary based on the size and diversity of the compared groups. The experiments showed that the hierarchical distance is more meaningful when calculated for similar sized groups of molecules.

As future works, the hierarchical distance could be extended to other contexts, to compare groups of different molecules that can be classified into a taxonomy. Also, we intend to investigate how the selected molecules can be used to fine tune their generators, in order to create a specific domain generation model.

Acknowledgments. This study was financed in part by the Coordenação de Aperfeiçoamento de Pessoal de Nível Superior - Brasil (CAPES) - Finance Code 001. The authors also thank the Brazilian research agencies FAPESP and CNPq for financial support.

References

1. Antoniou, G., Harmelen, F.v.: Web ontology language: owl. In: Handbook on ontologies, pp. 67–92. Springer (2004)
2. Bajusz, D., Rácz, A., Héberger, K.: Why is Tanimoto index an appropriate choice for fingerprint-based similarity calculations? J. Cheminform. **7** (2015)
3. Benhenda, M.: ChemGAN challenge for drug discovery: can AI reproduce natural chemical diversity? arXiv preprint arXiv:1708.08227 (2017)
4. Bickerton, G.R., Paolini, G.V., Besnard, J., Muresan, S., Hopkins, A.L.: Quantifying the chemical beauty of drugs. Nat. Chem. **4**(2), 90–98 (2012)
5. Bjerrum, E.J., Threlfall, R.: Molecular generation with recurrent neural networks (RNNs). arXiv preprint arXiv:1705.04612 (2017)
6. Brown, N., Ertl, P., Lewis, R., Luksch, T., Reker, D., Schneider, N.: Artificial intelligence in chemistry and drug design. J. Comput. Aided Mol. Des. **34**(7), 709–715 (2020). https://doi.org/10.1007/s10822-020-00317-x
7. Cao, D.S., Xu, Q., Hu, Q., Liang, Y.Z.: Manual for ChemoPy (2013)
8. Degtyarenko, K., et al.: ChEBI: a database and ontology for chemical entities of biological interest. Nucleic Acids Res. **36**(suppl 1), D344–D350 (2007)
9. DiMasi, J.A., Grabowski, H.G., Hansen, R.W.: Innovation in the pharmaceutical industry: new estimates of r&d costs. J. Health Econ. **47**, 20–33 (2016)
10. Elton, D.C., Boukouvalas, Z., Fuge, M.D., Chung, P.W.: Deep learning for molecular design-a review of the state of the art. Mol. Syst. Des. Eng. **4**(4), 828–849 (2019)
11. Evans, L., Phipps, R., Shanu-Wilson, J., Steele, J., Wrigley, S.: Methods for metabolite generation and characterization by NMR. In: Ma, S., Chowdhury, S.K. (eds.) Identification and Quantification of Drugs, Metabolites, Drug Metabolizing Enzymes, and Transporters (Second Edition), pp. 119–150. Elsevier, Amsterdam, second edition. (2020)
12. Foster, D.: Generative Deep Learning: Teaching Machines to Paint, Write, Compose, and Play. O'Reilly Media (2019)

13. Gaulton, A., et al.: The ChEMBL database in 2017. Nucleic Acids Res. **45**(D1), D945–D954 (2017)
14. Gómez-Bombarelli, R., et al.: Automatic chemical design using a data-driven continuous representation of molecules. ACS Cent. Sci. **4**(2), 268–276 (2018)
15. Gupta, A., Müller, A.T., Huisman, B.J., Fuchs, J.A., Schneider, P., Schneider, G.: Generative recurrent networks for de novo drug design. Mol. Inf. **37**(1–2), 1700111 (2018)
16. Jolliffe, I.: Principal component analysis. Encyclopedia of Statistics in Behavioral Science (2005)
17. Kadurin, A., Nikolenko, S., Khrabrov, K., Aliper, A., Zhavoronkov, A.: druGAN: an advanced generative adversarial autoencoder model for de novo generation of new molecules with desired molecular properties in silico. Mol. Pharm. **14**(9), 3098–3104 (2017)
18. Kim, S., et al.: PubChem 2019 update: improved access to chemical data. Nucleic Acids Res. **47**(D1), D1102–D1109 (2019)
19. Lipinski, C.A.: Lead-and drug-like compounds: the rule-of-five revolution. Drug Discov. Today Technol. **1**(4), 337–341 (2004)
20. Marwat, S.K., ur Rehman, F.: Medicinal and pharmacological potential of harmala (peganum harmala l.) seeds. In: Preedy, V.R., Watson, R.R., Patel, V.B. (eds.) Nuts and Seeds in Health and Disease Prevention, pp. 585–599. Academic Press, San Diego (2011)
21. Maziarz, K., et al.: Learning to extend molecular scaffolds with structural motifs. arXiv preprint arXiv:2103.03864 (2021)
22. McInnes, L., Healy, J., Melville, J.: UMAP: uniform manifold approximation and projection for dimension reduction. arXiv preprint arXiv:1802.03426 (2018)
23. Mitchell, J.B.: Machine learning methods in chemoinformatics. Wiley Interdisc. Rev. Comput. Mol. Sci. **4**(5), 468–481 (2014)
24. Olivecrona, M., Blaschke, T., Engkvist, O., Chen, H.: Molecular de-novo design through deep reinforcement learning. J. cheminform. **9**(1), 48 (2017)
25. Papamakarios, G., Pavlakou, T., Murray, I.: Masked autoregressive flow for density estimation. In: Advances in Neural Information Processing Systems, vol. 30 (2017)
26. Sanchez-Lengeling, B., Outeiral, C., Guimaraes, G.L., Aspuru-Guzik, A.: Optimizing distributions over molecular space. an objective-reinforced generative adversarial network for inverse-design chemistry (organic). ChemRxiv (2017)
27. Shi, C., Xu, M., Zhu, Z., Zhang, W., Zhang, M., Tang, J.: GraphAF: a flow-based autoregressive model for molecular graph generation. arXiv preprint arXiv:2001.09382 (2020)
28. Team, G.: GT4SD (Generative Toolkit for Scientific Discovery) (2022)
29. Vallender, S.: Calculation of the Wasserstein distance between probability distributions on the line. Theor. Probab. Appl. **18**(4), 784–786 (1974)
30. Vens, C., Struyf, J., Schietgat, L., Džeroski, S., Blockeel, H.: Decision trees for hierarchical multi-label classification. Mach. Learn. **73**(2), 185–214 (2008)
31. Wegner, J.K., et al.: Cheminformatics. Commun. ACM **55**(11), 65–75 (2012)
32. Weininger, D.: Smiles, a chemical language and information system. 1. introduction to methodology and encoding rules. J. Chem. Inf. Comput. Sci. **28**(1), 31–36 (1988)
33. Xiong, J., Xiong, Z., Chen, K., Jiang, H., Zheng, M.: Graph neural networks for automated de novo drug design. Drug Discovery Today **26**(6), 1382–1393 (2021)
34. You, J., Liu, B., Ying, Z., Pande, V., Leskovec, J.: Graph convolutional policy network for goal-directed molecular graph generation. In: Advances in neural information processing systems, vol. 31 (2018)

AutoMMLC: An Automated and Multi-objective Method for Multi-label Classification

Aline Marques Del Valle[1,2](✉), Rafael Gomes Mantovani[3], and Ricardo Cerri[2]

[1] Federal Institute of Education, Science and Technology of the South of Minas Gerais, Muzambinho, MG, Brazil
aline.valle@ifsuldeminas.edu.br

[2] Department of Computer Science, Federal University of São Carlos, Rodovia Washington Luis, km 235, São Carlos, SP, Brazil
cerri@ufscar.br

[3] Federal University of Technology - Paraná (UTFPR), Campus of Apucarana, Curitiba, PR, Brazil
rafaelmantovani@utfpr.edu.br

Abstract. Automated Machine Learning (AutoML) has achieved high popularity in recent years. However, most of these studies have investigated alternatives to single-label classification problems, presenting a need for more investigations in the multi-label classification scenario. From the AutoML point of view, the few studies on multi-label classification focus on automatically finding the best models based on mono-objective optimization. These tools train several multi-label classifiers in search of the one with the best performance in a single objective optimization process. In this work, we propose AutoMMLC, a new multi-objective AutoML method for multi-label classification, to find the best models that maximize the f-score measure and minimize the training time. Experiments were carried out with ten multi-label datasets and different versions of the proposed method using two multi-objective optimization algorithms: Multi-objective Random Search and Non-Dominated Sorting Genetic Algorithm II. We evaluated the Pareto front obtained by these methods through the hypervolume metric. The Wilcoxon test demonstrated that AutoMMLC versions had similar results for this metric. Multi-label Classification (MLC) algorithms were obtained from the Pareto frontiers through the Frugality Score and compared with the baseline algorithms. The Friedman test demonstrated that the MLC algorithms from AutoMMLC versions had equal performances to f-score and training time. Furthermore, they had better results than baseline algorithms for f-score and better results than most baseline algorithms for training time.

Keywords: Automated machine learning · Multi-label classification · Multi-objective optimization

M. C. Naldi and R. A. C. Bianchi (Eds.): BRACIS 2023, LNAI 14196, pp. 291–306, 2023.
https://doi.org/10.1007/978-3-031-45389-2_20

1 Introduction

Designing high-performance Machine Learning (ML) model is an arduous task that requires expert knowledge [7]. Automated Machine Learning (AutoML) is a research field that seeks to improve how ML applications are built by automating applications [20], creating automatically configured tools that perform well and are easy to use. However, AutoML is not only present in the search by ML models. It can also be applied to the pre-processing phases to the model interpretability phase, automating a specific sub-task of the ML pipeline or even the entire pipeline [18,20].

AutoML can solve Single-label Classification (SLC) and Multi-label Classification (MLC) problems. Each dataset instance is associated with a single label in SLC, while in MLC, each instance is associated with a subset of labels [9,17]. Most research dealing with AutoML and MLC focuses on automating searching for the best algorithms and hyperparameters used on the model induction task [12–14,18,19].

Given the search space, an optimization algorithm, and an evaluation criterion, AutoML trains and evaluates several models in search of the best model evaluated. As the search space comprises different classification algorithms and their corresponding hyperparameters, these studies deal with instances of Combined Algorithm Selection and Hyperparameter Optimization (CASH) optimization problems [7]. Furthermore, in these studies, the optimization algorithms for MLC are mono-objective, with an evaluation criterion usually related to the classifier performance.

However, we could evaluate multi-label classifiers considering more than one evaluation criterion [19]. Thus, we would have a multi-objective optimization problem, where a task involves more than one objective function to be maximized or minimized and produces a set of optimal solutions rather than a single solution, known as Pareto optimal solutions [3]. Multi-objective ML balances distinct evaluation criteria (objective functions) to improve model generalization and avoid models with local optima [8].

In this study, we propose a multi-objective AutoML method for MLC, denominated Automated Multi-objective Multi-label Classification (AutoMMLC), optimizing both the f-score performance measure and the training time of the classifiers. Thus, we want to automatically find models that maximize the f-score and have a lower computational time. We performed experiments evaluating the performance of AutoMMLC employing two different multi-objective algorithms: Multi-objective Random Search (MORS) [8] and Non-Dominated Sorting Genetic Algorithm II (NSGA-II) [2]. These algorithms were used because MORS serves as reasonable baselines, and NSGA-II is one of the most popular multi-objective evolutionary algorithms [8]. In addition, we evaluated the performance of AutoMMLC against other baselines simplest.

The main contributions of this paper are: i) the proposal of a new AutoML method for MLC, contributing to a subject still little explored, and ii) incorporating multi-objective criteria to search for the best ML classifiers. The remainder of this article is structured as follows: in Sect. 2, we review MLC, multi-objective optimization algorithms, and related works; in Sect. 3, we describe the experimental methodology adopted in this study; in Sect. 4, we present the results and discuss their implications; finally, Sect. 5 presents the findings and future works.

2 Background

2.1 Multi-label Classification

Given a label set, SLC associates a single label with each instance in the dataset. In MLC, an instance can be classified into two or more labels simultaneously. For example, textual data can be classified simultaneously with multiple labels, such as images, videos, music, and emotions. Since $L = \{l_j : j = 1, ..., q\}$ is the finite label set, MLC associates to each instance $\mathbf{x}_i$ of the dataset a set of labels $\mathbf{y}_i$ such that $\mathbf{y}_i \subseteq L$ [17]. According to Madjarov et al. [9], MLC algorithms can be characterized into three approaches: i) Problem Transformation (PT), which transforms multi-label problems into single-label problems and then employs traditional classification algorithms; ii) Algorithm Adaptation (AA), which creates algorithms that handle multiple labels simultaneously; and iii) Ensemble, which employs a set of MLCs as base classifiers.

2.2 Multi-objective Optimization

Multi-objective optimization aims to solve real problems with conflicting objectives, subject to possible constraints of the problem [3]. In the context of multi-objective optimization for ML problems, Karl et al. [8] formally define the problem according to Eq. 1.

$$\begin{array}{l} \min_{\lambda \in \Lambda} \; c(\lambda), \;\; \textit{subject to}: \\ k_1(\lambda) = 0, ..., k_n(\lambda) = 0 \\ \hat{k}_1(\lambda) \geq 0, ..., \hat{k}_n(\lambda) \geq 0 \end{array} \quad (1)$$

Since Λ is the set of possible ML candidate solutions, we want to evaluate each candidate solution λ. The λ evaluation is calculated from the m evaluation criteria of the ML models: $c_1 : \Lambda \rightarrow \mathbb{R}, \ldots, c_m : \Lambda \rightarrow \mathbb{R}$, with $m \in \mathbb{N}$. The objective is to find, among all the candidate solutions, the one with the minimum evaluation $c(\lambda)$, being $c : \Lambda \rightarrow \mathbb{R}^m$.

The multi-objective optimization result is a set of optimal solutions rather than a single solution [3]. This set is known as Pareto optimal solutions or Pareto frontier. They have different values for the objectives, where a solution can be good considering one of the objectives but poor concerning the others. It is up to the users to choose the best solution among the resulting solutions.

Most multi-objective algorithms use the concept of dominance to find Pareto optimal solutions. So for a minimization problem, solution λ dominates $\lambda^{'}$ if only $\forall i \in \{1, ..., m\} : c(\lambda_i) \leq c(\lambda_i^{'})$ and $\exists j \in \{1, ..., m\} : c(\lambda_j) < c(\lambda_j^{'})$ [8]. In other words, λ dominates $\lambda^{'}$ if the goal values of λ are equivalent to or better than the goal values of $\lambda^{'}$ and λ surpasses $\lambda^{'}$ in at least one of the objectives. Any other solution does not dominate Pareto optimal solutions. Based on the dominance concept, we describe below two algorithms:

- **Multi-objective Random Search (MORS)**: Random Search (RS) is one of the most popular mono-objective optimization algorithms. Given T trials, the mono-objective RS algorithm randomly produces candidate solutions and evaluates these solutions in search of the best one. According to Karl et al. [8], the multi-objective version of RS can be derived from RS and the concept of dominance. For this, we assess the T random samples and return the set of non-dominated solutions, the Pareto front.
- **Non-Dominated Sorting Genetic Algorithm II (NSGA-II)**: NSGA-II is an evolutionary algorithm [2]. Given an initial population, the population evolves over generations by crossover, mutation, and tournament selection operations. The fittest individuals, among parents and offspring, are selected for subsequent generations. Fitness is measured by non-domination rank and crowding distance density estimate. Thus, the populations are composed of individuals from the first Pareto frontiers with good dispersion.

We can find the quality of the Pareto frontier using distance-based (requires the true Pareto frontier) or volume-based (calculates the volume from the frontier to a point) indicators. Hypervolume is a volume-based quality indicator. For a minimization problem, it is the region bounded by the Pareto frontier points and by a reference point [6]. The larger the calculated hypervolume, the more the Pareto frontier minimizes the objectives of the problem.

2.3 Related Works

Most studies covering AutoML and MLC explored mono-objective optimization instances. Given a search space composed of single-label algorithms, multi-label algorithms, and hyperparameters [12–14,18,19], optimization algorithms explore the search space and evaluate the candidate multi-label classifiers in search of the best classifier. The studies used different evaluation criteria to evaluate candidate classifiers during optimization, aggregating different performance measures composing a single evaluation criterion or using one knowing evaluation criteria, such as Hamming Loss and F-score.

While handling AutoML for MLC as a mono-objective optimization problem, the related studies usually propose new algorithms and compare the results with other optimizers already used for MLC. In this sense, de Sá et al. [13] used Genetic Algorithms as the optimization algorithm, de Sá et al. [12] used Grammar-based Genetic Programming, and de Sá et al. [14] extended their previous works by also employing Bayesian Optimization as an optimization algorithm. Other authors explored the search space with Networks of Hierarchical Tasks [19], used Hyperband Optimization, Bayesian Optimization, and the combination of Bayesian Optimization and Hyperband as optimization algorithms [18].

The works focus on automating the search for ML models. Some suggest employing AutoML in the data pre-processing and post-processing steps [12,13]. All works employ one evaluation criterion in the optimization process. There are suggestions to study the evaluation criteria in the application domain to use the most appropriate ones [14] and employ multi-objective optimization to improve the generalization of the models [19]. The authors cite the need to improve the search space [13,18] using Meta Learning, for example [18].

3 Experimental Methodology

3.1 Datasets

We evaluated AutoMMLC in ten multi-label datasets traditionally used in studies on MLC [9,18] from different application domains: biology, multimedia (audio and music), and text. Table 1 presents the main characteristics of the datasets used. The number of instances in the collection ranges from [194, 7.395], the number of features ranges from [19, 1.836], and the number of labels from [6, 374]. All these datasets were available for download from the Mulan website[1].

Table 1. MLC datasets used in experiments. For each dataset, it is presented: its domain, number of instances (#I), number of features (#F), and number of labels (#L) These statistics values are shown before and after the preprocessing.

Name	Domain	Original			Preprocessed		
		#I	#F	#L	#I	#F	#L
bibtex	text	7395	1836	159	7380	1836	159
birds	multimedia	645	260	19	337	260	15
corel5k	multimedia	5000	499	374	4994	499	218
emotions	multimedia	593	72	6	593	72	6
enron	text	1702	1001	53	1675	1001	42
flags	multimedia	194	19	7	194	19	7
genbase	biology	662	1186	27	204	1186	12
medical	text	978	1449	45	940	1449	20
scene	multimedia	2407	294	6	2400	294	6
yeast	biology	2417	103	14	2417	103	14

3.2 Preprocessing

All datasets were split using a 10-fold cross-validation resampling strategy. To ensure the same proportion of instances of each class in the folds, we used the iterative stratification strategy [15]. It was coded using the *scikit-multilearn*

[1] https://mulan.sourceforge.net/datasets-mlc.html.

library [16]. Even after applying the iterative stratification, some folds ended without positive instances for some classes happened because some datasets originally had very few positive instances for the same labels. Thus, before the iterative stratification, we also preprocessed the datasets.

The data preprocessing step included data organization, cleaning, and data transformation. None of the selected datasets had missing values. So, in the preprocessing step, we treated the following situations:

- **Duplicated data**: we removed instances with duplicated values; and
- **Labels associated with few instances**: we also removed labels with ten or fewer positive instances. With this, we also excluded unlabeled instances.

In the data transformation step, we applied min-max normalization on all datasets to prevent different features from having outliers or features prevailing over others. The three right-most columns of Table 1 show the main statistics of the datasets after the preprocessing. One may note a change in the number of labels in some datasets: birds, corel5k, enron, genbase, and medical. In these datasets, the preprocessing also decreased the available number of instances. Once the preprocessing was finished, we generated the resampling 10-fold through iterative stratification, as mentioned before. The datasets and codes used in this work are freely available[2].

3.3 Search Space

During its execution, our AutoMMLC assesses different ML models to find the best model. The search space definition is essential in this process, as it defines the possible ML algorithms (or neural architectures) and hyperparameters values for inducing models. Figure 1 presents the search space of AutoMMLC, composed of 11 SLC algorithms and 10 MLC algorithms of the *scikit-learn* [10] and *scikit-multilearn* libraries [16].

The SLC algorithms included in the search space are K-Nearest Neighbors (KNN), Support Vector Machine (SVM), AdaBoost, Bernoulli Naïve Bayes (BNB), Decision Tree (DT), Extra Tree (ET), Gradient Boosting (GB), Logistic Regression (LR), Multinomial Naïve Bayes (MNB), Random Forest (RF), Multi-Layer Perceptron (MLP) and Stochastic Gradient Descent (SGD). The MLC algorithms included in the search space belong to the following categories:

- **Ensemble**: Random k labelsets Disjoint (RakelD);
- **PT**: Binary Relevance (BR) and Classifier Chains (CC); and
- **AA**: BR and KNN in version A (BRkNNa), BR and KNN in version B (BRkNNb), DT, Multi-Label KNN (MLkNN), Multi-Label Support Vector Machines (MLTSVM), Multi-Label Adaptive Resonance Associative Map (MLARAM), and RF.

[2] https://github.com/alinedelvalle/AutoMMLC.git.

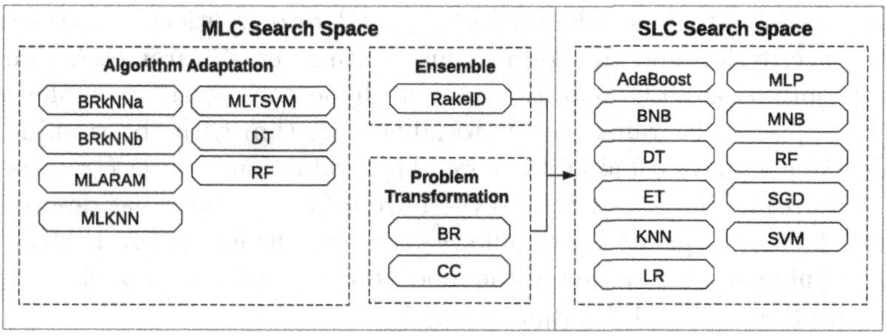

Fig. 1. Search space used in our experiments. It is composed of SLC and MLC algorithms. MLC algorithms belong to AA, Ensemble, and PT categories.

In Fig. 1, RakelD algorithm is an Label Powerset (LP) ensemble and has a base classifier. This base classifier is one of the SLC algorithms listed. The multi-label algorithms of the PT category also have a base classifier, which can be any of the 11 single-label algorithms included in the search space.

It is essential to mention that all these SLC and MLC algorithms can have tunable hyperparameters. Therefore, the search space also contains information about the possible values of each hyperparameter. In this work, we stored the possible values of the hyperparameters in arrays, which could assume discrete or continuous values. More information about the search space algorithms (their hyperparameters and values) is in Supplementary Material[3].

3.4 Evaluation Criteria and Objectives

The evaluation criteria in our ML model evaluation approach were the training time and the measure-based-example f-score (also known as F-measure). The selected criteria are related to the computational time and the performance of the models, respectively. F-score is the harmonic mean between precision and recall and can assume values between [0, 1]. Equation 2 presents the definition of f-score, where N is the number of instances in the dataset, $\mathbf{y}_i$ is the label set for the instance $\mathbf{x}_i$, and $h(\mathbf{x}_i)$ is the predicted label set for instance $\mathbf{x}_i$ [9,17].

$$f - score = \frac{1}{N}\sum_{1}^{N} \frac{2\,h(\mathbf{x}_i) \cup \mathbf{y}_i|}{|h(\mathbf{x}_i)| + |\mathbf{y}_i|} \tag{2}$$

Given that we want to maximize the f-score and minimize training time, we defined the objectives as training time and *1 - f-score.*

3.5 AutoMMLC$_{\text{MORS}}$

AutoMMLC$_{\text{MORS}}$ is the AutoMMLC method with the MORS algorithm. This method evaluates candidate solutions randomly sampled from the search space

[3] https://github.com/alinedelvalle/AutoMMLC/blob/main/BRACIS_Supplementary_Material.pdf.

and returns the set of non-dominated ones (Pareto frontier). It receives the termination criterion and the training and testing set as parameters. Then, it randomly obtains a MLC algorithm and its hyperparameters from the search space. In sequence, it trains this multi-label algorithm with the training data and calculates training time and f-score. The evaluation criteria (*training time* and *1 - f-score*) are necessary to find the Pareto front, obtained as described by Deb et al. [2]. Our implementation allows executing the method AutoML by fold and the training and evaluation of the models in parallel from a pool of threads, speeding up the AutoMMLC$_{\mathrm{MORS}}$.

3.6 AutoMMLC$_{\mathrm{NSGA}}$

Individual. In the AutoMMLC$_{\mathrm{NSGA}}$ optimization process, we represented candidate solutions (individuals) as an array of integers. They are created dynamically, and their structure is related to the search space. Figure 2 depicts examples of individuals and a reduced search space, considering only two candidates MLC algorithms (MLkNN and BR). This illustrative figure will be used to explain the individual. Each array position corresponds to a gene, and the first position specifies the MLC algorithm. We can get the MLC algorithms from the search space ([MLkNN, BR]). Thus, the possible values in the first gene are zero (MLkNN) or one (BR), which are indexes into the array of available MLC algorithms.

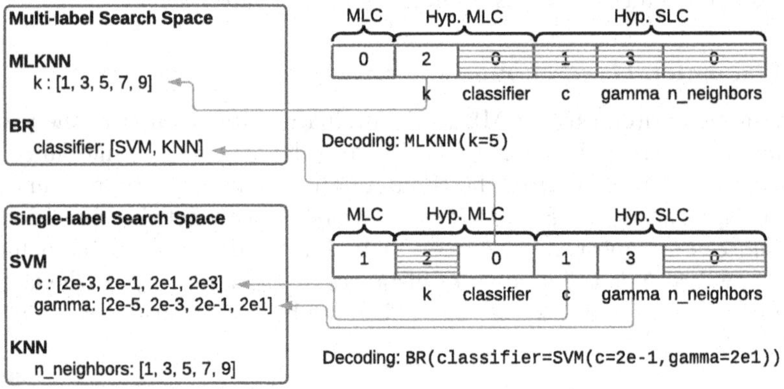

Fig. 2. In the Figure, there are two individuals and reduced search spaces for the single-label and multi-label algorithms.

The following individual's genes are the hyperparameters of the MLC algorithms in the multi-label search space. They are allocated in the individual in the same order they appear in the multi-label search space. Thus, they refer to the hyperparameters (Fig. 2): k and *classifier*. The other individual's genes are the hyperparameters of the SLC algorithms in the single-label search space.

They are also allocated in the individual in the same order as in the single-label search space. Therefore, the other positions refer to the hyperparameters (Fig. 2): *c*, *gamma* and *n_neighbors*.

The search space contains the possible values that a hyperparameter can assume. A hyperparameter can take on a finite number of possible values. Therefore, the hyperparameter space is discrete, so it was represented using arrays in search space. Thus, the value of a gene in the individual indicates one of the positions (indices) of this hyperparameter array. For example, the *gamma* hyperparameter in Fig. 2 has a set of possible values ($[2e-5, 2e-3, 2e-1, 2e1]$). The value of the gene referring to this hyperparameter in the individual is 3, which means $gamma = 2e1$. Following the described logic, individuals are randomly initialized based on the search space.

Figure 2 shows the decoding of the individual. The individual with the first gene equal to 0 has MLkNN as a MLC algorithm. This algorithm has hyperparameter k with hyperparameter space: $[1, 3, 5, 7, 9]$. The individual's gene referring to this hyperparameter contains a value of 2. Therefore, we have the MLkNN algorithm with k equal to 5. The hatched genes in Fig. 2 have values, but they are not used in the MLkNN algorithm.

On the other hand, when the individual's first gene is 1, the MLC algorithm is BR. This algorithm has the hyperparameter $classifier$ with hyperparameter space: $[SVM, KNN]$. The individual's gene referring to this hyperparameter contains a value of 0. Therefore, we have the BR algorithm with the $classifier$ equal to SVM. We must also decode the SVM base classifier ($c = 2e-1$ and $gamma = 2e1$).

AutoMMLC$_{\text{NSGA}}$ Implementation. AutoMMLC$_{\text{NSGA}}$ is the AutoMMLC method with the NSGA-II algorithm. In the implementation, we used the NSGA-II algorithm from Pymoo, a multi-objective optimization framework in Python [1]. For this, we defined the search space and developed classes that inherit from the Pymoo framework classes:

- **Sampling**: we specified how to create an initial population. Our algorithm selects the initial population randomly, selecting a valid value for each gene in the search space.
- **Problem**: in this class, we initialized the number of objectives and defined how to evaluate the candidate solution. This implementation involved decoding the individual in a MLC algorithm and assessing the correspondent classification model, measuring the training time, and calculating the f-score from the model. This process was similar to the defined MORS and was implemented in parallel.
- **Mutation**: we developed the mutation operator, which verifies whether individuals may or may not mutate with a 5% probability. The mutation alters one of the individual's genes with a valid search space value.

In the crossover, we employed uniform crossover available in the Pymoo framework, where each child's gene can be from one of the parents with a probability of 50%. The individuals resulting from crossover and mutation are valid, as the search space guarantees the consistency of the generated individuals.

3.7 Baseline Algorithms

The multi-label search space algorithms were used as baseline algorithms for comparison. We adopted the SVM algorithm as the base classifier of the MLC algorithms of the Problem Transformation approach. SVM was chosen because it is traditionally used as a base classifier [9]. For the hyperparameters penalty of the MLTSVM algorithm, we use the value $2e^{-3}$ (one of the values of our search space). We adopted the default hyperparameters of the implementations available in the textitScikit-learn [10] and *Scikit-Multilearn* [16] libraries for the other hyperparameters.

3.8 Settings for Running

We executed the AutoMMLC$_{\text{NSGA}}$ with a population of size 100 and 100 generations, totaling 10.000 candidate solutions evaluated. The mutation rate was set to 5%. We adopted early stopping, so if there was no change in the average of the objectives (training time and f-score) for ten consecutive generations, AutoMMLC$_{\text{NSGA}}$ ended. The budget size for AutoMMLC$_{\text{MORS}}$ was the same, i.e., it evaluated 10.000 random candidate solutions.

AutoMMLC$_{\text{NSGA}}$ and AutoMMLC$_{\text{MORS}}$ were implemented in parallel with a pool of 10 threads. The runtime of the MLC algorithms was limited to 20 minutes. Limiting the runtime of these algorithms is expected and commonly adopted in the AutoML context [12–14,18,19]. Given the complexity of some multi-label algorithms and the dimensionality of some datasets, our goal with the runtime limit was to ensure AutoML results promptly. If the algorithm runtime exceeds this pre-established limit threshold, the objectives receive their maximum values, 1 for *1 - f-score* and 20 min for training time. These values were also the reference point for the hypervolume calculation. The runtime of the baseline algorithms was also limited to 20 min.

In addition, we also stored all the evaluated candidate solutions and their corresponding objectives' values. Those values were stored for AutoMMLC$_{\text{NSGA}}$ and AutoMMLC$_{\text{MORS}}$. Thus, before considering a new candidate solution, we checked if it was among the solutions already evaluated. If so, we used the objectives already calculated, avoiding training the same classifier (algorithm and hyperparameters) more than once. With this requirement, we would like to decrease the total runtime required by the AutoMMLC.

3.9 Statistical Analysis

To compare the AutoMMLC$_{\text{NSGA}}$ and AutoMMLC$_{\text{MORS}}$ algorithms, we consider the hypervolume obtained from the Pareto frontiers of each fold of the datasets. We use the Wilcoxon non-parametric test [4] to evaluate statistically the hypervolumes with a significance level of 5%. Our null hypothesis was that AutoMMLC versions performed equally, producing equivalent hypervolumes.

We selected Pareto frontier algorithms using Frugality Score to compare AutoMMLC$_{\text{NSGA}}$ and AutoMMLC$_{\text{MORS}}$ with the baseline algorithms. The frugality Score is a measure for evaluating algorithms that combines a measure of performance and resources [5]. In this work, we combined the f-score with the training time, where the f-score is penalized with training time, as shown in Eq. 3.

$$Frugality = f\_score - \frac{0.5}{1 + \frac{1}{training\_time}} \tag{3}$$

The result of AutoMMLC$_{\text{NSGA}}$ and AutoMMLC$_{\text{MORS}}$ on a fold/dataset is the Pareto frontier. We calculated the Frugality Score for the Pareto frontier MLC algorithms and selected the highest score. Thus, we have a MLC algorithm representing the results of AutoMMLC in each fold/dataset. We used two Friedman tests [4] with a significance level of 5% to compare the MLC algorithms representing AutoMMLC$_{\text{NSGA}}$ and AutoMMLC$_{\text{MORS}}$ with the baseline algorithms to f-score and training time. Our null hypotheses were: the MLC algorithms had the same f-scores, and the MLC algorithms had the same training time. If the null hypotheses were rejected, the Nemenyi post-hoc test was applied, where the performances of two MLC algorithms are significantly different if the corresponding mean ranks differ by at least one critical difference value.

4 Results and Discussion

4.1 AutoMMLC$_{\text{NSGA}}$ and AutoMMLC$_{\text{MORS}}$

AutoMMLC$_{\text{NSGA}}$ and AutoMMLC$_{\text{MORS}}$ were run as described in the experimental setup (Sect. 3.8). Figure 3 presents the average hypervolume (k-folds) for the ten datasets obtained while evaluating candidate solutions. Hypervolume calculation occurred every 100 evaluations for AutoMMLC$_{\text{MORS}}$ and at the end of each generation for AutoMMLC$_{\text{NSGA}}$. AutoMMLC$_{\text{NSGA}}$ converges before 10.000 evaluations (or 100th generation), and its execution ends. Convergence means that after determinate generation, there were no changes in the values of the objectives and, consequently, in the value of the hypervolume. In AutoMMLC$_{\text{MORS}}$, there is also this tendency. As a result, the hypervolume stops increasing after a certain number of evaluations. However, this method has always evaluated the 10.000 candidate solutions.

AutoMMLC$_{\text{MORS}}$ evaluates more candidate solutions than AutoMMLC$_{\text{NSGA}}$, the runtime of this method is more. Both AutoMMLC methods consult candidate solutions that have already been evaluated, contributing to the execution time of methods. Furthermore, the methods stop training the MLC algorithms when the training time exceeds 20 minutes. This fact occurs more frequently in AutoMMLC$_{\text{MORS}}$ and increases your runtime. The evolutionary process of AutoMMLC$_{\text{NSGA}}$ contributes to the lower number of interrupted training since candidate solutions subject to interruption cease to be part of the population in the first generations.

AutoMMLC$_{\text{NSGA}}$ and AutoMMLC$_{\text{MORS}}$ results were the multi-objective algorithms results, the Pareto frontier. Thus, we could evaluate it by calculating the hypervolume. Figure 4 shows the boxplot of hypervolumes calculated

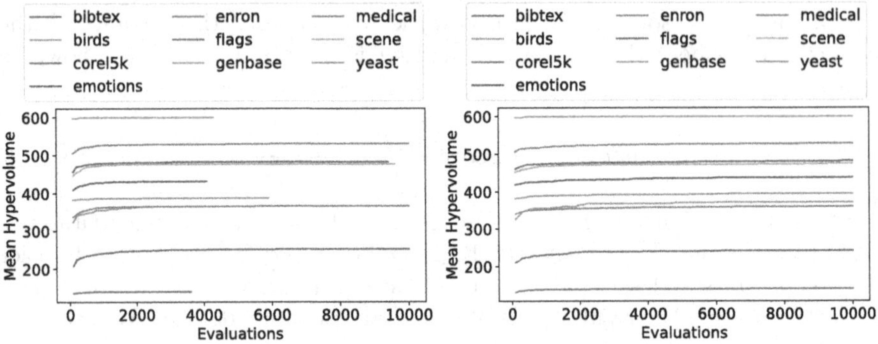

(a) $AutoMMLC_{NSGA}$ (b) $AutoMMLC_{MORS}$

Fig. 3. Convergence graphs of MLC methods. Each curve represents the average hypervolume of the 10-folds calculated as the solutions are evaluated.

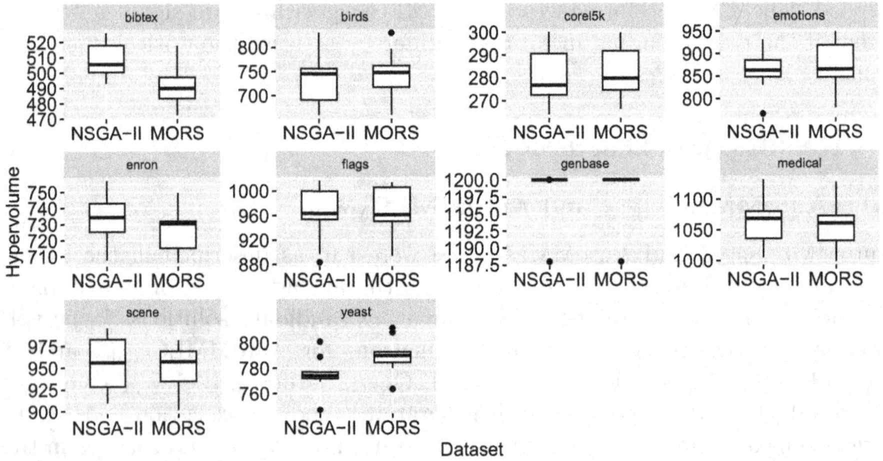

Fig. 4. Boxplot of hypervolumes. The hypervolumes were calculated from the Pareto frontier resulting from $\text{AutoMMLC}_{\text{NSGA}}$ and $\text{AutoMMLC}_{\text{MORS}}$. Each subplot brings the hypervolumes of the 10-fold from the datasets for both versions of the AutoMMLC.

for the 10-fold of each dataset, considering the results of the AutoMMLC with NSGA-II and MORS algorithms. As we wanted to minimize the objective values, the larger the hypervolume, the better the obtained Pareto frontier. Analyzing Fig. 4, we cannot assume which multi-objective algorithm got the best results. Then we did the Wilcoxon test. Our null hypothesis was that the hypervolumes obtained in the AutoMMLC versions were the same. The null hypothesis was accepted for all datasets. That is, $\text{AutoMMLC}_{\text{NSGA}}$ and $\text{AutoMMLC}_{\text{MORS}}$ had equal results for the hypervolume.

The Pareto frontiers resulting from $\text{AutoMMLC}_{\text{NSGA}}$ and $\text{AutoMMLC}_{\text{MORS}}$ comprise a set of MLC algorithms, one of the ten MLC algorithms in the search

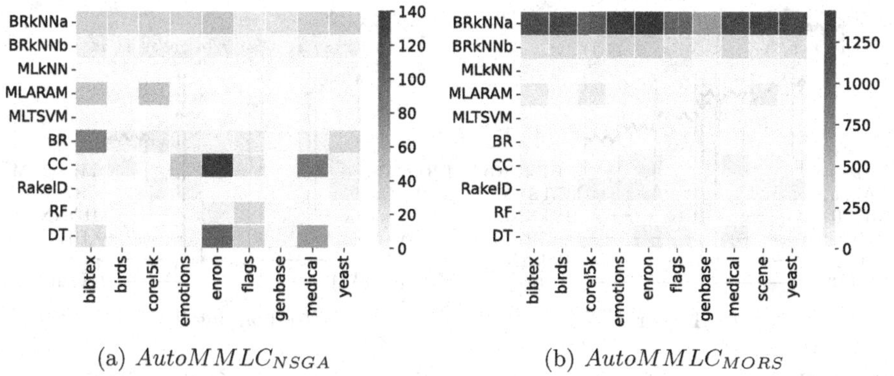

Fig. 5. Heat map of selected MLC algorithms by AutoMMLC. We counted the number of times the MLC algorithms were selected after executing the 10-folds of a dataset.

space (Sect. 3.3). Figure 5 shows heat maps with the occurrences of each MLC algorithm in the $\text{AutoMMLC}_{\text{NSGA}}$ and $\text{AutoMMLC}_{\text{MORS}}$ solutions after running the algorithms with the ten datasets. In the graphs of Fig. 5, there is a scale difference. This difference occurred because $\text{AutoMMLC}_{\text{MORS}}$ randomly selected from the search space many repeated MLC algorithms (as well as their hyperparameters), which ended up being part of the Pareto frontier. Among the most repeatable MLC algorithms is BRkNNa, followed by BRkNNb.

Except for the replications produced by $\text{AutoMMLC}_{\text{MORS}}$, both methods produced frontiers with different MLC algorithms. In the medical dataset, for example, the Pareto frontiers resulting from $\text{AutoMMLC}_{\text{MORS}}$ (in the 10-folds) had the MLC algorithms BRkNNa, BRkNNb, BR, CC, RakelD, and Decision Tree. These MLC algorithms also composited the Pareto frontiers resulting from $\text{AutoMMLC}_{\text{NSGA}}$ for the medical dataset. The two versions of AutoMMLC never included MLTSVM and MLkNN algorithms in their solutions.

4.2 Comparison of AutoMMLC with Baselines Algorithms

The Friedman test compared the MLC algorithms selected by the Frugality Score of $\text{AutoMMLC}_{\text{NSGA}}$ and $\text{AutoMMLC}_{\text{MORS}}$ solutions with the baseline algorithms to f-score and training time. The null hypothesis was that the MLC algorithms presented equal performances. The null hypothesis was rejected for the test with f-score and with training time.

Figure 6 presents the critical difference diagram for the Nemenyi test to the f-score. The analysis indicates no statistical difference between the two best MLC algorithms: algorithms from $\text{AutoMMLC}_{\text{NSGA}}$ and $\text{AutoMMLC}_{\text{MORS}}$. Figure 6 shows the critical difference diagram for the Nemenyi test to the training time. In this test, the MLC algorithms with the best training time are BRkNNb and BRkNNa, followed by algorithms from $\text{AutoMMLC}_{\text{NSGA}}$ and $\text{AutoMMLC}_{\text{MORS}}$.

The MLC algorithms representing $\text{AutoMMLC}_{\text{NSGA}}$ and $\text{AutoMMLC}_{\text{MORS}}$ do not have statistical differences for f-score and training time. However, these

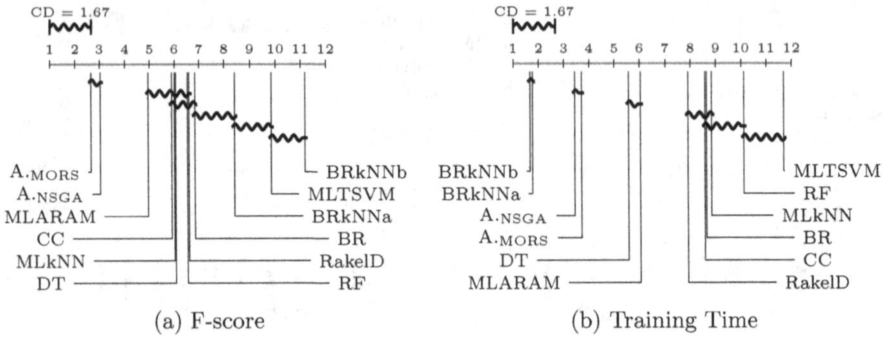

Fig. 6. Comparison of f-scores and training times values of MLC algorithms, according to the Nemenyi test. Groups of algorithms that are not significantly different are connected. $A._{NSGA}$ is AutoMMLC$_{\text{NSGA}}$ and $A._{MORS}$ is AutoMMLC$_{\text{MORS}}$.

algorithms are superior to baseline algorithms regarding f-score and superior to BR, CC, DT, MLARAM, MLkNN, MLTSVM, RakelD, and Random Forest algorithms regarding training time.

5 Conclusions

This work presented the AutoMMLC, a new multi-objective AutoML method for MLC, which seeks solutions that maximize the f-score and minimize the training time of the classifiers. AutoMMLC was developed using NSGA-II and MORS optimization algorithms. We ran both versions of the AutoMMLC with ten datasets, the same search space, and the settings for running. The Wilcoxon test indicated that AutoMMLC$_{\text{NSGA}}$ and AutoMMLC$_{\text{MORS}}$ were statistically equal results concerning hypervolume, but AutoMMLC$_{\text{NSGA}}$ was better runtime than AutoMMLC$_{\text{MORS}}$. Regarding baseline algorithms, AutoMMLC versions were statistically better than the f-score and had a lower training time than most baseline algorithms.

The results presented are still preliminary on a recent research topic that requires further exploration. Thus, in future work, we need to expand the analysis of data from AutoMMLC runs and the resulting Pareto frontiers. In addition, we need to study other ways to select Pareto frontier solutions to compare with other solutions already available in the literature. We can expand the comparisons, considering other AutoML solutions as baselines, be they mono-objective or multi-objective optimization. We can also produce new results by improving the search space. For this, we need to improve our representation of the search space to allow algorithms with deeper hierarchical levels. We can also add more algorithms to the search space, like the Multi-label Extension to Weka (MEKA) library algorithms [11]. Finally, more objectives could be used and have different weights in multi-objective optimization.

Acknowledgments. The authors would like to thank the Brazilian research agencies FAPESP, CAPES and CNPq for financial support.

References

1. Blank, J., Deb, K.: Pymoo: multi-objective optimization in python. IEEE Access **8**, 89497–89509 (2020)
2. Deb, K., Pratap, A., Agarwal, S., Meyarivan, T.: A fast and elitist multiobjective genetic algorithm: NSGA-II. IEEE Trans. Evol. Comput. **6**(2), 182–197 (2002)
3. Deb, K., Deb, K.: Multi-Objective Optimization, pp. 403–449. Springer, US, Boston, MA (2014)
4. Demšar, J.: Statistical comparisons of classifiers over multiple data sets. J. Mach. Learn. Res. **7**, 1–30 (2006)
5. Evchenko, M.M.: Frugal learning:applying machine learning with minimal resources (2016)
6. Fonseca, C., Paquete, L., Lopez-Ibanez, M.: An improved dimension-sweep algorithm for the hypervolume indicator. In: 2006 IEEE International Conference on Evolutionary Computation, pp. 1157–1163 (2006)
7. He, X., Zhao, K., Chu, X.: AutoML: a survey of the state-of-the-art. Knowl.-Based Syst. **212**, 106622 (2021)
8. Karl, F., et al.: Multi-objective hyperparameter optimization - an overview (2022)
9. Madjarov, G., Kocev, D., Gjorgjevikj, D., Džeroski, S.: An extensive experimental comparison of methods for multi-label learning. Pattern Recogn. **45**(9), 3084–3104 (2012), best Papers of Iberian Conference on Pattern Recognition and Image Analysis (IbPRIA'2011)
10. Pedregosa, F., et al.: Scikit-learn: machine learning in Python. J. Mach. Learn. Res. **12**, 2825–2830 (2011)
11. Read, J., Reutemann, P., Pfahringer, B., Holmes, G.: MEKA: a multi-label/multi-target extension to weka. J. Mach. Learn. Res. **17**(21), 1–5 (2016)
12. de Sá, A.G.C., Freitas, A.A., Pappa, G.L.: Automated selection and configuration of multi-label classification algorithms with grammar-based genetic programming. In: Auger, A., Fonseca, C.M., Lourenço, N., Machado, P., Paquete, L., Whitley, D. (eds.) Parallel Problem Solving from Nature - PPSN XV, pp. 308–320. Springer International Publishing, Cham (2018)
13. de Sá, A.G.C., Pappa, G.L., Freitas, A.A.: Towards a method for automatically selecting and configuring multi-label classification algorithms. In: Proceedings of the Genetic and Evolutionary Computation Conference Companion, pp. 1125–1132. GECCO 2017, Association for Computing Machinery, New York, NY, USA (2017)
14. de Sá, A.G.C., Pimenta, C.G., Pappa, G.L., Freitas, A.A.: A robust experimental evaluation of automated multi-label classification methods. In: Proceedings of the 2020 Genetic and Evolutionary Computation Conference, pp. 175–183. GECCO 2020, Association for Computing Machinery, New York, NY, USA (2020)
15. Sechidis, K., Tsoumakas, G., Vlahavas, I.: On the stratification of multi-label data. In: Gunopulos, D., Hofmann, T., Malerba, D., Vazirgiannis, M. (eds.) Machine Learning and Knowledge Discovery in Databases, pp. 145–158. Springer, Berlin Heidelberg, Berlin, Heidelberg (2011)
16. Szymanski, P., Kajdanowicz, T.: Scikit-multilearn: a scikit-based python environment for performing multi-label classification. J. Mach. Learn. Res. **20**(1), 209–230 (2019)

17. Tsoumakas, G., Katakis, I., Vlahavas, I.: Mining Multi-Label Data, pp. 667–685. Springer, US, Boston, MA (2010)
18. Wever, M., Tornede, A., Mohr, F., Hüllermeier, E.: AutoML for multi-label classification: overview and empirical evaluation. IEEE Trans. Pattern Anal. Mach. Intell. **43**(09), 3037–3054 (2021)
19. Wever, M.D., Mohr, F., Tornede, A., Hüllermeier, E.: Automating multi-label classification extending ML-Plan. In: 6th ICML Workshop on Automated Machine Learning, Long Beach, CA, USA (2019)
20. Zöller, M.A., Huber, M.F.: Benchmark and survey of automated machine learning frameworks. J. Artif. Int. Res. **70**, 409–472 (2021)

Merging Traditional Feature Extraction and Deep Learning for Enhanced Hop Variety Classification: A Comparative Study Using the UFOP-HVD Dataset

Pedro Castro[1], Gabriel Fortuna[2], Pedro Silva[1], Andrea G. C. Bianchi[1], Gladston Moreira[1], and Eduardo Luz[1](✉)

[1] Computing Department, Federal University of Ouro Preto (UFOP), Ouro Preto, MG 35402-163, Brazil
pedro.hnc@aluno.ufop.edu.br, {silvap,andrea,gladston,eduluz}@ufop.edu.br

[2] Brazuca Lúpulos, Petrópolis, RJ 25610-080, Brazil
https://csilab.ufop.br/

Abstract. Accurately identifying plant species and varieties is crucial across various disciplines, such as biology, medicine, and agronomy. While species identification is challenging, variety identification presents an even greater difficulty. Conventional identification methods, although effective, often require specialized and costly equipment, making them less accessible. In this work, we explore the problem of hop variety classification, comparing traditional feature extraction methods with deep learning approaches using the UFOP-HVD dataset. We address two research questions: whether traditional techniques can achieve competitive results given the limited number of images and whether combining traditional techniques and deep learning can improve the current state-of-the-art. Our findings indicate that traditional techniques yield competitive results for hop variety identification, offering advantages such as interpretability, reduced computational costs, and potential integration into mobile devices. Moreover, we introduce an ensemble method that improves the accuracy from 77.16% to 81.90%, establishing a new state-of-the-art for the UFOP-HVD dataset. These results demonstrate the potential of merging traditional methods with deep learning for challenging hop variety classification tasks, providing an initial baseline for future research.

Keywords: Hop variety · Plant recognition · Deep Learning · Ensemble · CNN

Supported by Universidade Federal de Ouro Preto, (FAPEMIG, grants APQ-01518-21, APQ-01647-22), CAPES and (CNPq, grants 307151/2022-0, 308400/2022-4).

M. C. Naldi and R. A. C. Bianchi (Eds.): BRACIS 2023, LNAI 14196, pp. 307–322, 2023.
https://doi.org/10.1007/978-3-031-45389-2_21

1 Introduction

Accurately identifying plant species and varieties holds significant value across numerous disciplines. Within biology, it serves as both a subject of investigation and an instrumental tool for teaching botany and promoting biodiversity conservation among high school and undergraduate students [20]. In medicine, plant identification enhances the quality and safety of herbal remedies, consequently reducing poisoning and intoxication incidents [8]. Forensic experts utilize this knowledge to bolster criminal investigations and support drug enforcement efforts [14]. In agronomy, the precise identification of plant varieties can optimize planting efficiency [5], as well as aid in detecting weeds detrimental to crop development.

Numerous methods exist for the identification of plant species and varieties, such as gas chromatography [1], mass spectrometry [18], electrochemical fingerprinting [12], molecular analysis [21], among other techniques. Although these methods effectively assist in recognizing plant varieties, they necessitate specialized and costly equipment, often utilized within laboratory settings.

The application of machine learning for plant recognition through image analysis has garnered interest due to its user-friendliness and accessibility [4]. In the machine learning approach, the only required equipment is a device capable of capturing plant images (or their components), such as a smartphone or a standard camera, which individuals can utilize without prior experience in plant classification. Although some researchers have demonstrated the feasibility of classifying plant species using computer vision with remarkable accuracy, a more challenging subclass of the problem persists: distinguishing varieties within the same species. Plants of the same species often exhibit strikingly similar appearances across leaves, stems, and flowers [19]. Nevertheless, in certain applications, it is crucial to classify plant varieties, as exemplified by hops used in beer production. Hops impart flavor, aroma, and bitterness; depending on the variety, the beverage's characteristics are significantly affected. Chemical processes differentiate hop varieties in the industry, but these methods are expensive and time-consuming.

In this context, researchers have made available a dataset, the UFOP-HVD [6][1], containing images of 12 hop varieties. In [7], deep learning techniques, specifically convolutional networks, were explored for hop variety classification. However, the dataset is limited, containing only 1,592 images (4,159 hop leaves). As a result, two research questions emerge: RQ1, given the limited number of images in the UFOP-HVD dataset, which may not favor deep learning techniques, is it possible to achieve better or more competitive results using traditional feature extraction techniques? And RQ2, can the combination of traditional techniques and deep learning improve the current state-of-the-art for the UFOP-HVD dataset?

The experimental findings of this study indicate that traditional techniques yield competitive results and offer several advantages, such as easier inter-

[1] Available at https://doi.org/10.6084/m9.figshare.14933178.

pretability, reduced computational costs, and the potential for seamless integration into mobile devices, compared to deep learning techniques. Nevertheless, merging traditional methods with deep learning has shown potential to yield significant advancements in classification. This was particularly evident in the tested hop varieties dataset, where an improvement in accuracy was observed from 77.16% to 81.90%, as demonstrated by our experimental findings.

Given that, the dataset under investigation is relatively new, this work also establishes an initial baseline for the problem, encompassing traditional and deep learning approaches.

2 Related Works and Dataset

Humulus lupulus L., commonly referred to as hops, is a climbing plant whose flowers are a principal ingredient in beer production worldwide. Hops impart flavor, bitterness, and aroma to the beverage [3], and play an important role in stabilization. Over 250 cataloged varieties of this plant exist [17], with distinguishing features including alpha-acids, beta-acids, and essential oils [25]. These components endow each beer with unique characteristics, potentially influencing its classification.

Numerous methods exist for identifying hop varieties based on their acids and essential oils [11,13,24]. However, these techniques may be inaccessible or unavailable to farmers. Machine learning and computer vision-based methods offer promising alternatives, given the extensive literature on plant species recognition [4]. For hops, the focus is on classifying the variety, which is a taxonomic level below species. As [19] states, intraspecific varieties share common genotype or phenotype traits. In the case of hops, visually observable morphological features are quite similar, as demonstrated in Fig. 1. To address this problem with computer vision, Brazilian researchers have provided an image dataset to facilitate the development of machine-learning techniques for the classification of hop varieties.

In [7], an end-to-end method for hop variety classification using the UFOP-HVD was proposed. The researchers investigated several CNN architectures: ResNets, EfficientNets, and InceptionNets. They explored image classification with and without leaf segmentation, with and without data augmentation, and proposed an ensemble architecture called Multi-Cropped-Full, which combined six models. For the leaf-only classification problem, termed "cropped classification" by the authors, a ResNet50 architecture with a 50% dropout achieved the best results, attaining 78% accuracy. The ensemble model, Multi-cropped-FULL, utilized multiple leaves from the same image and the entire image as input, achieving 81% accuracy. With the application of data augmentation techniques, the Multi-cropped-FULL model reached 95% accuracy. In the present study, we explore traditional machine learning techniques using the dataset in the "cropped" configuration, wherein each detected leaf image (an image that can have multiple hop leaves) is transformed into a new input for the problem. In this cropped evaluation format, only the main leaf should be considered. Given the limited number of samples, it is hypothesized that traditional machine-learning techniques should yield results comparable to those reported in [7].

2.1 UFOP Hop Varieties Dataset

The UFOP Hop Varieties Dataset (UFOP-HVD) [6] comprises 1,592 images of young and adult hop leaves across 12 varieties. The images were captured without controlling lighting, focus, distance, or angle and have resolutions ranging from $1{,}040 \times 520$ to $4{,}096 \times 3{,}072$. The dataset allocates 70% of the images for training, 15% for validation, and 15% for testing, establishing a protocol to be adhered to by authors. Representative samples from the dataset can be viewed in Fig. 1.

Fig. 1. Examples of the 12 hop varieties used in this work: (a) Cascade; (b) Centennial; (c) Cluster; (d) Comet; (e) Hallertau Mittelfrueh; (f) Nugget; (g) Saaz; (h) Sorachi Ace; (i) Tahoma; (j) Triple Pearl; (k) Triumph; (l) Zeus.

Each image in the dataset is accompanied by an XML file that specifies the plant variety and includes one or more leaves labeled using bounding boxes. These bounding boxes are defined by x and y coordinates for the upper-left and lower-right corners. Leaves that are not sharp or are significantly occluded have been excluded by the dataset author to maintain data quality.

Table 1 presents the number of images per class for each set. The photographs were captured using Motorola Moto G7, Samsung Galaxy A11, and Apple iPhone 11 devices.

Table 1. Class/variety division of training, validation, and test set size.

Variety	Train	Validation	Test	Total
Cascade	78	16	16	**110**
Centennial	89	18	18	**125**
Cluster	73	15	15	**103**
Comet	121	25	25	**171**
Hallertau Mittelfrueh	99	21	21	**141**
Nugget	82	17	17	**116**
Saaz	122	25	25	**172**
Sorachi Ace	137	28	28	**193**
Tahoma	85	17	17	**119**
Triple Pearl	84	17	17	**118**
Triumph	71	15	15	**101**
Zeus	87	18	18	**123**
Total	**1,128**	**232**	**232**	**1,592**

3 Methodology

In this study, hop classification is not conducted using the entire image. Instead, only the main leaf, which possesses the largest area among all leaves, is utilized. As illustrated in Fig. 2, the main leaf is marked with a red bounding box, while the remaining leaves are outlined in yellow. The initial step involves evaluating the area of all leaves in the image and extracting the largest one. Subsequently, the cropped image is resized such that its longest side measures L pixels. This parameter is further investigated in Sect. 4. This standardization process is essential to account for varying image resolutions. Lastly, the image is converted to grayscale in preparation for feature extraction algorithms.

3.1 Proposed Method

Figure 3 showcases the method, unifying all steps. The original image is preprocessed through cropping, resizing, and grayscale conversion. Next, GLCM, LBP, DAISY, and KASE feature extractors are applied. DAISY and KASE descriptors are normalized between 0 and 1, and fed to the Bag of Visual Words to create feature vectors. These vectors are normalized and concatenated into one, then sent to the SVM classifier.

To assess the impact of using traditional techniques on convolutional networks, a simple ensemble [10] method for combining the outputs of both is evaluated. In this study, the ensemble sums the predictions of the models to generate a final prediction, as demonstrated in Fig. 4.

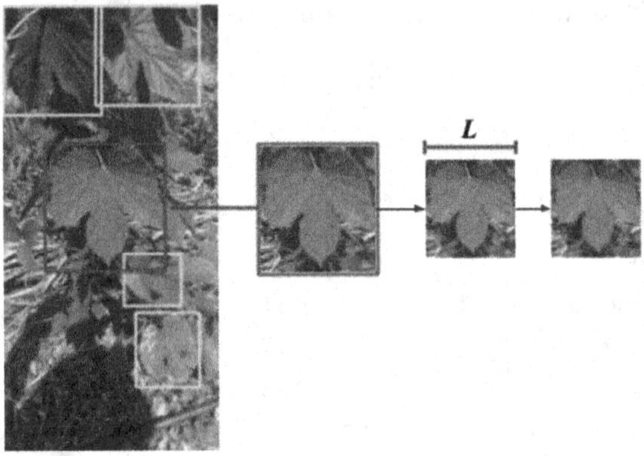

Fig. 2. Pre-processing containing the steps of cropping the main leaf, resizing, and converting to grayscale.

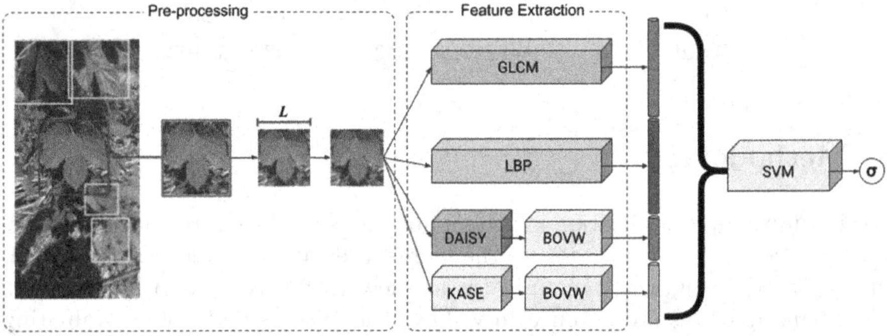

Fig. 3. Full proposed method with pre-processing, feature extraction, and classification steps.

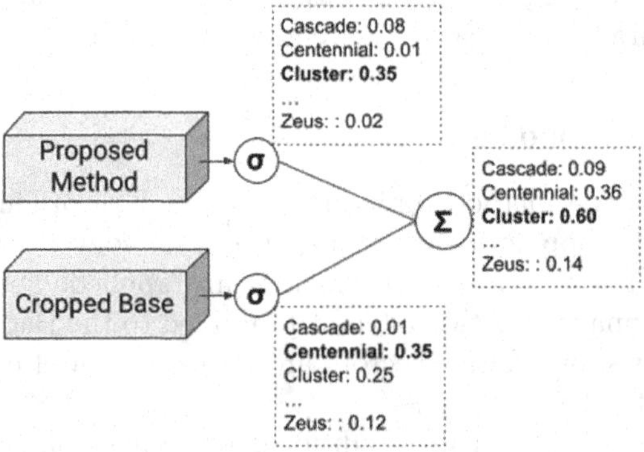

Fig. 4. Ensemble applied to the outputs of the proposed method and Cropped Base convolutional network. In this example, the proposed method indicated Cluster as the most likely class, while Cropped Base favored Centennial. In the summation of the predictions, the Cluster class prevailed.

3.2 Traditional Machine Learning Techniques

Feature extraction is vital for image classification in machine learning, as it helps mitigate errors caused by angle, lighting, and scale variations. This study investigates five robust feature extraction techniques resilient to these variations and employs a Support Vector Machine (SVM) with Radial Basis Function (RBF) for the classification step.

Gray Level Co-Occurrence Matrix (GLCM). [15] proposes a texture feature extraction method based on second-order statistics. A $p \times p$ matrix P is generated, where p represents the number of possible image intensities. Pairs of intensities are counted based on location parameters, such as distance and angle. Then, the co-occurrence matrix is normalized so that the sum of all its elements is 1. This matrix is usually sparse, and there are a few ways to better adapt it for classification. Haralick *et al.* [15] suggested 28 attributes generated from the matrix P, and those applied in this work are contrast, dissimilarity, homogeneity, angular second moment, energy, and correlation. The scalars obtained as a result form a vector of attributes.

Local Binary Pattern. The Local Binary Pattern (LBP) is a texture analysis feature extractor [22]. For each image pixel, it generates a value representing the pixel and its surroundings by determining the distance and number of neighbors. A binary code is formed by comparing the central pixel with its neighbors, and then converted to a decimal value. The original image pixels are replaced by these numerals, forming a new matrix that can be used directly as a feature or converted to a histogram as an attribute vector. Parameters, such as increasing the radius to the central pixel, can be adjusted. To reduce possible variations, [23] proposed a uniformity and circularity model. Uniform patterns have a maximum of 2-bit inversions, and circularity allows codes with the same shifted sequence to be marked as equivalent, reducing the number of combinations.

DAISY. [28] introduces a local region descriptor generator initially designed for stereo matching, but applicable for classification. It works with histograms of oriented gradients and can be applied to entire images or fragments around points of interest. The algorithm defines a central pixel c and calculates the gradient histogram in o orientations. R rings are generated around c with radius r_i. For each ring R_i, h points are uniformly distributed from 0 to 360°, and histograms with o directions are calculated. The total number of histograms H is $R \times h + 1$. The Gaussian filter size increases proportionally with r_i. Histograms are independently normalized and concatenated, resulting in a feature vector of size $H \times o$.

KASE. [2] is a method for detecting regions of interest and generating rotation and scale invariant descriptors. It involves building image gradients at various scales, calculating Hessian determinants on each scale, and using a sliding window to find local maxima for identifying points of interest. Descriptors are created for rectangular regions centered on each point, using first-order derivatives in horizontal and vertical directions. The 64-attribute descriptor is made rotation invariant by calculating the main orientation in a local neighborhood and rotating the descriptor towards that angle.

Bag of Visual Words. Direct use of DAISY and KASE descriptors in classifiers presents challenges due to varying input sizes and local rotation and scale invariance. The Bag of Visual Words (BOVW) [9,29] method overcomes these issues by grouping similar descriptors and generating a histogram representing a visual word vocabulary. This creates a constant-size feature vector suitable for classifiers. Clustering algorithms, specifically K-means, are used to group similar descriptors in this work.

3.3 Deep Learning Techniques

In this study, the Cropped Base convolutional networks outlined in [7] were replicated, employing the blocks from InceptionV3 [26], ResNet50 [16], and EfficientNetB3 [27] architectures. The purpose of reproducing these networks is to facilitate a comparison with traditional feature extraction and classification methods.

4 Experiments and Results

To address the research questions posed in this study, the experiments are divided into two groups. First, we investigate the feature extraction techniques with an SVM classifier, followed by tests with the re-implemented convolutional networks from the proposed work in [7], under the same setup, using UFOP-HVD data. Next, we evaluate the combination (ensemble) of multiple techniques. The evaluation metric employed is accuracy. In addition to experimenting with the models, this section also presents data on the interpretability of the base models. The source code for loading and preparing the data from the database is available at https://anonymous.4open.science/r/anonymous-8291.

4.1 RQ1: Can Better or More Competitive Results Be Achieved Using Traditional Feature Extraction Techniques?

Table 2 displays the classification accuracy using features generated by GLCM. Each row corresponds to a different configuration, and the 'Id' column identifies the configuration for reference. For all configurations, the co-occurrence angles of the pixels were 0°, 45°, 90°, 135°, 180°, 225°, 270°, and 315°. The parameter L, introduced earlier, corresponds to the length of the longest side of the resized image and can assume values of 400, 500, and 600 pixels. These values determine the model instances, which will be named L400, L500, and L600, respectively. The variable parameter was the distance between the co-occurring pixels. It can be observed that L400 and L500 instances benefited from the use of longer distances, while the L600's performance declined under these conditions. The accuracy contribution from greater distances suggests the need to further increase them to investigate the gain limit.

Table 2. Accuracy of GLCM feature extractor for different image sizes.

Id	Distance	Accuracy		
		$L = 400$	$L = 500$	$L = 600$
#1	1, 3, 5, 7	0.4267	0.4353	**0.4569**
#2	1, 3, 5, 7 e 9	0.4181	0.4310	**0.4569**
#3	1, 3, 5, 7, 9 e 11	0.4224	0.4397	0.4267
#4	1, 3, 5, 7, 9, 11 e 13	**0.4310**	**0.4526**	0.4440

Table 3 shows the classification results with LBP attributes. The variable parameter was the radius length, and the number of points was set as 8 times the radius value. The feature extraction method was the uniform pattern. L400 and L600 instances performed better with larger radii, indicating potential exploration of increasing these parameters. L500, which achieved the best accuracy for the attribute, did not exhibit noticeable patterns concerning the radius. LBP demonstrated the best individual performance among all extractors.

Table 3. Accuracy of LBP feature extractor for different image sizes.

Id	Radius	Accuracy		
		$L = 400$	$L = 500$	$L = 600$
#1	1, 2 e 3	0.4871	**0.5560**	0.5345
#2	1, 2, 3, 4 e 5	0.5345	0.5302	0.5259
#3	1, 2, 3, 4, 5, 6 e 7	0.5345	**0.5560**	0.5302
#4	1, 2, 3, 4, 5, 6, 7, 8 e 9	**0.5517**	0.5345	**0.5474**

The DAISY feature extractor has the most adjustable parameters and, consequently, more available configurations. The accepted values for the descriptor radius were 9, 18, and 36, which may contain 2 or 3 rings, as shown in Table 4. Additionally, for each instance, a different step parameter was used: 25, 33, and 40 for L400, L500, and L600, respectively. All settings employed a fixed number of 8 orientations and 16 histograms. Besides the DAISY descriptors parameterization, the number of K-means clusters, which form the basis for Bag of Visual Words, varied between 25, 50, 100, and 200 clusters. L500 performed better with larger clusters, while L400 and L600 performed optimally with 100.

Table 4. Top 10 accuracy of DAISY feature extractor for different image sizes.

Id	Clusters	Radius	Rings	Accuracy		
				$L = 400$	$L = 500$	$L = 600$
#14	100	9	3	**0.3621**	0.3276	0.2543
#15	100	18	2	0.3060	0.3017	0.3621
#16	100	18	3	0.3276	0.3147	0.3491
#17	100	36	2	0.3362	0.3448	**0.3836**
#18	100	36	3	0.3147	0.3405	0.3233
#19	200	9	2	0.3190	0.2974	0.3491
#20	200	9	3	0.3147	0.2802	0.2845
#21	200	18	2	0.3448	0.3319	0.3534
#22	200	18	3	0.3405	0.3060	0.3147
#23	200	36	2	0.3491	**0.3534**	0.3276
#24	200	36	3	0.3405	**0.3534**	0.3621

Lastly, Table 5 presents the accuracy for attributes obtained with KASE. The modified parameters were the number of clusters for the K-means algorithm and the presence or absence of contrast enhancement. When contrast enhancement was applied, a factor of 2 was used. The L500 instance achieved the best performance. In all cases, increased contrast contributed to an improvement in the hit rate. This is possible because the number of regions of interest detected by KASE increases, generating more descriptors that can help better discriminate one leaf from another. Moreover, all instances performed better with more clusters. It is crucial to investigate whether increasing the contrast factor or the number of clusters may lead to an even greater gain in accuracy.

Table 5. Accuracy of KASE feature extractor for different image sizes.

Id	Clusters	Contrast	Accuracy		
			$L = 400$	$L = 500$	$L = 600$
#1	25	No	0.3103	0.3103	0.3060
#2	25	Yes	0.3190	0.3233	0.3621
#3	50	No	0.3017	0.3879	0.3190
#4	50	Yes	0.3362	0.4009	0.4009
#5	100	No	0.3491	0.3534	0.3534
#6	100	Yes	0.3621	0.3664	0.4397
#7	200	No	0.3836	0.3491	0.4181
#8	200	Yes	**0.4310**	**0.4612**	**0.4440**

Combined Attributes. To explore different image features, the attributes generated by GLCM, LBP, DAISY, and KASE were combined into a single feature vector. Tests were conducted on the validation set using the SVM classifier with the parameter $C = 30$. Several combinations were performed with each extractor configuration. However, to reduce the total number of solutions, a heuristic was followed. Initially, all pairs composed of GLCM and LBP attributes were tested. Then, the pair with the best accuracy was chosen for the next step, where it was combined with the remaining features (DAISY and KASE). The tests were executed again, and this time, the tuple with the highest hit rate was chosen as the base model. This process was performed for the L400, L500, and L600 instances, and the base models returned are named MB400, MB500, and MB600, respectively. Table 6 displays the chosen configuration of each attribute for the base models, ensuring maximum performance. The total attributes of the final feature vector are also recorded.

Table 6. Base models generated from concatenating GLCM, LBP, DAISY, and KASE attributes.

Model	L	GLCM	LBP	DAISY	KASE	Attributes
MB400	400	#1	#2	#6	#2	372
MB500	500	#2	#3	#17	#2	603
MB600	600	#2	#3	#6	#1	528

The best models on the validation were executed on the test set to obtain the final accuracy. Table 7 presents these accuracies and compares them with those achieved by the Cropped Base models from the work of [7]. In [7], the Cropped Bases were applied to the validation set. Therefore, here we rerun them on the test set for comparison purposes. Cropped Base also uses the main cropped leaf at a 300 × 300 resolution. It has variants with the InceptionV3, ResNet50, and EfficientNetB3 neural networks, subject to a dropout rate of 0.2, 0.5, or 0.8, as seen in parentheses in the Table 7 'Model' column. MB400 and MB600 models outperformed InceptionV3 with dropout rates of 0.2 and 0.5. MB500 only surpassed InceptionV3 with a dropout rate of 0.2. However, it is worth noting the significant reduction in the proposed models' size in MB. The base models are at least 10 times smaller than the CNN-base model, making them suitable for use on devices with limited storage and memory capacity.

4.2 RQ2: Can the Combination of Techniques Improve the Current State-of-the-Art?

The final test setup focuses on the ensemble of the proposed models and the Cropped Bases. Table 8 displays the accuracies of the ensemble models in the 'Model' column with MB400, MB500, MB600, and their combinations. It can be observed that the ensembles improved the accuracy of all Cropped Bases.

Table 7. Comparison of the accuracy of the base models with the Cropped Bases applied to the test set.

Model	Size (MB)	Accuracy
Cropped Base (InceptionV3/0.2)	187.55	0.5948
Cropped Base (InceptionV3/0.5)	187.55	0.6336
Cropped Base (InceptionV3/0.8)	187.55	0.7026
Cropped Base (ResNet50/0.2)	200.71	0.7629
Cropped Base (ResNet50/0.5)	200.71	0.7586
Cropped Base (ResNet50/0.8)	200.71	0.7414
Cropped Base (EfficientNetB3/0.2)	99.65	0.7759
Cropped Base (EfficientNetB3/0.5)	99.65	0.7716
Cropped Base (EfficientNetB3/0.8)	99.65	0.7759
MB400	8.87	0.6594
MB500	13.03	0.6034
MB600	9.68	0.6465

The integration with MB400 showed greater gains in the convolutional networks with dropout rates of 0.2 and 0.5. The network achieving the highest success rate was EfficientNetB3, with a dropout rate of 0.5, increasing from 77.16% to 81.90% accuracy when its predictions were combined with the outputs of the MB400 and MB500 models simultaneously. It is also noted that the union of base models can enhance performance, as is the case with MB400 and MB600, which produced a 68.10% accuracy together.

Figure 5 presents the confusion matrix of the ensemble of models MB400, MB500, and EfficientNetB3 (dropout rate of 0.5). A significant improvement in many varieties is observed. Cascade achieved 100% accuracy, and five other varieties had an accuracy of at least 85%. Despite some improvement, Nugget had the lowest performance, while Cluster remained among the worst.

4.3 Interpretability

This study used variable permutation to evaluate the importance of GLCM, LBP, KASE, and DAISY features for the MB400, MB500, and MB600 models. LBP was the most significant feature, contributing over 50% to the final prediction (See Fig. 6). GLCM also played a substantial role, while MB500 relied more on DAISY and MB600 on KASE. MB400 had minimal influence from KASE and DAISY. Analyzing the feature importance by class revealed potential bottlenecks and opportunities for improving the models, such as removing or replacing KASE attributes for MB400, increasing LBP configurations for MB500, and adjusting GLCM parameters for MB600.

Table 8. Accuracy of an ensemble of the proposed method and the Cropped Base models.

Model	Accuracy						
	MB400	MB500	MB600	MB400+MB500	MB400 + MB600	MB500 + MB600	MB400+ MB500 MB600
Cropped Base (InceptionV3/0.2)	0.7026	0.6940	0.7241	0.7414	0.7457	0.7371	0.7457
Cropped Base (InceptionV3/0.5)	0.7629	0.7328	0.7500	0.7457	0.7500	0.7543	0.7500
Cropped Base (InceptionV3/0.8)	0.7198	0.7328	0.7328	0.7543	0.7629	0.7629	0.7716
Cropped Base (ResNet50/0.2)	0.7845	0.7845	0.7759	0.7759	0.7888	0.7888	0.7845
Cropped Base (ResNet50/0.5)	0.7802	0.7931	0.7802	0.8103	0.7888	0.7802	0.7845
Cropped Base (ResNet50/0.8)	0.7543	0.7543	0.7500	0.7457	0.7543	0.7500	0.7629
Cropped Base (EfficientNetB3/0.2)	0.8147	0.8017	0.7931	0.8017	0.7974	0.8103	0.7931
Cropped Base (EfficientNetB3/0.5)	0.7845	0.8017	0.7974	0.8190	0.8017	0.8103	0.7888
Cropped Base (EfficientNetB3/0.8)	0.7716	0.7716	0.7802	0.7672	0.7888	0.7931	0.7845
MB400	-	0.6552	0.6810	-	-	-	-
MB500	0.6552	-	0.6509	-	-	-	-
MB600	0.6810	0.6509	-	-	-	-	-

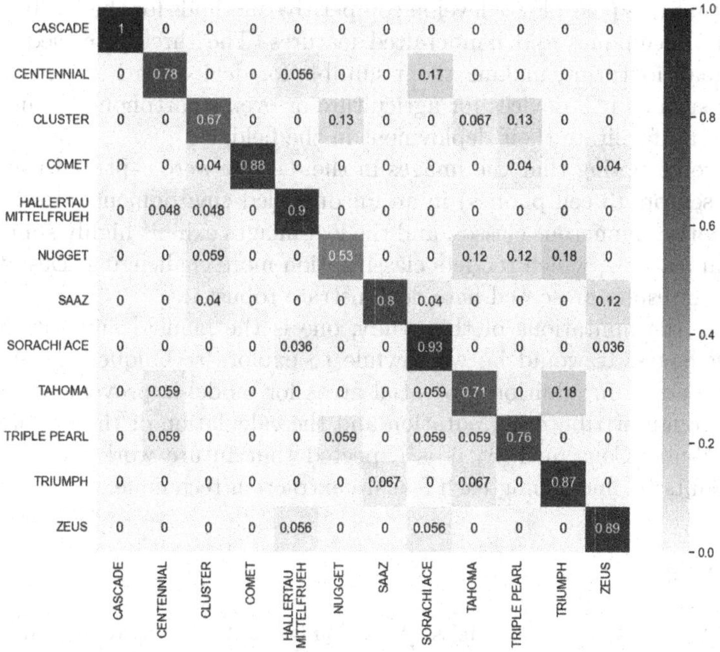

Fig. 5. Confusion matrix of ensemble of models MB400, MB500 and EfficientNetB3 (dropout of 0.5).

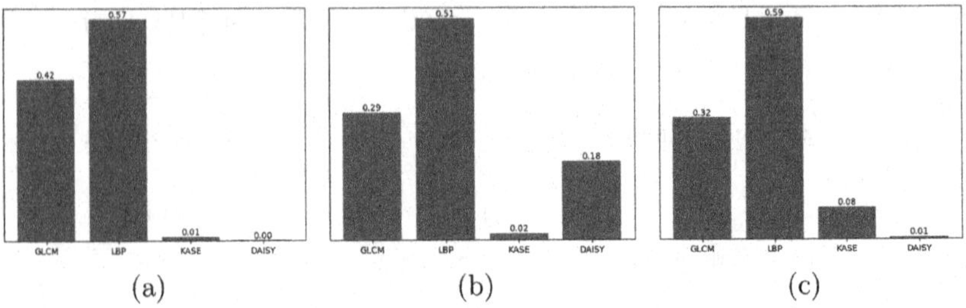

Fig. 6. Importance of GLCM, LBP, KASE, and DAISY features for each base model: (a) MB400; (b) MB500; (c) MB600.

5 Conclusion

In this study, we examine the problem of hop variety classification, comparing traditional feature extraction methods with deep learning approaches. We also introduce an ensemble method, achieving a new state-of-the-art for the UFOP-HVD Dataset (improving the accuracy from 77.16% to 81.90%). Our findings indicate that it is possible to develop competitive methods for this problem using traditional techniques and handcrafted features. The three proposed methods have a small footprint, making them suitable for devices with limited memory capacity, such as IoT devices for agriculture or even smartphones. This characteristic could facilitate their deployment in the field.

It is worth noting that the images in the dataset were captured using three different sensors (3 cell phones) in an uncontrolled environment. Moreover, the data comprises numerous classes, and the leaf images exhibit highly similar morphological features, which renders classification more challenging. Despite these factors, the results presented here demonstrate robustness.

Among the limitations of this study, one is the manual annotation of the bounding boxes. It would be worthwhile to explore techniques for automatic object detection. In addition, potential areas for model improvement were presented through attribute permutation and the calculation of the importance of each attribute. Consequently, it is expected that future work will yield even better results by fine-tuning each feature extraction technique.

References

1. Afifi, S.M., El-Mahis, A., Heiss, A.G., Farag, M.A.: Gas chromatography-mass spectrometry-based classification of 12 fennel (foeniculum vulgare miller) varieties based on their aroma profiles and estragole levels as analyzed using chemometric tools. ACS Omega **6**(8), 5775–5785 (2021)
2. Alcantarilla, P.F., Bartoli, A., Davison, A.J.: KAZE features. In: Fitzgibbon, A., Lazebnik, S., Perona, P., Sato, Y., Schmid, C. (eds.) ECCV 2012. LNCS, vol. 7577, pp. 214–227. Springer, Heidelberg (2012). https://doi.org/10.1007/978-3-642-33783-3_16

3. Astray, G., Gullón, P., Gullón, B., Munekata, P.E., Lorenzo, J.M.: Humulus lupulus L. As a natural source of functional biomolecules. Appl. Sci. **10**(15), 5074 (2020)
4. Azlah, M.A.F., Chua, L.S., Rahmad, F.R., Abdullah, F.I., Wan Alwi, S.R.: Review on techniques for plant leaf classification and recognition. Computers **8**(4), 77 (2019)
5. Bhattarai, U., Karkee, M.: A weakly-supervised approach for flower/fruit counting in apple orchards. Comput. Ind. **138**, 103635 (2022)
6. Castro, P., Luz, E., Moreira, G.: Dataset for hop varieties classification. Data Brief **38**, 107312 (2021)
7. Castro, P.H.N., Moreira, G.J.P., da Silva Luz, E.J.: An end-to-end deep learning system for hop classification. IEEE Latin America Trans. **20**(3), 430–442 (2021)
8. Chen, S., et al.: A renaissance in herbal medicine identification: from morphology to DNA. Biotechnol. Adv. **32**(7), 1237–1244 (2014)
9. Csurka, G., Dance, C., Fan, L., Willamowski, J., Bray, C.: Visual categorization with bags of keypoints. In: Workshop on Statistical Learning in Computer Vision, ECCV, pp. 1–2 (2004)
10. Dietterich, T.G.: Ensemble methods in machine learning. In: Kittler, J., Roli, F. (eds.) MCS 2000. LNCS, vol. 1857, pp. 1–15. Springer, Heidelberg (2000). https://doi.org/10.1007/3-540-45014-9_1
11. Duarte, L.M., Adriano, L.H.C., de Oliveira, M.A.L.: Capillary electrophoresis in association with chemometrics approach for bitterness hop (humulus lupulus L.) classification. Electrophoresis **39**(11), 1399–1409 (2018)
12. Fan, B., et al.: Development of an electrochemical technology for ten clematis SPP varieties identification. Int. J. Electrochem. Sci. **15**, 10212–10220 (2020)
13. Farag, M.A., Mahrous, E.A., Lübken, T., Porzel, A., Wessjohann, L.: Classification of commercial cultivars of humulus lupulus L. (hop) by chemometric pixel analysis of two dimensional nuclear magnetic resonance spectra. Metabolomics **10**(1), 21–32 (2014)
14. Ferri, G., Alù, M., Corradini, B., Beduschi, G.: Forensic botany: species identification of botanical trace evidence using a multigene barcoding approach. Int. J. Legal Med. **123**(5), 395–401 (2009)
15. Haralick, R.M., Shanmugam, K., Dinstein, I.H.: Textural features for image classification. IEEE Trans. Syst. Man Cybern. SMC **3**(6), 610–621 (1973)
16. He, K., Zhang, X., Ren, S., Sun, J.: Deep residual learning for image recognition. In: Proceedings of the IEEE Conference on Computer Vision and Pattern Recognition, pp. 770–778 (2016)
17. Healey, J.: The Hops List: 265 Beer Hop Varieties From Around the World (2016)
18. Jakab, A., Héberger, K., Forgács, E.: Comparative analysis of different plant oils by high-performance liquid chromatography-atmospheric pressure chemical ionization mass spectrometry. J. Chromatogr. A **976**(1–2), 255–263 (2002)
19. Jenks, M.A.: Plant nomenclature. Purdue University-Department of Horticulture and Landscape Architecture, Disponível (2011)
20. Killermann, W.: Research into biology teaching methods. J. Biol. Educ. **33**(1), 4–9 (1998)
21. Lotti, C., Ricciardi, L., Rainaldi, G., Ruta, C., Tarraf, W., De Mastro, G.: Morphological, biochemical, and molecular analysis of L. Open Agric. J. **13**(1) (2019)
22. Ojala, T., Pietikäinen, M., Harwood, D.: A comparative study of texture measures with classification based on featured distributions. Pattern Recogn. **29**(1), 51–59 (1996)

23. Ojala, T., Pietikainen, M., Maenpaa, T.: Multiresolution gray-scale and rotation invariant texture classification with local binary patterns. IEEE Trans. Pattern Anal. Mach. Intell. **24**(7), 971–987 (2002)
24. Shellie, R.A., Poynter, S.D., Li, J., Gathercole, J.L., Whittock, S.P., Koutoulis, A.: Varietal characterization of hop (humulus lupulus L.) by GC-MS analysis of hop cone extracts. J. Sep. Sci. **32**(21), 3720–3725 (2009)
25. Steenackers, B., De Cooman, L., De Vos, D.: Chemical transformations of characteristic hop secondary metabolites in relation to beer properties and the brewing process: a review. Food Chem. **172**, 742–756 (2015)
26. Szegedy, C., Vanhoucke, V., Ioffe, S., Shlens, J., Wojna, Z.: Rethinking the inception architecture for computer vision. In: Proceedings of the IEEE Conference on Computer Vision and Pattern Recognition, pp. 2818–2826 (2016)
27. Tan, M., Le, Q.: EfficientNet: rethinking model scaling for convolutional neural networks. In: International Conference on Machine Learning, pp. 6105–6114. PMLR (2019)
28. Tola, E., Lepetit, V., Fua, P.: DAISY: an efficient dense descriptor applied to wide-baseline stereo. IEEE Trans. Pattern Anal. Mach. Intell. **32**(5), 815–830 (2009)
29. Yang, J., Jiang, Y.G., Hauptmann, A.G., Ngo, C.W.: Evaluating bag-of-visual-words representations in scene classification. In: Proceedings of the International Workshop on Workshop on Multimedia Information Retrieval, pp. 197–206 (2007)

Feature Selection and Hyperparameter Fine-Tuning in Artificial Neural Networks for Wood Quality Classification

Mateus Roder[1], Leandro Aparecido Passos[1(✉)], João Paulo Papa[1], and André Luis Debiaso Rossi[2]

[1] Department of Computing, São Paulo State University, Av. Eng. Luiz Edmundo Carrijo Coube, 14-01, Bauru 17033-360, Brazil
{mateus.roder,leandro.passos,joao.papa}@unesp.br

[2] Department of Production Engineering, Paulo State University, Rua Geraldo Alckmin, 519 - Vila Nossa Sra. de Fatima, Itapeva 18409-010, Brazil
andre.rossi@unesp.br

Abstract. Quality classification of wood boards is an essential task in the sawmill industry, which is still usually performed by human operators in small to median companies in developing countries. Machine learning algorithms have been successfully employed to investigate the problem, offering a more affordable alternative compared to other solutions. However, such approaches usually present some drawbacks regarding the proper selection of their hyperparameters. Moreover, the models are susceptible to the features extracted from wood board images, which influence the induction of the model and, consequently, its generalization power. Therefore, in this paper, we investigate the problem of simultaneously tuning the hyperparameters of an artificial neural network (ANN) as well as selecting a subset of characteristics that better describes the wood board quality. Experiments were conducted over a private dataset composed of images obtained from a sawmill industry and described using different feature descriptors. The predictive performance of the model was compared against five baseline methods as well as a random search, performing either ANN hyperparameter tuning and feature selection. Experimental results suggest that hyperparameters should be adjusted according to the feature set, or the features should be selected considering the hyperparameter values. In summary, the best predictive performance, i.e., a balanced accuracy of 0.80, was achieved in two distinct scenarios: (i) performing only feature selection, and (ii) performing both tasks concomitantly. Thus, we suggest that at least one of the two approaches should be considered in the context of industrial applications.

Keywords: Artificial Neural Network Optimization · Wood Quality Classification · Sawmill Problem · Hyperparameter Tuning · Feature Selection

The authors are grateful to FAPESP grants #2016/06538-0, #2018/02822-1 and #2019/07825-1.

M. C. Naldi and R. A. C. Bianchi (Eds.): BRACIS 2023, LNAI 14196, pp. 323–337, 2023.
https://doi.org/10.1007/978-3-031-45389-2_22

1 Introduction

Industries have experienced many technological advances in recent years, resulting in more complex processes, systems, and products. As a consequence, the management of integrated manufacturing processes and operation analyses are crucial to delivering high-quality products to clients. In this scenario, the quality of raw materials is also paramount for high quality manufactured products. However, imperfection detection, as well as quality classification of raw materials, such as woods in sawmill companies, usually is still performed by trained human operators [1]. Notwithstanding, the process is inherently subjective, since it is a visual analysis, and these experts may suffer from fatigue after a long working period performing repetitive activities. Consequently, it is expected the increased number of incorrect classifications [25].

These disadvantages stimulated the scientific community towards the implementation of visual inspection systems, aiming to perform defect and quality classification autonomously. Therefore, this work focuses exclusively on Machine Learning (ML) techniques that have been successfully employed for these tasks on wood boards [6]. Recently, some studies investigated the performance of different ML techniques to classify the quality of wood surface [17,21]. These studies have used data generated from images captured in a real sawmill company and classified by a specialist in three levels of quality, according to the company rules, i.e., zero defect is found in the wood piece (A), only small defects, such as knots, are found (B), and defects that compromise the quality of the product, such as groups of knots and exposed pith (C). Figure 1 depicts some examples of wood images classified at each level.

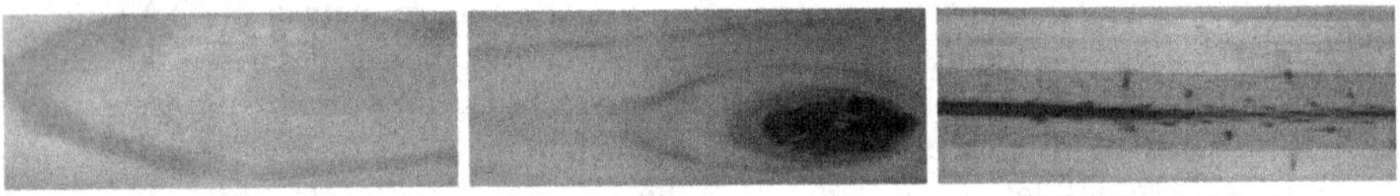

Wood quality A. Wood quality B. Wood quality C.

Fig. 1. Three different qualities of wood boards (A, B, and C) according to company's rule.

Despite the success obtained in these studies, each ML algorithm has its inherent tendency towards data specificity, which influences the model's induction and, thus, its predictive performance. Therefore, one can adjust such tendencies through a proper hyperparameter (HP) selection. The task of finding the best HP values is known as *hyperparameter tuning* and usually aims at improving the model's predictive performance while keeping the model as simple as possible. Although some HP values may fit sufficiently well different kinds of problems, it is a common practice to search for the hyperparameter that provides the best solutions concerning each problem at hand [17,21].

Besides, another challenge in the context of this work is extracting the images' more representative features, i.e., the features that best describes the problem, since the more descriptive they are, the higher the effectiveness of the technique. Regarding classification tasks, features are usually extracted through image descriptors, such as statistical measures from the Gray Level Co-occurrence Matrix (GLCM) [8] and Local Binary Patterns (LBP) [13]. However, many of these features may be correlated to each other or do not add any relevant information for the ML technique. Therefore, the process of selecting a subset of these features, referred to as feature selection (FS), can be applied to select the most descriptive ones. Moreover, since features are specific for each problem, FS has to be carried out for each data set separately.

Considering that both the HP tuning and the task of feature selection relies upon each other, they should be performed simultaneously to generate more robust models, concerning generalization purposes. In general, ML techniques require tuning more than one HP, since the tackled problems generally are described by many features, thus implying on large search spaces. In this context, metaheuristics approaches are commonly employed to solve such problems by randomly initializing a collection of candidate solutions, which interact among themselves and perform a directed exploration of the search space toward the results that best fit a desirable target function with an acceptable computational cost. Such approaches are commonly employed to solve problems related to ML techniques hyperparameter tuning [5,14,19] and feature selection [15], among others [4,22].

In this paper, we investigate the problem of FS and Artificial Neural Network (ANN) hyperparameter tuning applied in the context of wood boards quality classification. Experiments were carried out using the population-based metaheuristic Particle Swarm Optimization (PSO) [9,20] to simultaneously perform both tasks over a Multilayer Perceptron (MLP) ANN. Moreover, the results compared against five distinct baselines, as well as a random search, confirms the relevance of the proposed approach. We hypothesize that the predictive performance of ANN models can be improved since they depend on the HP values and the set of features used to describe the problem.

Therefore, the main contributions of this paper are twofold: (i) to propose a method capable of simultaneously selecting the hyperparameters that best performs over an MLP network as well as selecting the subset of features that best describe each image sample, and (ii) to foster the scientific community regarding material and wood quality classification. The remainder of this paper is presented as follows. Section 2 defines the problem of hyperparameter tuning and feature selection, and provides a brief description of some related works. Section 3 presents the main concepts of ANN and PSO. The experimental methodology employed to evaluate the effects of FS and MLP hyperparameter tuning over the models' performance is described in Sect. 4. Results are presented and discussed in Sect. 5, and finally conclusions are presented in Sect. 6.

2 Problem Definition and Related Work

Hyperparameter tuning and feature subset selection are two widely employed tasks carried out in the data mining context, aiming to improve models' predictive performance as well as simplifying them. Therefore, this section formalizes the problem of simultaneously performing these two tasks. Besides, it presents an overview of studies related to wood quality classification and the importance of HP tuning and FS.

2.1 Problem Definition

The problem investigated in this paper consists of tuning the HP values of an MLP Artificial Neural Network algorithm, as well as selecting a subset of features that are relevant for the problem of wood quality classification towards the improvement of the models' predictive performance.

Let A be an MLP algorithm that comprises the hyperparameter space Λ. For each hyperparameter setting $\lambda \in \Lambda$, let A_λ represent the learning algorithm A that employs the hyperparameter setting λ. Also, consider $D = \{(x_1, y_1), (x_2, y_2), \dots, (x_n, y_n)\}$ a dataset composed of n instances, such that $\boldsymbol{x} \in \mathbb{R}^m$ is a feature vector and y is the target value. Moreover, one can define κ as the subset of feature from $\boldsymbol{x}$. Finally, let A^κ be an algorithm A trained with a subset of features κ [12].

Therefore, the main goal of HP tuning is to finding $\lambda^* = \arg\min_{\lambda \in \Lambda} M(A_\lambda, D) \in \Lambda$ that minimizes some loss function, such as the misclassification rate using the algorithm A over instances not used for training purposes. Moreover, one can estimate the misclassification rate $M(A, D)$ achieved by A when trained and tested on D through a stratified multi-fold cross-validation resampling method.

Similarly to HP tuning, the goal of feature subset selection is to find $\kappa^* = \arg\min_{\kappa \subseteq \boldsymbol{x}} M(A^\kappa, D) \subseteq \boldsymbol{x}$ which minimizes the loss function achieved by A when trained on D.

Therefore, the aim of combining hyperparameter tuning and feature subset selection is to find the hyperparameter setting $\lambda^* \in \Lambda$ and the feature subset $\kappa^* \subseteq \boldsymbol{x}$ which has the lowest misclassification rate among all HP settings and features subsets, i.e., $(\kappa^*, \lambda^*) = \arg\min_{\lambda \in \Lambda, \kappa \subseteq \boldsymbol{x}} M(A_\lambda^\kappa, D)$.

2.2 Related Work

One of the first studies to investigate the problem of wood quality control using an automated visual inspection system based on machine learning techniques was accomplished by [16]. Since then, many others have investigated this problem aiming to improve predictive performance by analyzing and selecting different features, as well as optimizing HP values of ML techniques [21,24].

Tiryaki et al. [24] employed an ANN for modeling the wood surface roughness in the machining process. The study highlights some variables that have impact and influence in the surface roughness, such as wood species, the feed rate, the number of the cutter, and the cutting depth. The model's predictive performance was good enough to allow its application in the wood industry in order to optimize effort, time, and energy.

Others addressed the problem of combining FS and HP tuning for ML techniques over different applications. As previously mentioned, some techniques are more sensitive to HP tuning and FS than others. Besides, some optimization methods, such as metaheuristic approaches based on evolutionary algorithms and swarm intelligence [10], for instance, have successfully accomplished the task.

In this context, Roder et al. [21] used the PSO algorithm to tune the HP of an ANN applied for wood quality classification aiming to enhance the model's predictive performance. The authors employed the GLCM [8] to extract features from the same dataset used by [2], which is composed of five statistical measures for two angles, i.e., 0 and 90°: entropy, energy, maximum intensity, inverse difference moment, and correlation. Experimental results obtained up to 6% of accuracy gain concerning the ANN classification, and corroborate the necessity of tuning the ANN hyperparameters; stating the efficiency of PSO for such a task.

3 Theoretical Background

This section briefly introduces the main concepts of the techniques employed in this work, i.e., the Multilayer Perceptron neural network and the Particle Swarm Optimization algorithm, as well as the process of wood image feature extraction.

3.1 MultiLayer Perceptron ANN

MLPs are composed of an input and an output layer, as well as one or more hidden layers of neurons, which can be fully or partially connected. A neural network is called fully connected when each neuron from a given layer is connected to all neurons of the next one. Similarly, it is said to be partially connected when some neurons of adjacent layers are not connected. These connections are represented by a weight matrix, which is usually adjusted through gradient-based learning algorithms. With a single hidden layer, an MLP is capable of representing a large number of functions, thus sufficient for the purpose of this study. Figure 2 depicts the model architecture.

The conventional algorithm to train an MLP network is the backpropagation, which is composed of the forward pass, which exposes the input data to a series of linear operations followed by non-linear activations, and the backward phases, which is responsible for propagating the output error and update the network weights.

3.2 Particle Swarm Optimization

Particle Swarm Optimization [11] is a global optimization technique based on the social behavior of birds, fishes, and insects, among others. The method comprises

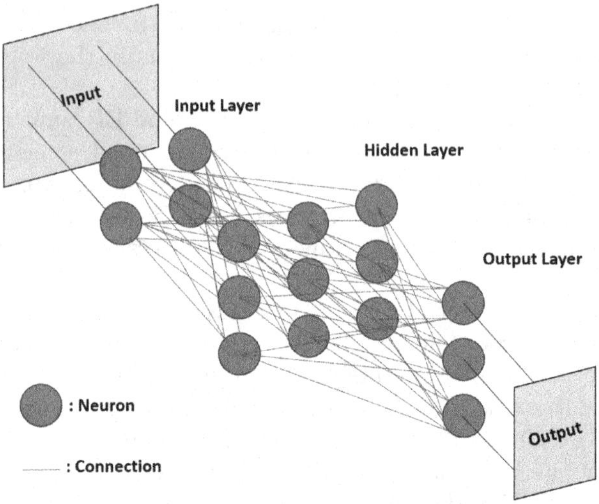

Fig. 2. MultiLayer Perceptron representation.

a swarm composed of a set of individuals capable of sharing information among themselves concerning their positions in the search space, as well as the relative quality of this position, denoted by a fitness function.

In short, each PSO particle is represented by its current position, velocity, and the best position found during the training process. The position of a particle i is represented by a point in a D-dimensional space, given by $\Psi_i = \{\psi_{i1}, \psi_{i2}, \ldots, \psi_{iD}\}$. Further, the particle velocity is defined by $\boldsymbol{v}_i = \{v_{i1}, v_{i2}, \ldots, v_{iD}\}$, and finally the best position found by this particle is represented by $\boldsymbol{p}_i = \{p_{i1}, p_{i2}, \ldots, p_{iD}\}$. Besides, the best position found among all particles is represented by $\boldsymbol{p}_g$.

A particle will move in a particular direction depending on its current position, velocity, and best position. Additionally, it also depends on the best position found by the other particles in the swarm. Therefore, the position of a particle $\psi_{ij}(t+1) = \psi_{ij}(t) + v_{ij}(t)$ is computed for each dimension $j \in \{1, 2, \ldots, D\}$ at time step t. Further, the velocity $v_{ij}(t)$ is updated using the following equation:

$$v_{ij}(t+1) = \Gamma \cdot v_{ij}(t) + \varphi_1 \cdot r_1 \cdot (p_{ij} - x_{ij}(t)) + \varphi_2 \cdot r_2 \cdot (p_{gd} - x_{ij}(t)), \qquad (1)$$

where Γ denotes the inertial weight, introduced by [23] to balance the global and local search, r_1 and r_2 are two independent values uniformly distributed in the range $[0, 1]$, and φ_1 and φ_2 are acceleration constants.

Such a representation of a particle is suitable for hyperparameter fine-tuning, which considers real- and integer-valued numbers. However, it is not adequate for feature selection tasks since it requires a categorical or binary representation. Therefore, this work employs a variation of the method, adapted for feature selection, as discussed in Sect. 4.

3.3 Feature Extractors

Haralick et al. [8] proposed a set of mathematical tools to extract statistical features from images using a Gray Level Co-occurrence Matrix. The GLCM is capable of describing the frequency of occurrences in grayscale transitions for an image in a pixel-by-pixel fashion. Further, the features are extracted considering the relationship of each pixel with its neighbors over four different angles, i.e., 0, 45, 90, and 135°, where 0 and 90^circ are the most employed ones. Moreover, the model is capable of extracting a total of 14 statistical features, named Angular Second Moment, Contrast, Correlation, Variance, Inverse Difference Moment, Sum Average, Sum Variance, Sum Entropy, Entropy, Difference Entropy, Information Measures of Correlation (1 and 2) and the Maximal Correlation Coefficient.

Another well-known texture descriptor is the Local Binary Pattern [13], which converts the image to a gray-scale level and performs a pixel-by-pixel comparison over the entire image considering a selected number of neighbors. In this comparison, the central pixel of the square formed by its neighbor assumes the value 1 if it is greater or equal to its neighbors, or 0 otherwise. Such value is stored in an array, which is further employed for converting the binary intensity code to a decimal value for the pixel at hand.

Afterward, the same computation is performed for all pixels compounding the image. Further, it is generated a histogram of the distribution of the values, which will compose the final vector describing the image. Notice the method was initially proposed with the number of neighborhood fixed in 3×3. Later, some changes allowed LBP to deal with a larger number of neighborhoods, resulting in a non-square structure, usually employing a circular pattern since it only requires the definition of the radius, instead of the NxN arrangement.

4 Methodology

This section presents the methodology concerning the material and methods employed during the experiments. It briefly describes the datasets, the process of feature extraction, the modeling of the hyperparameter fine-tuning and feature selection processes using PSO, the methods used for evaluation purpose, and the baselines considered for comparison.

4.1 Dataset

This work employs a dataset D composed of features extracted from 374 instances of wood board images obtained in a Brazilian sawmill [25]. As stated in the problem definition, each instance $D = (\boldsymbol{x}_i, y_i)$ is composed of an m-dimensional feature vector $\boldsymbol{x}_i \in \mathbb{R}^m$, whose features were extracted using both GLCM and LBP, as described in Sect. 4.2, and a target value y_i, denoting the wood quality. Each sample's target value is established according to rules defined by the sawmill company, where "A" stands for a high-quality standard and comprises 144 instances, "B" denotes an intermediate quality and comprises 177 instances, and "C" represents lower quality, comprising 53 samples.

4.2 Image Texture Descriptors

The feature set was obtained by joining the features extracted from two texture descriptors, namely, statistical measures extracted from GLCM and the LBP. While the statistical measures have the advantage of enabling the interpretation and comprehension of the image characteristics through different measures, LBP is robust in the treatment of gray-scale images, with a good performance for scale changes caused by illumination [13].

Concerning the GLCM, this paper employed 0° and 90° to extract six measures: angular second momentum, energy, contrast, correlation, dissimilarity, and homogeneity, resulting in 12 characteristics for each image. Besides, LPB uses 24 neighbors as well as a radius of size 3, resulting in 26 characteristics for each image. Thus, joining GLCM and LBP features resulted in 38 predictive attributes.

4.3 MLP Hyperparameter Tuning and Feature Selection Using PSO

This work employed a fully connected network composed of a single hidden layer for the task of classification. Further, it also employed the PSO algorithm to fine-tuning the three principal hyperparameters of the model, namely the number of neurons in the hidden layer γ, the learning rate η, and the momentum term μ.

As mentioned previously, the PSO algorithm was employed to perform a combined task of hyperparameter tuning and feature selection, hereafter referred to as HP-FS-PSO. Considering the former task, PSO decision variables are modeled admitting one integer values to represent the number of units in the hidden layer, as well as two real numbers to represent the learning rate and momentum.

Further, since the task of feature selection does not assume continuous representation, PSO requires some modifications to work properly in this context. The main change is made on the position representation, which must be treated as a result of probability analysis from particles' velocity to decide what features are relevant to the context [9]. Therefore, considering the task of feature selection, each particle's decision variable is represented by a binary value, where 1 means to consider a feature whereas 0 means to discard a feature.

Since the particle's decision variables employed for feature selection assume binary values, it is necessary to binarize each position, ψ_{ij} such that $\psi_{ij} = 1$ if $s(v_{ij}) > r_3$ and 0 otherwise. Notice r_3 is the threshold, a real number generated randomly in the range $[0, 1]$, and $s(\cdot)$ is the logistic function.

Concerning the MLP hyperparameters' search configuration, the hidden layer size is optimized in the range $[2, 60]$, the learning rate and momentum assume values in the range $[0, 1]$, and the feature subset selection is defined by a binary variable, i.e., assuming either 0 or 1. Additionally, it also presents the MLP hyperparameter default values used by Weka [7]. The number of neurons in the hidden layer is defined as $\gamma = \frac{(\text{NA+NC})}{2}$, where NA is the number of attributes and NC is the number of classes. Thus, the default value for our problem, which has 38 attributes and 3 classes, is 20 neurons.

Further, the PSO algorithm also has its own hyperparameters, defined as follows: Number of Particles $N = 30$, Acceleration Constant 1 $\varphi_1 = 1.494$, Acceleration Constant 2 $\varphi_2 = 1.494$, Inertia Weight $\omega = 0.729$, and Maximum Velocity υ. Notice tuning such hyperparameters would lead to a "never-ending" problem. Therefore, these values were empirically selected based on similar works [9,21]. Besides, the maximum velocity υ varies according to the upper limit of its respective hyperparameter.

The optimization process is performed until meeting the stop criterion, i.e., the maximum number of iterations, which was set in 300.

4.4 Evaluation

Metaheuristic algorithms guide their search for the best solutions according to the outcome of a fitness function. In this work, this measure is obtained from the predictive performance of the ANNs over a validation subset.

Moreover, since the dataset used in this work is composed of an imbalanced class distribution, i.e., there is a considerably smaller number of examples labeled as "C" class. Therefore, measures that acknowledge this imbalance are more suitable for the task. Thus, in this study, we considered the Balanced Accuracy (BAC) [3].

The ANN training and evaluation were performed using a nested stratified k-fold cross-validation (CV) re-sampling method. Such an approach splits the dataset into $k = 10$ partitions, where one of them is used to test, and the remaining folds are employed for training purposes. In this context, the training folds are used to train the model during the optimization process, i.e., finding the best MLP hyperparameters and the best subset of features. Therefore, PSO assesses the average BAC considering the fitness value over the validation set. Further, the best set of hyperparameters and features found in this process is them applied to train the model and induct the prediction of the testing set samples' labels. Such a process guarantees that the data used to evaluate the model is never used in the model training steps and, consequently, in the MLP hyperparameter tuning and feature selection processes.

Finally, due to the stochastic process of PSO, the optimization process was repeated during 10 runs, aiming to perform a statistical analysis through the Wilcoxon signed-rank test [26] with 5% of significance.

4.5 Baselines

In order to evaluate and compare the results obtained by PSO, six baselines methods were compared in the context of the combined MLP hyperparameter tuning and feature selection:

- Method 1 **(M1)** : Default hyperparameter values defined by Weka and the whole set of features;
- Method 2 **(M2)** : MLP hyperparameter tuned using PSO and the whole set of features;

- Method 3 **(M3)** : Default MLP hyperparameter values defined by Weka and feature subset selected by PSO;
- Method 4 **(M4)** : Default MLP hyperparameter values defined by Weka and dimensionality reduction performed by Principal Components Analysis (PCA);
- Method 5 **(M5)** : MLP hyperparameter values tuned by PSO and dimensionality reduction performed by PCA;
- Random Search **(RS)** : Random selection of MLP hyperparameter values and feature subset. This approach considered the same number of solutions evaluated by PSO.

The methodology adopted in this work aims at analyzing the PSO performance from different x perspectives. First, M1 is the baseline for both tasks, i.e., the MLP hyperparameter using default parameters provided by Weka considering the whole set of features. Further, M2 allows analyzing PSO influence for the task of MLP hyperparameter tuning, with no feature selection, while M3 investigates the opposite, i.e., PSO influence to the task of feature selection with using MLP default hyperparameters. Moreover, M4 employs the default hyperparameter with a dimensionally reduced feature subset performed by PCA, while M5 combines MLP hyperparameters tuning using PSO with PCA. Finally, RS represents the analysis of random combinations of MLP hyperparameter values and selected features.

The experiments carried out in this work were coded using Python[1] and R [18]. Further, the feature extraction task was implemented in Python using the Scikit-image[2] package, while the MLP network was developed using the RWeka package in R, which is an interface to Weka. Finally, PSO was also implemented in R.

5 Experimental Results

This section presents the predictive performance of the ANNs assessed for the HP-FS-PSO method, as well as the baselines techniques. Notice the results are also provided with the p-values considering the Wilcoxon signed-rank test compared to the HP-FS-PSO method as the reference for statistical analysis purposes. Further, values presented in bold stand for the most accurate result overall.

5.1 Optimization Evaluation

Regarding the optimization performance over the validation set, one can observe in Table 1 that HP-FS-PSO and M3 have obtained the best results, achieving a BAC average of 0.850. Such techniques provided an improvement of around 10% compared to M1, which represents an ANN using default HP values and

[1] https://www.python.org/.
[2] http://scikit-image.org/.

the whole set of features. Therefore, the most important finding from these results is that the optimization of MLP hyperparameters, and even more a proper selection of the more suitable features, have a strong influence in the induction of MLP models applied to wood boards quality classification. On the other hand, results also show that performing only one task may also be enough to increase performance since M3 performed only the task of feature selection, while M2, which obtained an average BAC of 0.834, performing only the task of MLP hyperparameter tuning. Although by a narrow margin, in this case, the set of features was more expressive for the model predictive performance than tuning the network hyperparameters.

This behavior suggests that a user should, at least, select a set of features for the hyperparameter values defined *a priori*. Besides the BAC values, Table 1 also provides the p-values considering the Wilcoxon signed-rank test compared to the HP-FS-PSO method as the reference. These p-values support our previous observations, considering only M3 and the random search obtained a significance level higher than $\alpha = 0.05$.

Table 1. Average BAC and the standard deviation concerning the task of MLP hyperparameter tuning and feature selection, considering the validation set over 10 executions. Notice the "p-values" are compared against the HP-FS-PSO reference.

	Validation (BAC± SD)	p-value
M1	0.752 ± 0.008	0.002
M2	0.834 ± 0.015	0.008
M3	**0.850 ± 0.015**	0.557
M4	0.739 ± 0.011	0.002
M5	0.825 ± 0.022	0.006
RS	**0.845** ± 0.012	0.131
HP-FS-PSO	**0.850** ± 0.011	Ref

Notice the positive behavior of the random search approach, which is somehow expected since the model presents itself as more sensitive to a proper selection of the features instead of the network hyperparameter tuning, which is expected to be a more straightforward task due to the binary nature of the search space. Further, the data dimensionality reduction performed by PCA showed to be inadequate for this problem, as denoted by methods M4, which obtained BAC average results lower than using a default configuration, and M5. The main reason lies in the fact that PCA may not be able to describe sufficiently well the problem due to its linear nature.

For a better understanding of each method's behavior during PSO convergence, Fig. 3 depicts the optimization performance (a) and the evolution of the BAC values (b) during 300 iterations. Figure 3 (a) considers the average values over 10 runs, where each iteration in the RS curve reflects the average evaluation among 30 executions, i.e., the same number of assessments performed by PSO considering 30 particles. Finally, M1 and M4 are represented by fixed lines since no optimization was performed over such approaches. Figure 3(b) corroborates our claim that HP-FS-PSO performed a guided search through the MLP hyper-parameters and features spaces, improving its performance over the iterations. Notice PSO can reduce the number of iterations required for finding reasonable BAC values since it obtained relatively high accuracies (around 0.840) after 80 iterations only.

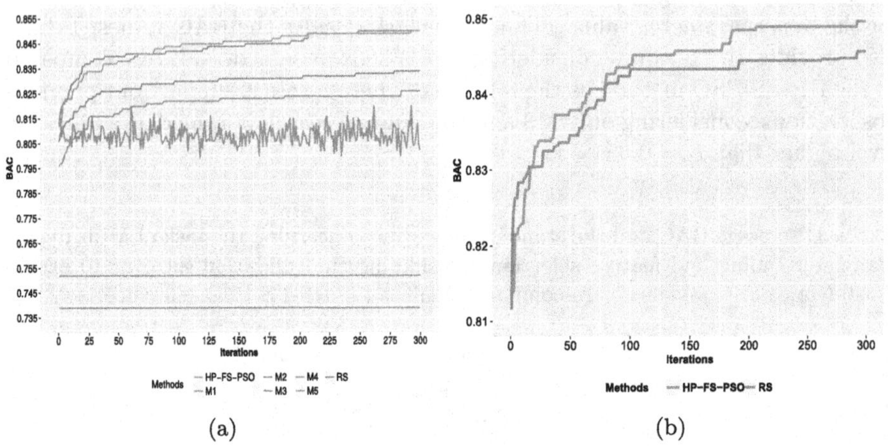

Fig. 3. PSO convergence considering the evaluation dataset (a) and evolution of the BAC values of HP-FS-PSO and RS. In this case, the best BAC value found by RS up to a iteration is kept in the next iterations (b).

Besides, Fig. 3(b) depicts the HP-FS-PSO performance compared against a random search considering the best results over each iteration, instead of an average. One can observe the random search performed slightly better during the first 30 iterations. Afterward, the HP-FS-PSO surpassed RS and kept this advantage until reaching the 300 iterations. As previously mentioned, it is possible to note that there is a considerable improvement of BAC for both methods in the first 100 iterations, and then there is a slowdown in the BAC growth. Such a piece of information is of extreme relevance for industrial applications, since it may save time and effort during the task of tuning the model's hyperparameter selecting the best subset of features.

5.2 Classification

This section investigates the ANN generalization power by evaluating the predictive performance of the model over the testing set, considering the best set of hyperparameters and a subset of features found during the optimization process. Table 2 presents the BAC values obtained in this context. Notice the values presented in bold stand for the most accurate approach overall.

Table 2. Average BAC and the standard deviation concerning the task of MLP wood quality classification considered the testing samples and the best set of hyperparameters and subfeatures found during the optimization process for each approach. Notice the "*p*-values" are compared against the HP-FS-PSO reference.

	Test (BAC± SD)	p-value
M1	0.755 ± 0.077	0.275
M2	0.779 ± 0.042	0.547
M3	**0.798** ± 0.087	0.722
M4	0.705 ± 0.057	0.010
M5	0.773 ± 0.071	0.232
RS	0.745 ± 0.080	0.131
HP-FS-PSO	0.794 ± 0.062	Ref

In general, the results of test data are in agreement with those of the validation set. The baseline method M3 led to the best BAC value, followed closely by HP-FS-PSO (differing only in the third decimal place). These performances are again superior to M1, supporting the idea of MLP hyperparameter tuning and feature selection influence.

However, differently from Table 1, HP-FS-PSO was not statistically better than M3, as observed in the p-value > 0.05. Therefore, the improvement obtained during the validation steps was not enough to provide a statistical difference, considering the test data. The high standard deviation over the testing set explains such behavior, which was considerably smaller regarding the optimization steps.

6 Conclusion

This paper analyzed the compound problem of ANN hyperparameters tuning and feature selection for the quality classification of wood boards in the sawmill industry. Experiments showed that a solution based on PSO led to satisfactory results compared to baseline methods. According to a statistical test, results show a significant difference during the optimization task but not for the generalization phase.

These experimental results suggest that MLP hyperparameter tuning and feature selection are essential to obtain models with higher predictive performance. Also, one can notice that these tasks are interdependent since the hyperparameter values should be adjusted according to a subset of features and vice versa. Consequently, for the problem investigated in this work, performing only one of them was enough to reach substantial gain. Finally, the accuracy obtained in this study supports employing machine learning models for industrial implementation, contributing to overall cost reduction and improvement in competitiveness. Regarding future works, we intend to perform a transfer learning from a CNN trained using a dataset composed of a more substantial number of wood image samples. Besides, we are willing to investigate and compare different image descriptors, non-linear data reduction techniques, and deep learning models.

Acknowledgments. The authors are grateful to FAPESP grants #2016/06538-0, #2018/02822-1, #2019/07825-1, and #2023/10823-6

References

1. Abdullah, A., Ismail, N.K.N., Kadir, T.A.A., Zain, J.M., Jusoh, N.A., Ali, N.M.: Agar wood grade determination system using image processing technique. In: Proceedings of the International Conference on Electrical Engineering and Informatics Institut Teknologi Bandung (2007)
2. Affonso, C., Rossi, A.L.D., Vieira, F.H.A., Carvalho, A.C.P.L.F.: Deep learning for biological image classification. Expert Syst. Appl. **85**, 114–122 (2017). https://doi.org/10.1016/j.eswa.2017.05.039
3. Brodersen, K.H., Ong, C.S., Stephan, K.E., Buhmann, J.M.: The balanced accuracy and its posterior distribution. In: 2010 20th International Conference on Pattern Recognition, pp. 3121–3124, August 2010. https://doi.org/10.1109/ICPR.2010.764
4. Cao, Y., et al.: A new intelligence fuzzy-based hybrid metaheuristic algorithm for analyzing the application of tea waste in concrete as natural fiber. Comput. Electron. Agric. **190**, 106420 (2021)
5. De Souza, L.A., et al.: Fine-tuning generative adversarial networks using metaheuristics-a case study on Barrett's esophagus identification. In: Bildverarbeitung für die Medizin, pp. 205–210 (2021)
6. Gu, I.Y.H., Andersson, H., Vicen, R.: Automatic classification of wood defects using support vector machines. In: Bolc, L., Kulikowski, J.L., Wojciechowski, K. (eds.) ICCVG 2008. LNCS, vol. 5337, pp. 356–367. Springer, Heidelberg (2009). https://doi.org/10.1007/978-3-642-02345-3_35
7. Hall, M., Frank, E., Holmes, G., Pfahringer, B., Reutemann, P., Witten, I.H.: The WEKA data mining software: an update. SIGKDD Explor. Newsl. **11**(1), 10–18 (2009). https://doi.org/10.1145/1656274.1656278
8. Haralick, R., Shanmugam, K., Distein, I.: Textual features for image classification. IEEE Trans. Syst. Man Cybern. SMC **3**(6), 610–621 (1973)
9. Kennedy, J., Eberhart, R.C.: A discrete binary version of the particle swarm algorithm. In: 1997 IEEE International Conference on Systems, Man, and Cybernetics. Computational Cybernetics and Simulation, vol. 5, pp. 4104–4108, October 1997. https://doi.org/10.1109/ICSMC.1997.637339
10. Kennedy, J., Eberhart, R.: Swarm Intelligence. Morgan Kaufmann Publishers (2001)
11. Kennedy, J., Eberhart, R.: Particle swarm optimization. In: Proceedings of the IEEE International Conference on Neural Networks, vol. 4, pp. 1942–1948. Perth, Australia (1995)

12. Luo, G.: A review of automatic selection methods for machine learning algorithms and hyper-parameter values. Netw. Model. Anal. Health Inform. Bioinforma. **5**(1), 1–16 (2016). https://doi.org/10.1007/s13721-016-0125-6
13. Ojala, T., Pietikainen, M., Harwood, D.: Comparative study of texture measures with classification based on feature distributions. Pattern Recogn., 51–59 (1996)
14. Passos, L.A., Paulo Papa, J.: Fine-tuning infinity restricted Boltzmann machines. In: 2017 30th SIBGRAPI Conference on Graphics, Patterns and Images (SIBGRAPI), pp. 63–70. IEEE (2017)
15. Pereira, C.R., Passos, L.A., Rodrigues, D., de Souza, A.N., Papa, J.P.: JADE-based feature selection for non-technical losses detection. In: Tavares, J.M.R.S., Natal Jorge, R.M. (eds.) VipIMAGE 2019. LNCVB, vol. 34, pp. 141–156. Springer, Cham (2019). https://doi.org/10.1007/978-3-030-32040-9_16
16. Pham, D.T., Alcock, R.J.: Automatic detection of defects on birch wood boards. Proc. Inst. Mech. Eng. Part E J. Process Mech. Eng. **210**(1), 45–52 (1996). https://doi.org/10.1243/0954408991529852
17. Qi, C., Fourie, A., Chen, Q.: Neural network and particle swarm optimization for predicting the unconfined compressive strength of cemented paste backfill. Constr. Build. Mater. **159**, 473–478 (2018). https://doi.org/10.1016/j.conbuildmat.2017.11.006
18. R Core Team: A Language and Environment for Statistical Computing. R Foundation for Statistical Computing, Vienna, Austria (2014)
19. Roder, M., Passos, L.A., de Rosa, G.H., de Albuquerque, V.H.C., Papa, J.P.: Reinforcing learning in deep belief networks through nature-inspired optimization. Appl. Soft Comput. **108**, 107466 (2021)
20. Roder, M., de Rosa, G.H., Passos, L.A., Papa, J.P., Rossi, A.L.D.: Harnessing particle swarm optimization through relativistic velocity. In: 2020 IEEE Congress on Evolutionary Computation (CEC), pp. 1–8. IEEE (2020)
21. Roder, M., Rossi, A.L.D., de Oliveira Affonso, C.: Boosting machine learning techniques for wood quality classification by particle swarm optimization. In: Encontro Nacional de Inteligência Artificial e Computacional. Sociedade Brasileira de Computação (2017)
22. Rodrigues, D., de Rosa, G.H., Passos, L.A., Papa, J.P.: Adaptive improved flower pollination algorithm for global optimization. In: Yang, X.-S., He, X.-S. (eds.) Nature-Inspired Computation in Data Mining and Machine Learning. SCI, vol. 855, pp. 1–21. Springer, Cham (2020). https://doi.org/10.1007/978-3-030-28553-1_1
23. Shi, Y., Eberhart, R.: A modified particle swarm optimizer. In: 1998 IEEE International Conference on Evolutionary Computation Proceedings. IEEE World Congress on Computational Intelligence (Cat. No. 98TH8360), pp. 69–73, May 1998. https://doi.org/10.1109/ICEC.1998.699146
24. Tiryaki, S., Malkoçoğlu, A., Özşahin, Ş.: Using artificial neural networks for modeling surface roughness of wood in machining process. Constr. Build. Mater. **66**, 329–335 (2014). https://doi.org/10.1016/j.conbuildmat.2014.05.098
25. Vieira, F.H.A.: Image processing through machine learning for wood quality classification. Ph.D. thesis, Faculdade de Engenharia de Guaratinguetá (FEG), UNESP (2016)
26. Wilcoxon, F.: Individual comparisons by ranking methods. Biometrics Bull. **1**(6), 80–83 (1945)

A Feature-Based Out-of-Distribution Detection Approach in Skin Lesion Classification

Thiago Carvalho[1(✉)], Marley Vellasco[1], José Franco Amaral[2], and Karla Figueiredo[2]

[1] Pontifical Catholic University of Rio de Janeiro, Rio de Janeiro, Brazil
tmedeiros@aluno.puc-rio.br, marley@ele.puc-rio.br
[2] Rio de Janeiro State University, Rio de Janeiro, Brazil
franco@eng.uerj.br, karlafigueiredo@ime.uerj.br

Abstract. When dealing with Deep Learning applications in open-set problems, accurately classifying known classes seen in the training phase is not the only aspect to be taken into account. In such a context, detecting Out-of-Distribution (OOD) samples plays an important role as an auxiliary task, generally solved by OOD detection methods. For medical applications, detecting unknown samples may in classification problems can be beneficial for many aspects, such as a better understanding of the diagnosis and probably a more adequate treatment. In this article, we evaluate a feature space-based approach, named as OpenPCS-Class, for OOD detection in medical applications, more specifically skin lesion classification. We compare the OpenPCS-Class against important OOD detection methods, evaluating different model architectures and OOD datasets. The OpenPCS-Class outperformed other methods at 48.4% and 5.3% in terms of FPR95 and AUROC, respectively.

Keywords: out-of-distribution detection · deep learning · feature space

1 Introduction

Deep Learning (DL) models have found widespread use in various applications, ranging from autonomous driving [19] and pest detection to speech recognition [26]. Despite its outstanding results in tasks related to computer vision and natural language processing, accuracy is not the only subject to be taken into consideration in a DL deployment [10]. Depending on the problem, other aspects may also become important, such as the explainability and the capability to handle samples from unknown classes [34].

The DL models are known to learn generally in closed-set assumptions, and such out-domain restrictions are reflected in their inefficiency in explicitly showing ignorance about input samples from unseen classes. As a result, a DL model trained in such a setup is often unable to identify an unknown class data as

M. C. Naldi and R. A. C. Bianchi (Eds.): BRACIS 2023, LNAI 14196, pp. 338–352, 2023.
https://doi.org/10.1007/978-3-031-45389-2_23

unknown, which leads to problems of model overconfidence [32]. The overconfidence has several natures, such as unidentified overfitting problems, bias, or even the choice of the softmax function for the model's output layer, making directly identifying unknown samples more difficult [34]. Therefore, the model needs to be robust and able to handle Out-of-distribution (OOD) samples, which can come in various forms depending on the problem.

For medical applications, OOD detection is an important auxiliary task to improve the ability to detect unseen classes in an open-set problem. For example, when classifying an unseen rare skin lesion using a DL model to classify skin lesions, it would be preferable to identify it as unknown instead of erroneously classifying it as one of the known classes [25,36]. Therefore, the OOD detection task has drawn attention to a wide range of applications, such as histopathology [22], X-ray [3], and magnetic resonance images [14] classification problems.

OOD Detection can be considered a recent field of research in the area of DL, having one of the main objectives to improve the ability of models to recognize unknown samples. In other words, an OOD detection algorithm should be able to identify whether an input can be considered known or unknown. The most straightforward option for OOD detection is to use activations from the model's output layer, as this is closest to the final inference result [9]. These strategies typically rely on logits or softmax outputs to compute confidence scores, which are then used to differentiate between known and unknown classes.

More recently, researchers have explored using the feature space of the model to identify unknown samples, based on the assumption that the feature space can be useful for OOD detection, as intermediate layers capture different levels of semantic features [20]. One of the methods is named Open Principal Component Score (OpenPCS), which uses a low dimensional feature space representation from Principal Component Analysis (PCA) to fit class-wise Gaussian distributions to identify whether data is known or unknown. This approach was first implemented for semantic segmentation problems, but it can be extended for multi-class classification, named as OpenPCS-Class [5]. However, the feature-space approach for OOD detection, especially the OpenPCS-Class, is still under-explored for many applications.

In this article, we evaluate the OpenPCS-Class for OOD detection in skin lesion classification problems. The objective is to evaluate the capability of a Gaussian-based approach using the feature space to identify unseen classes in this medical application, which is usually a complex task with numerous OOD classes related to unknown skin lesions. The contribution of this work is three-fold:

1. We evaluate the OpenPCS-Class method for OOD detection in skin lesion problems. We use different OOD data to evaluate the approach, ranging from samples of unseen classes of skin lesions to different medical problems.
2. We compare the results with traditional and state-of-the-art methods for OOD detection. We assess how these methods behave in the presence of different OOD classes and additional ID data.

3. We also evaluate these models in different model architectures to investigate the model's contribution to OOD detection using different space representations.

2 Related Works

Detecting OOD samples is crucial in building reliable Deep Learning models that need to operate effectively in an open-set scenario. In medical applications, such strategies allow DL models to enhance the robustness of such results in a critical task. These works are generally concentrated in semantic segmentation and image classification tasks [4]. Karimi et al. [13] proposed a spectral analysis of the intermediate features of DL models to enhance the robustness of the segmentation task in multiple organs by quantifying the uncertainty of the segmentation result. Wollek et al. [30] evaluated some state-of-the-art OOD detection methods in several medical application tasks related to the image classification problem, discussing the advantages and drawbacks of such methods in identifying unknown samples closer to the training classes.

Due to the relevance of this topic in DL applications to guarantee safety and robustness, there are a plethora of new strategies related to OOD detection. One of the most common methods for Out-of-Distribution (OOD) detection involves using the softmax output as an OOD score, known as Maximum Softmax Probability (MSP) [11]. The MSP is based on the idea that unknown class samples would generate lower confidence scores for each known class, which are then used to distinguish ID and OOD data. This method was evaluated in a wide range of problems, including medical applications. Zhang et al. [37], for example, evaluated the effectiveness of the MSP method in OOD detection for diabetic retinopathy detection and chest radiography-related problems. However, using softmax output can sometimes lead to overconfident scores on unknown data, which is inappropriate for OOD detection [33].

To avoid the issues associated with the softmax, the feature space can also distinguish between known and unknown samples. Lee et al. [15] proposed a method that uses the information from the feature space to detect OOD samples, assuming that the feature representation can be fitted into Gaussian distributions. In this case, the class-conditional Gaussian distributions are obtained and the score is computed as the Mahalanobis distance from a test sample to the closest class-conditional distribution [24]. This OOD detection method was applied in different medical applications related to image analysis, such as malaria parasitized cells classification [28], lung cancer classification [2], and skin lesion classification [25].

Despite its efficiency in the OOD detection task, the feature space is generally a high-dimensional representation, which can be often an inefficient representation with high redundancy and lead to a harder fit of the OOD detection method [31]. To alleviate the problem of high dimensionality in intermediate representations, Oliveira et al. [21] proposed a method for OOD detection, called OpenPCS, using PCA to reduce the dimensionality of the feature space. The low-dimensionality representation is then used to adjust class-conditional Gaussian distributions, and the score is calculated by finding the maximum likelihood between a sample's intermediate representation and the class-conditional distributions. More recently, Carvalho et al. [5] proposed an extension of its method for multi-class classification problems, named as OpenPCS-Class. This method was successfully evaluated in benchmark problems, but the OpenPCS-Class is still unexplored in different applications, including medical image analysis.

3 Detecting Unseen Samples Using Feature Space

In this section, we describe in detail the OpenPCS-Class method strategy. We also briefly introduce the OOD detection problem in skin lesion problems, motivating the applicability of this work. The code of this work is publicly available[1]

3.1 Open Principal Component Score for Image Classification

The Open Principal Component Score (OpenPCS) is a method that uses intermediate features for OOD detection in a semantic segmentation task. Originally, this method could be applied only to Fully Convolutional Networks (FCN), which can be prohibitively for direct utilization of OpenPCS for different DL tasks.

The OpenPCS-Class can be seen as the extension of the OpenPCS method for classification tasks. This method discards the need of a FCN but retains the main characteristics of using a combination of intermediate features in a low-level representation. For a better comprehension of the method, Fig. 1 displays the method overview for an image classification problem.

The OpenPCS-Class is an OOD detection method that can combine features from different layers to distinguish whether a sample belongs to a known or unknown class. For each model layer l, we transform the activation map $a^{(l)}$ to the corresponding activation vector $h^{(l)}$ by using a reduction method (e.g., average pooling). Therefore, we always obtain its feature vector independently from the layer specification.

One of the main abilities of the OpenPCS method is the capability to combine the feature representation from different layers, which is a user-defined parameter. For classification tasks, the features are combined by concatenating their vectors, resulting in a feature vector h. The drawback of such an approach is the high dimensionality of the feature vector h. To alleviate this issue, we apply the PCA to obtain a better representation in a low dimension.

To fit the eigenvectors and eigenvalues for the PCA, we follow a class-wise approach. Therefore, we use the collection of feature vectors related to each of the known classes to fit the parameters of the PCA, creating a specific dimensionality reduction for each of the known classes, according to Eq. 1.

$$h_c^* = h \cdot v_c \tag{1}$$

[1] Code available at https://github.com/mdrs-thiago/skin-lesion-ood-detection.

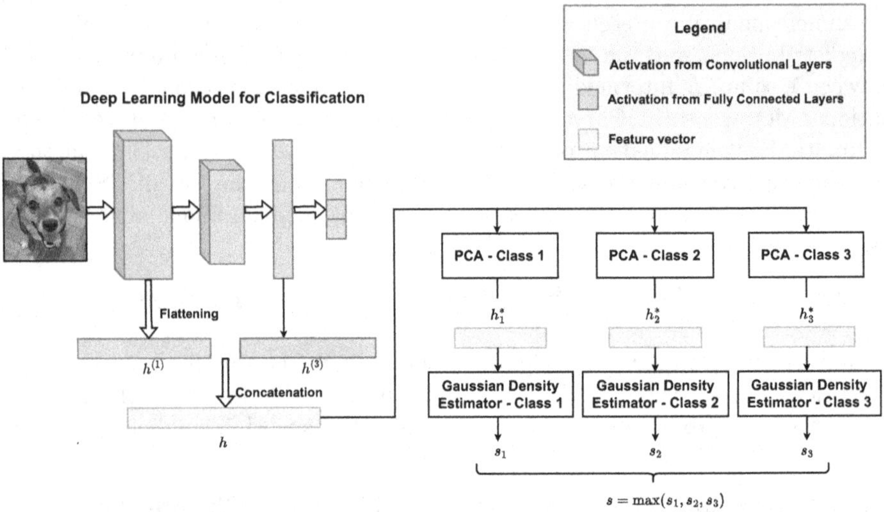

Fig. 1. OpenPCS-Class overview for image classification

where h is the feature representation, v_c is the eigenvector with the highest eigenvalues for dimensionality reduction for class c, and h_c^* is the feature vector after transformation for class c. Therefore, depending on the class c, the resulting low-dimension feature vector h_c^* can be different. It is important to note that we can obtain different low-dimensional representations for the same feature vector h, depending on the class c that we are evaluating.

The OOD score is computed by estimating how likely the feature vector is to each of the known class. In the literature, the Gaussian density estimator was successfully used to quantify the OOD-ness [18,21]. Therefore, we adopted the Gaussian density estimator to fit and compute its corresponding likelihood. Mathematically, the OOD score for each class c is computed according to Eq. 2

$$G_c(h_c^*) = \frac{1}{\sqrt{2\pi\sigma_c^2}} \exp\left(-\frac{(h_c^* - \mu_c)^2}{2\sigma_c^2}\right) \tag{2}$$

where μ_c and σ_c represents the mean and standard deviation for a known class c, and $G_c(a_c^*)$ represents the probability density of h_c^* generated by G_c. The final OOD score is the maximum log-likelihood over all known classes, as defined in Eq. 3.

$$s = \max_{c=1}^{n} \log\left[G_c\left(h_c^*\right)\right] \tag{3}$$

where n is the number of classes. In summary, to detect if a sample can be considered as OOD, we obtain its feature vector representation and apply a class-wise dimensionality reduction, and for each low-dimensionality representation vector, we calculate the log-likelihood to its corresponding class. The OOD score can be viewed as the maximum log-likelihood over all of the classes. For an

ID sample, the likelihood would be lower for all of the classes, except for the corresponding class, which yields a high s score. As for the OOD sample, the likelihood tends to be lower for all class-wise distributions, so the score s also is lower. Thereby, the ID and OOD samples can be distinguished by setting a threshold value for s.

3.2 OOD Detection in Skin Lesion Classification

With the world constantly evolving, new medical pathologies are frequently discovered through diagnosis. However, the identification of novel or rare diseases can be troublesome for DL-based automated diagnosis, potentially leading to incorrect classification and inappropriate treatment [36]. In such cases, OOD detection methods can play an essential role in identifying whether a new sample belongs to any of the known classes of the problem, thus providing an auxiliary task for DL-based approaches.

Specifically for dermatological-related tasks, OOD detection strategies can be handy to identify samples from unseen classes during the training phase. For instance, consider a deep learning problem aimed at automatically identifying the three most common skin lesions, as illustrated in Fig. 2. In an open-set scenario, the trained model may encounter unseen skin lesions, so it would be preferable to detect them as unknown instead of erroneously classifying them as the closer known class. Ideally, the OOD detection method should be capable of correctly classifying these samples as OOD, but this may depend on the chosen strategy [6]. Especially when ID and OOD samples are visually similar, it can be challenging to distinguish between known and unknown classes [30].

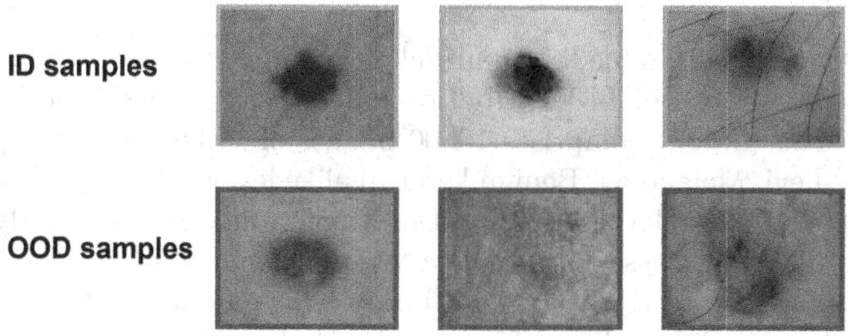

Fig. 2. Examples of In-Distributions and Out-of-Distributions samples of skin lesions

This article evaluates the feature space-based OOD detection method in different scenarios. In some experiments, we verify the capability of the OpenPCS-Class to detect near-OOD samples, typically skin lesions images taken in the same settings as the ID samples. We also evaluate some OOD detection approaches in the same problem (skin lesions), but under different conditions to evaluate the changing of such OOD detection strategies. Finally, we also evaluate these models in far-OOD detection samples, but related to medical applications.

4 Experiments

This section presents the experimental protocol for our case studies in the OOD detection task. In this work, we focused on the skin lesion classification problem, selecting different medical-related samples as OOD.

4.1 OOD Methods

We evaluated three robust methods commonly employed in this area to assess the OOD detection results. One of them is the Maximum Softmax Probability (MSP) method [11], which is a traditional approach that utilizes the softmax probability vector to identify unknown samples. By computing the maximum probability value of all classes, MSP assumes that a lower MSP score suggests that the model is less accurate about the predicted class, which could indicate an OOD sample.

Another method we selected is Energy-Based Out-of-Distribution detection (EBO) [16], a more sophisticated technique that uses the output space to calculate the OOD score. EBO computes the entropy of the logit and employs it as an OOD score to distinguish between OOD and ID samples.

We also opted for a feature space-based method for OOD detection to provide a more insightful discussion of the OpenPCS-Class approach. The Mahalanobis OOD detection method [12] measures the OOD score as the Mahalanobis distance of class-conditional Gaussian distribution in the feature space. In this case, if compared to the Gaussian distributions, OOD samples are expected to be further away from ID samples.

4.2 Datasets

For the OOD detection in medical multi-class classification, we have utilized the HAM10000 dataset as our In-Distribution dataset (D_{in}) for skin lesion classification [27]. This dataset comprises 10000 images of seven distinct skin lesions: Melanocytic nevi, Melanoma, Benign keratosis-like lesions, Basal cell carcinoma, Actinic keratoses, Vascular lesions, and Dermatofibroma. For this study, we have selected the first four classes as our ID classes, which contain 6705, 1113, 1099, and 514 samples, respectively. In addition, we have designated the remaining three classes as OOD samples to form the basis of our first case study. For the other experiments, we maintain the same D_{in} and ID classes, changing the OOD samples.

The dataset from the second case study consists of a wide range of skin lesion images taken in different parts of the body. However, a significant issue with this dataset is that some classes overlap with those found in D_{in}. Therefore, to ensure a fair comparison of the OOD detection task, we remove the overlapping classes from D_{out}.

The third selected D_{out} is related to the monkeypox classification problem [1]. The dataset contains images of monkeypox lesions and different skin lesions (e.g., chickenpox), given that the problem originally was built as a binary classification to identify whether a lesion can be considered monkeypox. Therefore, we used all images from this dataset as OOD samples in the third experiment.

For the fourth case study, we have manually selected images of rare skin lesions that do not belong to any of the classes in D_{in}. In this experiment, we have included additional ID images obtained from different circumstances than those found in the HAM10000 dataset. This collection of images will enable us to gain practical insights into the identification of unknown and uncommon classes in skin lesion classification and evaluate how the OOD detection methods perform when presented with different ID samples.

4.3 Metrics

To compare the methods, we selected three metrics to evaluate the OOD detection task in multi-classification problems [35].

AUROC (Area Under Recall Operating Curve) summarizes the Recall Operating Curve (ROC) as calculating the area under the curve. As ROC is usually used in a binary classification problem, to evaluate the OOD detection task using this metric, we consider only ID and OOD classes, independently from the fine-grained classes. Mathematically, the AUROC can be approximated as evaluating the True Positive Rate (TPR) and False Positive Rate (FPR) at discrete threshold values, presented in Eq. 4

$$\text{AUROC} = \sum_{i=1}^{n-1} \frac{1}{2}(x_{i+1} - x_i)(y_i + y_{i+1}) \tag{4}$$

where n is the number of thresholds, x_i and y_i are the false positive and true positive rates, respectively, at the i-th threshold.

AUPR (Area Under Precision-Recall Curve) is a metric that summarizes the Precision-Recall trade-off for different threshold values for a specific class. This metric is highly important for imbalance problems, which may be the case for our experiments. Therefore, we calculate the AUPR for the OOD class.

FPR95 indicates the False Positive Rate (FPR) when the True Positive Rate (TPR) is 95%. Typically, the FPR95 describes how likely the method could erroneously classify as unknown at a reasonably high TPR. Therefore, the lower the FPR95 is, the better the OOD detection method. Unlike the other ones, this is a dependent, since we define a cutoff value to classify as known or unknown.

4.4 Experimental Details

To evaluate our proposed approach, we used the same experimental procedure in all experiments. For the D_{in}, we split the dataset proportionally into training (60%), validation(20%), and test (20%) sets. We fit the OOD detection methods using the training set and, for all experiments, we use the test samples from D_{in} and the whole D_{out} to evaluate the separability between ID and OOD samples, respectively. During the testing phase, we randomly selected 500 samples from each set of D_{in} and D_{out} (when applicable) and computed the average metrics over ten runs. We also used the Wilcoxon signed-test rank to verify the statistical significance between the best result metric and all others. In Sect. 5, we denote an average result with a statistical difference using an underscore in the tables.

We also assess the impact of the OOD detection methods in different model architectures. As our problem is related to the image classification, we selected three models, Vision Transformer (ViT) model [7], ConvNeXT [17], and ResNet [8]. For the first two architectures, we used the pre-trained weights on ImageNet1k and finetuned the classification layer in the D_{in} problem. For the ResNet model architectures, we trained from scratch, following a similar training procedure as presented in the literature [29].

5 Discussion and Results

This section contains the results of the four case studies in medical applications. It is important to note that the selected OOD detection methods are similar in the experimental setup (i.e., it does not require any model retraining and just one forward pass is needed to identify OOD samples), but use different approaches to detect unseen classes.

For the first experiment, Table 1 summarizes the results for different OOD detection methods and architectures.

Table 1. OOD Detection Results for Experiment 1

Model	Method	AUROC ↑	AUPR ↑	FPR95 ↓
ViT	MSP	0.5755 ± 0.0536	0.8988 ± 0.0307	0.8754 ± 0.0322
	EBO	0.6150 ± 0.0331	0.9123 ± 0.0259	0.8521 ± 0.0111
	Mahalanobis	0.6564 ± 0.0403	0.8786 ± 0.1001	0.8955 ± 0.0291
	OpenPCS-Class	**<u>0.7636</u>** ± 0.0095	**<u>0.9607</u>** ± 0.0163	**<u>0.8132</u>** ± 0.0140
ResNet	MSP	0.6112 ± 0.0309	0.8301 ± 0.2550	0.8818 ± 0.0107
	EBO	0.5765 ± 0.0701	0.7537 ± 0.0612	0.9212 ± 0.0303
	Mahalanobis	0.6119 ± 0.0354	0.8110 ± 0.0273	0.8887 ± 0.0296
	OpenPCS-Class	**<u>0.6900</u>** ± 0.0201	**0.8571** ± 0.0109	**<u>0.8082</u>** ± 0.0114
ConvNeXT	MSP	0.6089 ± 0.0214	0.8178 ± 0.0203	**0.8904** ± 0.0215
	EBO	0.5742 ± 0.0300	0.7946 ± 0.0398	0.9512 ± 0.0316
	Mahalanobis	0.6217 ± 0.0115	0.8398 ± 0.0201	0.9092 ± 0.0124
	OpenPCS-Class	**0.6501** ± 0.0100	**0.8570** ± 0.0200	0.9110 ± 0.0197

The first experiment is a more challenging for discriminating wheter a sample belongs to a known or unknown class. This idea is reflected in the OOD detection metric results, showing a lower AUROC score for all of the methods, if compared to the other experiments. Even so, we noticed that OpenPCS-Class outperformed all three methods in terms of AUROC and AUPR, independently from the model architecture. Also, the FPR95 shows that our approach can enhance the OOD detection task considering a real-world scenario, considering the threshold that yields a TPR at 95%. In that case, the OpenPCS-Class can lower the FPR in this condition up to 7.1% (using ResNet model and MSP method).

The model architecture plays an important role in OOD detection. In this experiment, the ViT model increased the capability to detect OOD samples, at least for the OpenPCS-Class method. Especially for feature-based approaches, the model used can impact directly the results, given that different model architectures can yield feature activations.

The second experiment can be considered an easier task in OOD detection if compared to the first one. Although the classes from D_{out} are similar in the first two experiments, the images were obtained in different body parts, which can facilitate the OOD detection task. The results for the second experiment can be observed in Table 2.

Table 2. OOD Detection Results for Experiment 2

Model	Method	AUROC ↑	AUPR ↑	FPR95 ↓
ViT	MSP	0.6904 ± 0.0215	0.8012 ± 0.0194	0.9817 ± 0.0101
	EBO	0.6871 ± 0.0193	0.8305 ± 0.0143	0.9890 ± 0.0100
	Mahalanobis	0.9640 ± 0.0153	0.9876 ± 0.0095	0.1020 ± 0.0044
	OpenPCS-Class	**0.9847** ± 0.0085	**0.9978** ± 0.0003	**0.0656** ± 0.0006
ResNet	MSP	0.6908 ± 0.0148	**0.9352** ± 0.0102	**0.8066** ± 0.0052
	EBO	0.5786 ± 0.0109	0.8671 ± 0.0098	0.9632 ± 0.0032
	Mahalanobis	0.6578 ± 0.0102	0.9049 ± 0.0094	0.8832 ± 0.0091
	OpenPCS-Class	**0.7129** ± 0.0099	0.9168 ± 0.0092	0.8759 ± 0.0013
ConvNeXT	MSP	0.5413 ± 0.0120	0.8565 ± 0.0132	0.9197 ± 0.0091
	EBO	0.5149 ± 0.0093	0.8808 ± 0.0104	0.9732 ± 0.0019
	Mahalanobis	0.9868 ± 0.0054	0.9971 ± 0.0010	0.0425 ± 0.0009
	OpenPCS-Class	**0.9904** ± 0.0023	**0.9949** ± 0.0022	**0.0219** ± 0.0012

In this experiment, the approaches based on the feature space had a better OOD detection capability, if compared to those who use the output space. In fact, the feature space can contain low-level and high-level feature information, which can help to detect unknown classes in different contexts. On the other hand, the output space does not contain such kind of information, which may help to understand the difference between those approaches. Therefore, these strategies directly impact the scores generated, as illustrated in Fig. 3.

The main objective of OOD detection is to yield scores that could be easy to distinguish between ID and OOD samples. To evaluate the distributions obtained in Fig. 3, we conduct a Welch t-test [38], which rejected the hypothesis that the ID and OOD distributions have equal means ($p < 0.05$) only for the OpenPCS-Class.

The OpenPCS-Class, in this experiment, outperformed all three methods for OOD detection (decreased 48.7% in terms of FPR95, if compared to Mahalanobis and ConvNeXT). However, there is a slight difference between the OpenPCS-Class and Mahalanobis methods, depending on the model architecture. For

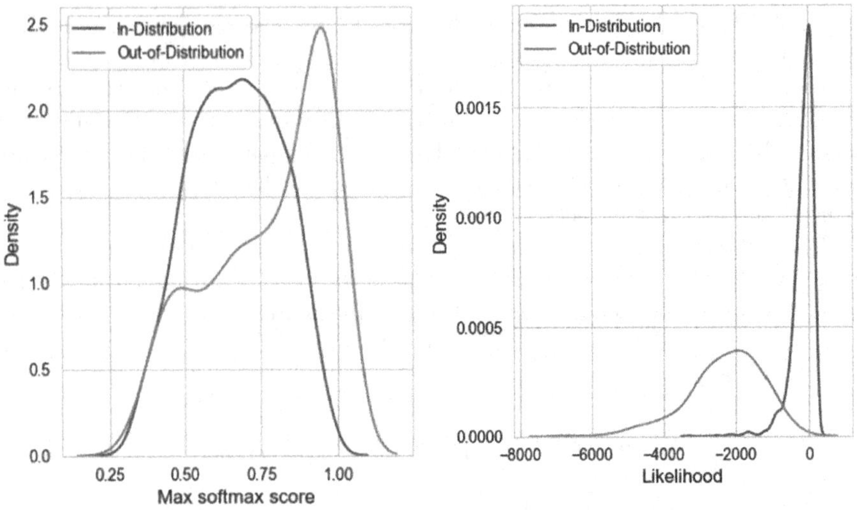

(a) (b)

Fig. 3. ID and OOD score distributions (a) Maximum softmax score from MSP (b) Maximum likelihood from OpenPCS-Class

Table 3. OOD Detection Results for Experiment 3

Model	Method	AUROC ↑	AUPR ↑	FPR95 ↓
ViT	MSP	0.7024 ± 0.0149	0.9779 ± 0.0094	0.8627 ± 0.0195
	EBO	0.6262 ± 0.0544	0.9506 ± 0.0130	0.9802 ± 0.0102
	Mahalanobis	0.9661 ± 0.0154	0.9886 ± 0.0058	0.1275 ± 0.0054
	OpenPCS-Class	**0.9889** ± 0.0017	**0.9994** ± 0.0004	**0.0392** ± 0.0051
ResNet	MSP	0.7638 ± 0.0340	0.9839 ± 0.0093	0.7941 ± 0.0216
	EBO	0.7413 ± 0.0219	0.9106 ± 0.0104	0.9212 ± 0.0200
	Mahalanobis	0.7916 ± 0.0184	0.9838 ± 0.0093	**0.7843** ± 0.0122
	OpenPCS-Class	**0.8325** ± 0.0146	**0.9847** ± 0.0088	0.8039 ± 0.0148
ConvNeXT	MSP	0.6563 ± 0.0109	0.9683 ± 0.0123	0.8922 ± 0.0100
	EBO	0.6693 ± 0.0154	0.9417 ± 0.0130	0.9781 ± 0.0098
	Mahalanobis	0.9784 ± 0.0106	0.9973 ± 0.0010	0.0790 ± 0.0012
	OpenPCS-Class	**0.9924** ± 0.0013	**0.9976** ± 0.0013	**0.0490** ± 0.0015

transformer-based models, both methods obtained, AUROC and AUPR metrics closer to one. This result corroborates recent findings that models based on Transformer architecture can enhance the robustness of OOD detection [23].

The third experiment uses skin lesions pathologies that are more different from those presented in D_{in}. The results are presented in Table 3.

Although the D_{out} in the third study case contains images from skin lesions, we noticed that all methods, independently from the model architecture, enhanced the OOD detection metrics, if compared to the previous experiments. As the OOD samples are related to diseases like monkeypox and chickenpox,

Table 4. OOD Detection Results for Experiment 4

Model	Method	AUROC ↑	AUPR ↑	FPR95 ↓
ViT	MSP	0.6901 ± 0.0216	0.9721 ± 0.0094	0.9048 ± 0.0105
	EBO	0.6137 ± 0.0311	0.9543 ± 0.0194	0.9802 ± 0.0210
	Mahalanobis	0.9611 ± 0.0093	0.9973 ± 0.0033	0.2698 ± 0.0197
	OpenPCS-Class	**0.9795** ± 0.0102	**0.9977** ± 0.0010	**0.0873** ± 0.0027
ResNet	MSP	0.7482 ± 0.0155	0.9779 ± 0.0107	0.8175 ± 0.0092
	EBO	0.6310 ± 0.0218	0.9175 ± 0.0091	0.9944 ± 0.0320
	Mahalanobis	0.7379 ± 0.0119	0.9606 ± 0.0088	0.8016 ± 0.0055
	OpenPCS-Class	**0.7878** ± 0.0107	**0.9677** ± 0.0083	**0.7698** ± 0.0080
ConvNeXT	MSP	0.5927 ± 0.0100	0.9457 ± 0.0095	0.8889 ± 0.0073
	EBO	0.7370 ± 0.0142	0.9019 ± 0.0122	0.8390 ± 0.0110
	Mahalanobis	0.9798 ± 0.0094	0.9966 ± 0.0015	0.1111 ± 0.0035
	OpenPCS-Class	**0.9852** ± 0.0064	**0.9980** ± 0.0010	**0.0873** ± 0.0024

the images are more dissimilar to those presented to the model in the training phase (using D_{in}), so there is a low confidence score in the output space, and the feature representation from OOD samples are more dissimilar to those presented from ID samples, resulting in an easier OOD detection task.

In this experiment, the feature-based approaches also obtained a considerably high capability to detect samples visually dissimilar to those presented in D_{in}. For Transformer-based approaches, Mahalanobis and OpenPCS-Class obtained comparable results, given that they could almost differentiate total ID and OOD samples. However, for the ResNet architecture, the difference between those approaches is more significant, showing better performance for the OpenPCS-Class method (increased 5.1% if compared to the Mahalanobis detector).

For the last case study, Table 4 displays the results for OOD detection using the same experimental protocol as the previous experiments.

In this experiment, we observed that feature-based approaches performed comparably better in detecting a wide range of pathologies as OOD. Even in the presence of new images for ID classes that are slightly different from those presented in the D_{in}, the OpenPCS-Class outperformed other methods in all three evaluation metrics. Therefore, even with visually different ID samples, the feature-space approaches obtained better results in the OOD detection task.

Although we only present the distributions for the second experiment, we used the Welch t-test for all experiments in this section. For all experiments using the transformer-based architectures, we noted that the ID and OOD distributions can be easily distinguished (i.e., it contains different means) for the OpenPCS-Class.

6 Conclusions

In this work, we evaluated the OpenPCS-Class for a new domain of application for OOD detection, more specifically for skin lesion problems. The feature space-based approaches, in general, obtained a superior OOD detection when the OOD samples are visually more dissimilar to ID ones, corresponding to the latter three experiments of this work.

Compared to all the methods evaluated in the experiments, the OpenPCS-Class outperformed in all scenarios regarding AUROC, and 9 (out of 12) in terms of average FPR95. More interestingly, the transformer-based models were more suitable for the OpenPCS-Class method, which always obtained superior OOD detection results.

Going forward, we aim to evaluate the OpenPCS-Class in different medical classification problems, to get a better perspective of feature space-based models in OOD detection application problems.

Acknowledgements. This work was supported in part by the Coordenação de Aperfeiçoamento de Pessoal de Nível Superior - Brasil (CAPES) - Finance Code 001, Conselho Nacional de Desenvolvimento e Pesquisa (CNPq) under Grants 140254/2021-8 and 308717/2020-1, and Fundação de Amparo à Pesquisa do Rio de Janeiro (FAPERJ)

References

1. Ali, S.N., et al.: Monkeypox skin lesion detection using deep learning models: a preliminary feasibility study. arXiv preprint arXiv:2207.03342 (2022)
2. Berger, C., Paschali, M., Glocker, B., Kamnitsas, K.: Confidence-based out-of-distribution detection: a comparative study and analysis. In: Sudre, C.H., et al. (eds.) UNSURE/PIPPI -2021. LNCS, vol. 12959, pp. 122–132. Springer, Cham (2021). https://doi.org/10.1007/978-3-030-87735-4_12
3. Calderon-Ramirez, S., Yang, S., Elizondo, D., Moemeni, A.: Dealing with distribution mismatch in semi-supervised deep learning for COVID-19 detection using chest X-ray images: a novel approach using feature densities. Appl. Soft Comput. **123**, 108983 (2022)
4. Cao, T., Huang, C.W., Hui, D.Y.T., Cohen, J.P.: A benchmark of medical out of distribution detection. arXiv preprint arXiv:2007.04250 (2020)
5. Carvalho, T., Vellasco, M., Amaral, J.F.: Out-of-distribution detection in deep learning models: a feature space-based approach. In: International Joint Conference on Neural Networks (2023)
6. Cho, W., Park, J., Choo, J.: Training auxiliary prototypical classifiers for explainable anomaly detection in medical image segmentation. In: Proceedings of the IEEE/CVF Winter Conference on Applications of Computer Vision, pp. 2624–2633 (2023)
7. Dosovitskiy, A., et al.: An image is worth 16 ×16 words: transformers for image recognition at scale. arXiv preprint arXiv:2010.11929 (2020)
8. He, K., Zhang, X., Ren, S., Sun, J.: Deep residual learning for image recognition. In: Proceedings of the IEEE Conference on Computer Vision and Pattern Recognition, pp. 770–778 (2016)
9. Hendrycks, D., et al.: Scaling out-of-distribution detection for real-world settings. In: International Conference on Machine Learning, pp. 8759–8773. PMLR (2022)
10. Hendrycks, D., Carlini, N., Schulman, J., Steinhardt, J.: Unsolved problems in ML safety. arXiv preprint arXiv:2109.13916 (2021)
11. Hendrycks, D., Gimpel, K.: A baseline for detecting misclassified and out-of-distribution examples in neural networks. arXiv preprint arXiv:1610.02136 (2016)

12. Kamoi, R., Kobayashi, K.: Why is the mahalanobis distance effective for anomaly detection? arXiv preprint arXiv:2003.00402 (2020)
13. Karimi, D., Gholipour, A.: Improving calibration and out-of-distribution detection in deep models for medical image segmentation. IEEE Trans. Artif. Intell. **4**, 383–397 (2022)
14. Lambert, B., Forbes, F., Doyle, S., Tucholka, A., Dojat, M.: Improving uncertainty-based out-of-distribution detection for medical image segmentation. arXiv preprint arXiv:2211.05421 (2022)
15. Lee, K., Lee, K., Lee, H., Shin, J.: A simple unified framework for detecting out-of-distribution samples and adversarial attacks. In: Advances in neural information processing systems, vol. 31 (2018)
16. Liu, W., Wang, X., Owens, J., Li, Y.: Energy-based out-of-distribution detection. Adv. Neural. Inf. Process. Syst. **33**, 21464–21475 (2020)
17. Liu, Z., Mao, H., Wu, C.Y., Feichtenhofer, C., Darrell, T., Xie, S.: A convnet for the 2020s. In: Proceedings of the IEEE/CVF Conference on Computer Vision and Pattern Recognition, pp. 11976–11986 (2022)
18. Martinez, J.A.C., Oliveira, H., dos Santos, J.A., Feitosa, R.Q.: Open set semantic segmentation for multitemporal crop recognition. IEEE Geosci. Remote Sens. Lett. **19**, 1–5 (2021)
19. Muhammad, K., et al.: Vision-based semantic segmentation in scene understanding for autonomous driving: recent achievements, challenges, and outlooks. IEEE Trans. Intell. Transp. Syst. **23**, 22694–22715 (2022)
20. Nunes, I., Pereira, M.B., Oliveira, H., Santos, J.A.D., Poggi, M.: Fuss: Fusing superpixels for improved segmentation consistency. arXiv preprint arXiv:2206.02714 (2022)
21. Oliveira, H., Silva, C., Machado, G.L., Nogueira, K., Dos Santos, J.A.: Fully convolutional open set segmentation. Mach. Learn. **112**, 1733–1784 (2021)
22. Pawlowski, N., Glocker, B.: Abnormality detection in histopathology via density estimation with normalising flows. In: Medical Imaging with Deep Learning (2021)
23. Podolskiy, A., Lipin, D., Bout, A., Artemova, E., Piontkovskaya, I.: Revisiting Mahalanobis distance for transformer-based out-of-domain detection. In: Proceedings of the AAAI Conference on Artificial Intelligence, vol. 35, pp. 13675–13682 (2021)
24. Ren, J., Fort, S., Liu, J., Roy, A.G., Padhy, S., Lakshminarayanan, B.: A simple fix to mahalanobis distance for improving near-OOD detection. arXiv preprint arXiv:2106.09022 (2021)
25. Roy, A.G., et al.: Does your dermatology classifier know what it doesn't know? detecting the long-tail of unseen conditions. Med. Image Anal. **75**, 102274 (2022)
26. Swetha, P., Srilatha, J.: Applications of speech recognition in the agriculture sector: a review. ECS Trans. **107**(1), 19377 (2022)
27. Tschandl, P., Rosendahl, C., Kittler, H.: The HAM10000 dataset, a large collection of multi-source dermatoscopic images of common pigmented skin lesions. Sci. Data **5**(1), 1–9 (2018)
28. Uwimana, A., Senanayake, R.: Out of distribution detection and adversarial attacks on deep neural networks for robust medical image analysis. arXiv preprint arXiv:2107.04882 (2021)
29. Wightman, R., Touvron, H., Jégou, H.: ResNet strikes back: an improved training procedure in timm. arXiv preprint arXiv:2110.00476 (2021)
30. Wollek, A., Willem, T., Ingrisch, M., Sabel, B., Lasser, T.: A knee cannot have lung disease: out-of-distribution detection with in-distribution voting using the medical example of chest X-ray classification. arXiv preprint arXiv:2208.01077 (2022)

31. Wright, J., Ma, Y.: High-Dimensional Data Analysis with Low-Dimensional Models: Principles, Computation, and Applications. Cambridge University Press (2022)
32. Wu, Y., et al.: Revisit overconfidence for OOD detection: reassigned contrastive learning with adaptive class-dependent threshold. In: Proceedings of the 2022 Conference of the North American Chapter of the Association for Computational Linguistics: Human Language Technologies, pp. 4165–4179 (2022)
33. Wu, Y., et al.: Disentangling confidence score distribution for out-of-domain intent detection with energy-based learning. arXiv preprint arXiv:2210.08830 (2022)
34. Yang, J., Zhou, K., Li, Y., Liu, Z.: Generalized out-of-distribution detection: a survey. arXiv preprint arXiv:2110.11334 (2021)
35. Ye, N., et al.: OOD-bench: quantifying and understanding two dimensions of out-of-distribution generalization. In: Proceedings of the IEEE/CVF Conference on Computer Vision and Pattern Recognition, pp. 7947–7958 (2022)
36. Zadorozhny, K., Thoral, P., Elbers, P., Cinà, G.: Out-of-distribution detection for medical applications: guidelines for practical evaluation. In: Shaban-Nejad, A., Michalowski, M., Bianco, S. (eds.) Multimodal AI in Healthcare. Studies in Computational Intelligence, vol. 1060, pp. 137–153. Springer, Cham (2022). https://doi.org/10.1007/978-3-031-14771-5_10
37. Zhang, O., Delbrouck, J.-B., Rubin, D.L.: Out of distribution detection for medical images. In: Sudre, C.H., et al. (eds.) UNSURE/PIPPI -2021. LNCS, vol. 12959, pp. 102–111. Springer, Cham (2021). https://doi.org/10.1007/978-3-030-87735-4_10
38. Zimmerman, D.W., Zumbo, B.D.: Rank transformations and the power of the student T test and welch T'test for non-normal populations with unequal variances. Can. J. Exp. Psychol. **47**(3), 523 (1993)

A Framework for Characterizing What Makes an Instance Hard to Classify

Maria Gabriela Valeriano[1,2(✉)], Pedro Yuri Arbs Paiva[1], Carlos Roberto Veiga Kiffer[2], and Ana Carolina Lorena[1]

[1] Instituto Tecnológico de Aeronáutica, Praça Marechal Eduardo Gomes, São José dos Campos, Brazil
{valeriano,paiva,aclorena}@ita.br
[2] Universidade Federal de São Paulo, Rua Botucatu, São Paulo, Brazil
carlos.kiffer@unifesp.br

Abstract. The health domain has been largely benefited by Machine Learning solutions, which can be used for building predictive models to support medical decisions. But, for increasing the reliability of these systems, it is important to understand when the models are prone to failures. In this paper, we investigate what can we learn from the instances of a dataset which are hard to classify by Machine Learning models. Different reasons may explain why one or a set of instances are misclassified, despite the predictive model used. They can be either noisy, anomalous or placed in overlapping regions, to name a few. Our framework works at two levels: the original base dataset and a meta-dataset built to reflect the hardness level of the instances. A two-dimensional hardness embedding is assembled, which can be visually inspected to determine sets of instances to scrutinize better. We show some analysis that can be undertaken in this hardness space that allow to characterize why some of the instances are hard to classify, with case studies on health datasets.

Keywords: Explainability · Data-center · Instance-hardness

1 Introduction

Machine Learning (ML) models are regarded as promising solutions for revolutionizing healthcare [1]. These data-driven techniques may leverage knowledge from the large volumes of data continuously gathered by health systems and agents [2]. As a recent example, different predictive models were built during the COVID-19 pandemic to assist diagnosis and prognosis of patients using hospital data [3,4]. Nonetheless, there are issues still preventing the widespread usage of ML predictive models in health decision making and planning. Some concerns are the risk of bias and inappropriate or incomplete model performance evaluation [5].

Given that health databases are not collected with the objective of data analysis in the first place, the Data Scientist and ML practitioner must deal

M. C. Naldi and R. A. C. Bianchi (Eds.): BRACIS 2023, LNAI 14196, pp. 353–367, 2023.
https://doi.org/10.1007/978-3-031-45389-2_24

with these issues by relying on their own experience or on ad-hoc procedures, which can bias the achieved results. Data-centric frameworks have recently been proposed to aid these professionals in better assessing data quality and taking potential corrective measures, which will result in more reliable and trustful predictive models [6,7].

One fruitful direction is to assess and monitor learning performance for each dataset instance and prevent relying only on averages over an entire dataset, as it is common practice [8]. Furthermore, examining the performance at the instance level can reduce biases in the evaluation of ML models, as specific important groups of instances can be mistaken by the ML models, despite the obtainment of an overall high average predictive performance. This makes it important to know which instances are systematically misclassified by ML models and why they are so, a concern that has led to the recent literature of *instance hardness* analysis [9].

According to Smith et al. [9], *instance hardness* can be measured as the average misclassification error of a pool of diverse classifiers when predicting the label of such instance. This metric allows identifying instances in a dataset that are inherently difficult to have their label predicted, despite of the classification technique used. However, going beyond, the literature also presents *hardness meta-features* which prospect possible reasons why the instance is hard to classify [9,10].

Recently, a framework for relating the classification performance of classifiers of distinct biases and the values of different hardness meta-features was framed [8,11]. The relationship of this information is used to produce a 2-D hardness embedding where the instances are linearly distributed according to their hardness level. Here this framework is extended for gaining data-centric insights by taking advantage of the organization of the dataset observations in the hardness embedding. These analyzes take place at two levels: the original input features and the meta-features. Inspecting general trends in the input features of the hard instances of a dataset can be particularly informative and allow to involve the data domain experts in the ML pipeline. On the other hand, looking at the hardness meta-features allows Data Scientists to understand structural problems with such instances, which may be preventing the obtainment of a better predictive accuracy. This dual interplay allows for taking into account the main stakeholders involved in data analysis for decision support.

In particular, our framework is applied here to datasets from the health domain, which is a critically requester reliable predictive models. As case studies, we consider a set of COVID prognosis datasets from Brazil, one of the countries hit hardest by the pandemic. The classification problem consists in predicting whether a COVID-hospitalized patient will develop into an aggravated condition or not, information that can support decision making at both clinical and management levels at hospitals. Our framework contributes on the following ways:

- **Providing a principled approach for identifying hard instances in a dataset:** taking advantage of the hardness embedding of a dataset, where

easy and hard instances are interposed, different types of analyzes are made possible. One of them is to define hardness and easiness footprints, which are dense regions of the representation space which encompass more hard and easy instances, respectively. This allows to define a principled strategy for choosing sets of instances to be further examined;
- **Providing meaningful insights on data quality and structure:** the hardness profile of a dataset can be inspected for gaining insights about where and why ML models are failing;
- **Being actionable by different stakeholders:** we combine two levels of knowledge. While inspecting relationships of the input features that lead to increased hardness levels can be a valuable tool for better data understanding and auditing with the aid of domain experts, the meta-features allow to prospect structural problems in the data which can be informative for Data Science and ML practitioners.

This paper is structured as follows: Section 2 presents some related work. Section 3 presents our formulation. Section 4 presents some experimental results. Section 5 concludes the paper with some discussions.

2 Related Work

ML techniques are data-hungry and a common mistake is to consider that more data will always result in more accurate predictive models. But some issues present in real data, such as noise, sparse regions and outliers, can impair the ML system's performance, trustfulness and acceptance. Therefore, data quality plays a central role in developing reliable ML systems.

Since predictions inside health scenarios involve highly delicate matters, data quality assessment and cleaning are even more critical. Some recent works have used data-centered approaches to formulate strategies allowing to: assess how difficult it is to predict the class of each instance in a dataset [6,8,13]; understand why some instances are frequently misclassified, while also trying to explain the behavior of classifiers for such instances [8,14]; predict test instances that will be reliably classified or not [7]; remove hard or ambiguous instances in a data sculpting strategy to improve ML algorithms performance [6,13]; and identifying anomalous regions of the input feature space offering useful explanations and insights about data quality as well as model performance [6,8,15]. But there are still gaps to be filled that may help increasing the acceptance of such tools. One of them is providing more interpretable and actionable insights to the different stakeholders involved in data analysis.

In order to increase the trustfulness and acceptance of AI systems, it is imperative that their decisions are transparent and can be scrutinized by different stakeholders. This has lead to the increase in the *Explainable Artificial Intelligence* (XAI) area [16]. Many XAI platforms and strategies are currently available and most of them focus on evaluating how the predictive input features influence the results of the ML models, such as permutation tests [17] and Shapley values [18]. But there are other layers of explainability that are not so commonly explored in the literature, despite having potential to offer rich insights for better understanding the strengths and weaknesses of the ML models.

One of them is working at a meta-level where general properties of the data that may lead to impairments in a good predictive performance are identified and characterized [9,19]. Taking steps towards this direction, we can extract local rules based on at most two input features for subsets of instances a given model struggles to classify correctly, that is, instances the model regards as hard to classify [15]. Our work follows a similar direction, but with a more general approach for finding and characterizing the hard instances from a dataset. While combining the outputs of many models, we also leverage on meta-features able to describe why some instances are hard to classify, giving insights at different yet complementary perspectives.

3 Formulation

Figure 1 presents an overview of the methodology followed in our paper. Given a dataset $\mathcal{D}$, a set $\mathcal{F}$ of hardness meta-features describing the level of difficulty in classifying each instance in $\mathcal{D}$ according to different perspectives is extracted. Algorithmic performance $\mathcal{P}$ is registered for each individual instance of the dataset for a pool of classification algorithms $\mathcal{A}$. Joining $\mathcal{F}$ and $\mathcal{P}$, a hardness embedding is produced where the instances are placed so as to present linear trends of difficulty level, from bottom right (easy) to top left (hard). Next two types of analyzes are performed. One of them involves defining regions concentrating hard and easy to classify instances. They are called hardness and easiness footprints, respectively. By extracting patterns from the opposing footprints, we are able to obtain insights on instance hardness at different levels and perspectives. The second approach involves a visual inspection of the hardness embedding for identifying observations lying in regions of interest to be inspected.

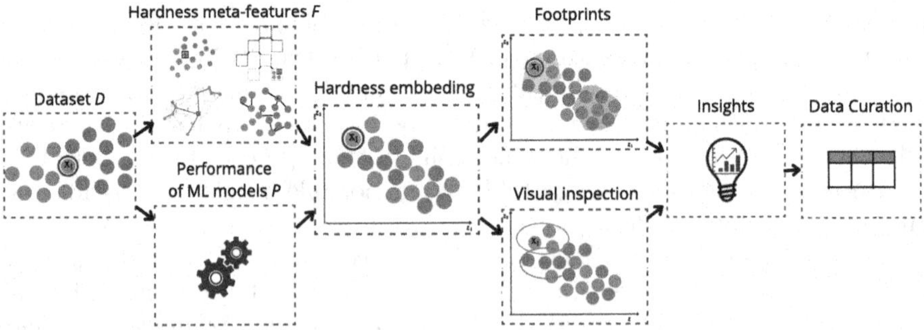

Fig. 1. Overview of the proposed framework.

3.1 Instance Hardness Analysis

Let us define $\mathcal{D}$ formally as containing n pairs of labeled observations $(\mathbf{x}_i, y_i)$. Each $\mathbf{x}_i \in \mathcal{X}$ is an instance described by m input features and is labeled in $y_i \in \mathcal{Y}$, where $\mathcal{Y}$ is a discrete and non-ordered set of classes. In addition, let $h : \mathcal{X} \rightarrow \mathcal{Y}$ denote a classification hypothesis, that is, a ML predictive model generated from $\mathcal{D}$. The Instance hardness of an instance $\mathbf{x}_i$ can be defined as the probability of misclassification when it is subject to a pool of different learning algorithms, that is:

$$IH_{\mathcal{A}}(\mathbf{x}_i, y_i) = 1 - \frac{1}{|\mathcal{A}|} \sum_{j=1}^{|\mathcal{A}|} p(y_i|\mathbf{x}_i, h_j(\mathcal{D})), \tag{1}$$

where $p(y_i|\mathbf{x}_i, h_j(\mathcal{D}))$ is the probability the j-th ML model in the pool attributes $\mathbf{x}_i$ to its expected class y_i. The intuition is that if a pool of diverse classifiers is unable to predict the expected class of $\mathbf{x}_i$ with a high probability (proxy to a high confidence), then this instance is intrinsically hard to classify.

In turn, *hardness meta-features* (HM) can be used for prospecting possible reasons why an instance is hard to classify or not [9,10]. Summarizing, given an instance $\mathbf{x}_i$:

- kDN (k-Disagreeing Neighbors) gives the percentage of k nearest neighbors of $\mathbf{x}_i$ with label different from y_i;
- $N1I$ (fraction of nearby instances of different classes) gives the fraction of instances that do not belong to class y_i which are connected to $\mathbf{x}_i$ in a Minimum Spanning Tree built from $\mathcal{D}$;
- $N2I$ (ratio of the intra-class and extra-class distances) takes the ratio of the distance $\mathbf{x}_i$ has to the nearest neighbor from its class y_i to the distance $\mathbf{x}_i$ has to its nearest enemy (nearest instance from another class $y_j \neq y_i$);
- $LSCI$ (Local Set Cardinality) measures the size of the set containing instances from class y_i which are closest to $\mathbf{x}_i$ than to its nearest enemy (this set is named local set of the instance);
- LSR (Local Set Radius) gives the radius of the previous set;
- U (Usefulness) considers the number of local sets an instance belongs to;
- H (Harmfulness) takes the number of instances $\mathbf{x}_i$ is nearest enemy of;
- CL (Class Likelihood) measures the likelihood an instance belongs to its class y_i;
- CLD (Class Likelihood Difference) measures the difference between the likelihood an instance belongs to the class y_i to the likelihood it belongs to any other class $y_j \neq y_i$;
- $F1I$ (Fraction of features in overlapping areas) gives the percentage of input features lying in overlapping regions of the classes;
- DCP (Disjunct Class Percentage) gives the percentage of examples of class y_i which are placed in a same disjunct as $\mathbf{x}_i$, where disjuncts are defined by a decision tree algorithm;
- TD (Tree Depth) gives the depth where the instance is classified in a decision tree model, which can be either pruned (TD_P) or unpruned (TD_U).

All measures are standardized, by definition, in the [0, 1] interval, so that larger values are attributed to instances hard to classify according to the measured criterion.

3.2 Hardness Embedding

The hardness embedding is built from a composition of several sets. All individual instances $\mathbf{x}_i \in \mathcal{D}$ compose the Instance Set $\mathcal{I}$. The Algorithm Set $\mathcal{A}$ comprises a portfolio of classification algorithms of distinct biases. The Performance Set $\mathcal{P}$ records the predictive performance obtained by each algorithm in $\mathcal{A}$ for every instance $\mathbf{x}_i$. The Feature Set $\mathcal{F}$ contains the HM extracted from the instances $\mathbf{x}_i$. Therefore, for each instance $\mathbf{x}_i \in \mathcal{I}$ and for each algorithm $\alpha_j \in \mathcal{A}$, a feature vector $\mathbf{f}(\mathbf{x}_i) \in \mathcal{F}$, and an algorithm performance metric $p_m(\alpha_j, \mathbf{x}_i)$ are measured. The process is repeated for all the instances in $\mathcal{I}$ and algorithms in $\mathcal{A}$, generating a meta-dataset $\mathcal{M} = \{\mathcal{I}, \mathcal{F}, \mathcal{A}, \mathcal{P}\}$.

The algorithms included in the pool $\mathcal{A}$ are: Bagging, Gradient Boosting (GB), Support Vector Machine (SVM, with both linear and RBF kernels), Multilayer Perceptron (MLP), Logistic Regression (LR) and Random Forest (RF). They present distinct learning mechanisms and are commonly adopted to solve ML classification tasks in the literature, specially when tabular datasets are concerned. To assess the performance of the algorithms, a five-fold CV strategy is used. Their performance $\mathcal{P}$ is evaluated using the log-loss error per instance.

The meta-dataset $\mathcal{M}$ is next subject to a meta-feature selection step so that only the most informative meta-features are preserved. This selection is performed using a Neighborhood Component Feature Selection (NCFS) algorithm [21], where we seek to find meta-features more related to the predictive performance of the classifiers in $\mathcal{A}$.

The resulting meta-dataset is then projected into a 2-D embedding presenting linear trends in the meta-features and algorithmic performance measures, also named *Instance Space* (IS), as described in [8,11]. The result is a projection matrix able to map each instance to the 2-D hardness embedding based on the selected meta-features values. A further rotation step is also included, so that the hard instances are placed in the upper left quadrant and the easier instances are in the lower right quadrant of the hardness embedding.

Figure 2 presents examples of hardness embeddings of three datasets for predicting the outcome of COVID patient severity. The first and third datasets contain the symptoms and comorbidities of citizens undergoing diagnosis tests. In the second example, patients of an hospital from the São Paulo metropolitan area (Brazil) are prognosed based on a set of routine laboratory tests. In all cases, classes are represented by different shapes. The colors give the instance hardness level of the instances as computed by Eq. 1. The higher the IH value, the redder the color; the lower the IH value, the bluer the color. There is a mix of patients that are easy or hard to classify per class, with both triangles and circles in the upper left or lower right regions of the plot.

Whilst this plot can be colored according to all meta-features $\mathcal{F}$ values, performance measures $\mathcal{P}$ values of the classification algorithms $\mathcal{A}$ and even by the

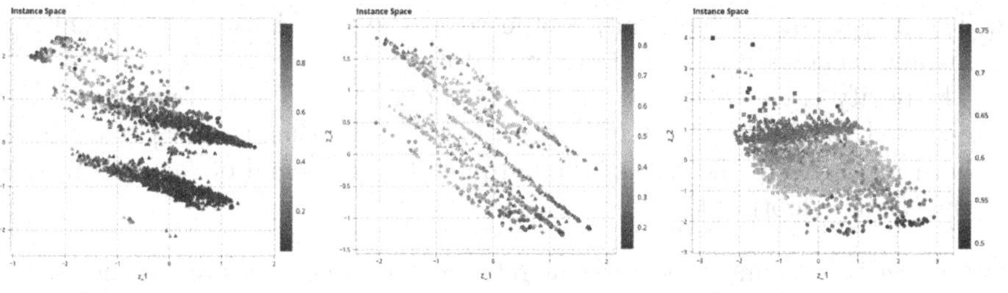

(a) *Severity*. (b) *Hospital 1*. (c) *Hospitalization*.

Fig. 2. Hardness embedding of three datasets with contrasting profiles: (a) *Severity*; (b) *Hospital 1*; (c) *Hospitalization*.

original input features values of the dataset $\mathcal{D}$, selections of specific regions can also be saved and scrutinized.

3.3 Instance Easiness and Instance Hardness Footprints

A *footprint* is defined in [22,23] as a region in the instance space where an algorithm is expected to perform well based on inference from empirical performance analysis. Here we use such a concept in order to define regions from the hardness embedding which concentrate more easy or hard instances. They are named *easiness* and *hardness* footprints, respectively.

In order to construct these footprints, first the instances must be categorized as either easy or hard. This can be done by imposing different thresholds on IH values. In fact, we consider three categories: easy, hard and others. The later category encompasses instances that have intermediate hardness levels. We focus our analysis on contrasting easy and hard instances. Ultimately, the objective is to understand why some instances are hard to classify. For segmenting the instances, we consider a threshold T, which must be higher than 0 and lower than 0.5. Since the IH values are bounded in the [0, 1] interval, we have:

- $\mathbf{x}_i$ is considered easy if $IH_{\mathcal{A}}(\mathbf{x}_i, y_i) \leq T$;
- $\mathbf{x}_i$ is considered hard if $IH_{\mathcal{A}}(\mathbf{x}_i, y_i) \geq 1 - T$.

Then for each set of instances, easy or hard, the DBSCAN algorithm [24] is used to identify high density clusters in the instance (hardness) space. α-shapes are used to construct hulls which enclose all the points within the clusters [25]. For each cluster hull, a Delaunay triangulation creates a partition, and those triangles that do not satisfy a minimum purity (defined as the percentage of easy or hard instances enclosed within it, which is 0.7 by default) requirement are removed. The union of the remaining triangles gives the corresponding easiness or hardness footprint. As in [22], it is also possible to extract the following objective measures from the footprints:

1. The area of the footprint in the 2-D hardness embedding, normalized by the total area of the space;

2. The density of the footprint, computed as the ratio between the number of instances enclosed by the footprint and its area;
3. The purity of the footprint, which corresponds to the percentage of instances enclosed by the footprint from a given category, that is, the percentage of hard instances in the case of the hardness footprint and of easy instances in the easiness footprint.

Taking the hardness footprint as reference, a large area implies the dataset has many instances hard to classify, so that they occupy a large portion of the hardness embedding. A large density means such area is dense. And a large purity is observed when most of the instances enclosed in the hardness footprint are indeed hard. Conversely, it is possible to define similar concepts for the easiness footprint and the easy instances. The easiness and hardness footprints' areas can be regarded as an indicative of the hardness profile of the dataset. When the hardness area is large, the dataset and the underlying classification problem can be considered more difficult to solve. In contrast, easy datasets/problems will show a large easiness area.

4 Experiments

Four datasets from the health domain, involving the prognosis of individuals diagnosed for COVID-19 in Brazil, were used in our experiments. Datasets and their description are available in a public repository[1]. Summarizing, we have:

1. ***Hospital 1 and Hospital 2***: prognosis of COVID-19 severity for patients hospitalized at two distinct private hospitals from the São Paulo metropolitan area, using some standard laboratory tests;
2. ***Hospitalization and Severity***: prediction of the hospitalization length when a citizen from São José dos Campos-SP is hospitalized with a positive COVID diagnosis (*Hospitalization*) and if he/she will evolve to a severe condition (*Severity*), based on initial symptoms and comorbidities reported.

4.1 Footprint Analysis

First, we analyze the easiness and hardness footprints of three datasets with distinct hardness profiles. They are *Severity*, *Hospital 1*, and *Hospitalization* datasets with easy, intermediate, and hard profiles, respectively. Following the same order, the average accuracy rate achieved by the seven classifiers for each of these datasets was: 0.900 (std 0.003), 0.673 (std 0.030), and 0.398 (std 0.022).

To determine such footprints, we did an experimental procedure varying the T value from 0.3 to 0.495, with steps of 0.02. Next, we monitored the purity values of the resulting footprints. We chose the T values leading to the highest purity values for both easiness and hardness footprints. Purest footprints are preferred, as they ensure most of the instances the footprint encloses are indeed

[1] (https://github.com/gabivaleriano/explaining_healthdata).

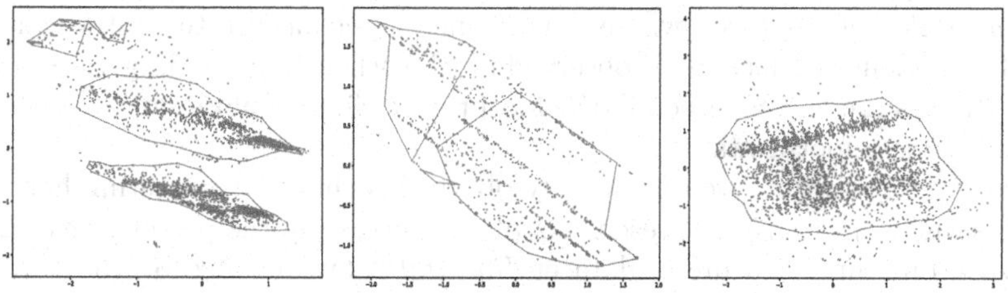

(a) *Severity.* (b) *Hospital 1.* (c) *Hospitalization.*

Fig. 3. Easy and hard footprints of datasets with contrasting profiles: (a) *Severity*, (b) *Hospital 1*; (c) *Hospitalization*. T value = 0.048.

easy/hard. Figure 3 shows the obtained footprints: hardness in red and easiness in blue. As expected, the hardness footprint dimensions increase according to the dataset's difficulty level. For instance, when the T value reaches 0.45, all instances are considered hard in the *Hospitalization* dataset.

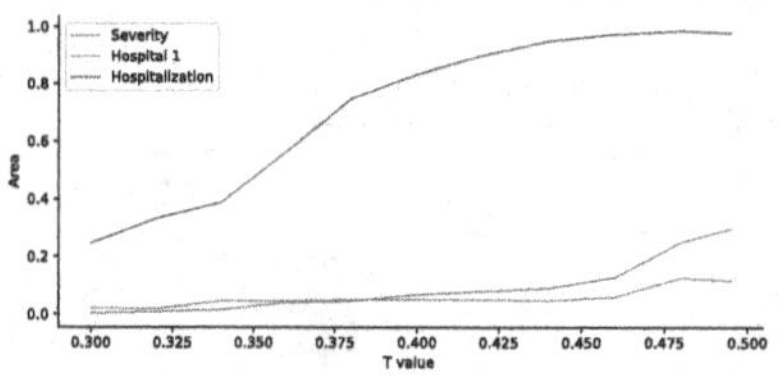

(a) Area

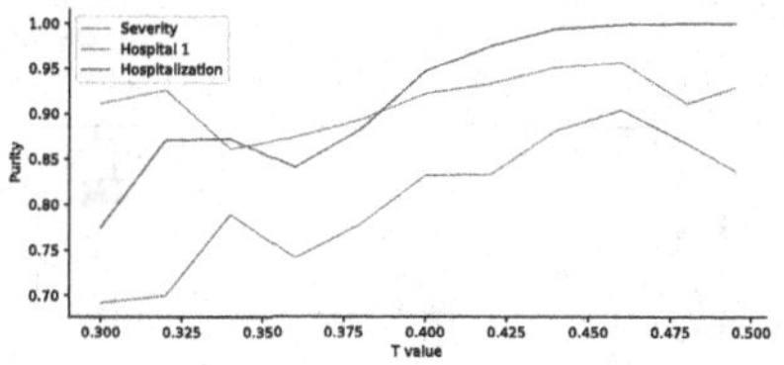

(b) Purity

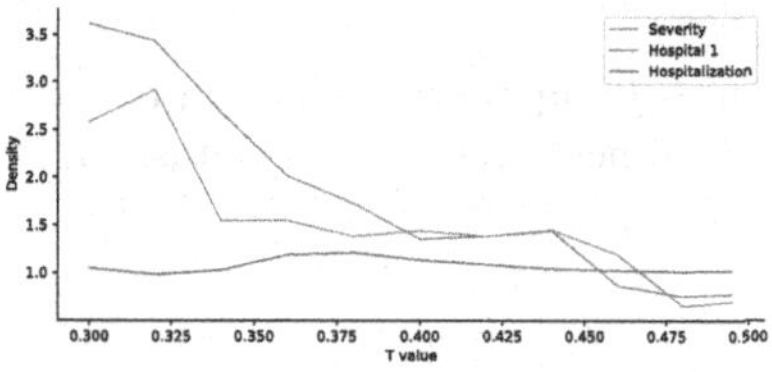

(c) Density

Fig. 4. Areas, purities and densities of the hardness footprints for increasing T values.

Figure 4 presents the variations of the area, density and purity values of the hardness footprint for increasing T values in the datasets. As expected, when T increases, the footprints' **areas** grow, as more instances obey the inequalities involving the instance hardness values. But this increase is less accentuated for the datasets with easy and intermediate hardness profiles. **Purity** is always over 0.7 and varies less for the *Severity* dataset, while presenting an increasing tendency in the other problems. In contrast, whilst **density** values are stable in the *Hospitalization* dataset, where the instances are concentrated in the center of the IS, a tendency of decrease is observed for the other datasets (larger areas with less instances are encompassed in the footprints). Similar plots can be generated for the easiness footprints.

From the footprints we can already obtain insights by contrasting hard and easy instances. In Fig. 5 we present boxplots contrasting the percentage of Lymphocytes, one input feature of dataset *Hospital 1* for instances in the hardness (left boxplots) and easiness (right boxplots) footprints. Colors matches the original class, where red stands for patients that developed to a severe condition and blue is the opposite class. Clearly, there is an inversion on the expected features' values when we compare the plots. In fact, a reduction in the proportion of lymphocytes is expected for severe cases and not the opposite [20]. This is already an indicative that hard instances are not following the expected pattern from the class they belong to. Figure 5b presents boxplots of the kDN meta-feature values for the hardness (left) and easiness (right) footprints. Easy instances have a low kDN value (are surrounded by elements sharing their class label), while hard instances have a high kDN value (are close to elements from the opposite class), despite the class they belong to. Therefore, hard instances are contrasting their ground truth labels and are contained in overlapping areas.

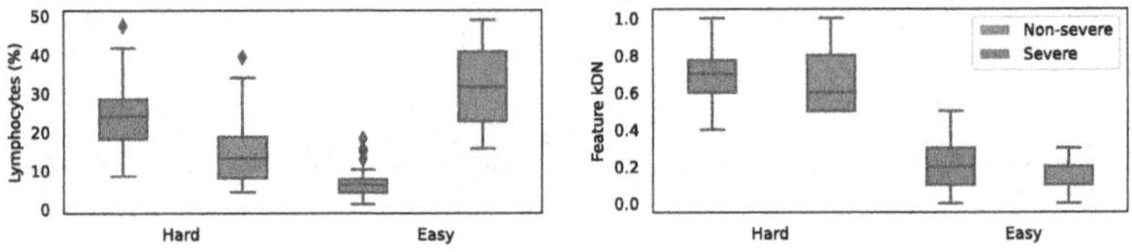

(a) Lymphocytes (%). (b) kDN.

Fig. 5. Plots of the input features Lymphocytes (%) and of meta-feature kDN for the hard instances (left) and easy instances (right) from dataset *Hospital 1*.

4.2 Data Sculpting

Now we move our analysis to a practical usage of our framework in a data sculpting strategy. The idea is to remove the instances of the hardness footprint, assuming they are probably noisy and incorrect. The results of this procedure are illustrated in Fig. 6. Increasing thresholds of T are considered, which means that we start removing instances with the highest hardness levels. The average AUC performance of the seven classifiers from the pool $\mathcal{A}$ is shown in the y-axis, with standard deviations in gray. The performance achieved with no sculpting (keeping all instances) is shown as a dotted line.

In Fig. 6a, we train and test the classifier over the same dataset (*Hospital 1*). As expected, removing the hard instances increases the accuracy achieved. In Fig. 6b, the sculpting takes place at instances from dataset *Hospital 1*, used

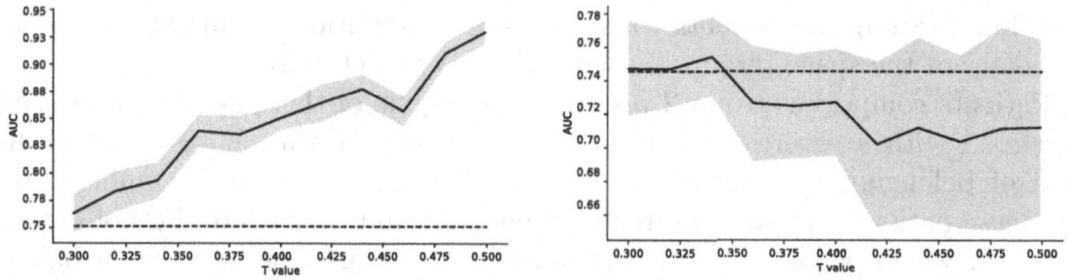

(a) Training and testing with *Hospital 1*. (b) Training with *Hospital 1* and testing with *Hospital 2*

Fig. 6. Average model performance scored by AUC in a data sculpting process. The dotted line is the performance without removing any instance from *Hospital 1*.

for training, but test takes place at a different dataset, *Hospital 2*. This second analysis already entails a higher difficulty level, since it presupposes the profiles of patients from *Hospital 1* are predictive of the conditions of the patients from *Hospital 2*. In this case, there is only a subtle increase of accuracy over the original baseline (the dotted line), for T around 0.30 and 0.34. But the results are detrimental for other T values. That is, in this case the benefit occurs only when a small fraction of the hardest training instances are removed. The standard deviation values are also usually high, indicating a large variation of results.

As in [6] when randomly removing ambiguous instances, removing hard instances in a dataset always increases model performance for the same dataset (Fig. 6a). However, inside health scenarios the removal of hard instances can be dangerous since they are not necessarily noisy or incorrect. Instead they can represent sub populations with atypical yet correct features' values. So the models might become incapable to correctly classify these types of instances. This might explain why when sculpting data from *Hospital 1* and testing the results in data from *Hospital 2* the accuracy decreases after a threshold around 0.34. Therefore, a more informed data sculpting process is needed.

4.3 Analyzing Groups of Data with a Domain Expert

Here we inspect closely groups of hard instances in a dataset and show that more guided insights can be devised. This is possible taking advantage of the different possibilities of visual inspection of the hardness embedding, which can have the instances colored according to all features, meta-features and algorithmic performance values.

In spite of the easy profile of the *Severity* dataset (Fig. 2a), there are still instances with a high level of difficulty placed in the upper left corner of the hardness embedding. Most of the patients with a severe condition are placed in the bottom of the space, indicating this class is fairly easy to classify (group 1). In the top left of the space there are non-severe patients very hard to classify (group 2) and also some severe cases hard to classify in the top of the hardness embedding (group 3). Patients without a severe condition are mostly placed in the center of the space (group 4), being also easy to classify.

Patients composing group 2 present at least one of three attributes heavily correlated with a severe condition. As a consequence, they have a low likelihood of belonging to their registered class (high CL values). To understand why these patients did not require an extended hospitalization or did not evolve to death, we inspected the raw databases, obtaining some valuable insights. From the 90 instances, 17 were mislabeled due to data preprocessing failures and should be corrected. The remaining were hospitalized but released before ten days of hospitalization. Between then, nine patients were under 35 years old, which can explain the quick recovery. For others, the short hospitalization could be explained either because of disease recovery despite the apparent severity of the case, or by an external factor, like reduced hospital capacity for receiving such patients. The complete analysis of the other groups can be found in our repository.

In the *Hospital 1* dataset, the nature of the attributes is different, since the blood tests have distinct reference values according to sex and age and can also change in the presence of other diseases and medication. These are some reasons why the predictive models perform poorly in this dataset. The IS also offers insights which help to understand why, how, and where the ML models are failing. Comparing meta-feature values for the severe individuals hard to classify reveals six types of hard instances. One of these groups is composed of eight instances labeled as non-severe, although they have the opposite profile. Indeed, they have a low likelihood of belonging to their registered class (CL) and are surrounded by elements from a different class ($N1$). Each individual in this group is a man between fifty and sixty years old, with low percentage of lymphocytes and C-reactive protein count as well as high percentages of neutrophils. Investigating the original database it is possible to follow up the values of these same blood tests in the next few days. This inspection reveals that, besides their initial condition, all of them recovered before 14 days of hospitalization (the proxy for considering a severe patient in this dataset). Therefore, this group is composed of outlier patients with a faster recovery than would be initially expected.

Another group is composed of 30 patients that also present a clinical condition similar to the severe class, but are labeled as non-severe patients. All meta-features are high for these patients (CL, DCP and $N1$), revealing they are difficult to classify according to different perspectives. Analyzing the original database, we found an unexpected pattern. Some of these patients are released without recovering (still presenting low percentage of lymphocytes and C-reactive protein counts) and sometimes in a worse condition. We could not recover the real reason for hospital discharge. These patients might have been transferred between hospitals, although being registered as medical releases in the system. These are more suitable candidates for data sculpting, since they might be noisy, as justified by the high values for their meta-features. These analysis illustrate how our framework can support the insertion of expert knowledge to improve data quality and model performance, in a more guided data sculpting process.

5 Discussions and Conclusions

Data-centric analysis have been taking an important role in ML, providing means to assure more trustfulness to the area. In this paper, we obtain interpretations on reasons why some observations of a dataset can be considered hard to classify. This is done considering both original input features and also meta-features which describe possible abstract structural reasons explaining why some instances are hard to classify. The choice of the instances to be examined is based on a projection of the dataset into a hardness embedding showing linear trends of classification difficulty according to different perspectives. We show the value of our proposal in gaining insights about data from the health domain.

Our framework was devised to the analysis of datasets in a tabular format, which are very abundant in the health domain. But plenty of non-structured data, such as image and text, are also gathered in this domain (eg. X-ray images). This does not prevent applying our framework to non-structured data. For instance, it is possible to extract structured representations from non-structured data by imputing them to some deep learning trained models [26].

For obtaining the hardness embedding, many meta-features and algorithmic performance measures must be extracted from the dataset first. This clearly implies in a computational cost which cannot be disregarded. Running the analysis on a MacBook Pro OS computer (M1 processor with 8 GB of memory), the average time taken to obtain the hardness embeddings for our datasets was around four minutes. In general, the hardness meta-features have at most a quadratic asymptotic cost on the number of observations the dataset has, mostly because some of them require building a distance matrix between all pairs of observations of the dataset. But the main cost is usually incurred by the cross-validation training-testing of the classifiers in the pool $\mathcal{A}$. Computational cost can be saved by reducing the amount of classifiers considered, although we can argue that the classifiers chosen here are common representatives everyone tests when tabular data are concerned, at least as baselines.

Another limitation of our current approach is that data must be labeled in advance. We shall investigate strategies to allow attributing a expected hardness level to unlabeled test instances. Also, applying the analysis to imbalanced datasets must be done with care, since the minority class observations will tend to be pointed as hard just because they are outnumbered. And missing values must also be dealt with in advance. Both issues can be solved in the future by including data re-sampling and missing value imputation strategies inside the framework.

Other types of analysis can also be devised and investigated. For instance, we can test the effects of including or excluding features from a dataset. As more informative features are present, we expect the easiness footprint to grow and the hardness footprint to shrink. Other analysis possible is to establish when a prediction should better be discarded and be entrusted to a specialist instead and to devise a human-in-the-loop procedure.

Acknowledgements. This study was financed in part by the Coordenação de Aperfeiçoamento de Pessoal de Nível Superior - Brasil (CAPES) - Finance Code 001. The authors also thank the financial support of FAPESP (grant 2021/06870-3) and CNPq.

References

1. Anderson, D., Bjarnadottir, M.V., Nenova, Z.: Machine learning in healthcare: operational and financial impact. In: Babich, V., Birge, J.R., Hilary, G. (eds.) Innovative Technology at the Interface of Finance and Operations, vol. 11, pp. 153–174. Springer, Cham (2022). https://doi.org/10.1007/978-3-030-75729-8_5
2. Imrie, F., Cebere, B., McKinney, E.F., van der Schaar M.: AutoPrognosis 2.0: democratizing diagnostic and prognostic modeling in healthcare with automated machine learning. arXiv preprint arXiv:2210.12090 (2022)
3. de Moraes, B.A.F., Miraglia, J., Donato, T., Filho, A.: Covid-19 diagnosis prediction in emergency care patients: a machine learning approach. MedRxiv, 2020-04 (2020)
4. Fernandes, F.T., de Oliveira, T.A., Teixeira, C.E., de Moraes Batista, A.F., Dalla Costa, G., Chiavegatto Filho, A.D.P.: A multipurpose machine learning approach to predict covid-19 negative prognosis in São Paulo, Brazil. Sci. Rep. **11**(1), 1–7 (2021)
5. Wynants, L., et al.: Prediction models for diagnosis and prognosis of covid-19: systematic review and critical appraisal. BMJ **369** (2020). https://doi.org/10.1136/bmj.m1328
6. Seedat, N., Crabbe J., van der Schaar, M.: Data-SUITE: data-centric identification of in-distribution incongruous examples. arXiv preprint arXiv:2202.08836 (2022)
7. Seedat, N., Crabbe J., Bica, I., van der Schaar, M.: Data-IQ: characterizing subgroups with heterogeneous outcomes in tabular data. arXiv preprint arXiv:2210.13043 (2022)
8. Paiva, P.Y.A., Moreno, C.C., Smith-Miles, K., Valeriano, M.G., Lorena, A.C.: Relating instance hardness to classification performance in a dataset: a visual approach. Mach. Learn., 1–39 (2022)
9. Smith, M.R., Martinez, T., Giraud-Carrier, C.: An instance level analysis of data complexity. Mach. Learn. **95**(2), 225–256 (2014)
10. Arruda, J.L.M., Prudêncio, R.B.C., Lorena, A.C.: Measuring instance hardness using data complexity measures. In: Cerri, R., Prati, R.C. (eds.) BRACIS 2020. LNCS (LNAI), vol. 12320, pp. 483–497. Springer, Cham (2020). https://doi.org/10.1007/978-3-030-61380-8_33
11. Paiva, P.Y.A., Smith-Miles, K., Valeriano, M.G., Lorena, A.C.: PyHard: a novel tool for generating hardness embeddings to support data-centric analysis. arXiv preprint arXiv:2109.14430 (2021)
12. Valeriano, M.G., et al.: Let the data speak: analysing data from multiple health centers of the São Paulo metropolitan area for covid-19 clinical deterioration prediction. In: 2022 22nd IEEE International Symposium on Cluster, Cloud and Internet Computing (CCGrid), pp. 948–951. IEEE (2022)
13. Zheng, K., Chen, G., Herschel, M., Ngiam, K.Y., Ooi, B.C., Gao, J.: PACE: learning effective task decomposition for human-in-the-loop healthcare delivery. In: Proceedings of the 2021 International Conference on Management of Data, pp. 2156–2168 (2021)

14. Houston, A., Cosma, G., Turner, P., Bennett, A.: Predicting surgical outcomes for chronic exertional compartment syndrome using a machine learning framework with embedded trust by interrogation strategies. Sci. Rep. **11**(1), 1–15 (2021)
15. Prudêncio, R.B., Silva Filho, T.M.: Explaining learning performance with local performance regions and maximally relevant meta-rules. In: Xavier-Junior, J.C., Rios, R.A. (eds.) Brazilian Conference on Intelligent Systems, pp. 550–564. Springer, Cham (2022). https://doi.org/10.1007/978-3-031-21686-2_38
16. Gunning, D., Stefik, M., Choi, J., Miller, T., Stumpf, S., Yang, G.-Z.: XAI-explainable artificial intelligence. Sci. Rob. **4**(37), eaay7120 (2019)
17. Ojala, M., Garriga, G.C.: Permutation tests for studying classifier performance. J. Mach. Learn. Res. **11**(6) (2010)
18. Ghorbani, A., Zou, J.: Data Shapley: equitable valuation of data for machine learning. In: International Conference on Machine Learning, pp. 2242–2251. PMLR (2019)
19. Lorena, A.C., Garcia, L.P., Lehmann, J., Souto, M.C., Ho, T.K.: How complex is your classification problem? A survey on measuring classification complexity. ACM Comput. Surv. **52**(5), 1–34 (2019)
20. Jafarzadeh, A., Jafarzadeh, S., Nozari, P., Mokhtari, P., Nemati, M.: Lymphopenia an important immunological abnormality in patients with covid-19: possible mechanisms. Scand. J. Immunol. **93**(2), e12967 (2021)
21. Amankwaa-Kyeremeh, B., Greet, C., Zanin, M., Skinner, W., Asamoah, R.K.: Selecting key predictor parameters for regression analysis using modified Neighbourhood Component Analysis (NCA) algorithm. In: Proceedings of 6th UMaT Biennial International Mining and Mineral Conference, pp. 320–325 (2020)
22. Smith-Miles, K., Tan, T.T.: Measuring algorithm footprints in instance space. In: 2012 IEEE Congress on Evolutionary Computation, pp. 1–8. IEEE (2012)
23. Muñoz, M.A., Villanova, L., Baatar, D., Smith-Miles, K.: Instance spaces for machine learning classification. Mach. Learn. **107**(1), 109–147 (2018)
24. Khan, K., Rehman, S.U., Aziz, K., Fong, S., Sarasvady, S.: DBSCAN: past, present and future. In: The Fifth International Conference on the Applications of Digital Information and Web Technologies, pp. 232–238. IEEE (2014)
25. Edelsbrunner, H.: Alpha shapes-a survey. Tessellations Sci. **27**, 1–25 (2010)
26. Najafabadi, M.M., Villanustre, F., Khoshgoftaar, T.M., Seliya, N., Wald, R., Muharemagic, E.: Deep learning applications and challenges in big data analytics. J. Big Data **2**(1), 1–21 (2015)

Physicochemical Properties for Promoter Classification

Lauro Moraes(✉), Eduardo Luz, and Gladston Moreira

Universidade Federal de Ouro Preto, Ouro Preto-MG, Brazil
{lauromoraes,eduluz,gladston}@ufop.edu.br

Abstract. The accurate identification of promoter regions in DNA sequences holds significant importance in the field of bioinformatics. While this problem has garnered substantial attention in the literature, it remains unresolved. Several researchers have achieved notable outcomes by employing diverse machine-learning techniques to predict promoter regions. However, only a few have thoroughly explored the utilization of features derived from the physicochemical properties of DNA across various organism types. This study investigates the advantages of incorporating these features in the training of machine-learning models. The research evaluates and compares the performance of multiple metrics on diverse datasets encompassing both prokaryotic and eukaryotic organisms. The state-of-the-art CNNProm method is employed as the baseline for our experiments. The models and source code associated with this study can be accessed at the following URL of the project's repository: https://anonymous.4open.science/r/bracis-paper-1458/.

Keywords: Bioinformatics · Machine Learning · Physicochemical properties · Promoter classification

1 Introduction

The identification of gene products and their location within DNA sequences that have not been experimentally characterized, commonly known as gene finding, is a central topic of interest in computational biology. Prediction of promoter sequences and transcriptional start points can help signal a transcript's approximate start, thereby identifying one end of a gene. This information is particularly useful in DNA sequences derived from higher eukaryotes, where coding regions are isolated segments embedded within a non-coding DNA background [28].

A promoter region in DNA is a noncoding sequence of DNA located upstream of a gene that is responsible for regulating the expression of that gene [17]. The promoter region contains binding sites for transcription factors, which are proteins that bind to the promoter and recruit RNA polymerase to initiate transcription. The specific pattern of binding sites in the promoter determines the level of gene expression and the conditions under which the gene is active [5].

High-resolution promoter recognition in DNA is an important area of research in bioinformatics. It involves the use of algorithms to find the transcription start

M. C. Naldi and R. A. C. Bianchi (Eds.): BRACIS 2023, LNAI 14196, pp. 368–382, 2023.
https://doi.org/10.1007/978-3-031-45389-2_25

site (TSS) of a gene without the need for laborious and costly experimental techniques such as aligning expressed sequence tags (ESTs), complementary DNAs (cDNAs) or messenger RNAs (mRNAs) to the entire genome [33]. The TSS is a specific genomic location within the promoter region where RNA polymerase binds and initiates transcription. It marks the starting point of the transcription process, where the DNA sequence is transcribed into RNA. Defining the TSS position as a reference, the upstream region is the DNA sequence located before the transcription start site, and the downstream region is situated after the TSS.

Identifying the promoter region allows us to study the regulation of gene expression and understand how different genes are controlled in various tissues, conditions, and stages of development [31]. In addition, by identifying the promoter regions for a set of genes, one can infer information about the molecular mechanisms that control gene expression and how these mechanisms may change in response to different stimuli. So, these algorithms can efficiently delimit regions involved in transcriptional regulation, guiding further experimental work, given the lower cost associated with computational approaches [33].

The identification of promoter regions is often performed by computational methods, such as promoter prediction algorithms, that analyze the DNA sequence and look for characteristic features such as TATA boxes [20] and CpG islands [10] and other short conserved sequences. However, multiple groups of genes don't contain these features. Methods for identifying promoters have been widely adopted in bioinformatics, as it allows for the generation of comprehensive maps of promoter regions in a genome which can be used for further analysis, such as studying gene regulation in response to environmental cues or identifying novel regulatory elements. Despite recent advances in promoter recognition algorithms, accurately identifying promoters remains challenging due to the diversity and complexity of these sequences in genome [34].

Methods for promoter prediction have been established to recognize promoter regions within the DNA sequences of both prokaryotic and eukaryotic organisms. These methods harness various machine learning algorithms for their functioning. The support vector machines (SVM) [9] were used in [26] to evaluate the strength in a dataset of *Escherichia coli* Trc promoter. [1] applied a genetic algorithm to calibrate the SVM hyperparameters using a human promoter dataset. [3] used datasets of three higher eukaryotes, Saccharomyces cerevisiae, *A. thaliana*, and human, to train Convolutional Neural Network (CNN) [24] with Long Short Term Memory (LSTM) [18] and Random forest (RF) [4] models. A study of promoter prediction on bacterial datasets was performed by [8] using RF models. In [25] a stacked ensemble of LightGBM [21], XGBoost [6], AdaBoost [14], GBDT [15], and RF models were used on a *Escherichia coli* dataset. The work of [30] proposed some CNN models to identify promoters on different organisms datasets. In [27] the same datasets were used to train a capsule neural network (CapsNet) [29].

The existing literature of promoter prediction lacks comprehensive investigations regarding the influence of DNA physicochemical properties on the accuracy of promoter prediction across diverse organisms. Consequently, the primary objective of this study is to examine the efficacy of features derived from DNA physicochemical properties in training machine learning models for predicting promoter regions in both prokaryotic and eukaryotic organisms. To assess the performance of our models, we utilize the datasets introduced by Umarov et al. (2017) [30], which encompass a wide range of organisms. For each property, we train and compare multiple machine learning models, as well as ensembles of these models, to identify the most effective properties. To evaluate the overall predictive capacity of these models, we compare their performance against the state-of-the-art method for these datasets, namely CNNProm [29]. The results indicate that the models based solely on properties do not yield optimal performance. Still, they may serve as a foundational step for future research by combining them with other techniques and models to enhance predictive accuracy.

2 Materials and Methods

2.1 Benchmark Datasets

The seven datasets utilized in this study were sourced from the supplementary materials made available by [30] and can be accessed on the corresponding GitHub repository[1]. Regrettably, the dataset pertaining to "Human TATA" could not be accessed through the repository and therefore was excluded from the analysis in this work.

The eukaryotic organism sequences, spanning 251 *bp* (base pairs) with a transcription start site (TSS) located at the 200th position, were sourced exclusively from the EPD database [12]. The dataset includes *Arabidopsis* TATA and *Arabidopsis* non-TATA sequences from a plant species, as well as Mouse TATA, Mouse non-TATA, and Human non-TATA sequences from mammalian species. Non-promoter sequences in these datasets are comprised of random gene fragments located after the first exons.

The bacterial promoter sequences of *Bacillus subtilis* were obtained from DBTBS [19], while the *Escherichia coli s*70 sequences were acquired from RegulonDB [16]. These sequences, which comprise 81 *bp*, contain the transcriptional start site (TSS) at the 60th position. Conversely, the non-promoter sequences for these prokaryotic organisms include the reverse sequences of random fragments extracted from protein-coding genes.

Figure 1 shows the number of promoters and non-promoters in each dataset. Notably, all datasets have more non-promoter samples than promoter samples. This characteristic can potentially affect the model learning process as they tend to classify the data into the majority class [22]. Besides, the prokaryotic datasets have fewer samples compared to eukaryotic datasets.

[1] https://github.com/solovictor/CNNPromoterData.

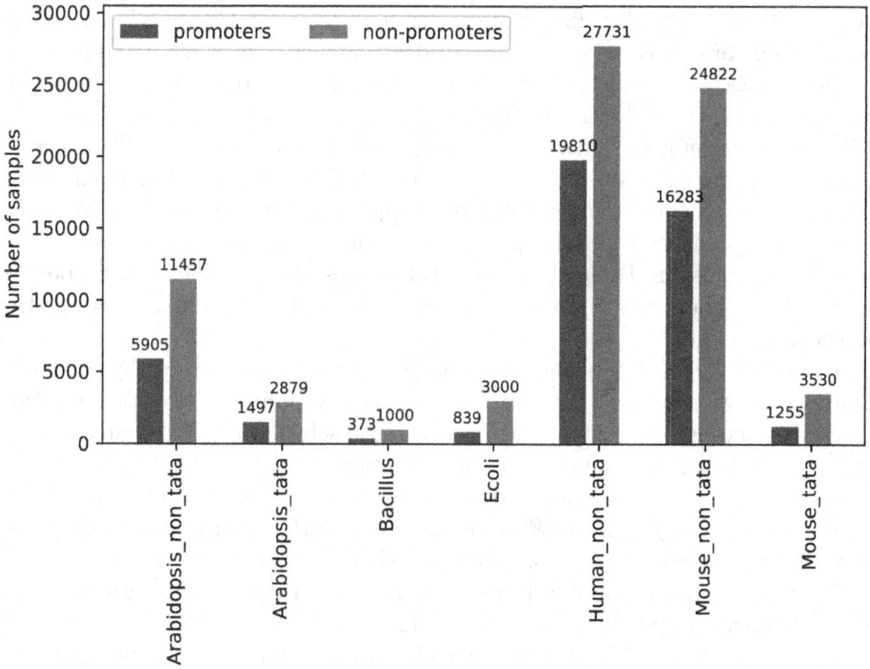

Fig. 1. The number of promoters samples and non-promoters samples present in each benchmark dataset.

2.2 Feature Representation

Adenine (A), cytosine (C), guanine (G), and thymine (T) are the four nucleotides that comprise the alphabet $V = \{A, C, G, T\}$ forming the basis of DNA. A DNA sample of length $l_S \in \mathbb{N}^*$ base pairs is represented by $S \in V^{ls}$, which can be written as a sequence $S = s_1...s_i...s_{l_S}$, where s_i is the i-th nucleotide of the sequence S.

To convert the DNA sequence into a numeric format, we can use the physicochemical properties of dinucleotides or trinucleotides. A dinucleotide represents a pair of adjacent nucleotides in a DNA sequence, while a trinucleotide represents three adjacent nucleotides. These properties capture the physicochemical characteristics of nucleotides and the structural features of DNA [7]. For example, the properties may include measures of hydrogen bond formation, base stacking, base pairing, or flexibility of the DNA helix.

K-mers are substrings of length **k** that are extracted from a longer DNA or protein sequence. This bioinformatics concept can be used in sequence analysis to identify motifs, patterns, or features characteristic of a specific biological function. Considering the nucleotide alphabet V, there is $|V|^k$ possible mers with length k.

As the name suggests, dinucleotides are two adjacent nucleotides that occur in a DNA sequence. It can be seen as a special case of k-mers with $k = 2$.

There are $4^2 = 16$ possible dinucleotides, such as "AT", "CG", "TA", etc. While trinucleotides are three adjacent nucleotides that occur in a DNA sequence. It is a special case of k-mers with $k = 3$. There are $4^3 = 64$ possible trinucleotides, such as "ATG", "CGT", "TAA", etc.

There are several physicochemical properties in the literature. The work of [7] used 38 properties for dinucleotides and 12 for trinucleotides, making the mapping values available for each in the supplementary data section. We used these data to convert the dataset sequences and create the input features for our tested models. In Table 1, we show the names of all of these 50 properties that we used. There is one column for the dinucleotide names and one for the trinucleotide names.

To convert the DNA sequence into dinucleotide or trinucleotide physicochemical properties, we can use a function that maps each dinucleotide or trinucleotide to a corresponding numerical value based on the selected physicochemical property. This mapping function can be represented as $f : V^k \rightarrow \mathbb{R}$.

A sequence S with l_s nucleotides is converted to a sequence of $l_s - k + 1$ property values. So, $f_{d_m}(S) \rightarrow \mathbb{R}^{l_s - 1}$ is the function that maps each dinucleotide to its related physicochemical property m, where $1 \leq m \leq 38$. While $f_{t_n}(S) \rightarrow \mathbb{R}^{l_s - 2}$ define a similar mapping function to convert trinucleotides into the m-th physicochemical property, where $1 \leq n \leq 12$.

Using a sliding window strategy, with length 2 *bp* for dinucleotides and 3 *bp* for trinucleotides, and a step size of 1 *bp*, we apply the corresponding mapping function to get all physicochemical property numeric values and create the feature vectors that we use in this work to train machine learning models. Figure 2 shows an example of converting a nucleotide sequence of length 8 *bp* into two numeric vectors to be used as features. On the left, we use a sliding window of 2 *bp* to extract all 7 dinucleotides and then apply the mapping function to convert them to numeric values using the physicochemical property "Shift". On the right, we apply a similar strategy with a sliding window of 3 *bp* to create a mapped vector of length 6 using the trinucleotide physicochemical property "Nucleosome".

With this approach, we can describe the profiles of properties in each dataset. For example, Fig. 3 illustrates all 12 trinucleotide average profiles in the Human_non_tata dataset. There is one graph for each trinucleotide property. Each point presents the average property value in a specific position of the analyzed sequences. The positions in the x-axis give negative values for the upstream part of sequences and positive ones for the downstream portion. Position zero is the TSS. We can see the different patterns in the promoter and non-promoter samples. The non-promoter's averages tend not to show significant variations, while the promoter's averages tend to vary around the TSS position.

2.3 K-Fold Cross-validation Evaluation

K-fold cross-validation [13] is a popular technique used in machine learning and statistical modeling for assessing model performance. It involves dividing the dataset into k subsets or folds, using one fold as the validation set and the

Table 1. List of all physicochemical properties names. There are 38 dinucleotides and 12 trinucleotides.

Number	Dinucleotides	Trinucleotides
1	Base stacking	Bendability (DNase)
2	Protein-induced deformability	Bendability (consensus)
3	B-DNA twist	Trinucleotide GC content
4	Dinucleotide GC content	Nucleosome positioning
5	A-philicity	Consensus-roll
6	Propeller twist	Consensus-rigid
7	Duplex stability (free energy)	DNase I
8	Duplex stability (disrupt energy)	DNase I-rigid
9	DNA denaturation	MW-daltons
10	Bending stiffness	MW-kg
11	Protein-DNA twist	Nucleosome
12	Stabilizing energy of Z-DNA	Nucleosome-rigid
13	Aida_BA_transition	
14	Breslauer_dG	
15	Breslauer_dH	
16	Breslauer_dS	
17	Electron interaction	
18	Hartman_trans_free_energy	
19	Helix-coil_transition	
20	Ivanov_BA_transition	
21	Lisser_BZ_transition	
22	Polar_interaction	
23	SantaLucia_dG	
24	SantaLucia_dH	
25	SantaLucia_dS	
26	Sarai_flexibility	
27	Stability	
28	Stacking_energy	
29	Sugimoto_dG	
30	Sugimoto_dH	
31	Sugimoto_dS	
32	Watson-Crick_interaction	
33	Twist	
34	Tilt	
35	Roll	
36	Shift	
37	Slide	
38	Rise	

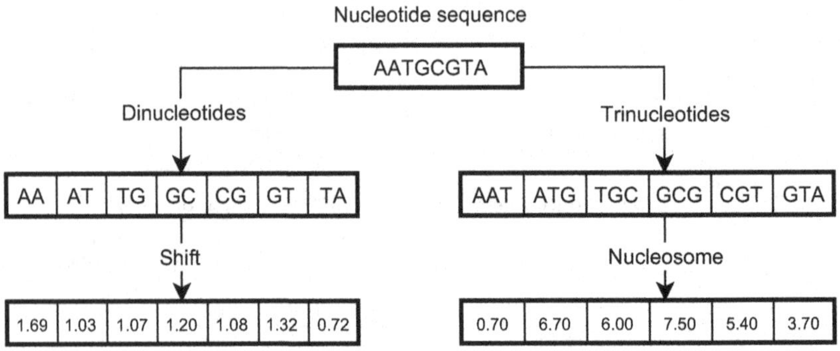

Fig. 2. Illustration of the conversion process of a sequence. On the left, the dinucleotides are mapped to the physicochemical property "Shift". On the right, the trinucleotides are mapped to the physicochemical property "Nucleosome".

remaining k-1 folds as the training set. This process is repeated k times, with each fold serving as the validation set once. By repeatedly training and evaluating the model on different subsets of the data, k-fold cross-validation provides a robust estimate of the model's performance on new, unseen data, helping to mitigate the risk of overfitting.

To address the issue of imbalanced target variables, a modification of k-fold cross-validation called stratified k-fold cross-validation [11] is often employed. Stratified k-fold cross-validation ensures that each fold maintains a similar distribution of classes as the overall dataset. It involves dividing the dataset into k folds and adjusting the fold split to preserve the class distribution in both the training and validation sets. This technique is particularly useful when dealing with imbalanced datasets, as it ensures that the model is evaluated on all classes rather than being biased towards the most frequent class.

The performance metrics obtained from each fold in k-fold cross-validation are typically averaged to obtain an overall estimate of the model's performance. This technique applies to various evaluation metrics and provides a more reliable assessment of the model's generalization ability. By employing k-fold cross-validation, researchers can confidently evaluate their models and make informed decisions regarding model selection and performance comparison in the field of machine learning and statistical modeling.

2.4 Evaluation Measures

The problem of Promoter Classification can be defined as a binary classification task, where the predicted class variable can take on two values: 1 (one) if the DNA sequence is predicted to be a promoter and 0 (zero) otherwise. This classification task leads to four possible outcomes: a true positive (TP), which occurs when the model correctly identifies a promoter sequence; a true negative (TN), which occurs when the model correctly identifies a non-promoter sequence; a

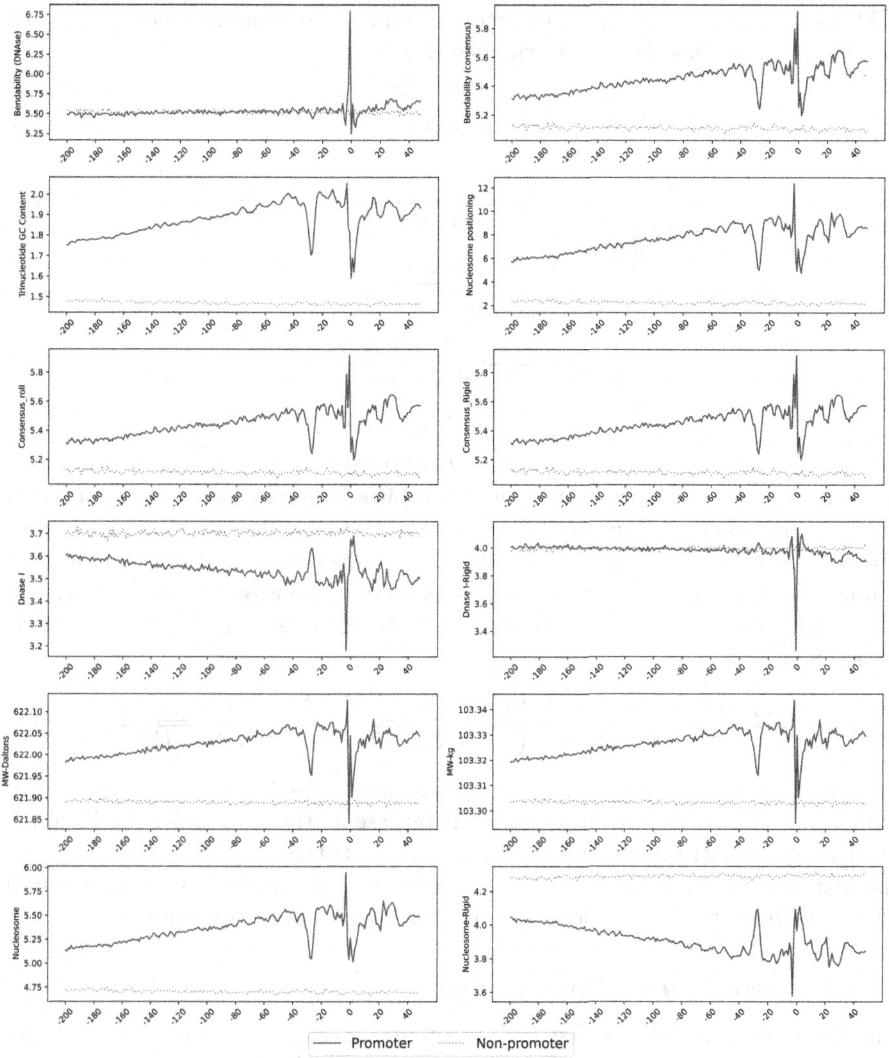

Fig. 3. Average profiles of twelve features along promoter and non-promoter sequences of the 251*bp* dataset Human_non_tata. Each graph is related to a different trinucleotide's physicochemical properties.

false positive (FP), which occurs when the model incorrectly identifies a non-promoter sequence as a promoter; and a false negative (FN), which occurs when the model incorrectly identifies a promoter sequence as a non-promoter.

Precision (Prec) measures the relevance of the predicted true positives (Eq. 1). Sensitivity, also known as Recall (Sn), measures the proportion of true positives (Eq. 2). Specificity (Sp) measures the proportion of true negatives (Eq. 3). Accuracy (Acc) reflects the proportion of correct predictions (Eq. 4). The F1-score

(F1) metric, a harmonic mean of precision and recall, combines both measures and assesses the classifier's performance (Eq. 5).

$$Prec = \frac{TP}{TP + FP} \tag{1}$$

$$Sn = \frac{TP}{TP + FN} \tag{2}$$

$$Sp = \frac{TN}{TN + FP} \tag{3}$$

$$Acc = \frac{TP + TN}{TP + TN + FP + FN} \tag{4}$$

$$F1 = 2 \times \frac{Prec \times Sn}{Prec + Sn} \tag{5}$$

Matthews Correlation Coefficient (MCC) is a balanced metric that considers all four components of the confusion matrix and provides a balanced measure ranging from -1 to +1, with +1 indicating perfect classification, 0 indicating random classification, and -1 indicating completely opposite classification (Eq. 6). It is more robust than other metrics in handling class imbalance and biased datasets [35]

$$MCC = \frac{TP \times TN - FP \times FN}{\sqrt{(TP + FP)(TP + FN)(TN + FP)(TN + FN)}} \tag{6}$$

The Area Under the Curve (AUC) is a metric used to evaluate the performance of a binary classification model. It represents the area under the Receiver Operating Characteristic (ROC) curve, which plots the true positive rate against the false positive rate for different classification thresholds. AUC ranges from 0 to 1, with higher values indicating better classification performance.

2.5 Hyperparameter Optimization

The Tree-structured Parzen Estimator (TPE) [2] is a Bayesian optimization algorithm that efficiently explores the search space to find optimal hyperparameters for machine learning models. TPE models the relationship between hyperparameters and performance metrics by maintaining two probability density functions (PDFs) representing successful and unsuccessful configurations. It utilizes these PDFs to guide the search towards regions of the search space more likely to yield improved performance.

During the tuning process, TPE samples hyperparameter configurations based on its probability distributions, evaluate model performance using cross-validation and updates the PDFs accordingly. This iterative approach enables the algorithm to explore and refine the search space, ultimately leading to the identification of hyperparameter configurations that optimize model performance. TPE's combination of exploration and exploitation strategies makes it a powerful tool for effectively and efficiently optimizing machine learning models.

2.6 Ensemble Methods

The voting ensemble method [23] combines predictions from multiple individual models to make a final prediction. It involves aggregating the outputs of different models, either by majority voting (for classification problems) or by averaging (for regression problems). By combining the predictions of multiple models, the voting ensemble could enhance the overall prediction accuracy and robustness.

The stacking ensemble method [32] leverages the predictions of multiple base models to train a meta-model, which learns to make predictions based on the outputs of the base models. The predictions, along with the original target variable, are then used to train a meta-model that learns to combine the base models' outputs. The stacking ensemble can capture complex interactions among the models and exploit their complementary strengths, leading to improved prediction performance.

3 Experiments and Results

In order to assess the performance of physicochemical properties in the context of promoter prediction, a series of experiments were conducted. The evaluation encompassed the utilization of individual properties as well as combinations of the most promising ones across seven diverse datasets representing various organism types. To ascertain the effectiveness of the proposed methods within the domain of the problem, the obtained results were compared against a baseline that is regarded as state-of-the-art on these specific datasets.

3.1 Experiment Workflow

The implemented processing workflow encompasses the transformation of raw DNA sequences within a given dataset into physicochemical property values. Subsequently, one or more properties are selected, and their corresponding values are concatenated to form the feature set. The evaluation process involves a 5-fold cross-validation, which partitions the dataset into training and test subsets. Within the training subset, a further 5-fold cross-validation is conducted to compare various machine learning methods and identify the optimal one for hyperparameter tuning. Finally, the tuned model is evaluated using the test dataset.

To conduct our experiments, we utilized PyCaret 3[2], an advanced machine learning library. PyCaret offers a comprehensive range of eighteen classification methods, which were employed in our study (Table 2).

3.2 Individual Properties' Performances

Using the 5-fold cross-validation method, we compared several machine learning models to observe the prediction power of each physicochemical property.

[2] Official page: https://pycaret.gitbook.io/.

Table 2. Metrics results of the best models with only one property. We show the property and the results of the evaluated model in each dataset.

Dataset	Property	Acc	AUC	Sn	Prec	F1	MCC
*Arabidopsis*_non_tata	dinuc18	0.91	0.96	0.84	0.89	0.86	0.79
*Arabidopsis*_tata	dinuc18	0.96	0.99	0.95	0.95	0.95	0.92
Bacillus subtilis	dinuc18	0.88	0.93	0.71	0.83	0.76	0.69
*Escherichia coli s*70	dinuc23	0.91	0.95	0.71	0.84	0.77	0.71
Human_non_tata	dinuc15	0.86	0.92	0.78	0.88	0.83	0.72
Mouse_non_tata	dinuc02	0.92	0.97	0.88	0.91	0.89	0.82
Mouse_tata	dinuc01	0.87	0.97	0.87	0.87	0.87	0.78

Then, we chose the best one that uses that property. After, we tuned the hyperparameters of the selected models using the same training data and the 5-fold cross-validation. Figure 4 presents the results from the tuned models. The bars show the MCC scores' maximum, minimum, and average from evaluating these models. Each dataset has fifty models, one for each property, among dinucleotides and trinucleotides. The graph suggests that the models trained on prokaryotic datasets get worse results than those from the eukaryotic datasets. The labels on the x-axis of the graph refer to Table 1. Although we tested eighteen different models, all the best models were the Light Gradient Boosting Machine.

3.3 Performance Comparisons

We utilized CNNProm [30] and MCC scores baseline for comparing the performance of the models under evaluation, given our belief that it represents the state-of-the-art solution for this problem. The corresponding results can be found in Table 3. The scores attributed to CNNProm are as reported by its authors. Columns labeled as "Single" denote the best-performing individual property model. The columns "Vote5", "Vote10", "Stack5", and "Stack10" represent results from ensemble models that employed either the voting or stacking method, incorporating the top 5 or 10 properties, respectively.

The performance of the "Single" model demonstrated relatively lower average results when compared with other models. Interestingly, the outcomes derived from the ensemble models did not exhibit a significant discrepancy whether they incorporated the top 5 or 10 properties, insinuating that an increased number of properties does not substantially enhance the final model's efficacy. Likewise, no notable divergence was observed between the "Vote" and "Stack" ensemble methodologies.

When comparing the investigated models to the baseline model CNNProm, we observed similar performance in 3 out of 7 datasets. Specifically, in the *Arabidopsis*_non_tata dataset, our model exhibited a marginally lower performance by 1% compared to CNNProm. Conversely, in the *Arabidopsis*_tata dataset, our

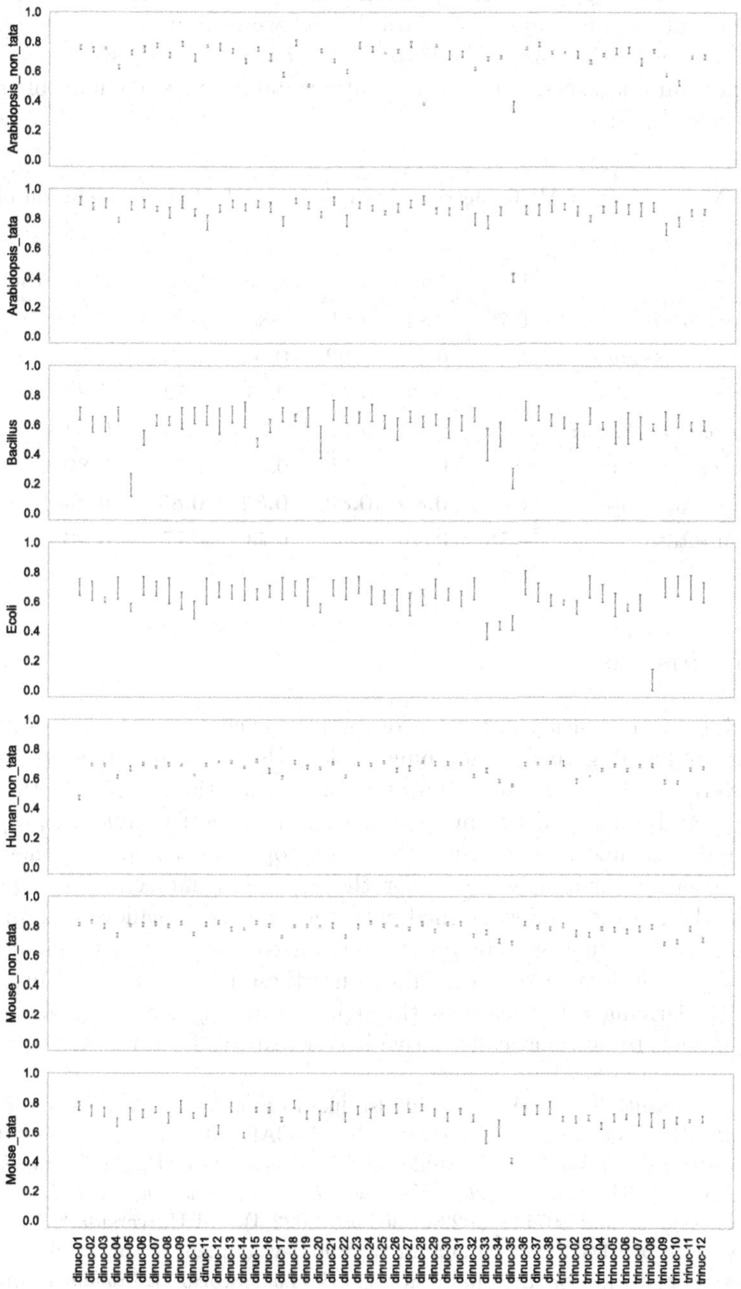

Fig. 4. MCC scores (y-axis) from 5-fold cross-validation of each physicochemical property (x-axis) model evaluated on the training subset for each dataset. The graph shows the maximum, minimum, and average computed scores.

model achieved a slightly higher performance by 1% in comparison. In addition, the results for the Mouse_non_tata dataset were equivalent to the baseline model. However, in the *Bacillus subtilis*, *Escherichia coli s*70 (procaryotes), and Human_non_tata datasets, our implementations did not yield comparable results to the baseline model.

Table 3. Mean values of MCC metrics obtained from 5-fold cross-validation on each dataset.

Dataset	Single	$Vote_5$	$Vote_{10}$	$Stack_5$	$Stack_{10}$	CNNProm
*Arabidopsis*_non_tata	0.79	0.84	0.84	0.85	0.85	**0.86**
*Arabidopsis*_tata	**0.92**	0.91	**0.92**	**0.92**	**0.92**	0.91
Bacillus subtilis	0.69	0.76	0.75	0.73	0.72	**0.86**
*Escherichia coli s*70	0.71	0.75	0.75	0.74	0.75	**0.84**
Human_non_tata	0.72	0.74	0.74	0.74	0.74	**0.89**
Mouse_non_tata	0.82	**0.83**	**0.83**	**0.83**	**0.83**	**0.83**
Mouse_tata	0.78	0.80	0.80	0.79	0.77	**0.93**

4 Conclusions

This work presents a study of the use of the physicochemical properties of DNA as features in training machine-learning models. Using different metrics, we evaluated several models using seven datasets of procaryotic and eucaryotic organisms. Our study analyzed the prediction performance of individual properties models and ensemble models using the best properties, comparing them with the state-of-art method CNNProm for the evaluated datasets. We concluded that even though the models trained with the properties achieved competitive prediction results, they were unable to surpass the established baseline. So, it is necessary to evaluate new strategies to use these physicochemical properties values, like deriving other features through feature engineering or using deep learning models to automatically derive better features from them.

Acknowledgment. The authors would also like to thank the *Coordenação de Aperfeiçoamento de Pessoal de Nível Superior* - Brazil (CAPES) - Finance Code 001, *Fundacão de Amparo à Pesquisa do Estado de Minas Gerais* (FAPEMIG, grants APQ-01518-21, APQ-01647-22), *Conselho Nacional de Desenvolvimento Científico e Tecnológico* (CNPq, grants 307151/2022-0, 308400/2022-4) and Universidade Federal de Ouro Preto (PROPPI/UFOP) for supporting the development of this study. We want to express our gratitude for the collaboration of the *Laboratório Multiusuários de Bioinformática* of *Núcleo de Pesquisas em Ciências Biológicas* (NUPEB/UFOP).

References

1. Arslan, H.: A new promoter prediction method using support vector machines. In: 2019 27th Signal Processing and Communications Applications Conference (SIU), pp. 1–4. IEEE (2019)
2. Bergstra, J., Bardenet, R., Bengio, Y., Kégl, B.: Algorithms for hyper-parameter optimization. In: Advances in Neural Information Processing Systems, vol. 24 (2011)
3. Bhandari, N., Khare, S., Walambe, R., Kotecha, K.: Comparison of machine learning and deep learning techniques in promoter prediction across diverse species. PeerJ Comput. Sci. **7**, e365 (2021)
4. Breiman, L.: Random forests. Mach. Learn. **45**, 5–32 (2001)
5. Cartharius, K., et al.: Matinspector and beyond: promoter analysis based on transcription factor binding sites. Bioinformatics **21**(13), 2933–2942 (2005)
6. Chen, T., Guestrin, C.: XGBoost: a scalable tree boosting system. In: Proceedings of the 22nd ACM SIGKDD International Conference on Knowledge Discovery and Data Mining, pp. 785–794 (2016)
7. Chen, W., Lei, T.Y., Jin, D.C., Lin, H., Chou, K.C.: PSEKNC: a flexible web server for generating pseudo k-tuple nucleotide composition. Anal. Biochem. **456**, 53–60 (2014)
8. Chevez-Guardado, R., Peña-Castillo, L.: Promotech: a general tool for bacterial promoter recognition. Genome Biol. **22**, 1–16 (2021)
9. Cortes, C., Vapnik, V.: Support-vector networks. Mach. Learn. **20**, 273–297 (1995)
10. Deaton, A.M., Bird, A.: CPG islands and the regulation of transcription. Genes Dev. **25**(10), 1010–1022 (2011)
11. Dietterich, T.G.: Approximate statistical tests for comparing supervised classification learning algorithms. Neural Comput. **10**(7), 1895–1923 (1998)
12. Dreos, R., Ambrosini, G., Cavin Périer, R., Bucher, P.: EPD and EPDNEW, high-quality promoter resources in the next-generation sequencing era. Nucleic Acids Res. **41**(D1), D157–D164 (2013)
13. Efron, B.: Estimating the error rate of a prediction rule: improvement on cross-validation. J. Am. Stat. Assoc. **78**(382), 316–331 (1983)
14. Freund, Y., Schapire, R.E.: A decision-theoretic generalization of on-line learning and an application to boosting. J. Comput. Syst. Sci. **55**(1), 119–139 (1997)
15. Friedman, J.H.: Greedy function approximation: a gradient boosting machine. Ann. Stat. 1189–1232 (2001)
16. Gama-Castro, S., et al.: Regulondb version 9.0: high-level integration of gene regulation, coexpression, motif clustering and beyond. Nucleic Acids Res. **44**(D1), D133–D143 (2016)
17. Goñi, J.R., Pérez, A., Torrents, D., Orozco, M.: Determining promoter location based on DNA structure first-principles calculations. Genome Biol. **8**(12), R263 (2007)
18. Hochreiter, S., Schmidhuber, J.: Long short-term memory. Neural Comput. **9**(8), 1735–1780 (1997)
19. Ishii, T., Yoshida, K.i., Terai, G., Fujita, Y., Nakai, K.: DBTBS: a database of bacillus subtilis promoters and transcription factors. Nucleic Acids Res. **29**(1), 278–280 (2001)
20. Juven-Gershon, T., Kadonaga, J.T.: Regulation of gene expression via the core promoter and the basal transcriptional machinery. Dev. Biol. **339**(2), 225–229 (2010)

21. Ke, G., et al.: LightGBM: a highly efficient gradient boosting decision tree. In: Advances in Neural Information Processing Systems, vol. 30 (2017)
22. Kotsiantis, S., Kanellopoulos, D., Pintelas, P., et al.: Handling imbalanced datasets: a review. GESTS Int. Trans. Comput. Sci. Eng. **30**(1), 25–36 (2006)
23. Kuncheva, L.I.: Combining pattern classifiers: methods and algorithms. John Wiley & Sons (2014)
24. LeCun, Y., Bottou, L., Bengio, Y., Haffner, P.: Gradient-based learning applied to document recognition. Proc. IEEE **86**(11), 2278–2324 (1998)
25. Li, F., et al.: Computational prediction and interpretation of both general and specific types of promoters in Escherichia coli by exploiting a stacked ensemble-learning framework. Brief. Bioinform. **22**(2), 2126–2140 (2021)
26. Meng, H., Ma, Y., Mai, G., Wang, Y., Liu, C.: Construction of precise support vector machine based models for predicting promoter strength. Quant. Biol. **5**, 90–98 (2017)
27. Moraes, L., Silva, P., Luz, E., Moreira, G.: CapsProm: a capsule network for promoter prediction. Comput. Biol. Med. **147**, 105627 (2022)
28. Pedersen, A.G., Baldi, P., Chauvin, Y., Brunak, S.: The biology of eukaryotic promoter prediction-a review. Comput. Chem. **23**(3–4), 191–207 (1999)
29. Sabour, S., Frosst, N., Hinton, G.E.: Dynamic routing between capsules. In: Advances in Neural Information Processing Systems, vol. 30 (2017)
30. Umarov, R.K., Solovyev, V.V.: Recognition of prokaryotic and eukaryotic promoters using convolutional deep learning neural networks. PLoS ONE **12**(2), e0171410 (2017)
31. Wasserman, W.W., Sandelin, A.: Applied bioinformatics for the identification of regulatory elements. Nat. Rev. Genet. **5**(4), 276–287 (2004)
32. Wolpert, D.H.: Stacked generalization. Neural Netw. **5**(2), 241–259 (1992)
33. Zeng, J., Zhu, S., Yan, H.: Towards accurate human promoter recognition: a review of currently used sequence features and classification methods. Brief. Bioinform. **10**(5), 498–508 (2009)
34. Zhang, M., et al.: Critical assessment of computational tools for prokaryotic and eukaryotic promoter prediction. Brief. Bioinform. **23**(2) (2022)
35. Zhu, Q.: On the performance of Matthews correlation coefficient (MCC) for imbalanced dataset. Pattern Recogn. Lett. **136**, 71–80 (2020)

Machine Learning Analysis

Critical Analysis of AI Indicators in Terms of Weighting and Aggregation Approaches

Renata Pelissari[1], Betania Campello[2], Guilherme Dean Pelegrina[2](✉), Ricardo Suyama[3], and Leonardo Tomazeli Duarte[2]

[1] School of Electrical Engineering and Computation at the University of Campinas, Campinas, Brazil

[2] School of Applied Sciences at the University of Campinas, Limeira, Brazil

b180839@dac.unicamp.br,guidean@unicamp.br,leonardo.duarte@fca.unicamp.br

[3] Center for Engineering, Modeling and Applied Social at Federal University of ABC, Santo André, Brazil

ricardo.suyama@ufabc.edu.br

Abstract. National Artificial Intelligence (AI) strategies have been implemented by several countries worldwide. These strategies aim to guide AI policy priorities and foster research, innovation, and development in AI. Alongside the development of national AI strategies, AI indices have emerged as tools to compare countries' AI development levels. This study focuses on a specific AI indicator, the Global AI Index (GAII) proposed by Tortoise Media, which ranks 62 countries based on their level of investment, innovation, and implementation of AI. The GAII computes a ranking of countries by aggregating sub-dimensions within these categories using a weighted sum approach, with subjective weight assignments. This paper critically analyzes the weighting and aggregation approaches used in the GAII, employing two techniques. Firstly, the Stochastic Multicriteria Acceptability Analysis (SMAA) is used to explore changes in rankings by varying the weights. Secondly, a non-additive aggregation model known as the Choquet integral is applied to consider potential interactions among dimensions. The findings indicate that the weights assigned to criteria strongly influence the final ranking. Additionally, there are interactions between AI dimensions that should be taken into account to address unbalanced achievements across dimensions. This study contributes to the development of more robust and objective methodologies for comparing countries' AI development levels.

Keywords: Global AI Index · Multicriteria decision analysis · SMAA

Work supported by São Paulo Research Foundation (FAPESP) under the grants #2020/09838-0 (BIOS - Brazilian Institute of Data Science), #2020/10572-5 and #2023/04159-6. L. T. Duarte and R. Suyama would like to thank the National Council for Scientific and Technological Development (CNPq, Brazil) for the financial support.

M. C. Naldi and R. A. C. Bianchi (Eds.): BRACIS 2023, LNAI 14196, pp. 385–399, 2023.
https://doi.org/10.1007/978-3-031-45389-2_26

1 Introduction

Launched in 2017, Canada's national Artificial Intelligence (AI) Strategy was the first in the world with the aim of guiding AI policy priorities at a country level. This helped Canada to guide AI policy definition and prioritize investments, to stimulate research, innovation and development of solutions in AI as well as its conscientious and ethical aspects [1]. Finland developed its national AI strategy also in 2017, closely followed by Japan, France, Germany and the United Kingdom in 2018. More than 30 other countries and regions have launched their national AI strategies as of 2021[1], including Brazil.

Accompanying the development of national AI strategies, AI indices started to be built in order to compare nations on their level of AI development. The Global AI Index proposed by Tortoise Media provides a ranking of 62 countries around the world in order to benchmark nations on their level of investment, innovation and implementation of AI [2]. The Global AI Vibrancy Tool[2] proposed by the Stanford Institute for Human-Centered Artificial Intelligence (Stanford HAI) provides a weighted ranking of 29 countries with the aim of identifying the countries that are leading AI race in terms of two main dimensions: Research and Development and Economy.

Each AI dimension used in the construction of an indicator is associated with a relative importance weight, intending to account for the fact that contributions of different dimensions to the level of AI development have varying degrees of impact levels. Tortoise Media adopted a weighting approach based on subjective assumptions, which may affect the composite scoring for each country and consequently their position in the ranking. On the other hand, Stanford HAI's approach, worrying about the subjectivity of dimension weights and its impact on the final ranking, adopts a user preference-based weighting methodology and provides an interactive tool that allows the user to change the dimension weights, obtaining different rankings depending on the preference declared by the user.

Despite the differences in the aforementioned weighting approaches, those indices are similar in the sense that they are built based on a linear aggregation of multiple dimensions–although different dimensions are adopted in each of them. Generally, when one assumes a linear aggregation, one does not consider interactions among criteria. Indeed, when defining the set of criteria, non-redundancy is a desirable property [3]. However, one frequently observes correlated criteria in real applications and, therefore, adopting an approach that models interaction between criteria could be useful to avoid biased results [4].

As pointed out in the composite indicator literature [5,6], indicators that take into account multiple dimensions should be aggregated and weighted accordingly, considering correlation and compensability issues among indicators and avoiding subjective weighting approaches. Robustness analysis should be undertaken to assess the impact of assumptions and hypotheses set when building the composite indicator in terms of the choice of weights and the aggregation method (among

[1] https://hai.stanford.edu/news/state-ai-10-charts. Accessed date: 16 May 2023.
[2] https://aiindex.stanford.edu/vibrancy/. Accessed date: 16 May 2023.

others). As a consequence, it is likely that changes in these assumptions may be required, thus ultimately leading to different final decisions.

Therefore, motivated by the requirements that usually stands behind the development of robust indicators, this paper presents a critical analysis of AI indicators when comparing countries in terms of weighting and aggregation approaches. We attempt to answer the following 3 research questions:

1. Do the criteria weights influence the resulting AI ranking of countries?
2. Is the hypothesis of interactions between AI dimensions true?
3. Does the use of a non-linear aggregator in order to consider the interaction between criteria influence the resulting AI ranking of countries?

In the conducted analysis we consider the Global AI Index proposed by Tortoise because it presents a ranking with a greater number of countries and considers a greater number of dimensions in the analysis, and also due to the ease of data acquisition. In order to answer those questions, we apply the MCDA (Multiple Criteria Decision Aid) methods SMAA and Choquet Integral.

The first question is answered by conducting robustness analyses through the application of the Stochastic Multicriteria Acceptability Analysis (SMAA) [7,8]. SMAA is based on an inverse weight space analysis in order to describe the criteria weights that make each country the most "preferred" one. Therefore, it does not require weights to be pre-defined. SMAA can also be used with different decision models besides the weighted sum. The result given by SMAA–among other descriptive measures–is the probability of a country occupying each of the positions in the ranking. Since SMAA considers simultaneously the uncertainty in all parameters, it is particularly useful for robustness analysis. We apply SMAA with the weighted sum varying the weights and conduct a comprehensive analysis of the ranking variation, comparing the results with the Tortoise ranking. We consider two scenarios: weight information totally missing and weights following the preference order adopted in the Tortoise Index.

In order to answer Question 2, we show that the AI dimensions are statistically redundant. Finally, to answer Question 3, we evaluate the use of a non linear aggregator, called Choquet integral [9], which takes into account interaction among criteria. We also compare the ranking obtained by means of the Choquet integral with the Tortoise ranking.

This paper is organized as follows. In Sect. 2, we introduce the Multiple Criteria Decision Aid problem. Section 3 presents an overview of the methodology used in Tortoise to derive the Global AI Index. In Sect. 4, we provide the theoretical background on the SMAA methodology and the Choquet integral. Section 5 presents the methodology adopted in this paper. Results are presented and discussed in Sect. 6. We conclude this paper in Sect. 7.

2 Multiple Criteria Decision Aid

Multiple Criteria Decision Aid (MCDA) is an area of research concerned with mathematical and computational design tools that can be used either by an individual decision-maker (DM) or a group of DMs, to evaluate a finite number of decision alternatives regarding a set of performance criteria, which are determined according to the decision context [10]. These approaches make it possible to reduce the subjectivity inherent in decision-making processes while considering the preferences of the DM(s).

The DM is who has the power over the decision and is responsible for setting the model parameters. The set of alternatives $\mathcal{A} = \{a_1, a_2, \ldots, a_m\}$ is a finite set of m elements, with $m \geq 2$, being all of them considered possible solutions for the studied problem. Decision criteria $G = \{g_1, g_2, \ldots, g_n\}$ are qualitative or quantitative attributes used to evaluate the different alternatives. In MCDA we assume that there are at least two criteria ($n \geq 2$). For each criterion g_j is given a relative importance w_j called criterion weight, $j = 1, \ldots, n$. We denote $\mathbf{w} = (w_1, w_2, \ldots, w_n)$ the criteria weight vector.

Each alternative a_i is evaluated according to each criterion g_j, for $i = 1, \ldots, m$ and $j = 1, \ldots, n$, representing the performance of the alternative in relation to that criterion. $g_j(a_i)$ denotes the performance of the alternative a_i in relation to the criterion g_j, for $i = 1, \ldots, m$ and $j = 1, \ldots, n$.

The different MCDA methods aim to solve a decision-making problem with m alternatives $\mathcal{A} = \{a_1, \ldots, a_m\}$, evaluated according to n criteria $\{g_1, \ldots, g_n\}$. Without loss of generality, it is assumed that all criteria must be maximized. Thus, the decision problem is defined as (1)

$$\max\{g_1(a), \ldots, g_n(a)/a \in \mathcal{A}\}. \tag{1}$$

A utility function $u(a_i, \mathbf{w})$ is then applied as the aggregation procedure in order to obtain a utility value–or score–that, in the context of the ranking problematic, is used to order the set of alternatives obtaining a final ranking.

3 Tortoise's Approach to Deriving the Global AI Index: An Overview of the Methodology

The Global AI Index (GAII) proposed by Tortoise Media [2] aims at ranking 62 countries, represented by $[a_1, \ldots, a_{62}]$, based on their level of development in articial intelligence, which is measured by combining three categories of indicators:

Implementation: This category evaluates how artificial intelligence is being implemented by businesses, governments, and communities. It comprises three dimensions: Talent, Infrastructure, and Operating Environment;

Innovation: This category measures technological advancements and methodological breakthroughs that indicate a greater potential for artificial intelligence in the future. This pillar is divided into two dimensions: Research and Development;

Investment: This category assesses the financial and procedural commitments made towards artificial intelligence, and is composed of two dimensions: Commercial Ventures and Government Strategy.

Scores for each of the seven dimensions–also called criteria–are obtained by aggregating several sub-criteria. These scores are displayed on the Tortoise website[3]. Then, a Total Score (TS) for each country is obtained through a weighted sum (WS), calculated as

$$\mathrm{TS}_i^{WS} = \frac{1}{\sum_{j=1}^{n} w_j} \sum_{j=1}^{n} w_j g_j(a_i), \forall i, \qquad (2)$$

where w_j represents the weight assigned to criterion g_j, and $g_j(a_i)$ is the score of country a_i over criterion g_j, for $i = 1, \cdots, 62$ and $j = 1, \cdots, 7$. The weights used by Tortoise to obtain the total score are: Talent ($w_1 = 5$), Infrastructure ($w_2 = 3$), Operating Environment ($w_3 = 2$), Research ($w_4 = 5$), Development ($w_5 = 3$), Government strategy ($w_6 = 1$), and Commercial ventures ($w_7 = 5$). The scores of the top six countries according to the Tortoise GAII are illustrated in Table 1.

Table 1. Scores for each AI dimension and the total score (TS) of the top six countries determined by Tortoise GAII.

Country	Tal.	Infra.	OP	Res.	Dev.	GS	Com.	TS
USA	100	94.02	64.56	100	100	77.39	100	95.38
China	16.51	100	91.57	71.42	79.97	94.87	44.02	61.57
UK	39.65	71.43	74.65	36.52	5.03	82.82	18.91	41.53
Canada	31.29	77.06	93.94	30.67	25.78	100	14.88	40.85
Israel	35.77	67.59	82.44	32.63	27.96	43.91	27.33	40.58
Singapore	39.39	84.34	3.15	37.67	22.55	79.82	15.07	39.47

One may note in Table 1 that the USA has emerged as the leader in the GAII, scoring the maximum possible points in four out of the seven criteria: Talent, Infrastructure, Operating Environment, and Government Strategy. As a result, the USA received the highest overall score among the 62 countries ranked by the GAII.

This approach provides an understanding of a country's AI capacity by considering multiple factors, and the weights assigned to each criterion reflect their relative importance in determining a country's overall score. However, two main drawbacks related to Tortoise GAII can be highlighted. Firstly, the assignment of weights can be considered subjective, as noted by Tortoise itself [2]. Secondly, the adoption of the weighted sum as an aggregator is based on the unverified hypothesis of independence between the criteria.

3 https://www.tortoisemedia.com/intelligence/global-ai/. Accessed date: 16 May 2023.

4 Preliminaries

In this section, we present the building blocks of our proposal: the SMAA algorithm and the Choquet integral, including a description on the adjustment of the Choquet integral parameters.

4.1 SMAA

SMAA (Stochastic Multicriteria Acceptability Analysis) is a simulation-based method for discrete multicriteria decision makings problems where model parameters are uncertain, imprecise, or, specifically in the case of criteria weights, partially or totally missing [7,11]. Uncertain information is represented by probability distributions. Throughout a Monte-Carlo simulation process, values for the uncertain variables are sampled from their distributions, and alternatives are evaluated by applying the decision model that can be a weighted average, for instance, or any other aggregation procedure. In a ranking problem, SMAA determines all possible rankings for alternatives and quantifies the results in terms of probabilities. Usually, the recommended solution is the ranking with the highest probability.

Different SMAA variants have been proposed in the literature [11]. The basis for most of those variants is the SMAA-2 version proposed in [8]. SMAA-2 computes three descriptive measures. The rank acceptability index, denoted by b_i^s, describes the probability of an alternative a_i being in s-th position of the rank. It ranges between 0 and 1, and the closer b_i^s is to 1, the greater the probability of a_i being in position s. Central weight vector w_i^c describes Decision-Maker (DM) preferences supporting alternative a_i being ranked first. Central weight vectors allow an inverse decision-making approach: DM can learn the weights that lead an alternative to rank first instead of previously defining them and building a solution for the problem. The confidence factor p_i^c is the probability of an alternative being the preferred one with the criteria weights expressed by its central weight vector.

As probabilities for all possible solutions are provided by SMAA, this is a methodology that describes how robust the model is subject to different uncertainties in the input data, being so a useful tool for robustness analysis. The rank acceptability index can support this analysis. For instance, alternatives with high acceptability for the best ranks are candidates for occupying the best places in the rank, while alternatives with large acceptability for the worst ranks should be avoided in the best positions even if they would have fairly high acceptability for the best ranks. If none of the alternatives receives high acceptability indices for the best ranks, it indicates a need to measure the criteria, preferences or both more accurately.

4.2 Choquet Integral

The (discrete) Choquet integral (CI) [9] is a non-linear aggregation function that takes into account interaction among criteria. It is defined as follows:

$$TS_i^{CI} = \sum_{j=1}^{n} \left[g_{(j)}(a_i) - g_{(j-1)}(a_i)\right] \mu\left(\{(j), \ldots, (n)\}\right), \tag{3}$$

where $(1), \ldots, (n)$ indicates a permutation of the indices j such that $0 = g_{(0)}(a_i) \leq g_{(1)}(a_i) \leq g_{(j)}(a_i) \leq \ldots \leq g_{(n)}(a_i)$ and $\mu(\cdot)$ represents the set of parameters known as capacity coefficients. A capacity $\mu : 2^N \rightarrow \mathbb{R}_+$, where $N = \{1, 2, \ldots, n\}$ is the set of criteria, is a set function that satisfies the following axioms:

(a) Normalization: $\mu(\emptyset) = 0$ and $\mu(N) = 1$,
(b) Monotonicity: $\forall A \subseteq B \subseteq N, \mu(S) \leq \mu(T)$.

An interesting aspect of the Choquet integral is that the capacity coefficients are associated with the Shapley values [13], i.e., a well-known solution concept from game theory. In multicriteria decision making, the Shapley value of a criterion j, represented by ϕ_j, indicates its marginal contribution on the aggregation procedure. The linear relation between μ and ϕ_j is given as follows:

$$\phi_j = \sum_{A \subseteq N \setminus \{j\}} \frac{(n - |A| - 1)! \, |A|!}{n!} \left[\mu(A \cup \{j\}) - \mu(A)\right], \tag{4}$$

where $|A|$ represents the cardinality of subset A. A property associated with the Shapley values that are useful when learning the capacity coefficients and interpreting the obtained parameters is that $\phi_j \geq 0$, $\forall j = 1$. Therefore, the importance assigned to each criterion is at least zero and the higher the ϕ_j, the higher criterion j contributes to the aggregation.

Besides the marginal contributions, one may also interpret the interaction between criteria. In this case, the Shapley interaction index between criteria j, j' is given by [14,15]

$$I_{j,j'} = \sum_{A \subseteq N \setminus \{j,j'\}} \frac{(n - |A| - 2)! \, |A|!}{(n-1)!} \left[\mu(A \cup \{j, j'\}) - \mu(A \cup \{j\}) - \mu(A \cup \{j'\}) + \mu(A)\right]. \tag{5}$$

Each $I_{j,j'}$ can be interpreted as the interaction degree between criteria j, j'. If $I_{j,j'} < 0$, there is a negative interaction (or redundant effect) between criteria j, j'. If $I_{j,j'} > 0$, there is a positive interaction (or complementary effect) between criteria j, j'. In the case where $I_{j,j'} = 0$, there is no interaction criteria j, j', and they act independently.

Although there exists a generalization of Shapley index to any coalition of criteria (see [16] for further details), for the scope of this paper, we will restrict this parameters to singletons and pairs of criteria. Indeed, in this study, we consider a particular case of the Choquet integral, which is based on the notion of 2-additive capacity [16]. A 2-additive capacity implies that only exists interactions between pairs of criteria. In other works, we consider that the interaction among 3 or more criteria must be zero. Based on a 2-additive capacity, the Choquet integral can be defined by means of the Shapley values and Shapley interaction indices as follows [16]:

$$TS_i^{2adCI} = \sum_{I_{j,j'}>0} \min\{g_j(a_i), g_{j'}(a_i)\}I_{j,j'} + \sum_{I_{j,j'}<0} \max\{g_j(a_i), g_{j'}(a_i)\}|I_{j,j'}| + \sum_{j=1}^{n} g_j(a_i)(\phi_j - \frac{1}{2}\sum_{j'\neq j}|I_{j,j'}|). \quad (6)$$

An interesting aspect of Eq. (6) in comparison with Eq. (3) is that one reduces the number of parameters from 2^n to $n(n+1)/2$. Surely, we lose flexibility to model all kinds of interaction among criteria. However, the 2-additive Choquet integral offers a good trade-off between flexibility and model complexity [17,18].

When one assumes a 2-additive capacity, one may redefine the axioms of a capacity in terms of the Shapley values and interaction indices as

$$\sum_{j=1}^{n} \phi_j = 1 \quad (7)$$

and

$$\phi_j - \frac{1}{2}\sum_{j'\neq j}|I_{j,j'}| \geq 0, \forall j \in N. \quad (8)$$

4.3 An Unsupervised Approach to Learn the Choquet Integral Parameters

Once one adopts the Choquet integral as the aggregation function to calculate the scores and rank the alternatives, one needs to define its parameters. This task in the Choquet integral is quite complicated, as one has several parameters to be defined. However, one may adopt a strategy that can automatically adjust some parameters without defining them subjectively.

Inspired by [19], we consider in this paper a non-supervised approach to automatically adjust the Shapley interaction indices. The goal is to define the interaction index $I_{j,j'}$ as close as possible from the negative of a similarity measure $\rho_{j,j'}$ between pairs of criteria, such as the correlation coefficient between them [20]. The idea behind this approach is to mitigate, for instance, biased results provided by correlated criteria. Suppose that two criteria are positively correlated. If we do not take this data structure characteristic into account, when one aggregates the evaluations provided by these criteria, one may sum twice the same information. Therefore, as the Choquet integral can model interaction between criteria, one may define a negative interaction index (which models a redundant effect) to positively correlated criteria. This will reduce the impact of criteria correlations.

The optimization problem used to automatically adjust the Shapley interaction indices is given as follows:

$$\begin{array}{ll} \min\limits_{I_{j,j'}, \forall j,j' \in N} & \sum_{j,j'} \left(I_{j,j'} + \rho_{j,j'} \right)^2 \\ \text{s.t.} & \phi_j - \frac{1}{2} \sum_{j' \neq j} \pm I_{j,j'} \geq 0, \; \forall j \in N, \\ & \sum_j \phi_j = 1 \end{array} \tag{9}$$

where $\pm$ in the first constraint avoids the use of absolute values. Note that, in this optimization problem, we do not find the Shapley values. Indeed, they should be (subjectively) predefined. Moreover, as it is a quadratic problem, it can be easily tackled by most of the available solvers.

5 Methodology

In order to answer the 3 research questions set in this study, we conducted four analyses of the GAII proposed by Tortoise [2] considering the same 62 countries as alternatives and the same seven AI dimensions as criteria: Talent (g_1), Infrastructure (g_2), Operating (g_3), Environment (g_4), Research (g_5), Development (g_6), Commercial Ventures (g_7) and Government Strategy (g_8). We take the scores displayed in Tortoise's website[4] as criteria performance.

In the first analysis, we apply the weighted sum of Eq. (2) to obtain a total score for each country, as in GAII, assuming weight information is totally missing. Throughout SMAA application, weights are randomly generated and the rank acceptability index is given as result. The weighted sum is also applied with SMAA in the second analysis, but weights follow an ordinal preference. We assume the same order of preference as the one adopted in the construction of the Tortoise index [2], i.e., w_1 (Talent) $= w_4$ (Research) $= w_7$ (Commercial ventures) $> w_2$ (Infrastructure) $= w_5$ (Development) $> w_3$ (Operating Environment) $> w_6$ (Government strategy). In this case, throughout SMAA application, weights are randomly generated by respecting the constraint imposed by these preferences. A comprehensive analysis based on the rank acceptability index is then conducted, comparing these two results with the Tortoise ranking.

The third analysis consists in verify the redundancies in the dataset, measured by the correlation coefficient between pairs of criteria. In the fourth analysis, we evaluate the use of the 2-additive Choquet integral to aggregate the criteria information. For this purpose, we apply the non-supervised approach, presented in Sect. 4.3, to obtain the interaction indices $I_{j,j1}$ that are as close as possible to the negative of the correlation coefficients. As the Shapley values ϕ_j, $j = 1, \ldots, n$, we assume the same weights as in the Tortoise analysis. However, we normalize them as follows:

$$\phi_j = \frac{w_j}{\sum_{j=1}^{n} w_j}. \tag{10}$$

[4] https://www.tortoisemedia.com/intelligence/global-ai/ - Accessed date: 16 May 2023.

Based on the obtained interaction indices, we apply the Choquet integral expressed in Eq. (6). The obtained scores are used to construct the new ranking. We then compare the ranking provided by the Choquet integral with respect to the Tortoise one. We verify if an approach that models criteria interactions may lead to a different ranking of countries.

6 Results and Discussion

In this section, we present and discuss the results of the conducted analysis.

6.1 Probabilistic Country Rankings: A Weight Sum Method and SMAA Perspective

The first analysis refers to the application of the weighted sum and SMAA with randomly generated weights. Table 2 shows the obtained rank acceptability indices, with the top six countries ranked according to Tortoise GAII. We have highlighted the highest percentage in each column to indicate the country with the highest probability of being in that particular position. Results indicate that the USA and China are the two most acceptable alternatives, having the highest acceptability for the first two ranks. In particular, the USA has an acceptability score of 93.94% for the first rank and 6% for the second rank, which adds up to almost 100%. As a result, the USA seems to be a robust choice for being in the first position, regardless of criteria weights. Similarly, China appears as a robust choice for the second position, indicating that the first and second positions obtained in the Tortoise ranking come across as being robust choices.

From the third position onwards, the ranking acceptability indices indicate a ranking significantly different from the Tortoise GAII. This is evident in the case of the UK, which holds the third position in Tortoise GAII but has only a 6.59% probability of being ranked third when the weights are random generated, and a potential 46.19% chance of assuming the fourth position. Moreover, no robust choice can be made for the third position as the highest acceptability index achieved was only 24.34% by Israel.

We have elaborated Fig. 1, which presents the same information as Table 2, but in the form of a heat map that shows all possible combinations of country and position, sorted according to the Tortoise GAII. As before, it is evident that the USA and China have a high probability of ranking first and second, respectively. In the middle of the heat map, the probabilities for other countries to occupy a specific position are low, and towards the end positions, the probabilities increase, albeit not robustly. This first analysis shows that the interchanges in the criteria weights have a significant impact on the ranking.

The second analysis to the application of weighted sum and SMAA considering ordinal weights and no longer completely random weights. Table 3 presents the rank acceptability indices of this analysis. The first remark of the results is that there is a 100% chance that USA and China be in first and second place, respectively. Despite a high probability of the UK ranking third at 63.04%, there

Table 2. Rank acceptability indices with randomly generated weights of the top six countries of the Tortoise GAII.

		1st	2nd	3rd	4th	5th	6th
1st	USA	93.94	6.00	0.00	0.06	0.00	0.00
2nd	China	4.84	92.32	0.47	1.88	0.01	0.15
3rd	UK	0.43	0.75	6.59	46.19	4.34	4.37
4th	Canada	0.26	0.37	23.52	29.09	6.21	5.76
5th	Israel	0.16	0.16	24.34	10.53	8.80	10.77
6th	Singapore	0.11	0.15	15.14	6.00	9.22	11.11

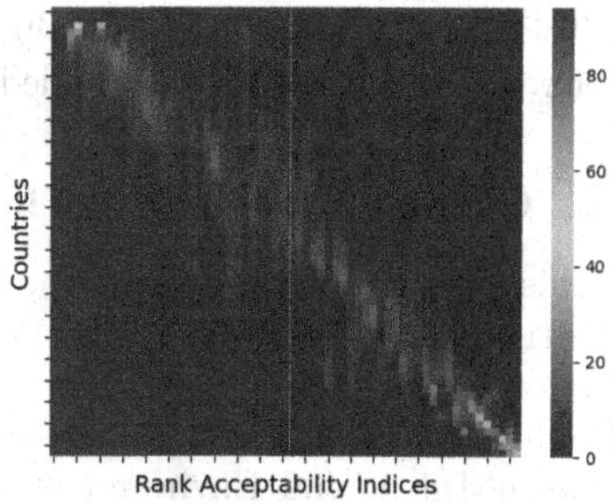

Fig. 1. Rank acceptability indices (%) with randomly generated weights.

remains a significant chance of 31.49% that it could instead be positioned fifth. It is noteworthy that Singapore ranks sixth in the Tortoise Index, despite having a high probability (76.49%) of ranking fourth. Conversely, Canada, which ranks fourth in the Tortoise GAII, has only a 6.91% probability of being in that position.

Table 3. Rank acceptability indices with ordinal weights of the top six countries of the Tortoise GAII.

		1st	2nd	3rd	4th	5th	6th
1st	USA	100.0	0.0	0.00	0.00	0.00	0.00
2nd	China	0.0	100.0	0.00	0.00	0.00	0.00
3rd	UK	0.0	0.0	63.04	2.24	31.49	2.31
4th	Canada	0.0	0.0	34.88	6.92	52.27	5.00
5th	Israel	0.0	0.0	1.99	10.80	12.18	72.69
6th	Singapore	0.0	0.0	0.09	76.49	3.00	15.08

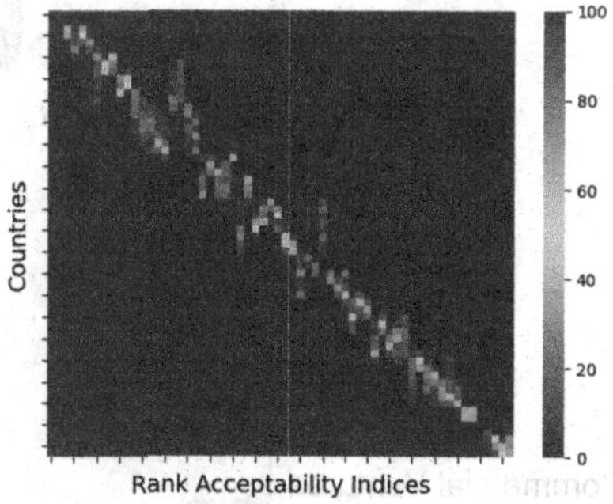

Fig. 2. Rank acceptability indices (%) with ordinal weights.

Figure 2 illustrates the same values presented in Table 3 in the form of a heat map, showing all possible combinations of countries and positions. As expected, maintaining the order of relevant weights results in a more robust solution. However, Fig. 2 also reveals that this robust solution may differ from Tortoise GAII at certain points.

These results show that criteria weights do influence the obtained AI ranking of countries. Therefore, it is essential to conduct sensitivity analysis to evaluate the impact of weight changes and avoid unfair treatment of countries by assuming subjective weights. For example, while UK was ranked third in Tortoise GAII, when random weights were used (even with order preferences), it had a low probability of being ranked in that position. Similarly, Singapore ranked sixth in Tortoise GAII but had a 76.49% probability of being ranked fourth when preference weight order was applied.

6.2 Country Rankings: A Choquet Perspective

In this subsection, we present the results using the Choquet integral. As first investigation, we calculated the Pearson correlation coefficients between pairs of criteria. The obtained values are presented in Fig. 3. There are clearly a lot of redundancies within this dataset. For instance, there are strong correlations between Talent and Environment ($\rho = 0.8103$), Talent and Commercial Ventures ($\rho = 0.7951$), Environment and Research ($\rho = 0.8459$), Environment and Commercial Ventures ($\rho = 0.8474$), and Research and Commercial Ventures ($\rho = 0.7759$).

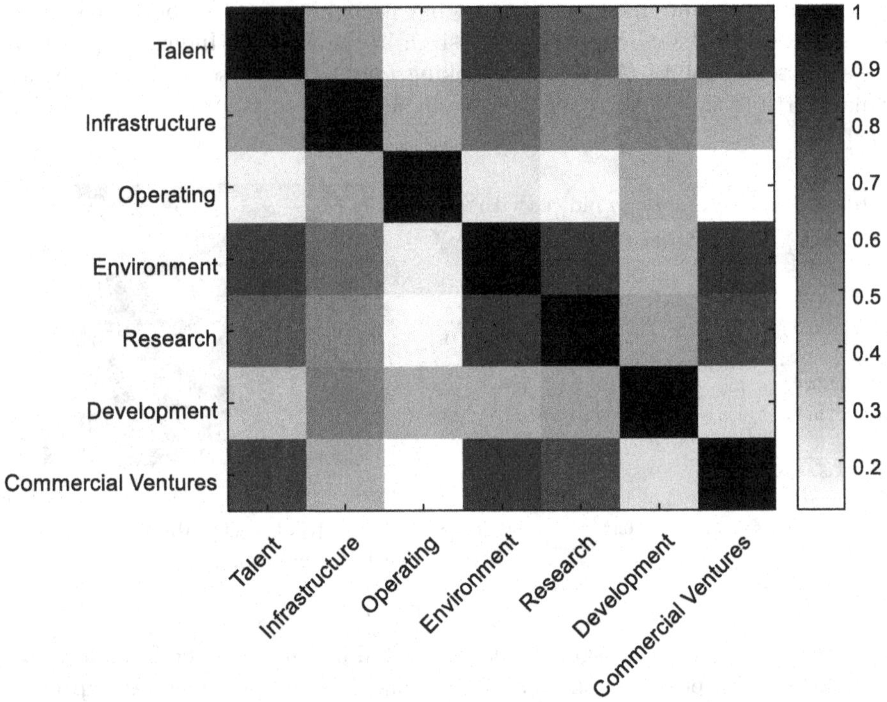

Fig. 3. Correlations coefficients among AI dimensions.

Aiming at mitigating the effect of positively correlated criteria in the aggregation procedure, we searched for interaction indices as close as possible from the negative of the correlation coefficients that can be applied with the Choquet integral. As the Shapley values we assumed the same weights as in the Tortoise analysis but normalized by Eq. (10). By solving the optimization problem (9), we achieved $I_{1,2} = -0.006$, $I_{1,4} = -0.153$, $I_{1,5} = -0.060$, $I_{1,7} = -0.198$,

$I_{2,3} = -0.145$, $I_{2,4} = -0.037$, $I_{2,6} = -0.062$, $I_{3,6} = -0.022$, $I_{4,5} = -0.100$, $I_{4,7} = -0.128$, $I_{5,7} = -0.091$ (the remaining interaction indices are practically zero). Based on these parameters, the 2-additive Choquet integral leads to the total scores presented in Fig. 4 (for the top 6 countries according to Tortoise GAII).

Table 4. Scores for each AI dimension and the total score (TS) of the top six countries determined by the 2-additive Choquet.

Country	Tal.	Infra.	OP	Res.	Dev.	GS	Com.	TS
USA	100	94.02	64.56	100	100	77.39	100	98.28
China	16.51	100	91.57	71.42	79.97	94.87	44.02	75.77
UK	39.65	71.43	74.65	36.52	5.03	82.82	18.91	47.65
Canada	31.29	77.06	93.94	30.67	25.78	100	14.88	47.43
Israel	35.77	67.59	82.44	32.63	27.96	43.91	27.33	45.46
Singapore	39.39	84.34	3.15	37.67	22.55	79.82	15.07	49.56

An interesting remark is that Singapore achieved the third position in the ranking. Note the redundant effect between Infrastructure and Operating modeled by the Choquet integral ($I_{2,3} = -0.145$) in order to mitigate the positive correlation between them ($\rho_{2,3} = 0.4130$). This helps to explain how Singapore achieved a better score in comparison with UK and could move from the 6th to the 3rd position (in comparison with Tortoise GAII). Indeed, by looking at Eq. (6), the part associated with a negative interaction index takes the maximum between the scores. Therefore, the bad performance of Singapore in Operating was overcome by the very good score on Infrastructure (which is higher than the score of UK).

7 Conclusion

This paper presented a critical analysis of AI indicators for comparing countries. We apply the SMAA methodology and the Choquet integral to analyze the Tortoise GAII in terms of criteria weights and aggregation procedure. The SMAA analysis results in rank acceptability indices for all countries, which can be used for deriving robust conclusions. More specifically, the rank acceptability index allows quantifying the amount of instability in the results induced by uncertain criteria weights. By applying the Choquet integral, we explored how the solutions change when a non-linear function is assumed.

Regarding the first hypothesis that the criteria weights influence the resulting AI ranking, we have observed that even when randomly varying the weights while adhering to the same ordinal preference as used in Tortoise GAII, the positions with the highest probabilities do not always align with those presented in the

Tortoise ranking. Furthermore, for certain ranking positions, it may not be even possible to find a robust choice. It can be concluded that the decision regarding weight determination will strongly influence the final ranking.

The hypothesis of interaction between criteria was verified, indicating that an aggregation procedure than the weighted sum shall be applied in the construction of such indicators when considering the same AI dimensions as criteria. This was confirmed after comparing the ranking determined by the 2-additive Choquet with the Tortoise ranking, since compensations between criteria are made through interaction indices, resulting in changes in rankings.

It is important to note that the rank acceptability index given by SMAA is able to provide only a rough ranking of the alternatives because there is no objective way to combine acceptability indices for different ranks to reach a complete ranking [12]. This characteristic naturally opens up possibilities for future studies, such as the use of the SMAA pairwise winning index to propose a new AI ranking of the countries. A wider outlook would also include the proposal of new measures in order to compare the probability matrix with the given ranking. As another future study, we are planning to apply the SMAA methodology also with the Choquet integral in order to conduct a robustness analysis with respect to model structure.

References

1. CIFAR. Pan-Canadian AI Strategy Impact Assessment Report. Canada's AI Ecosystem: Government Investment Propels Private Sector Growth. University of Toronto (2020). https://cifar.ca/wp-content/uploads/2020/11/Pan-Canadian-AI-Strategy-Impact-Assessment-Report.pdf. Accessed 2 May 2023
2. Tortoise. The Global AI Index: Methodology. Tortoise Media (2021). https://www.tortoisemedia.com/wp-content/uploads/sites/3/2021/12/Global-AI-Index-Methodology-3.0-211201-v2.pdf. Accessed 4 May 2023
3. Keeney, R.L., Raiffa, H.: Decisions with Multiple Objectives. Wiley, New York (1976)
4. Pelegrina, G.D., Duarte, L.T., Grabisch, M., Romano, J.M.T.: Dealing with redundancies among criteria in multicriteria decision making through independent component analysis. Comput. Ind. Eng. **169**, 108171 (2022)
5. Greco, S., Ishizaka, A., Tasiou, M., Torrisi, G.: On the methodological framework of composite indices: a review of the issues of weighting, aggregation, and robustness. Soc. Indic. Res. **141**, 61–94 (2019)
6. Nardo, M., Saisana, M., Saltelli, A., Tarantola, S., Hoffmann, A., Giovannini, E.: Handbook on Constructing Composite Indicators: Methodology and User Guide. OECD, Paris (France) (2008)
7. Lahdelma, R., Hokkanen, J., Salminen, P.: SMAA - stochastic multiobjective acceptability analysis. Eur. J. Oper. Res. **106**, 137–143 (1998)
8. Lahdelma, R., Salminen, P. SMAA-2: stochastic multicriteria acceptability analysis for group decision making. Oper. Res. **49** (2001)
9. Choquet, G.: Theory of capacities. Annales de l'Institut Fourier **5**, 131–295 (1954)
10. Par Jean-Charles Pomerol and Sergio Barba-Romero: Multicriterion Decision in Management: Principles and Practice. Kluwer Academic Publishers, Massachusetts (2000)

11. Pelissari, R., Oliveira, M.C., Amor, S.B., Kandakoglu, A., Helleno, A.L.: SMAA methods and their applications: a literature review and future research directions. Ann. Oper. Res. **293**, 433–493 (2020)
12. Lahdelma, R., Salminen, P.: SMAA in robustness analysis. In: Doumpos, M., Zopounidis, C., Grigoroudis, E. (eds.) Robustness Analysis in Decision Aiding, Optimization, and Analytics. ISORMS, vol. 241, pp. 1–20. Springer, Cham (2016). https://doi.org/10.1007/978-3-319-33121-8_1
13. Shapley, L.S.: A Value for N-person Games. Annals of Mathematics Studies: Vol. 28. Contributions to the Theory of Games, vol. II, pp. 307–317 (1953)
14. Murofushi, T., Soneda, S.: Techniques for reading fuzzy measures (III): interaction index. In: 9th Fuzzy System Symposium, Sapporo, Japan, pp. 693–696 (1993)
15. Grabisch, M.: Alternative representations of discrete fuzzy measures for decision making. Int. J. Uncertain. Fuzziness Knowl.-Based Syst. **5**(5), 587–607 (1997)
16. Grabisch, M.: *k*-order additive discrete fuzzy measures and their representation. Fuzzy Sets Syst. **92**, 167–189 (1997)
17. Grabisch, M., Prade, H., Raufaste, E., Terrier, P.: Application of the Choquet integral to subjective mental workload evaluation. IFAC Proc. Volumes **39**(4), 135–140 (2006)
18. Mayag, B., Grabisch, M., Labreuche, C.: A characterization of the 2-additive Choquet integral through cardinal information. Fuzzy Sets Syst. **184**(1), 84–105 (2011)
19. Duarte, L.T.: A novel multicriteria decision aiding method based on unsupervised aggregation via the Choquet integral. IEEE Trans. Eng. Manag. **65**(2), 293–302 (2018)
20. Papoulis, A., Pillai, S.U.: Probability, Random Variables and Stochastic Processes. 4th edn. McGraw-Hill, New York (2002)

Estimating Code Running Time Complexity with Machine Learning

Ricardo J. Pfitscher[1(✉)], Gabriel B. Rodenbusch[1], Anderson Dias[2], Paulo Vieira[2], and Nuno M. M. D. Fouto[3]

[1] Federal University of Santa Catarina, Florianópolis , Brazil
ricardo.pfitscher@ufsc.br, gabriel.braun@grad.ufsc.br
[2] UniSociesc Campus Anita Garibaldi, Joinville, Brazil
[3] University of São Paulo, São Paulo, Brazil
nfouto@usp.br

Abstract. The running time complexity is a crucial measure for determining the computational efficiency of a given program or algorithm. Depending on the problem complexity class, it can be considered intractable; a program that solves this problem will consume so many resources for a sufficiently large input that it will be unfeasible to execute it. Due to Alan Turing's halting problem, it is impossible to write a program capable of determining the execution time of a given program and, therefore, classifying it according to its complexity class. Despite this limitation, an approximate running time value can be helpful to support development teams in evaluating the efficiency of their produced code. Furthermore, software-integrated development environments (IDEs) could show real-time efficiency indicators for their programmers. Recent research efforts have made, through artificial intelligence techniques, complexity estimations based on code characteristics (e.g., number of nested loops and number of conditional tests). However, there are no databases that relate code characteristics with complexity classes considered inefficient (*e.g.*, $O(c^n)$ and $O(n!)$), which limits current research results. This research compared three machine learning approaches (i.e., Random Forest, eXtreme Gradient Boosting, and Artificial Neural Networks) regarding their accuracy in predicting Java program codes' efficiency and complexity class. We train each model using a dataset that merges data from literature and 394 program codes with their respective complexity classes crawled from a publicly available website. Results show that Random Forest resulted in the best accuracy, being 90.17% accurate when predicting codes' efficiency and 89.84% in estimating complexity classes.

Keywords: Running time complexity prediction · Algorithm analysis · Machine Learning

1 Introduction

Computer programming is a primary task in a software development lifecycle. It essentially involves translating algorithms into functions, methods, and pro-

M. C. Naldi and R. A. C. Bianchi (Eds.): BRACIS 2023, LNAI 14196, pp. 400–414, 2023.
https://doi.org/10.1007/978-3-031-45389-2_27

cedures that perform tasks that, in conjunction, solve a specific problem. Before the programming action, analysts should determine the algorithms that apply to the referred situation. Such an analysis task requires technical knowledge in many areas, from business rules to computational resource efficiency. Regarding the latter, a relevant discipline is algorithm analysis. One result of its activities is a mathematical function that expresses the upper bound running time of a specific algorithm, a.k.a. running time complexity [10].

The complexity class of an algorithm or code directly impacts its resource usage efficiency. Such an impact on the software's future performance highlights the need for a precise analysis. Also, depending on its asymptotic behavior, the study may indicate that the program takes so long to run that it is considered intractable [6]. On the other hand, beginner developers may rely on inefficient solutions to solve coding problems without understanding the impacts on running time complexity. Determining the running time complexity is arduous and requires a deep understanding of code behavior. An analyst must consider multiple variables when analyzing a coded function, including the number of recursive calls, the size of iteration loops, and the number of nested loops. Also, analysts must consider how these variables relate and how the computer program uses computational resources. Consequently, experts can misevaluate the running-time function, promoting a search for tools and frameworks to aid in the complexity analysis process.

Alan Turing proved through the halting problem that estimating the code complexity is mathematically impossible [12]. Thus, recent research applied machine learning techniques and mathematical models to estimate running-time complexity functions [9,18]. With such approximation values, developers could get real-time feedback on the efficiency of their code. However, a relevant barrier that hinders the evolution and adoption of such models is the lack of datasets that correlate coding characteristics to the respective runtime complexity class.

This research aims to fill this gap by developing a learning method to predict the runtime complexity of a program code. To achieve this objective, we built a representative dataset with program codes, related characteristics and complexities, and compared machine learning approaches according to their accuracy in predicting efficiency (if the code is efficient or not) and the runtime complexity of program codes. The contributions of this paper are twofold: first, it makes available to the machine learning community a dataset containing 394 Java code files grouped into eight distinct complexity classes and each code having 16 metadata information; second, it shows that a Random Forest model trained with a dataset merged from the published work of Sikka et al. [18] and web crawled data can achieve an accuracy of 90.34% when predicting code efficiency and 89.26% when predicting complexity classes.

The remainder of this text is organized as follows, Sect. 2 discusses the related work and current limitations. Then, Sect. 3 describe the methodology followed to build datasets and develop the machine learning models. Next, Sect. 4 presents the complexity prediction results obtained with trained models and Sect. 5 discusses the limitations of classifications. Finally, the Sect. 6 presents this work's conclusions and future works.

2 Related Work

Given the limitations of writing a computer program to determine the running time of program source codes, a few works appear in the literature to address this problem. However, the recent advances in artificial intelligence propelled the development of models that estimate code complexity.

The seminal work of Hutter et al. [9] assesses machine learning approaches to predict the performance of algorithms. The results show that the proposed approaches based on Random Forests and approximate Gaussian processes are the ones that better predict the performance of parameterized algorithms used to solve NP-hard problems. The correlation coefficients of predictions reached values superior to 0.9. Although the research of Hutter et al. provided relevant results, our focus differs from theirs because we wish to estimate the runtime complexity regarding asymptotic order.

The research work of Sikka et al. [18] addressed the runtime complexity estimation problem using machine learning models. The paper has two main contributions. First, it publishes the Code Runtime Complexity Dataset (CoRCoD) composed of 932 code files belonging to 5 different classes of complexities, namely *constant*-time, *logarithmic*-time, *linear*-time, *linearithmic*-time, and *quadratic*-time. Second, it shows that the Random Forest model achieved the best results for predicting code complexity, with an accuracy of 71.84% using code features as attributes and 83.57% when Abstract Syntax Tree (AST) applies to generate code embedding used in training. Although the paper contributes a significant step for future research, we argue that the dataset does not have entries for inefficient codes. Thus, we address this problem by crawling a public website to publish a more comprehensive dataset; also, we evaluated more recent models, such as the eXtreme Gradient Boosting Trees (XGBT) and Artificial Neural Networks (ANN).

The efforts published by Sepideh Seifzadeh in a public blog [17] appear to be the results of ongoing research at IBM. The publication has two significant highlights: first, the developed models use The CodeNet Dataset with around 14M code samples for roughly 4K programming problems; second, they trained an ANN and a Light Gradient Boosting Machine (LGBM) using both code features and code graph representation to predict six classes of runtime complexity near-constant, linear, log-linear, polynomial, exponential, and factorial. The preliminary results show up to 80% of accuracy in predictions, which supports machine learning for addressing the problem of runtime complexity estimation. Despite the promising results, the problem remains open, with space for improvements; also, we accessed the CodeNet used in the results and do not find the tags that classify codes according to complexity classes, we suppose authors manually tagged the dataset, but it is not clear in the published article.

3 Material and Methods

This section presents the materials and methodology used to compare approaches to estimate code running time complexity. Section 3.1 explains the steps taken to consolidate a dataset with representativeness for training machine learning models, including a discussion regarding dataset balancing. Section 3.2 explains how we extracted attributes from codes and the assumptions leveraged for complexity classification, as also discusses the process used to select features for the machine learning models. Finally, Sect. 3.3 presents the machine learning models used to predict the codes' complexity and process used for comparisons. All the source codes we used in this paper are available on GitHub[1].

3.1 Dataset Consolidation

The first step toward developing this research project is dataset consolidation. This work relies on three datasets. The first one is the *reference* dataset published in the work of Sikka et al. [18], which consists of 931 Java code files, with 14 metadata information relative to the codes and their respective complexity. Considering that such a dataset does not contain codes with intractable complexity, this work builds a second one with data from a publicly available website, which we call *crawled* dataset (394 entries). The third one, the *merged* dataset (1325 entries), results from merging the reference and the crawled datasets.

We developed a web crawler to extract the information from the platform `geekforgeeks.org`, which contains a specific section to discuss fundamentals of algorithms[2]. Such section discusses several algorithms, their respective codes, and complexities. The crawler consists of a Python script that uses the selenium library to scratch the website and extract codes and complexity information and runs as follows. First, it accesses the main page, which contains a list of topics related to the fundamentals of algorithms. Then, the crawler runs a login process and computes the list of links to be visited on the page. Next, each page in the link list is accessed to verify if it contains content relevant to complexity prediction. In other words, the crawler searches if the page has algorithms' source codes and runtime complexity information.

Considering that the focus of this research is to extend the reference dataset, the crawler always selects Java programming language. Also, the page should provide text describing the time complexity of the code in any place near it. As soon as the crawler identifies that the page contains the relevant content, it copies each code and searches for the nearest complexity information. It is essential to highlight that many pages have multiple program codes; thus, we save each one and associate it with the runtime complexity information closer to the point where the program code is exhibited. Such nearest assumption involves a risk of misclassification, as the pages from the website do not follow a static standard; however, we understand that this is the best effort for automated computing of a dataset.

[1] https://github.com/ricardopfitscher/RuTiCo.

[2] https://www.geeksforgeeks.org/fundamentals-of-algorithms/.

The crawler process resulted in a dataset with 394 program codes and 78 distinct complexity classes. Many classes extracted from the Web pages do not follow the general form of asymptotic classes; for example, the classes $O(n|n|)$ (algorithm: compute the sum of digits in all numbers from 1 to n) and $O(m^2k + k^3logn)$ (algorithm: count ways to reach the nth stair) do not match any general case; for those cases, we manually evaluated each entry and defined their value according to the closest dominant class.

To produce a feasible scenario, instead of 78 complexities classes, we reduced the classification scope to eight categories: constant, logarithmic (includes double logarithmic and polylogarithmic), sublinear (fractional power), linear, linearithmic, quadratic, polynomial, and exponential (we also consider factorial as exponential). In addition, we split the program codes into two major categories: the program codes that are efficient (constant, logarithmic, sublinear, linear, and linearithmic) and the inefficient ones (quadratic, polynomial, and exponential). We assume as polynomial time, the asymptotic functions that are equal to or greater than $O(n^3)$. We understand this is a questionable assumption, as according to [10], polynomial time functions have the form of n^c for every c greater than 1. However, we argue that algorithms that run in $O(n^3)$ are much more inefficient than the ones that run in $O(n^2)$ time, and thus, supposing $O(n^3)$ is distinct from $O(n^2)$ is quite reasonable. Figure 1 depicts the resultant distribution of complexity classes in crawled and reference datasets.

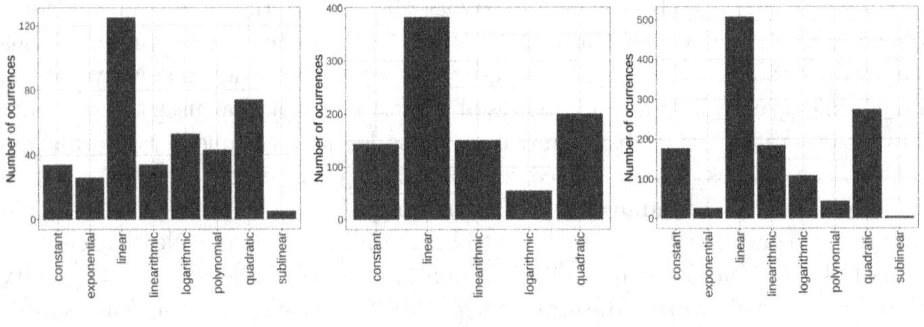

(a) crawled dataset (b) reference dataset (c) merged dataset

Fig. 1. Distribution of complexity classes in the available datasets

As shown in Fig. 1b, the reference dataset does not contain the exponential time and the polynomial time codes, which severally reduces the size of the inefficient class. Also, as one can notice, both datasets are imbalanced: in the crawled dataset (Fig. 1a) the smaller class (sublinear) contains five entries while the larger class (linear) contains 125 entries; in the reference dataset the smaller class (logarithmic) contains 55 entries while the larger class (also the linear) contains 383 entries. Such unbalancing also impacts the classes distribution in the merged dataset (Fig. 1c), making the linear and quadratic classes more representative. Regarding efficiency, the distribution in each dataset is as follows:

63.70% of codes are from efficient classes in the crawled dataset, 78.51% in the reference dataset, and 74.18% in the merged dataset.

Considering that an imbalanced dataset can impact the abstraction capability of machine learning models [14] and that our dataset can be considered small compared to what is considered in current state-of-the-art, we rely on the SMOTE tool to balance the data on each dataset. The SMOTE (Synthetic Minority Oversampling Technique) [4] is a well-known algorithm for solving imbalanced classification problems. The general idea of this method is to artificially generate new samples of the minority class using the nearest neighbors of these cases. Such synthetic data would be generated between the random data and the randomly selected k-nearest neighbor; the procedure is then repeated until the minority class has a size equal to or close to the majority. Due to applying SMOTE, all the resultant complexity classes hold a similar number of occurrences.

3.2 Features Extraction and Selection

We wrote the code features extraction program in Python ("`crawler.py` file on the repository") and relied on the `javalang` library to extract the features described in Table 1 for each Java code in the datasets. The second column in the table describes if the former work of Sikka et al. [18] relies on the feature for their prediction. For the cases where column values are with "no*", the feature was only used in [18] for the manual classification of algorithms and not for the machine learning models. For the case of recursive calls, we measured the number of recursive calls in the code instead of if there is a recursive call in the code.

As we established the set of complexity classes and balanced the datasets, the next step is to perform a feature selection to determine the most relevant characteristics to distinguish the codes according to their complexity classes. Multiple approaches exist for determining the most relevant features of random forest models. According to Speiser et al. [19] for datasets with many predictors, the methods implemented in the R packages `varSelRF` and `Boruta` [11] are preferable due to computational efficiency. Thus, in this work, we rely on the Boruta package to define which code attributes are more relevant to predict the complexity classes and the efficiency of a given code (please read the "`classificators.R`" file in the repository). Boruta is a feature selection algorithm that relies on Random Forest to output a variable importance measure (VIM); the method's rationality consists of progressively eliminating irrelevant features by comparing original attributes' importance with importance achievable at random until the test is stable. Table 2 depicts the resultant feature functions from Boruta process.

An analysis of the feature selection results depicted in Table 2 permits some relevant considerations: (i) the efficiency estimation requires fewer features than the complexity class estimation; (ii) the crawled dataset also uses fewer features for both predicting efficiency and complexity classes and; (iii) the number of switches (attribute number 3) in codes is irrelevant for the predictions.

Table 1. Features extracted from codes and respective descriptions

No.	Feature	Sikka et al. [18]	Description
1	num_if	yes	Number of if statements in the code
2	num_else	no	Number of else statements in the code
3	num_switch	yes	Number of switch statements
4	num_loof	yes	Number of loops, including for and while statements
5	num_break	yes	Number of break statements
6	num_priority	no*	Number of priority queues instantiated
7	num_binSearch	no*	Number of calls for a binary search
8	num_minMax	no*	Number data of calls for min() or max() functions
9	num_sort	no*	Number of calls for the sort() function
10	num_hash_map	no*	Number of hash maps instantiated
11	num_hash_set	no*	Number of hash sets instantiated
12	num_recursive	distinct	Number of recursive calls for a given function
13	num_nested_loop	yes	Depth of nested loops
14	num_vari	yes	Number of variables declared
15	num_method	yes	Number of methods declared
16	num_state	yes	Number of statements

Table 2. Features resultant from Boruta feature selection process in each dataset according to the dependent variable

Dataset	Dependent variable	Features selected (No. in Table 1)
Crawled	Efficiency	1,4,8,12,13,14,15,16
Crawled	Complexity class	1,2,4,7,8,11,12,13,14,15,16
Reference	Efficiency	1,2,4,5,6,8,9,10,12,13,14,15,16
Reference	Complexity class	1,2,4,5,6,7,8,9,10,12,13,14,15,16
Merged	Efficiency	1,2,4,7,8,9,10,12,13,14,15,16
Merged	Complexity class	1,2,4,5,6,7,8,9,10,11,12,13,14,15,16

3.3 Model Training and Validation

The last step of this research consists of comparing supervised machine learning models according to their ability to predict the running-time complexity class of a program code. To this end, we first use the merged dataset to find out which method have the best accuracy to predict both efficiency (two-class prediction problem) and complexity class (multiclass prediction). Considering the results provided by the work of [18], which found that Random Forest had the best accuracy among eight classification algorithms, we included eXtreme Gradient Boosting Trees (XGBT) and ANN in the comparisons because the previous work do not considered that approaches. We conducted the implementations of Random Forest both in R and in Python (files "`classificators.R`" and "`classificators.py`", respectively) and the other models only in Python.

XGBT is an effective and scalable tree-boosting system that combines novel sparsity-aware algorithms and weighted quantile sketch for approximate tree learning [5]. This system is an optimized version of the gradient boosting machine algorithm (GBDT), created by Friedman [7], which uses decision trees for classification. The traditional GBDT approach only deals with the first derivative in learning; XGBoost improves the loss function with Taylor expansion, reducing modeling complexities and the likelihood of model over-fitness [3].

Regarding ANNs, each implementation requires several configuration parameters, which we discuss in the following. We normalized data using the `StandardScaler` function from scikit-learn [13] to scale the features so that they have a mean of 0 and a standard deviation of 1. Such scaling is a typical pre-processing step to improve model performance [20]. We then established the hyperparameters based on the best random search results, which have been proven more efficient for hyper parametrization than trials on a grid [1]. When the corresponding dropout rate exceeds zero, a Dropout layer is added after the dense layer to prevent overfitting. The same applies to the batch normalization flag: when it is true, it adds a *BatchNormalization* layer after the dense layer to normalize the inputs and improve the convergence of the model [8]. Lastly, the model is optimized through Adam, defined by the random search, and trained for 500 epochs with a batch size equal to 64.

We used the `train_test_split` function from scikit-learn [13] to split the balanced dataset into train- and test-data. To prevent overfitting issues, we chose to maintain most of the function parameters as standard as possible. That way, we altered only the stratify. When different from None, the samples are stratified through *StratifiedKFold*, so that each set contains approximately the same percentage of samples of each target class as the complete set. Stratification has been found to improve upon standard cross-validation both in terms of bias and variance. The standard method of randomly distributing multi-label training samples can create issues when test subsets lack even a single positive example of a rare label. This, in turn, can lead to calculation problems for various multi-label evaluation measures [16].

After we established that Random Forest is the machine learning approach that provides the best results (see Sect. 4), we conducted a systematic process for learning and predicting complexity classes:

1. We used the crawled dataset for training the models with a sample of 70% and tested with 30% remaining data;
2. We compare our results to Sikka et al. [18] by running a cross-validation process with 70% of the reference dataset and validating using 30% of their data;
3. We evaluated the generalization capability of the crawled dataset by training the ML model with 70% of the crawled dataset and testing with 100% of the reference dataset, and;
4. We in-depth evaluate the confusion matrix of the Random Forest model using 70% of the merged dataset for training and 30% for testing.

4 Results

This section presents the results for predicting computer program code efficiency and complexity classes based on their attributes. First, we compare the accuracy of predictions for each model in the merged dataset to select the appropriate approach to perform our in-depth systematic evaluation process. Table 3 depicts these comparison results; the accuracy values show that the Random Forest is the best approach among the evaluated methods, predicting codes' efficiency with 90.17% accuracy and the complexity class with 89.84%. The XGBT provided an accuracy close to the Random Forest, and given the nondeterministic behavior of ANN it resulted in a range of accuracy, also smaller than the Random Forest. Such results follow the findings of [18], which pointed out the Random Forest as the algorithm that resulted in better accuracy for predicting complexity classes (71.84%). Thus, we will consider the Random Forest for our systematic evaluation process.

Table 3. Comparison of machine learning models to predict efficiency and complexity class on the merged dataset

Objective	Random Forest	XGBT	ANN
Efficiency	*90.17%*	90%	between 85.5% and 87.5%
Complexity class	*89.84%*	87.38%	between 84.5% and 85.5%

As we established Random Forest as the reference ML approach, we now assess its accuracy to predict the efficiency of codes using our systematic evaluation process. Table 4 contains prediction results for the efficiency of available algorithm codes. Two main conclusions arise from the analysis of the use of Random Forest to predict the efficiency of source codes: i) when the training and test data are from the same dataset, the model can achieve up to 80% of accuracy (80.13% for crawled dataset, 93.85% for reference dataset, and 90.17% for the merged dataset); ii) when train and test data are from distinct datasets, accuracy drops to 83.57% (trained with 70% of crawled dataset and 100% of reference dataset). Although this fall in accuracy seems a poor result, if we consider that the data are from distinct sources, the crawled dataset has a good generalization concerning efficiency classification. Also, by analyzing the F1-score's results (up to 80%), we can conclude that all the random forest models have a good balance between precision and recall, which means that they have a good fit to distinguish the efficiency of source codes.

After assessing the Random Forest models' ability to classify codes regarding their efficiency, we retrained each model to predict complexity classes. Similar to the efficiency classification, we leverage our four-step systematic process to compare the models. Table 5 depicts these results. An accuracy analysis shows that the Random Forest model properly predicts the complexity class of source codes from the crawled dataset, with an accuracy of 80%.

Table 4. Summary of efficiency prediction using Random Forest

Train set	Validation set	Accuracy	Precision	Recall	F1-score
70% crawled	30% of crawled dataset	80.13%	80.33%	80.13%	80.11%
70% reference	30% of reference dataset	93.85%	93.92%	93.85%	93.85%
70% crawled	100% of reference dataset	83.57%	83.91%	83.57%	83.72%
70% merged	30% of merged dataset	90.17%	90.22%	90.17%	90.17%

The results in Table 5 also show that the reference- and merged-based models achieved an accuracy superior to 88%, outstanding the finds of Sikka et al. [18], which presented an accuracy of 71.4% for predicting all classes. We claim that the resultant improvement justifies by both the balancing and the feature selection processes we leveraged. However, when the predictions ran in the reference dataset, the model trained with data from the crawled data had a lower accuracy, only 44.04%. This indicates that the model does not have enough generalization for distinguishing the complexity class when it only relies on the crawled dataset for training. It is worth mentioning that the trained model does not include any of the data in the reference dataset. To understand the misclassifications, we analyzed the confusion matrix of predictions of the crawled-based model on the reference dataset (Table 6).

Table 5. Summary of complexity class prediction using Random Forest model

Train set	Validation test	Accuracy
70% crawled	30% of crawled dataset	80.00%
70% reference	30% of reference dataset	88.35%
70% crawled	100% of reference dataset	**44.04%**
70% merged	30% of merged dataset	89.84%

The results depicted in Table 6 show that 112 predictions occurred for classes that even exist in the dataset (sublinear, polynomial, and exponential time), representing 12.03% of the total data, most of them pointing to inefficient classes (polynomial and exponential time). Such a result demonstrates that the crawled dataset does not have enough representativeness to allow machine learning models to generalize complexity classes. Also, we observed that the better predictions occurred for the linear and linearithmic categories (65.01% and 74% of accuracy), and the worst case occurred in the logarithmic class (09.09% of accuracy), which indicates that the characteristics of these classes require a more profound study. For such analysis, we assess the mean decrease accuracy measure - MDA [2] for each feature to understand their importance in predicting the complexity class on each model. Figure 2 presents such MDA results.

The analysis of top-3 MDA depicted in Fig. 2 allows us to conclude that the three most relevant features for code complexity classification in all the models are the number of variables (num_vari), the number of statements (num_state)

Table 6. Confusion matrix of predictions in the reference dataset with the Random Forest model trained using 70% of crawled dataset

	Actual class				
Predicted	constant	logarithmic	linear	linearithmic	quadratic
constant	49	0	4	3	1
logarithmic	13	5	35	5	0
sublinear	0	0	0	0	0
linear	57	*39*	249	*111*	53
linearithmic	8	4	31	15	6
quadratic	5	1	29	4	92
polynomial	4	0	8	3	31
exponential	7	6	27	9	17
Accuracy	**34.26%**	**09.09%**	**65.01%**	**74.00%**	**46.00%**

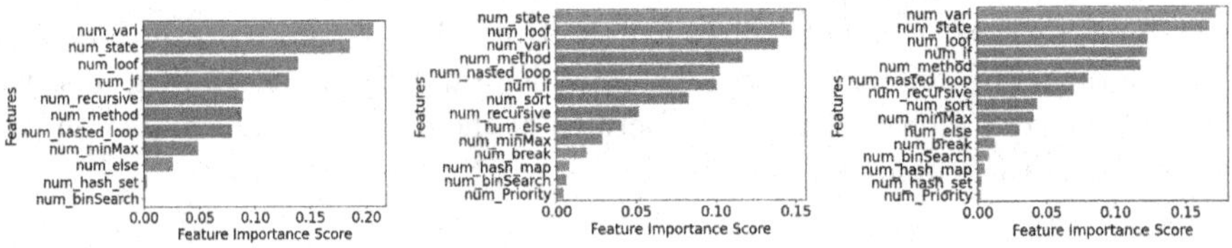

(a) crawled-based model (b) reference-based model (c) merged-based model

Fig. 2. Most relevant features according to MDA results in each model

and the number of loops (num_loof). The depth of nested loops and the number of recursive calls only appear in the top-5 relevant features in two of three models, which is surprising, as these metrics serve to compute recurrence functions in code complexity analysis. Another relevant aspect of MDA analysis is that the second most relevant feature varies significantly from one model to another, which also hinders the abstraction capability of the model based on the crawled dataset. Considering that the crawled-based and the reference-based models share the top-3 relevant features, we now assess the frequency density function of these features in both datasets to understand the reasons why models misclassify linear, linearithmic, and logarithmic classes (Fig. 3).

Two main considerations arise from the analysis of Fig. 3: first, the density function is different in the two datasets, both in terms of the number of occurrences and the behavior of the distribution; second, the linear and linearithmic classes have density distribution functions with similar behavior for the three evaluated metrics, independently of the dataset. While the first consideration explains why the crawled-based model cannot properly identify the classes of the reference dataset, the second justifies the lack of distinction between classes.

To have a clear view of the limitations of the merged-based model, we also depicted the confusion matrix of its predictions. The results in Table 7 show

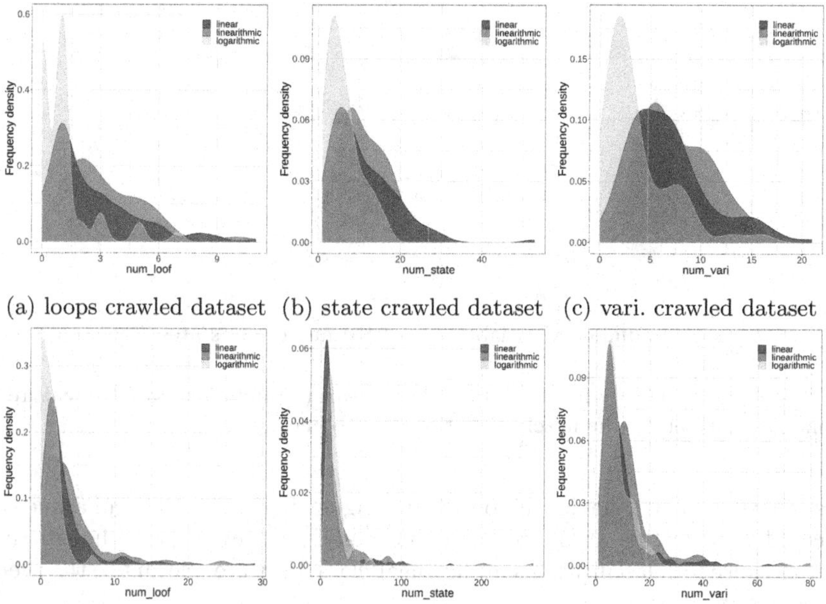

(a) loops crawled dataset (b) state crawled dataset (c) vari. crawled dataset

(d) loops ref. dataset (e) state ref. dataset (f) vari. ref. dataset

Fig. 3. Frequency density functions for the top-3 most relevant features in the crawled and reference datasets in the linear, linearithmic, and logarithmic classes.

Table 7. Confusion matrix of predictions in the validation part of merged dataset

	Actual class							
Predicted	constant	logarithmic	sublinear	linear	linearithmic	quadratic	polynomial	exponential
constant	137	4	0	12	1	1	0	0
logarithmic	6	140	0	10	2	1	0	1
sublinear	0	0	151	3	0	0	0	0
linear	5	8	1	99	9	5	1	0
linearithmic	0	0	0	14	137	2	0	0
quadratic	1	1	0	13	2	135	4	1
polynomial	0	0	0	1	0	9	147	0
exponential	1	0	0	1	2	0	0	150
Accuracy	**91.33%**	**91.05%**	**99.34%**	**64.70%**	**89.54%**	**88.23%**	**96.71%**	**98.68%**

accurate predictions for most of the class, with accuracies superior to 90%; the discrepant case occurred in the linear class, with 25 predictions pointing to faster time classes (i.e., constant, logarithmic, and sublinear) and 29 predictions to slower classes (i.e., linearithmic, quadratic, polynomial, and exponential). To understand this lack of accuracy, we plotted the frequency density function for the `num_vari` and `num_state` features in the merged dataset (Fig. 4).

The density functions depicted in Fig. 4 show that misclassifications mainly occurred because the number of variables and states in codes from the mispointed

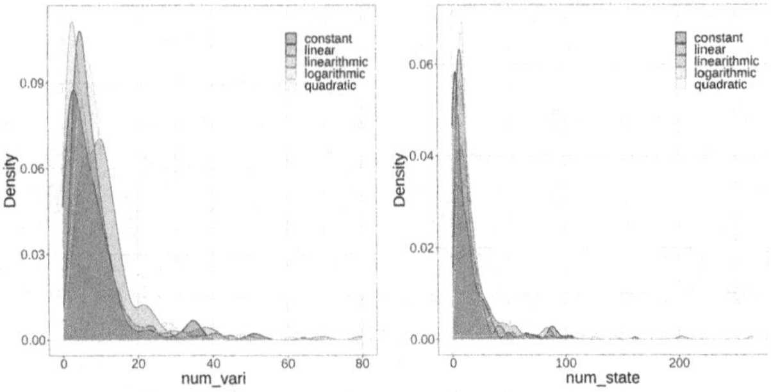

(a) number of variables (b) number of states

Fig. 4. Frequency density functions for the num_vari (a) and num_state (b) features in the merged dataset for the misclassifications in the linear class

classes have similar behavior, hindering distinguishing them. One can argue that we could remove this attribute from the models; however, even though these features cause misclassifications in the linear class, they are relevant to the overall model classification.

5 Discussion

Besides this paper showing promising results in estimating the code runtime complexity, outperforming the state-of-the-art, we point out the following limitations that will be addressed in future research:

- *The Web crawler may not collect the correct complexity of each code.* The dataset used for training and estimations resulted from a Web scrapping process. The collection process may imply biased information, as the crawler ran automatically. However, given the lack of datasets containing inefficient codes, we understand that the potential bias does not influence the comparison results, which shows that Random Forest achieved the best results. We will provide an in-depth analysis of the collected data in future works, including a manual verification.
- *Trustworthiness of* `geeksforgeeks.org`. One can argue that the information provided by geeksforgeeks.org may not be trustworthy. However, considering the lack of publicly available datasets containing inefficient codes and their respective complexity classes, we understand that the provided information is the best effort to have a comprehensive dataset containing the most relevant complexity classes.
- *Small dataset for machine learning purposes.* The merged dataset used for training the machine learning models contains 1325 entries, which is small compared to today's standards. However, considering the lack of comprehensive datasets and the absence of tags to classify the codes published in the CodeNet dataset, we consider our results as a relevant step towards the use of AI to predict code complexity.

- *The dataset balancing process may cause overfitting.* Santos et al. [15] discuss the impacts of balancing datasets before running the cross-validation process; they demonstrate that oversampling techniques may generate replicated entries in test and training sets, which implies in lack of generalization of ML models caused by overfitting. However, the SMOTE method generates synthetic entries based on a proximity function, without duplicating any entry. Thus, we argue that the lack of duplication reduces the overfitting issues.

6 Concluding Remarks

In this paper, we investigated the use of machine learning models to predict the runtime complexity class of computer program codes. Considering that the reference dataset published by Sikka et al. [18] does not contain most of the inefficient classes from the literature, we build a second one with data from a publicly available website. Next, we compare machine learning models to predict program codes' efficiency and complexity classes.

Results show the Random Forest as the best approach, predicting code efficiency with an accuracy of up to 80% and can classify the runtime complexity with an accuracy superior to 81% for a model trained with data from a merged dataset and 87.4% when the model is trained with the reference dataset. Such a result outperforms the ones found by Sikka et al. [18], which found an accuracy of 71.4% using a Random Forest Model. We argue that the feature selection process and balanced datasets supported the accuracy enhancement. However, the Random Forest Model's accuracy drops to less than 50% when we train it with the crawled dataset and try to predict the complexity classes in the reference dataset. We claim that this occurred for two primary reasons: first, we trained the model in a dataset that has more classes than the reference dataset; two, most of the classification errors occurred in the logarithmic and linearithmic categories, which we demonstrated to have a similar behavior regarding the top-3 most important features.

In future work, we aim to develop a complete complexity prediction framework containing a learning component deployed on the cloud and an extension for software IDEs. Research challenges related to the framework include but are not limited to: *i)* studying the applicability of other machine learning models to predict computer programs' efficiency and complexity class, including natural language processing and deep learning frameworks; and *ii)* building an even more comprehensive dataset of program codes and complexity classes mainly in polynomial and exponential classes with constant update. In addition, we aim to manually tag the dataset published by the CodeNet project, which will benefit future research and improve the accuracy of predictors.

References

1. Bergstra, J., Bengio, Y.: Random search for hyper-parameter optimization. J. Mach. Learn. Res. **13**(2) (2012)

2. Breiman, L.: Random forests. Mach. Learn. **45**, 5–32 (2001)
3. Chang, Y.C., Chang, K.H., Wu, G.J.: Application of extreme gradient boosting trees in the construction of credit risk assessment models for financial institutions. Appl. Soft Comput. **73**, 914–920 (2018)
4. Chawla, N.V., Bowyer, K.W., Hall, L.O., Kegelmeyer, W.P.: Smote: synthetic minority over-sampling technique. J. Artif. Intell. Res. **16**, 321–357 (2002)
5. Chen, T., Guestrin, C.: Xgboost: A scalable tree boosting system. In: Proceedings of the 22nd ACM SIGKDD International Conference on Knowledge Discovery and Data Mining, pp. 785–794 (2016)
6. Cormen, T.H., Leiserson, C.E., Rivest, R.L., Stein, C.: Introduction to algorithms. MIT press (2022)
7. Friedman, J.H.: Greedy function approximation: a gradient boosting machine. Annals of statistics, pp. 1189–1232 (2001)
8. Garbin, C., Zhu, X., Marques, O.: Dropout vs. batch normalization: an empirical study of their impact to deep learning. Multimed. Tools Appl. **79**, 12777–12815 (2020)
9. Hutter, F., Xu, L., Hoos, H.H., Leyton-Brown, K.: Algorithm runtime prediction: methods & evaluation. Artif. Intell. **206**, 79–111 (2014)
10. Kleinberg, J., Tardos, E.: Algorithm design. Pearson Education India (2006)
11. Kursa, M.B., Rudnicki, W.R.: Feature selection with the boruta package. J. Stat. Softw. **36**, 1–13 (2010)
12. Lucas, S.: The origins of the halting problem. J. Logical Algebraic Methods Programming **121**, 100687 (2021)
13. Pedregosa, F., et al.: Scikit-learn: machine learning in Python. J. Mach. Learn. Res. **12**, 2825–2830 (2011)
14. Ramyachitra, D., Manikandan, P.: Imbalanced dataset classification and solutions: a review. Int. J. Comput. Bus. Res. (IJCBR) **5**(4), 1–29 (2014)
15. Santos, M.S., Soares, J.P., Abreu, P.H., Araujo, H., Santos, J.: Cross-validation for imbalanced datasets: avoiding overoptimistic and overfitting approaches [research frontier]. IEEE Comput. Intell. Magaz. **13**(4), 59–76 (2018)
16. Sechidis, K., Tsoumakas, G., Vlahavas, I.: On the stratification of multi-label data. In: Gunopulos, D., Hofmann, T., Malerba, D., Vazirgiannis, M. (eds.) ECML PKDD 2011. LNCS (LNAI), vol. 6913, pp. 145–158. Springer, Heidelberg (2011). https://doi.org/10.1007/978-3-642-23808-6_10
17. Seifzadeh, S.: Ai for code: Predict code complexity using IBM's CodeNet Dataset, October 2021. https://community.ibm.com/community/user/ai-datascience/blogs/sepideh-seifzadeh1/2021/10/05/ai-for-code-predict-code-complexity-using-ibms-cod
18. Sikka, J., Satya, K., Kumar, Y., Uppal, S., Shah, R.R., Zimmermann, R.: Learning based methods for code runtime complexity prediction. In: Jose, J.M., Yilmaz, E., Magalhães, J., Castells, P., Ferro, N., Silva, M.J., Martins, F. (eds.) ECIR 2020. LNCS, vol. 12035, pp. 313–325. Springer, Cham (2020). https://doi.org/10.1007/978-3-030-45439-5_21
19. Speiser, J.L., Miller, M.E., Tooze, J., Ip, E.: A comparison of random forest variable selection methods for classification prediction modeling. Expert Syst. Appl. **134**, 93–101 (2019)
20. Thara, D., PremaSudha, B., Xiong, F.: Auto-detection of epileptic seizure events using deep neural network with different feature scaling techniques. Pattern Recogn. Lett. **128**, 544–550 (2019)

The Effect of Statistical Hypothesis Testing on Machine Learning Model Selection

Marcel Chacon Gonçalves[1] and Rodrigo Silva[2(✉)]

[1] Graduate Program on Computer Science, Universidade Federal de Ouro Preto, Ouro Preto, Brazil
marcel.chacon@aluno.ufop.edu.br

[2] Department of Computer Science, Universidade Federal de Ouro Preto, Ouro Preto, Brazil
rodrigo.silva@ufop.edu.br

Abstract. Statistical tests of hypothesis play a crucial role in evaluating the performance of machine learning (ML) models and selecting the best model among a set of candidates. However, their effectiveness in selecting models over larger periods of time remains unclear. This study aims to investigate the impact of statistical tests on ML model selection in sequential experiments. Specifically, we examine whether selecting models based on statistical tests leads to higher quality models after a significant number of iterations and explore the effect of the number of tests performed and the preferred statistical test for different experimental time horizons.

The study on binary classification problems reveals that the use of statistical tests should be approached with caution, particularly in challenging scenarios where generating improved models is difficult. The analysis demonstrates that statistical tests may impede progress and impose overly stringent acceptance criteria for new models, hindering the selection of high-quality models. The findings also indicate that the dominance of versions without statistical tests remained consistent, suggesting the need for further research in this area.

Although this study is limited by the number of datasets and the absence of pre-test assumption verification, it emphasizes the importance of understanding the impact of statistical tests on ML model selection.

Keywords: Machine Learning · Model Selection · Hypothesis testing

1 Introduction

In recent years, machine learning (ML) has become a popular tool for data analysis in various fields such as finance [1], healthcare [2], and marketing [8]. The ultimate goal of machine learning is to build predictive models that can accurately predict the target variable based on the input features.

Choosing the best model among a set of candidate models, however, can be a daunting task. One approach is to use statistical tests of hypothesis to compare the performance of different models [4].

M. C. Naldi and R. A. C. Bianchi (Eds.): BRACIS 2023, LNAI 14196, pp. 415–427, 2023.
https://doi.org/10.1007/978-3-031-45389-2_28

The statistical tests of hypothesis are widely used in the scientific community to evaluate the significance of a result or to compare the performance of different methods [7]. In machine learning, statistical tests are used to determine whether there is a significant difference between the performance of two or more models.

The use of statistical tests in ML model selection has become an important topic of research due to its impact on the performance of the final model [3–5,14,17]. The use bayesian statistics is studied in [3,5] while the use of frequentist tests is the focus of [4,14,17]. While these studies investigate the ability of statistical tests to choose the best model in a single experiment, this study aims to understand their effectiveness when they are applied over larger periods of time. More specifically, this study investigates whether the statistical tests commonly used for selecting the best ML model in a single experiment remain effective when applied in sequences of experiments. To achieve this, using a set a binary classification problems, the quality of the models selected by each test is examined after a significant number of iterations. Other parameters such as the number of iterations and the amount of data collected to feed the statistical tests are also investigated. In this context, the research questions investigated in this study can be laid down as follows:

1. The selection of machine learning models based on statistical tests of hypothesis lead to higher quality models after a large number of iterations?
2. What is the effect of the number of tests performed to acquire data to these statistical test of hypothesis?
3. Is there a preferred statistical test for different experimental time horizons?

The rest of the paper is organized as follows. In Sect. 2, we will provide a brief overview of statistical tests of hypothesis and their application in machine learning. In Sects. 3 and 4, we will describe the experimental setup and datasets used in this study. In Sect. 5, we will present and analyze the results of our experiments. Finally, in Sect. 6, we will conclude the paper with a discussion of the implications of our findings and future research directions.

2 Statistical Hypothesis Testing

In this section, we introduce the hypothesis tests used in this study. Given our objective of comparing two groups, which is the simplest scenario, i.e., comparing two algorithms, we have selected methods for this purpose. We have chosen the t-test as a representative parametric test, while the Mann-Whitney U test has been selected as the non-parametric option.

2.1 Paired T-Test for Two Related Samples

A paired t-test is a statistical test used to determine whether there is a significant difference between the means of two related samples [11]. While using it, each individual in one sample is paired with an individual in the other sample based on some common characteristic, such as before-and-after measurements of the same individual, or measurements from two different methods applied to the same subjects. It assumes that the two samples are normally distributed and have equal variances.

The null hypothesis of the paired t-test is that there is no difference between the means of the two samples. If the p-value returned by it, which is a representation of the probability of obtaining a statistic as extreme as the one observed, assuming that the null hypothesis is true, is less than a predefined significance level (e.g., 0.05), the null hypothesis can be rejected, and it usually concluded that there is a significant difference between the means of the two samples.

The `scipy.stats.ttest_rel` function is the method in the SciPy library [16] that performs a paired t-test for two related samples. It can be used in machine learning to compare the performance of two models on the same dataset. This information can be useful in model selection and can help to identify the best performing model for a given task. The test statistic t is computed as follows:

$$t = \frac{\overline{X}_D - \mu_0}{s_D/\sqrt{n}} \tag{1}$$

where, $\overline{X}_D$ and s_D are the average and the standard deviation of the difference between all pairs. μ_0 is the true mean of the difference under the null hypothesis. It is zero if we want to test whether the average of the difference is significant. The number of pairs is represented by n, which is also used to calculate the degrees of freedom as $n-1$.

2.2 Independent T-Test for Two Independent Samples

An independent t-test is a statistical test used to determine whether there is a significant difference between the means of two independent samples.

The `scipy.stats.ttest_ind` function is a method in the SciPy library [16] that performs an independent t-test for two independent samples. It assumes that the two samples are normally distributed and have equal variances.

The null hypothesis of the independent t-test is that there is no difference between the means of the two samples [11]. If the p-value returned by the ttest_ind function is less than a predefined significance level (e.g., 0.05), the null hypothesis can be rejected, and it can be concluded that there is a significant difference between the means of the two samples. The test t statistic is computed as:

$$t = \frac{\overline{X}_1 - \overline{X}_2}{s_p \times \sqrt{\frac{1}{n_1} + \frac{1}{n_2}}} \tag{2}$$

where

$$s_p = \sqrt{\frac{(n_1 - 1)s^2_{X_1} + (n_2 - 1)s^2_{X_2}}{n_1 + n_2 - 2}} \tag{3}$$

s_p is the pooled standard deviation of the two samples. It is defined in this way so that its square is an unbiased estimator of the common variance. For more details, see [12].

2.3 Mann-Whitney U Test

The Mann-Whitney U test is a non-parametric statistical test used to determine whether there is a significant difference between the distributions of two independent samples.

The `ss.mannwhitneyu` function is a method in the SciPy library [16] that performs the Mann-Whitney U test, also known as the Wilcoxon rank-sum test.

A very general formulation is to assume that:

- All the observations from both groups are independent of each other,
- The responses are at least ordinal (i.e., one can at least say, of any two observations, which is the greater),
- Under the null hypothesis H_0, the distributions of both populations are identical.
- The alternative hypothesis H_1 is that the distributions are not identical.

The function takes two arrays of different sizes as input. The arrays represent the two samples that are being compared. The function returns two values: the calculated U statistic and the associated p-value. The U statistic is a measure of the difference between the ranks of the two samples.

The null hypothesis of the Mann-Whitney U test is that there is no difference between the distributions of the two samples. If the p-value returned by the ss.mannwhitneyu function is less than a predefined significance level (e.g., 0.05), the null hypothesis can be rejected, and it can be concluded that there is a significant difference between the distributions of the two samples.

The Mann-Whitney U test is commonly used in machine learning to compare the performance of two models, especially when the assumptions of the t-test (such as normality and equal variances) are not met.

In details, let $X_1, \cdots, X_n$ be an independent and identically distributed (i.i.d.) sample from X and $Y_1, \cdots, Y_m$ an i.i.d. sample from Y and each sample independent from another. The corresponding Mann-Whitney U statistic is defined as:

$$U = \sum_{i=1}^{n} \sum_{j=1}^{m} S(X_i, Y_j) \tag{4}$$

with

$$S(X,Y) = \begin{cases} 1, \text{ if } X > Y \\ 0.5, \text{ if } X = Y \\ 0, \text{ if } X < Y \end{cases} \tag{5}$$

For large samples, assign numeric rank to all the observations, independent of the group they are. Then, U is given by $min(U_1, U_2)$, where: $U_1 = n_1 n_2 + \frac{n_1(n_1+1)}{2} - R_1$, and $U_2 = n_1 n_2 + \frac{n_2(n_2+1)}{2} - R_2$. R_i is the sum of the ranks from observations of the group i and n_i is the sample size of the group i.

3 The Test Bed

To evaluate the impact of using the aforementioned statistical tests for model selection over time, a simplified evolutionary algorithm is defined. This algorithm operates on a population of n model types, each having a set of associated hyper-parameters. The model with the highest performance, according to metric M, is designated as best model.

During each generation, a random model type is selected and its parameters undergo random mutations. The mutated model then competes with the model of the same type in the population. If a statistical test reveals a significant difference in performance at a significance level of $\alpha = 0.05$, the hyper-parameters of the model in the population are updated. In such cases, the mutated model also competes against the current best model using the same process.

Ultimately, the method returns the best model obtained through this evolutionary process. A pseudocode for the method is given in Algorithm 1.

Algorithm 1. Evolutionary Model Selection

```
Require: Population size n, metric M, confidence level α
Initialize population with n model types, each with their hyper-parameters
best_model ← None
function EVALUATE_POPULATION
    for each generation do
        model ← random_model(population)
        mutated_model ← mutate(model)
        performance_difference ← compare_performance(model, mutated_model)
        if statistical_test(performance_difference, α = 0.05) then
            update_hyper-parameters(model)
            if test((performance(mutated_model) == performance(best_model)))
  == False and (performance(mutated_model) > performance(best_model)) then
                update_population(mutated_model)
                best_model ← mutated_model
            end if
        end if
    end for
    return best_model
end function
best_model ← EVALUATE_POPULATION
return best_model
```

4 Experimental Setup

4.1 Datasets

Four binary classification problems defined over four datasets from the UCI repository [6] were selected for analysis. They are described below:

1. The Banking Dataset - Marketing Targets [13] (Banking) represents customer data in the banking industry for predicting conversion. It features a large sample size ($n \approx 45000$) and includes seventeen predictors, consisting of seven numeric and nine categorical variables (three of which are binary).
2. The Default of Credit Card Clients Dataset [18] (Credit) contains information about default payments of credit card clients in Taiwan from 2005. With a substantial sample size ($n \approx 30000$), it comprises 24 predictors, comprising fifteen numeric and nine categorical variables.
3. The Heart Disease Dataset [10] (Heart) integrates information from four databases: Cleveland, Hungary, Switzerland, and Long Beach V. It encompasses 76 attributes, but published experiments focus on a subset of 14 attributes. The "target" field denotes the presence of heart disease in patients. The dataset has a sample size of $n = 303$ and includes thirteen attributes, of which five are numeric and eight are categorical.
4. The Spambase Data Set [9] (Spambase) consists of a sample of $n = 4601$ emails classified as either "spam" or "non-spam." It contains fifty-seven numeric predictors, including metrics such as capital character frequency and the percentage of matching words.

The datasets are preprocessed by converting categorical variables to numerical values using LabelEncoder and normalizing the feature values using preprocessing normalize from scikit-learn. The preprocessed data is then split into training and testing sets using train_test_split from scikit-learn.

4.2 Models and Hyper-parameters

The following models were chosen for the study:

- Random Forest Classifier
- K-Nearest Neighbors Classifier
- Decision Tree Classifier
- XGBoost Classifier

The sampling functions used in the hyper-parameter mutation phase are denoted by:

- $\mathcal{S}_{[\mu_1,\mu_2,\min]}$: Samples from a Skellan Distribution defined with μ_1 and μ_2 and the center on the current value of the hyper-parameter. The samples truncated to min if its value is below min

- $\mathcal{N}_{[\sigma,\min,\max]}$: Samples from a Normal Distribution defined with μ on the current value of the hyper-parameter and σ. The sample is truncated to min if its value is below min or max if its value is above *max*
- $\mathcal{U}_{[v_1,v_2,...,v_k]}$: Samples the values $v_1, v_2, ..., v_k$ using a uniform distribution.

The hyper-parameters varied in each model and the related sampling functions are given below:

- Random Forest Classifier:
  • Number of estimators: $\mathcal{S}_{[10,10,3]}$
  • Maximum depth: $\mathcal{S}_{[2,2,2]}$
  • Maximum number of features: $\mathcal{U}_{['sqrt','log2',None]}$
- K-Nearest Neighbors Classifier:
  • Number of neighbors: $\mathcal{S}_{[1,1,3]}$
  • Weights: $\mathcal{U}_{['uniform','distance']}$
- Decision Tree Classifier:
  • Maximum depth: $\mathcal{S}_{[1,1,2]}$
  • Maximum number of features: $\mathcal{U}_{['sqrt','log2',None]}$
  • Criterion: $\mathcal{U}_{['gini','entropy','log\_loss']}$
- XGBoost Classifier:
  • Tree method: $\mathcal{U}_{['auto','exact','approx']}$
  • Maximum depth: $\mathcal{S}_{[1,1,1]}$
  • Booster: $\mathcal{U}_{['gbtree','dart']}$
  • Number of estimators: $\mathcal{S}_{[1,1,1]}$
  • Subsample: $\mathcal{N}_{[0.3,10-2,1]}$

If no sampling function is the defined, the mutation selects a random element from the set of possible values. Otherwise, it ramdomly samples from the defined distribution.

The F1 score [15] is used as the evaluation metric for the classification models. The F1 metric is a particular case of a general measure F_β. This measure is a weighting between precision and recall, indicated in cases of unbalanced data, which can generate more inaccurate results when only one of these two measures is used.

4.3 Statistical Tests

The statistical tests used by Algorithm 1 to compare the performance of different hyper-parameters are:

1. Paired t-test for two related samples (*ttest_rel*)
2. Two-sample t-test (*ttest_in*)
3. Mann-Whitney U test (*mannwhitneyu*)

They are performed on the set of evaluation metrics collected using k-fold cross-validation. As baseline, Algorithm 1 was also without any statistical test. In this case is simply selects the models with the best average performance as given by the cross-validation procedure. This procedure will be called *dummy_stats_test* for rest of the text.

4.4 Experimental Procedure

Algorithm 1 was executed 10 independent times for each combination of the following factor levels:

- Datasets: [Banking,Credit,Heart,Spambase]
- Statistical Tests: [Paired t-test for two related samples (*ttest_rel*), Two-sample t-test (*ttest_in*), Mann-Whitney U test (*mannwhitneyu*), Dummy (*dummy_stats_test*)]
- K folds: [10, 30, 50]
- Number of generations: [25, 100, 1000]

5 Results

This section summarizes the results obtained using Algorithm 1 for all tested conditions related to the research questions in Sect. 1. The results are presented as the percentage improvement in performance between the final best model ($F1_{final}$) and the initial best model ($F1_0$) which was computed by the equation below:

$$\%\text{improvement} = \frac{F1_{final} - F1_0}{F1_0} \tag{6}$$

5.1 Performance over Datasets

Figure 1 shows the box-plots for the *%improvement* obtained with Algorithm 1 combining all the possible configurations. It can be seen that the proposed procedure was more successful for the Spambase dataset with an average improvement around 7% reaching 17.5% improvement in some scenarios. Conversely, the improvements obtained for the other datasets were more modest with averages around 3%, 1.7% and 2.6% for the Banking, Credit and Heart datasets, respectively.

5.2 Performance of the Statistical Tests Varying Number of Folds

Figure 2 shows the box-plots for the *%improvement* obtained with Algorithm 1 when we vary the applied statistical test and the number of folds used in the cross-validations procedure. It can be seen that, in this scenario where the dataset effect is confounded increasing the number of folds had no effect in the *%improvement*.

Figure 3 presents the factors from Fig. 2 for each dataset, separately. Looking at the *Spambase* results, the dataset where Algorithm 1 performed the set, it can be seen that, even though there is still no significant difference in the average *%improvement*, increasing the number of folds contributed to reduce the variance of the results for the parametric tests. This phenomenon, however, could not be observed in the overall results, possibly because of the difficulty Algorithm 1 had to generate improved models.

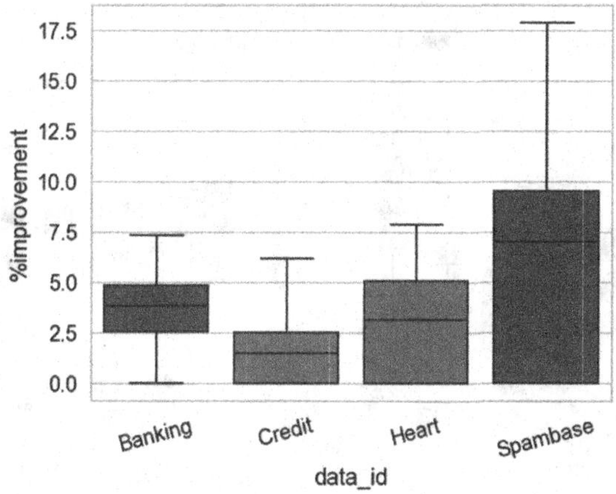

Fig. 1. % of improvement by dataset

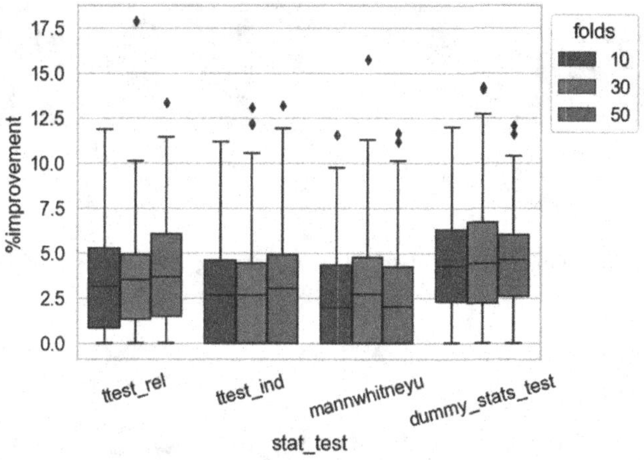

Fig. 2. % of improvement by Statistical Test varying the number of folds

5.3 Performance of the Statistical Tests Varying the Maximum Number of Generations

Figure 4 displays the performance Algorithm 1 in regards to the *%improvemnt* when varying the statistical test and the maximum number of generations (iterations) allowed. As expected, when the number of generations increases the *%improvement* also increases. These overall results do not show an interaction between the statistical test and the number of iterations since the difference in the performance among the different statistical tests and the dummy test remains almost constant as the number o generations increases.

Figure 5 displays the performance Algorithm 1 in regards to the *%improvemnt* for each dataset. For the datasets *Heart*, *Banking* and *Credit*,

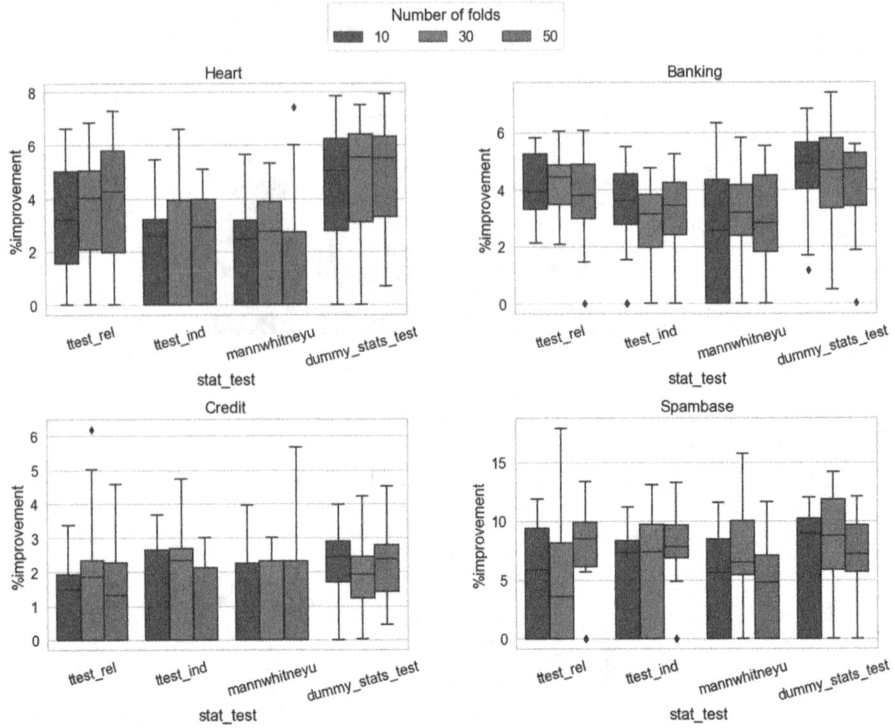

Fig. 3. % of improvement by Statistical Test varying the number of folds for each dataset

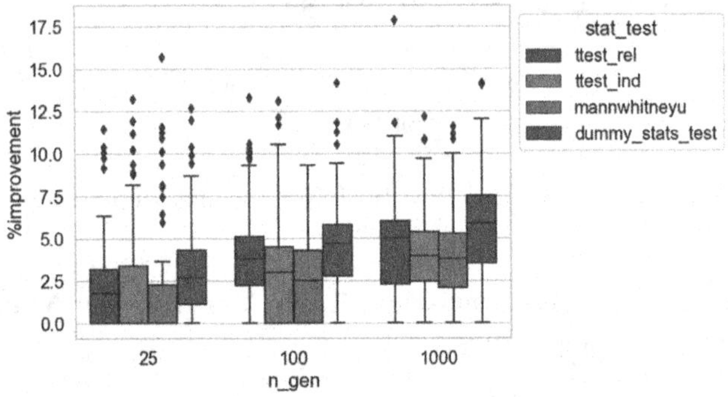

Fig. 4. % of improvement by Statistical Test and Maximum Number of Generations

even for a small number of iterations, or 25 or 100, the use of no statistical test (dummy) has led to better average performances of Algorithm 1. Among the versions with an actual statistical test, the paired t-test with relates samples (ttest_rel) has led to better or almost equal results on average.

For the *Spambase* where is was easy to generate improved models, the difference between the dummy and the other versions, only shows at the limit of 1000 generations.

Overall, these results suggest that in a challenging scenario where generating improved models is difficult, the use of statistical tests may hinder progress by imposing a an excessively stringent standard for accepting new models.

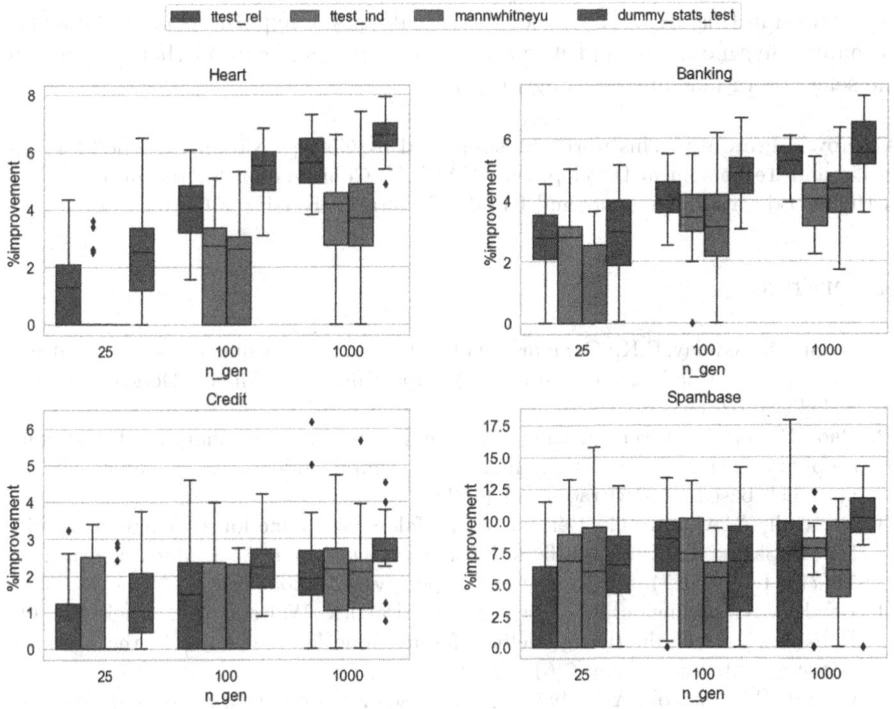

Fig. 5. % of improvement by Statistical Test and Maximum Number of Generations for each dataset

6 Conclusion

This study investigates the effectiveness of statistical tests commonly used for selecting the best machine learning (ML) model when applied over larger periods of time. The study aims to understand whether a selection procedure based on statistical tests leads to higher quality models after a significant number of iterations. Additionally, the impact of the number of tests (number of folds in cross-validation) performed to acquire data for these statistical tests and the preferred statistical test for different experimental time horizons are examined.

In summary, the analysis suggests that the use of statistical tests in ML model selection should be carefully considered, especially in challenging scenarios, as they may hinder progress and impose overly stringent criteria for accepting new models.

Although the number of datasets in this study is limited, and the proposed procedure did not verify the assumptions of the applied tests beforehand, it is noteworthy that the dominance of the versions without statistical tests remained consistent. This highlights the importance of conducting further research and exploration in this area. The use of statistical tests of hypothesis is fundamental in many scientific fields, and it is crucial to better understand their impact on the selection of machine learning models.

Acknowledgments. This work was supported by CNPq - National Council for Scientific and Technological Development, CAPES - Coordination for the Improvement of Higher Education Personnel and UFOP - Federal University of Ouro Preto.

References

1. Aygun, B., Gunay, E.K.: Comparison of statistical and machine learning algorithms for forecasting daily bitcoin returns. Avrupa Bilim ve Teknoloji Dergisi (21), pp. 444–454 (2021)
2. Bao, D., et al.: Discriminating between p16-negative oropharyngeal and non-oropharyngeal origins by their metastatic lymph nodes using machine learning approach based on MRI radiomics (2022)
3. Benavoli, A., Corani, G., Demšar, J., Zaffalon, M.: Time for a change: a tutorial for comparing multiple classifiers through bayesian analysis. J. Mach. Learn. Res. **18**(77), 1–36 (2017). http://jmlr.org/papers/v18/16-305.html
4. Bender, A., Schneider, N., Segler, M., Patrick Walters, W., Engkvist, O., Rodrigues, T.: Evaluation guidelines for machine learning tools in the chemical sciences. Nat. Rev. Chem. **6**(6), 428–442 (2022)
5. Corani, G., Benavoli, A.: A bayesian approach for comparing cross-validated algorithms on multiple data sets. Mach. Learn. **100**(2–3), 285–304 (2015)
6. Dua, D., Graff, C.: UCI machine learning repository (2017). http://archive.ics.uci.edu/ml
7. Fagerland, M.W.: t-tests, non-parametric tests, and large studies-a paradox of statistical practice? BMC Med. Res. Methodol. **12**(1), 1–7 (2012)
8. Hair, J.F., Jr., Sarstedt, M.: Data, measurement, and causal inferences in machine learning: opportunities and challenges for marketing. J. Market. Theory Practice **29**(1), 65–77 (2021)
9. Hopkins, M., Reeber, E., Forman, G., Suermondt, J.: Spambase. UCI Machine Learning Repository (1999). https://doi.org/10.24432/C53G6X
10. Janosi, A., Steinbrunn, W., Pfisterer, M., Detrano, R., M.D., M.: Heart Disease. UCI Machine Learning Repository (1988). https://doi.org/10.24432/C52P4X
11. Kim, T.K.: T test as a parametric statistic. Korean J. Anesthesiol. **68**(6), 540–546 (2015)
12. Morettin, P.A., Bussab, W.O.: Estatística básica. Saraiva Educação SA (2017)
13. Moro, S., Rita, P., Cortez, P.: Bank Marketing. UCI Machine Learning Repository (2012). https://doi.org/10.24432/C5K306

14. Trawiński, B., Smetek, M., Telec, Z., Lasota, T.: Nonparametric statistical analysis for multiple comparison of machine learning regression algorithms. Int. J. Appl. Math. Comput. Sci. **22**(4), 867–881 (2012)
15. Van Rijsbergen, C.J.: Information retrieval. (No Title) (1979)
16. Virtanen, P., et al.: SciPy 1.0 Contributors: SciPy 1.0: fundamental algorithms for scientific computing in python. Nature Methods **17**, 261–272 (2020). https://doi.org/10.1038/s41592-019-0686-2
17. Wong, T.T., Yeh, P.Y.: Reliable accuracy estimates from k-fold cross validation. IEEE Trans. Knowl. Data Eng. **32**(8), 1586–1594 (2019)
18. Yeh, I.C.: default of credit card clients. UCI Mach. Learn. Repository (2016). https://doi.org/10.24432/C55S3H

Author Index

M. C. Naldi and R. A. C. Bianchi (Eds.): BRACIS 2023, LNAI 14196, pp. 429–433, 2023.
https://doi.org/10.1007/978-3-031-45389-2

Z

Printed in the United States
by Baker & Taylor Publisher Services